Advanced Quantum Mechanics

T0211402

Franz Schwabl

Advanced Quantum Mechanics

Translated by Roginald Hilton and Angela Lahee

Fourth Edition
With 79 Figures, 4 Tables, and 104 Problems

 Springer

Professor Dr. Franz Schwabl
Physik-Department
Technische Universität München
James-Franck-Str. 2
85748 Garching, Germany
schwabl@physik.tu-muenchen.de

Translators:
Dr. Roginald Hilton
Dr. Angela Lahee

Title of the original German edition: *Quantenmechanik für Fortgeschrittene (QM II)*
(Springer-Lehrbuch)

© Springer-Verlag Berlin Heidelberg 2008

ISBN 978-3-642-09874-1   e-ISBN 978-3-540-85062-5

DOI 10.1007/978-3-540-85062-5

Cover design: eStudio Calamar, Girona/Spain

Printed on acid-free paper

9 8 7 6 5 4 3 2 1

springer.com

The true physics is that which will, one day,
achieve the inclusion of man in his wholeness
in a coherent picture of the world.

Pierre Teilhard de Chardin

To my daughter Birgitta

Preface to the Fourth Edition

In this latest edition new material has been added, which includes many additional clarifying remarks and cross references. The design of all figures has been reworked, the layout has been improved and unified to enhance the didactic appeal of the book, however, in the course of these changes I have attempted to keep intact its underlying compact nature. I am grateful to many colleagues for their help with this substantial revision. Again, special thanks go to Uwe Täuber and Roger Hilton for discussions, comments and many constructive suggestions. I should like to thank Dr. Herbert Müller for his generous help in all computer problems. Concerning the graphics, I am very grateful to Mr Wenzel Schürmann for essential support and to Ms Christina Di Stefano and Mr Benjamin Sánchez who undertook the graphical design of the diagrams.

It is my pleasure to thank Dr. Thorsten Schneider and Mrs Jacqueline Lenz of Springer for the excellent co-operation, as well as the le-tex setting team for their careful incorporation of the amendments for this new edition. Finally, I should like to thank all colleagues and students who, over the years, have made suggestions to improve the usefulness of this book.

Munich, June 2008   *F. Schwabl*

Preface to the First Edition

This textbook deals with advanced topics in the field of quantum mechanics, material which is usually encountered in a second university course on quantum mechanics. The book, which comprises a total of 15 chapters, is divided into three parts: I. Many-Body Systems, II. Relativistic Wave Equations, and III. Relativistic Fields. The text is written in such a way as to attach importance to a rigorous presentation while, at the same time, requiring no prior knowledge, except in the field of basic quantum mechanics. The inclusion of all mathematical steps and full presentation of intermediate calculations ensures ease of understanding. A number of problems are included at the end of each chapter. Sections or parts thereof that can be omitted in a first reading are marked with a star, and subsidiary calculations and remarks not essential for comprehension are given in small print. It is not necessary to have read Part I in order to understand Parts II and III. References to other works in the literature are given whenever it is felt they serve a useful purpose. These are by no means complete and are simply intended to encourage further reading. A list of other textbooks is included at the end of each of the three parts.

In contrast to Quantum Mechanics I, the present book treats relativistic phenomena, and classical and relativistic quantum fields.

Part I introduces the formalism of second quantization and applies this to the most important problems that can be described using simple methods. These include the weakly interacting electron gas and excitations in weakly interacting Bose gases. The basic properties of the correlation and response functions of many-particle systems are also treated here.

The second part deals with the Klein–Gordon and Dirac equations. Important aspects, such as motion in a Coulomb potential are discussed, and particular attention is paid to symmetry properties.

The third part presents Noether's theorem, the quantization of the Klein–Gordon, Dirac, and radiation fields, and the spin-statistics theorem. The final chapter treats interacting fields using the example of quantum electrodynamics: S-matrix theory, Wick's theorem, Feynman rules, a few simple processes such as Mott scattering and electron–electron scattering, and basic aspects of radiative corrections are discussed.

The book is aimed at advanced students of physics and related disciplines, and it is hoped that some sections will also serve to augment the teaching material already available.

This book stems from lectures given regularly by the author at the Technical University Munich. Many colleagues and coworkers assisted in the production and correction of the manuscript: Ms. I. Wefers, Ms. E. Jörg-Müller, Ms. C. Schwierz, A. Vilfan, S. Clar, K. Schenk, M. Hummel, E. Wefers, B. Kaufmann, M. Bulenda, J. Wilhelm, K. Kroy, P. Maier, C. Feuchter, A. Wonhas. The problems were conceived with the help of E. Frey and W. Gasser. Dr. Gasser also read through the entire manuscript and made many valuable suggestions. I am indebted to Dr. A. Lahee for supplying the initial English version of this difficult text, and my special thanks go to Dr. Roginald Hilton for his perceptive revision that has ensured the fidelity of the final rendition.

To all those mentioned here, and to the numerous other colleagues who gave their help so generously, as well as to Dr. Hans-Jürgen Kölsch of Springer-Verlag, I wish to express my sincere gratitude.

Munich, March 1999   *F. Schwabl*

Table of Contents

Nonrelativistic Many-Particle Systems

1. Second Quantization

In this first chapter, we shall consider nonrelativistic systems consisting of a large number of identical particles. In order to treat these, we will introduce a particularly efficient formalism, namely, the method of second quantization. Nature has given us two types of particle, bosons and fermions. These have states that are, respectively, completely symmetric and completely antisymmetric. Fermions possess half-integer spin values, whereas boson spins have integer values. This connection between spin and symmetry (statistics) is proved within relativistic quantum field theory (the spin-statistics theorem). An important consequence in many-particle physics is the existence of Fermi–Dirac statistics and Bose–Einstein statistics. We shall begin in Sect. 1.1 with some preliminary remarks which follow on from Chap. 13 of Quantum Mechanics[1]. In the subsequent sections of this chapter, we shall develop the formalism of second quantization, i.e. the quantum field theoretical representation of many-particle systems.

1.1 Identical Particles, Many-Particle States, and Permutation Symmetry

1.1.1 States and Observables of Identical Particles

We consider N identical particles (e.g., electrons, π mesons). The Hamiltonian

$$H = H(1, 2, \ldots, N) \tag{1.1.1}$$

is symmetric in the variables $1, 2, \ldots, N$. Here $1 \equiv \mathbf{x}_1, \sigma_1$ denotes the position and spin degrees of freedom of particle 1 and correspondingly for the other particles. Similarly, we write a wave function in the form

$$\psi = \psi(1, 2, \ldots, N). \tag{1.1.2}$$

The permutation operator P_{ij}, which interchanges i and j, has the following effect on an arbitrary N-particle wave function

---

[1] F. Schwabl, *Quantum Mechanics*, 4th ed., Springer, Berlin Heidelberg, 2007; in subsequent citations this book will be referred to as QM I.

$$P_{ij}\psi(\dots,i,\dots,j,\dots) = \psi(\dots,j,\dots,i,\dots). \tag{1.1.3}$$

We remind the reader of a few important properties of this operator. Since $P_{ij}^2 = 1$, the eigenvalues of P_{ij} are ± 1. Due to the symmetry of the Hamiltonian, one has for every element P of the permutation group

$$PH = HP. \tag{1.1.4}$$

The permutation group S_N which consists of all permutations of N objects has $N!$ elements. Every permutation P can be represented as a product of transpositions P_{ij}. An element is said to be even (odd) when the number of P_{ij}'s is even (odd).[2]

A few *properties*:

(i) If $\psi(1,\dots,N)$ is an eigenfunction of H with eigenvalue E, then the same also holds true for $P\psi(1,\dots,N)$.
   Proof. $H\psi = E\psi \Rightarrow HP\psi = PH\psi = EP\psi$.
(ii) For every permutation one has

$$\langle \varphi | \psi \rangle = \langle P\varphi | P\psi \rangle , \tag{1.1.5}$$

   as follows by renaming the integration variables.
(iii) The adjoint permutation operator $P^\dagger$ is defined as usual by

$$\langle \varphi | P\psi \rangle = \langle P^\dagger \varphi | \psi \rangle .$$

   It follows from this that

$$\langle \varphi | P\psi \rangle = \langle P^{-1}\varphi | P^{-1}P\psi \rangle = \langle P^{-1}\varphi | \psi \rangle \Rightarrow P^\dagger = P^{-1}$$

   and thus P is unitary

$$P^\dagger P = PP^\dagger = 1 . \tag{1.1.6}$$

(iv) For every symmetric operator $S(1,\dots,N)$ we have

$$[P, S] = 0 \tag{1.1.7}$$

   and

$$\langle P\psi_i | S | P\psi_j \rangle = \langle \psi_i | P^\dagger SP | \psi_j \rangle = \langle \psi_i | P^\dagger PS | \psi_j \rangle = \langle \psi_i | S | \psi_j \rangle . \tag{1.1.8}$$

   This proves that the matrix elements of symmetric operators are the same in the states ψ_i and in the permutated states $P\psi_i$.

---

[2] It is well known that every permutation can be represented as a product of cycles that have no element in common, e.g., (124)(35). Every cycle can be written as a product of transpositions,

   e.g.   (12)   odd
   $P_{124} \equiv (124) = (14)(12)$ even

   Each cycle is carried out from left to right ($1 \to 2, 2 \to 4, 4 \to 1$), whereas the products of cycles are applied from right to left.

(v) The converse of (iv) is also true. The requirement that an exchange of identical particles should not have any observable consequences implies that all observables O must be symmetric, i.e., permutation invariant. *Proof.* $\langle \psi | O | \psi \rangle = \langle P\psi | O | P\psi \rangle = \langle \psi | P^\dagger O P | \psi \rangle$ holds for arbitrary ψ. Thus, $P^\dagger O P = O$ and, hence, $PO = OP$.

Since identical particles are all influenced identically by any physical process, *all physical operators must be symmetric*. Hence, the states ψ and $P\psi$ are experimentally indistinguishable. The question arises as to whether all these $N!$ states are realized in nature.

In fact, the *totally symmetric* and *totally antisymmetric* states ψ_s and ψ_a do play a special role. These states are defined by

$$P_{ij}\psi_{\substack{s\\a}}(\dots,i,\dots,j,\dots) = \pm\psi_{\substack{s\\a}}(\dots,i,\dots,j,\dots) \tag{1.1.9}$$

for all P_{ij}.

It is an *experimental* fact that there are two types of particle, *bosons* and *fermions*, whose states are totally symmetric and totally antisymmetric, respectively. As mentioned at the outset, bosons have integral, and fermions half-integral spin.

Remarks:

(i) The symmetry character of a state does not change in the course of time:

$$\psi(t) = T\mathrm{e}^{-\frac{i}{\hbar}\int_0^t dt' H(t')}\psi(0) \Rightarrow P\psi(t) = T\mathrm{e}^{-\frac{i}{\hbar}\int_0^t dt' H(t')}P\psi(0) , \tag{1.1.10}$$

where T is the time-ordering operator.[3]

(ii) For arbitrary permutations P, the states introduced in (1.1.9) satisfy

$$P\psi_s = \psi_s \tag{1.1.11}$$

$$P\psi_a = (-1)^P \psi_a \text{ , with } (-1)^P = \begin{cases} 1 \text{ for even permutations} \\ -1 \text{ for odd permutations.} \end{cases}$$

Thus, the states ψ_s and ψ_a form the basis of two one-dimensional representations of the permutation group S_N. For ψ_s, every P is assigned the number 1, and for ψ_a every even (odd) element is assigned the number $1(-1)$. Since, in the case of three or more particles, the P_{ij} do not all commute with one another, there are, in addition to ψ_s and ψ_a, also states for which not all P_{ij} are diagonal. Due to noncommutativity, a complete set of common eigenfunctions of all P_{ij} cannot exist. These states are basis functions of higher-dimensional representations of the permutation group. These states are not realized in nature; they are referred to

---

[3] QM I, Chap. 16.

as parasymmetric states.[4]. The fictitious particles that are described by these states are known as paraparticles and are said to obey parastatistics.

1.1.2 Examples

(i) *Two particles*
Let $\psi(1,2)$ be an arbitrary wave function. The permutation P_{12} leads to $P_{12}\psi(1,2) = \psi(2,1)$.
From these two wave functions one can form

$$\begin{aligned}\psi_s &= \psi(1,2) + \psi(2,1) \quad \text{symmetric} \\ \psi_a &= \psi(1,2) - \psi(2,1) \quad \text{antisymmetric}\end{aligned} \tag{1.1.12}$$

under the operation P_{12}. For two particles, the symmetric and antisymmetric states exhaust all possibilities.

(ii) *Three particles*
We consider the example of a wave function that is a function only of the spatial coordinates

$$\psi(1,2,3) = \psi(x_1, x_2, x_3).$$

Application of the permutation P_{123} yields

$$P_{123}\,\psi(x_1, x_2, x_3) = \psi(x_2, x_3, x_1),$$

i.e., particle 1 is replaced by particle 2, particle 2 by particle 3, and particle 3 by particle 1, e.g., $\psi(1,2,3) = \mathrm{e}^{-x_1^2(x_2^2 - x_3^2)^2}$, $P_{12}\,\psi(1,2,3) = \mathrm{e}^{-x_2^2(x_1^2 - x_3^2)^2}$, $P_{123}\,\psi(1,2,3) = \mathrm{e}^{-x_2^2(x_3^2 - x_1^2)^2}$. We consider

$$\begin{aligned}P_{13}P_{12}\,\psi(1,2,3) &= P_{13}\,\psi(2,1,3) &= \psi(2,3,1) &= P_{123}\,\psi(1,2,3) \\ P_{12}P_{13}\,\psi(1,2,3) &= P_{12}\,\psi(3,2,1) &= \psi(3,1,2) &= P_{132}\,\psi(1,2,3) \\ (P_{123})^2\psi(1,2,3) &= P_{123}\,\psi(2,3,1) &= \psi(3,1,2) &= P_{132}\,\psi(1,2,3).\end{aligned}$$

Clearly, $P_{13}P_{12} \neq P_{12}P_{13}$.
S_3, the permutation group for three objects, consists of the following $3! = 6$ elements:

$$S_3 = \{1, P_{12}, P_{23}, P_{31}, P_{123}, P_{132} = (P_{123})^2\}. \tag{1.1.13}$$

We now consider the effect of a permutation P on a ket vector. Thus far we have only allowed P to act on spatial wave functions or inside scalar products which lead to integrals over products of spatial wave functions.
Let us assume that we have the state

$$|\psi\rangle = \sum_{x_1, x_2, x_3} \overbrace{|x_1\rangle_1 \, |x_2\rangle_2 \, |x_3\rangle_3}^{\text{direct product}} \psi(x_1, x_2, x_3) \tag{1.1.14}$$

---

[4] A.M.L. Messiah and O.W. Greenberg, Phys. Rev. **B 136**, 248 (1964), **B 138**, 1155 (1965).

with $\psi(x_1, x_2, x_3) = \langle x_1|_1 \langle x_2|_2 \langle x_3|_3 |\psi\rangle$. In $|x_i\rangle_j$ the particle is labeled by the number j and the spatial coordinate is x_i. The effect of P_{123}, for example, is defined as follows:

$$P_{123} |\psi\rangle = \sum_{x_1, x_2, x_3} |x_1\rangle_2 |x_2\rangle_3 |x_3\rangle_1 \, \psi(x_1, x_2, x_3) \, .$$

$$= \sum_{x_1, x_2, x_3} |x_3\rangle_1 |x_1\rangle_2 |x_2\rangle_3 \, \psi(x_1, x_2, x_3)$$

In the second line the basis vectors of the three particles in the direct product are once more written in the usual order, 1,2,3. We can now rename the summation variables according to $(x_1, x_2, x_3) \rightarrow P_{123}(x_1, x_2, x_3) = (x_2, x_3, x_1)$. From this, it follows that

$$P_{123} |\psi\rangle = \sum_{x_1, x_2, x_3} |x_1\rangle_1 |x_2\rangle_2 |x_3\rangle_3 \, \psi(x_2, x_3, x_1) \, .$$

If the state $|\psi\rangle$ has the wave function $\psi(x_1, x_2, x_3)$, then $P |\psi\rangle$ has the wave function $P\psi(x_1, x_2, x_3)$. The particles are exchanged under the permutation. Finally, we discuss the *basis vectors for three particles:* If we start from the state $|\alpha\rangle |\beta\rangle |\gamma\rangle$ and apply the elements of the group S_3, we get the six states

$$|\alpha\rangle |\beta\rangle |\gamma\rangle$$
$$P_{12} |\alpha\rangle |\beta\rangle |\gamma\rangle = |\beta\rangle |\alpha\rangle |\gamma\rangle \, , \quad P_{23} |\alpha\rangle |\beta\rangle |\gamma\rangle = |\alpha\rangle |\gamma\rangle |\beta\rangle \, ,$$
$$P_{31} |\alpha\rangle |\beta\rangle |\gamma\rangle = |\gamma\rangle |\beta\rangle |\alpha\rangle \, , \tag{1.1.15}$$
$$P_{123} |\alpha\rangle_1 |\beta\rangle_2 |\gamma\rangle_3 = |\alpha\rangle_2 |\beta\rangle_3 |\gamma\rangle_1 = |\gamma\rangle |\alpha\rangle |\beta\rangle \, ,$$
$$P_{132} |\alpha\rangle |\beta\rangle |\gamma\rangle = |\beta\rangle |\gamma\rangle |\alpha\rangle \, .$$

Except in the fourth line, the indices for the particle number are not written out, but are determined by the position within the product (particle 1 is the first factor, etc.). It is the particles that are permutated, not the arguments of the states.

If we assume that α, β, and γ are all different, then the same is true of the six states given in (1.1.15). One can group and combine these in the following way to yield invariant subspaces [5]:

Invariant subspaces:

Basis 1 (symmetric basis):

$$\frac{1}{\sqrt{6}} (|\alpha\rangle |\beta\rangle |\gamma\rangle + |\beta\rangle |\alpha\rangle |\gamma\rangle + |\alpha\rangle |\gamma\rangle |\beta\rangle + |\gamma\rangle |\beta\rangle |\alpha\rangle + |\gamma\rangle |\alpha\rangle |\beta\rangle + |\beta\rangle |\gamma\rangle |\alpha\rangle)$$

$$\tag{1.1.16a}$$

Basis 2 (antisymmetric basis):

$$\frac{1}{\sqrt{6}} (|\alpha\rangle |\beta\rangle |\gamma\rangle - |\beta\rangle |\alpha\rangle |\gamma\rangle - |\alpha\rangle |\gamma\rangle |\beta\rangle - |\gamma\rangle |\beta\rangle |\alpha\rangle + |\gamma\rangle |\alpha\rangle |\beta\rangle + |\beta\rangle |\gamma\rangle |\alpha\rangle)$$

$$\tag{1.1.16b}$$

---

[5] An invariant subspace is a subspace of states which transforms into itself on application of the group elements.

Basis 3:

$$\begin{cases} \frac{1}{\sqrt{12}}(2\,|\alpha\rangle\,|\beta\rangle\,|\gamma\rangle + 2\,|\beta\rangle\,|\alpha\rangle\,|\gamma\rangle - |\alpha\rangle\,|\gamma\rangle\,|\beta\rangle - |\gamma\rangle\,|\beta\rangle\,|\alpha\rangle \\ \qquad - |\gamma\rangle\,|\alpha\rangle\,|\beta\rangle - |\beta\rangle\,|\gamma\rangle\,|\alpha\rangle) \\ \frac{1}{2}(0 + 0 - |\alpha\rangle\,|\gamma\rangle\,|\beta\rangle + |\gamma\rangle\,|\beta\rangle\,|\alpha\rangle + |\gamma\rangle\,|\alpha\rangle\,|\beta\rangle - |\beta\rangle\,|\gamma\rangle\,|\alpha\rangle) \end{cases}$$

$$(1.1.16c)$$

Basis 4:

$$\begin{cases} \frac{1}{2}(0 + 0 - |\alpha\rangle\,|\gamma\rangle\,|\beta\rangle + |\gamma\rangle\,|\beta\rangle\,|\alpha\rangle - |\gamma\rangle\,|\alpha\rangle\,|\beta\rangle + |\beta\rangle\,|\gamma\rangle\,|\alpha\rangle) \\ \frac{1}{\sqrt{12}}(2\,|\alpha\rangle\,|\beta\rangle\,|\gamma\rangle - 2\,|\beta\rangle\,|\alpha\rangle\,|\gamma\rangle + |\alpha\rangle\,|\gamma\rangle\,|\beta\rangle + |\gamma\rangle\,|\beta\rangle\,|\alpha\rangle \\ \qquad - |\gamma\rangle\,|\alpha\rangle\,|\beta\rangle - |\beta\rangle\,|\gamma\rangle\,|\alpha\rangle)\,. \end{cases}$$

$$(1.1.16d)$$

In the bases 3 and 4, the first of the two functions in each case is even under P_{12} and the second is odd under P_{12} (immediately below we shall call these two functions $|\psi_1\rangle$ and $|\psi_2\rangle$). Other operations give rise to a linear combination of the two functions:

$$P_{12}\,|\psi_1\rangle = |\psi_1\rangle \;,\; P_{12}\,|\psi_2\rangle = -\,|\psi_2\rangle \;, \qquad\qquad (1.1.17a)$$

$$P_{13}\,|\psi_1\rangle = \alpha_{11}\,|\psi_1\rangle + \alpha_{12}\,|\psi_2\rangle \;,\; P_{13}\,|\psi_2\rangle = \alpha_{21}\,|\psi_1\rangle + \alpha_{22}\,|\psi_2\rangle \;, \qquad (1.1.17b)$$

with coefficients α_{ij}. In matrix form, (1.1.17b) can be written as

$$P_{13}\begin{pmatrix}|\psi_1\rangle\\|\psi_2\rangle\end{pmatrix} = \begin{pmatrix}\alpha_{11} & \alpha_{12}\\\alpha_{21} & \alpha_{22}\end{pmatrix}\begin{pmatrix}|\psi_1\rangle\\|\psi_2\rangle\end{pmatrix}\,. \qquad (1.1.17c)$$

The elements P_{12} and P_{13} are thus represented by 2×2 matrices

$$P_{12} = \begin{pmatrix}1 & 0\\0 & -1\end{pmatrix}\;,\; P_{13} = \begin{pmatrix}\alpha_{11} & \alpha_{12}\\\alpha_{21} & \alpha_{22}\end{pmatrix}\,. \qquad (1.1.18)$$

This fact implies that the basis vectors $|\psi_1\rangle$ and $|\psi_2\rangle$ span a two-dimensional representation of the permutation group S_3. The explicit calculation will be carried out in Problem 1.2.

1.2 Completely Symmetric and Antisymmetric States

We begin with the single-particle states $|i\rangle$: $|1\rangle$, $|2\rangle$, ... The single-particle states of the particles 1, 2, ..., α, ..., N are denoted by $|i\rangle_1$, $|i\rangle_2$, ..., $|i\rangle_\alpha$, ..., $|i\rangle_N$. These enable us to write the basis states of the N-particle system

$$|i_1,\ldots,i_\alpha,\ldots,i_N\rangle = |i_1\rangle_1 \ldots |i_\alpha\rangle_\alpha \ldots |i_N\rangle_N\,, \qquad (1.2.1)$$

where particle 1 is in state $|i_1\rangle_1$ and particle α in state $|i_\alpha\rangle_\alpha$, etc. (The subscript outside the ket is the number labeling the particle, and the index within the ket identifies the state of this particle.)

Provided that the $\{|i\rangle\}$ form a complete orthonormal set, the product states defined above likewise represent a *complete orthonormal system* in the space of N-particle states. The symmetrized and antisymmetrized basis states are then defined by

$$S_\pm |i_1, i_2, \ldots, i_N\rangle \equiv \frac{1}{\sqrt{N!}} \sum_P (\pm 1)^P P |i_1, i_2, \ldots, i_N\rangle \ . \tag{1.2.2}$$

In other words, we apply all $N!$ elements of the permutation group S_N of N objects and, for fermions, we multiply by (-1) when P is an odd permutation. The states defined in (1.2.2) are of two types: completely symmetric and completely antisymmetric.

Remarks regarding the properties of $S_\pm \equiv \frac{1}{\sqrt{N!}} \sum_P (\pm 1)^P P$:

(i) Let S_N be the permutation group (or symmetric group) of N quantities.

Assertion: For every element $P \in S_N$, one has $PS_N = S_N$.
Proof. The set PS_N contains exactly the same number of elements as S_N and these, due to the group property, are all contained in S_N. Furthermore, the elements of PS_N are all different since, if one had $PP_1 = PP_2$, then, after multiplication by P^{-1}, it would follow that $P_1 = P_2$.
Thus

$$PS_N = S_N P = S_N \ . \tag{1.2.3}$$

(ii) It follows from this that

$$PS_+ = S_+ P = S_+ \tag{1.2.4a}$$

and

$$PS_- = S_- P = (-1)^P S_- . \tag{1.2.4b}$$

If P is even, then even elements remain even and odd ones remain odd. If P is odd, then multiplication by P changes even into odd elements and vice versa.

$$PS_+ |i_1, \ldots, i_N\rangle = S_+ |i_1, \ldots, i_N\rangle$$
$$PS_- |i_1, \ldots, i_N\rangle = (-1)^P S_- |i_1, \ldots, i_N\rangle$$
Special case $P_{ij} S_- |i_1, \ldots, i_N\rangle = -S_- |i_1, \ldots, i_N\rangle \ .$

(iii) If $|i_1, \ldots, i_N\rangle$ contains single-particle states occurring more than once, then $S_+ |i_1, \ldots, i_N\rangle$ is no longer normalized to unity. Let us assume that the first state occurs n_1 times, the second n_2 times, etc. Since $S_+ |i_1, \ldots, i_N\rangle$ contains a total of $N!$ terms, of which $\frac{N!}{n_1! n_2! \ldots}$ are different, each of these terms occurs with a multiplicity of $n_1! n_2! \ldots$.

$$\langle i_1, \ldots, i_N | S_+^\dagger S_+ |i_1, \ldots, i_N\rangle = \frac{1}{N!} (n_1! n_2! \ldots)^2 \frac{N!}{n_1! n_2! \ldots} = n_1! n_2! \ldots$$

Thus, the *normalized Bose basis functions* are

$$S_+ |i_1, \ldots, i_N\rangle \frac{1}{\sqrt{n_1! n_2! \ldots}} = \frac{1}{\sqrt{N! n_1! n_2! \ldots}} \sum_P P |i_1, \ldots, i_N\rangle . \quad (1.2.5)$$

(iv) A further property of $S_\pm$ is

$$S_\pm^2 = \sqrt{N!} S_\pm , \qquad (1.2.6a)$$

since $S_\pm^2 = \frac{1}{\sqrt{N!}} \sum_P (\pm 1)^P P S_\pm = \frac{1}{\sqrt{N!}} \sum_P S_\pm = \sqrt{N!} S_\pm$. We now consider
an arbitrary N-particle state, which we expand in the basis $|i_1\rangle \ldots |i_N\rangle$

$$|z\rangle = \sum_{i_1, \ldots, i_N} |i_1\rangle \ldots |i_N\rangle \underbrace{\langle i_1, \ldots, i_N|z\rangle}_{c_{i_1, \ldots, i_N}} .$$

Application of $S_\pm$ yields

$$S_\pm |z\rangle = \sum_{i_1, \ldots, i_N} S_\pm |i_1\rangle \ldots |i_N\rangle c_{i_1, \ldots, i_N} = \sum_{i_1, \ldots, i_N} |i_1\rangle \ldots |i_N\rangle S_\pm c_{i_1, \ldots, i_N}$$

and further application of $\frac{1}{\sqrt{N!}} S_\pm$, with the identity (1.2.6a), results in

$$S_\pm |z\rangle = \frac{1}{\sqrt{N!}} \sum_{i_1, \ldots, i_N} S_\pm |i_1\rangle \ldots |i_N\rangle (S_\pm c_{i_1, \ldots, i_N}). \qquad (1.2.6b)$$

Equation (1.2.6b) implies that every symmetrized state can be expanded in
terms of the symmetrized basis states (1.2.2).

1.3 Bosons

1.3.1 States, Fock Space, Creation and Annihilation Operators

The state (1.2.5) is fully characterized by specifying the occupation numbers

$$|n_1, n_2, \ldots\rangle = S_+ |i_1, i_2, \ldots, i_N\rangle \frac{1}{\sqrt{n_1! n_2! \ldots}} . \qquad (1.3.1)$$

Here, n_1 is the number of times that the state 1 occurs, n_2 the number of
times that state 2 occurs, ... Alternatively: n_1 is the number of particles in
state 1, n_2 is the number of particles in state 2, ... The sum of all occupation
numbers n_i must be equal to the total number of particles:

$$\sum_{i=1}^{\infty} n_i = N. \qquad (1.3.2)$$

Apart from this constraint, the n_i can take any of the values $0, 1, 2, \ldots$. The factor $(n_1! n_2! \ldots)^{-1/2}$, together with the factor $1/\sqrt{N!}$ contained in S_+, has the effect of normalizing $|n_1, n_2, \ldots\rangle$ (see point (iii)). These states form a complete set of completely symmetric N-particle states. By linear superposition, one can construct from these any desired symmetric N-particle state.

We now combine the states for $N = 0, 1, 2, \ldots$ and obtain a complete orthonormal system of states for arbitrary particle number, which satisfy the orthogonality relation[6]

$$\langle n_1, n_2, \ldots | n_1', n_2', \ldots \rangle = \delta_{n_1, n_1'} \delta_{n_2, n_2'} \cdots \qquad (1.3.3a)$$

and the completeness relation

$$\sum_{n_1, n_2, \ldots} |n_1, n_2, \ldots\rangle \langle n_1, n_2, \ldots| = \mathbb{1} . \qquad (1.3.3b)$$

This extended space is the *direct sum* of the space with no particles (vacuum state $|0\rangle$), the space with one particle, the space with two particles, etc.; it is known as *Fock space*.

The operators we have considered so far act only within a subspace of fixed particle number. On applying $\mathbf{p}, \mathbf{x}$ etc. to an N-particle state, we obtain again an N-particle state. We now define *creation* and *annihilation operators*, which lead from the space of N-particle states to the spaces of $N \pm 1$-particle states:

$$a_i^\dagger |\ldots, n_i, \ldots\rangle = \sqrt{n_i + 1} |\ldots, n_i + 1, \ldots\rangle . \qquad (1.3.4)$$

Taking the adjoint of this equation and relabeling $n_i \to n_i'$, we have

$$\langle \ldots, n_i', \ldots | a_i = \sqrt{n_i' + 1} \langle \ldots, n_i' + 1, \ldots| . \qquad (1.3.5)$$

Multiplying this equation by $|\ldots, n_i, \ldots\rangle$ yields

$$\langle \ldots, n_i', \ldots | a_i |\ldots, n_i, \ldots\rangle = \sqrt{n_i} \, \delta_{n_i' + 1, n_i} .$$

Expressed in words, the operator a_i reduces the occupation number by 1.
Assertion:

$$a_i |\ldots, n_i, \ldots\rangle = \sqrt{n_i} |\ldots, n_i - 1, \ldots\rangle \quad \text{for } n_i \geq 1 \qquad (1.3.6)$$

and

$$a_i |\ldots, n_i = 0, \ldots\rangle = 0 .$$

---

[6] In the states $|n_1, n_2, \ldots\rangle$, the n_1, n_2 etc. are arbitrary natural numbers whose sum is not constrained. The (vanishing) scalar product between states of differing particle number is defined by (1.3.3a).

Proof:

$$a_i |\ldots, n_i, \ldots\rangle = \sum_{n_i'=0}^{\infty} |\ldots, n_i', \ldots\rangle \langle \ldots, n_i', \ldots | a_i |\ldots, n_i, \ldots\rangle$$

$$= \sum_{n_i'=0}^{\infty} |\ldots, n_i', \ldots\rangle \sqrt{n_i}\, \delta_{n_i'+1, n_i}$$

$$= \begin{cases} \sqrt{n_i} |\ldots, n_i - 1, \ldots\rangle & \text{for } n_i \geq 1 \\ 0 & \text{for } n_i = 0 \end{cases} .$$

The operator $a_i^\dagger$ increases the occupation number of the state $|i\rangle$ by 1, and the operator a_i reduces it by 1. The operators $a_i^\dagger$ and a_i are thus called *creation* and *annihilation operators*. The above relations and the completeness of the states yield the *Bose commutation relations*

$$[a_i, a_j] = 0, \quad [a_i^\dagger, a_j^\dagger] = 0, \quad [a_i, a_j^\dagger] = \delta_{ij} . \tag{1.3.7a,b,c}$$

Proof. It is clear that (1.3.7a) holds for $i = j$, since a_i commutes with itself. For $i \neq j$, it follows from (1.3.6) that

$$a_i a_j |\ldots, n_i, \ldots, n_j, \ldots\rangle = \sqrt{n_i}\sqrt{n_j} |\ldots, n_i - 1, \ldots, n_j - 1, \ldots\rangle$$

$$= a_j a_i |\ldots, n_i, \ldots, n_j, \ldots\rangle$$

which proves (1.3.7a) and, by taking the hermitian conjugate, also (1.3.7b). For $j \neq i$ we have

$$a_i a_j^\dagger |\ldots, n_i, \ldots, n_j, \ldots\rangle = \sqrt{n_i}\sqrt{n_j + 1} |\ldots, n_i - 1, \ldots, n_j + 1, \ldots\rangle$$

$$= a_j^\dagger a_i |\ldots, n_i, \ldots, n_j, \ldots\rangle$$

and

$$\left(a_i a_i^\dagger - a_i^\dagger a_i\right) |\ldots, n_i, \ldots, n_j, \ldots\rangle =$$

$$\left(\sqrt{n_i + 1}\sqrt{n_i + 1} - \sqrt{n_i}\sqrt{n_i}\right) |\ldots, n_i, \ldots, n_j, \ldots\rangle$$

hence also proving (1.3.7c).

Starting from the *ground state* $\equiv$ *vacuum state*

$$|0\rangle \equiv |0, 0, \ldots\rangle , \tag{1.3.8}$$

which contains no particles at all, we can construct all states: single-particle states

$$a_i^\dagger |0\rangle , \ldots ,$$

two-particle states

$$\frac{1}{\sqrt{2!}} \left(a_i^\dagger\right)^2 |0\rangle , a_i^\dagger a_j^\dagger |0\rangle , \ldots$$

and the general many-particle state

$$|n_1, n_2, \ldots\rangle = \frac{1}{\sqrt{n_1! n_2! \ldots}} \left(a_1^\dagger\right)^{n_1} \left(a_2^\dagger\right)^{n_2} \ldots |0\rangle . \tag{1.3.9}$$

Normalization:

$$a^\dagger |n-1\rangle = \sqrt{n}\, |n\rangle \tag{1.3.10}$$

$$\left\| a^\dagger |n-1\rangle \right\| = \sqrt{n}$$

$$|n\rangle = \frac{1}{\sqrt{n}} a^\dagger |n-1\rangle .$$

1.3.2 The Particle-Number Operator

The particle-number operator (occupation-number operator for the state $|i\rangle$) is defined by

$$\hat{n}_i = a_i^\dagger a_i . \tag{1.3.11}$$

The states introduced above are eigenfunctions of $\hat{n}_i$:

$$\hat{n}_i |\ldots, n_i, \ldots\rangle = n_i |\ldots, n_i, \ldots\rangle , \tag{1.3.12}$$

and the corresponding eigenvalue of $\hat{n}_i$ is the number of particles in the state i.

The operator for the total number of particles is given by

$$\hat{N} = \sum_i \hat{n}_i. \tag{1.3.13}$$

Applying this operator to the states $|\ldots, \hat{n}_i, \ldots\rangle$ yields

$$\hat{N} |n_1, n_2, \ldots\rangle = \left(\sum_i n_i\right) |n_1, n_2, \ldots\rangle . \tag{1.3.14}$$

Assuming that the particles do not interact with one another and, furthermore, that the states $|i\rangle$ are the eigenstates of the single-particle Hamiltonian with eigenvalues ϵ_i, the full Hamiltonian can be written as

$$H_0 = \sum_i \hat{n}_i \epsilon_i \tag{1.3.15a}$$

$$H_0 |n_1, \ldots\rangle = \left(\sum_i n_i \epsilon_i\right) |n_1, \ldots\rangle . \tag{1.3.15b}$$

The commutation relations and the properties of the particle-number operator are analogous to those of the harmonic oscillator.

1.3.3 General Single- and Many-Particle Operators

Let us consider an operator for the N-particle system which is a sum of single-particle operators

$$T = t_1 + t_2 + \ldots + t_N \equiv \sum_\alpha t_\alpha \,, \qquad (1.3.16)$$

e.g., for the kinetic energy $t_\alpha = \mathbf{p}_\alpha^2/2m$, and for the potential $V(\mathbf{x}_\alpha)$. For one particle, the single-particle operator is t. Its matrix elements in the basis $|i\rangle$ are

$$t_{ij} = \langle i| t |j\rangle \,, \qquad (1.3.17)$$

such that

$$t = \sum_{i,j} t_{ij} |i\rangle \langle j| \qquad (1.3.18)$$

and for the full N-particle system

$$T = \sum_{i,j} t_{ij} \sum_{\alpha=1}^{N} |i\rangle_\alpha \langle j|_\alpha \,. \qquad (1.3.19)$$

Our aim is to represent this operator in terms of creation and annihilation operators. We begin by taking a pair of states i, j from (1.3.19) and calculating their effect on an arbitrary state (1.3.1). We assume initially that $j \neq i$

$$\sum_\alpha |i\rangle_\alpha \langle j|_\alpha |\ldots, n_i, \ldots, n_j, \ldots\rangle$$

$$\equiv \sum_\alpha |i\rangle_\alpha \langle j|_\alpha S_+ |i_1, i_2, \ldots, i_N\rangle \frac{1}{\sqrt{n_1! n_2! \ldots}} \qquad (1.3.20)$$

$$= S_+ \sum_\alpha |i\rangle_\alpha \langle j|_\alpha |i_1, i_2, \ldots, i_N\rangle \frac{1}{\sqrt{n_1! n_2! \ldots}} \,.$$

It is possible, as was done in the third line, to bring the S_+ to the front, since it commutes with every symmetric operator. If the state j is n_j-fold occupied, it gives rise to n_j terms in which $|j\rangle$ is replaced by $|i\rangle$. Hence, the effect of S_+ is to yield n_j states $|\ldots, n_i + 1, \ldots, n_j - 1, \ldots\rangle$, where the change in the normalization should be noted. Equation (1.3.20) thus leads on to

$$= n_j \sqrt{n_i + 1} \frac{1}{\sqrt{n_j}} |\ldots, n_i + 1, \ldots, n_j - 1, \ldots\rangle$$

$$= \sqrt{n_j} \sqrt{n_i + 1} |\ldots, n_i + 1, \ldots, n_j - 1, \ldots\rangle \qquad (1.3.20')$$

$$= a_i^\dagger a_j |\ldots, n_i, \ldots, n_j, \ldots\rangle \,.$$

For $j = i$, the i is replaced n_i times by itself, thus yielding

$$n_i \left|\ldots, n_i, \ldots\right\rangle = a_i^\dagger a_i \left|\ldots, n_i, \ldots\right\rangle .$$

Thus, for any N, we have

$$\sum_{\alpha=1}^{N} |i\rangle_\alpha \langle j|_\alpha = a_i^\dagger a_j.$$

From this it follows that, for any single-particle operator,

$$T = \sum_{i,j} t_{ij} a_i^\dagger a_j , \tag{1.3.21}$$

where

$$t_{ij} = \langle i| t |j\rangle . \tag{1.3.22}$$

The special case $t_{ij} = \epsilon_i \delta_{ij}$ leads to

$$H_0 = \sum_{i} \epsilon_i a_i^\dagger a_i ,$$

i.e., to (1.3.15a).

In a similar way one can show that *two-particle operators*

$$F = \frac{1}{2} \sum_{\alpha \neq \beta} f^{(2)}(\mathbf{x}_\alpha, \mathbf{x}_\beta) \tag{1.3.23}$$

can be written in the form

$$F = \frac{1}{2} \sum_{i,j,k,m} \langle i, j| f^{(2)} |k, m\rangle a_i^\dagger a_j^\dagger a_m a_k, \tag{1.3.24}$$

where

$$\langle i, j| f^{(2)} |k, m\rangle = \int dx \int dy\, \varphi_i^*(x) \varphi_j^*(y) f^{(2)}(x, y) \varphi_k(x) \varphi_m(y) . \tag{1.3.25}$$

In (1.3.23), the condition $\alpha \neq \beta$ is required as, otherwise, we would have only a single-particle operator. The factor $\frac{1}{2}$ in (1.3.23) is to ensure that each interaction is included only once since, for identical particles, symmetry implies that $f^{(2)}(\mathbf{x}_\alpha, \mathbf{x}_\beta) = f^{(2)}(\mathbf{x}_\beta, \mathbf{x}_\alpha)$.

Proof of (1.3.24). One first expresses F in the form

$$F = \frac{1}{2} \sum_{\alpha \neq \beta} \sum_{i,j,k,m} \langle i, j| f^{(2)} |k, m\rangle |i\rangle_\alpha |j\rangle_\beta \langle k|_\alpha \langle m|_\beta .$$

We now investigate the action of one term of the sum constituting F:

$$\sum_{\alpha \neq \beta} |i\rangle_\alpha |j\rangle_\beta \langle k|_\alpha \langle m|_\beta |\ldots, n_i, \ldots, n_j, \ldots, n_k, \ldots, n_m, \ldots\rangle$$

$$= n_k n_m \frac{1}{\sqrt{n_k}\sqrt{n_m}} \sqrt{n_i+1}\sqrt{n_j+1}$$

$$|\ldots, n_i+1, \ldots, n_j+1, \ldots, n_k-1, \ldots, n_m-1, \ldots\rangle$$

$$= a_i^\dagger a_j^\dagger a_k a_m |\ldots, n_i, \ldots, n_j, \ldots, n_k, \ldots, n_m, \ldots\rangle \ .$$

Here, we have assumed that the states are different. If the states are identical, the derivation has to be supplemented in a similar way to that for the single-particle operators.

A somewhat shorter derivation, and one which also covers the case of fermions, proceeds as follows: The commutator and anticommutator for bosons and fermions, respectively, are combined in the form $[a_k, a_j]_\mp = \delta_{kj}$.

$$\sum_{\alpha \neq \beta} |i\rangle_\alpha |j\rangle_\beta \langle k|_\alpha \langle m|_\beta = \sum_{\alpha \neq \beta} |i\rangle_\alpha \langle k|_\alpha |j\rangle_\beta \langle m|_\beta$$

$$= \sum_{\alpha, \beta} |i\rangle_\alpha \langle k|_\alpha |j\rangle_\beta \langle m|_\beta - \underbrace{\langle k|j\rangle}_{\delta_{kj}} \sum_\alpha |i\rangle_\alpha \langle m|_\alpha$$

$$= a_i^\dagger a_k a_j^\dagger a_m - a_i^\dagger \underbrace{[a_k, a_j^\dagger]_\mp}_{a_k a_j^\dagger \mp a_j^\dagger a_k} a_m$$

$$= \pm a_i^\dagger a_j^\dagger a_k a_m = a_i^\dagger a_j^\dagger a_m a_k \ ,$$

$$(1.3.26)$$

for $\genfrac{}{}{0pt}{}{\text{bosons}}{\text{fermions}}$.

This completes the proof of the form (1.3.24).

1.4 Fermions

1.4.1 States, Fock Space, Creation and Annihilation Operators

For fermions, one needs to consider the states $S_- |i_1, i_2, \ldots, i_N\rangle$ defined in (1.2.2), which can also be represented in the form of a determinant:

$$S_- |i_1, i_2, \ldots, i_N\rangle = \frac{1}{\sqrt{N!}} \begin{vmatrix} |i_1\rangle_1 & |i_1\rangle_2 & \cdots & |i_1\rangle_N \\ \vdots & \vdots & \ddots & \vdots \\ |i_N\rangle_1 & |i_N\rangle_2 & \cdots & |i_N\rangle_N \end{vmatrix} \ . \qquad (1.4.1)$$

The determinants of one-particle states are called *Slater determinants*. If any of the single-particle states in (1.4.1) are the same, the result is zero. This is a statement of the Pauli principle: two identical fermions must not occupy the same state. On the other hand, when all the i_α are different, then this antisymmetrized state is normalized to 1. In addition, we have

$$S_- \,|i_2, i_1, \ldots\rangle = -S_- \,|i_1, i_2, \ldots\rangle \,. \tag{1.4.2}$$

This dependence on the order is a general property of determinants.

Here, too, we shall characterize the states by specifying their occupation numbers, which can now take the values 0 and 1. The state with n_1 particles in state 1 and n_2 particles in state 2, etc., is

$$|n_1, n_2, \ldots\rangle \,.$$

The state in which there are no particles is the vacuum state, represented by

$$|0\rangle = |0, 0, \ldots\rangle \,.$$

This state must not be confused with the null vector!

We combine these states (vacuum state, single-particle states, two-particle states, ...) to give a state space. In other words, we form the direct sum of the state spaces for the various fixed particle numbers. For fermions, this space is once again known as *Fock space*. In this state space a scalar product is defined as follows:

$$\langle n_1, n_2, \ldots | n_1', n_2', \ldots\rangle = \delta_{n_1, n_1'} \delta_{n_2, n_2'} \cdots \,; \tag{1.4.3a}$$

i.e., for states with equal particle number (from a single subspace), it is identical to the previous scalar product, and for states from different subspaces it always vanishes. Furthermore, we have the completeness relation

$$\sum_{n_1=0}^{1} \sum_{n_2=0}^{1} \ldots |n_1, n_2, \ldots\rangle \langle n_1, n_2, \ldots| = \mathbb{1} \,. \tag{1.4.3b}$$

Here, we wish to introduce creation operators $a_i^\dagger$ once again. These must be defined such that the result of applying them twice is zero. Furthermore, the order in which they are applied must play a role. We thus define the creation operators $a_i^\dagger$ by

$$S_- \,|i_1, i_2, \ldots, i_N\rangle = a_{i_1}^\dagger a_{i_2}^\dagger \ldots a_{i_N}^\dagger \,|0\rangle$$
$$S_- \,|i_2, i_1, \ldots, i_N\rangle = a_{i_2}^\dagger a_{i_1}^\dagger \ldots a_{i_N}^\dagger \,|0\rangle \,. \tag{1.4.4}$$

Since these states are equal except in sign, the anticommutator is

$$\{a_i^\dagger, a_j^\dagger\} = 0, \tag{1.4.5a}$$

which also implies the impossibility of double occupation

$$\left(a_i^\dagger\right)^2 = 0. \tag{1.4.5b}$$

The anticommutator encountered in (1.4.5a) and the commutator of two operators A and B are defined by

$$\begin{aligned}\{A, B\} &\equiv [A, B]_+ \equiv AB + BA \\ [A, B] &\equiv [A, B]_- \equiv AB - BA\,.\end{aligned} \tag{1.4.6}$$

Given these preliminaries, we can now address the precise formulation. If one wants to characterize the states by means of occupation numbers, one has to choose a particular ordering of the states. This is arbitrary but, once chosen, must be adhered to. The states are then represented as

$$|n_1, n_2, \ldots\rangle = \left(a_1^\dagger\right)^{n_1} \left(a_2^\dagger\right)^{n_2} \ldots |0\rangle\,, \quad n_i = 0, 1. \tag{1.4.7}$$

The effect of the operator $a_i^\dagger$ must be

$$a_i^\dagger |\ldots, n_i, \ldots\rangle = (1 - n_i)(-1)^{\sum_{j<i} n_j} |\ldots, n_i + 1, \ldots\rangle\,. \tag{1.4.8}$$

The number of particles is increased by 1, but for a state that is already occupied, the factor $(1 - n_i)$ yields zero. The phase factor corresponds to the number of anticommutations necessary to bring the $a_i^\dagger$ to the position i. The adjoint relation reads:

$$\langle\ldots, n_i, \ldots| a_i = (1 - n_i)(-1)^{\sum_{j<i} n_j} \langle\ldots, n_i + 1, \ldots|\,. \tag{1.4.9}$$

This yields the matrix element

$$\langle\ldots, n_i, \ldots| a_i |\ldots, n_i', \ldots\rangle = (1 - n_i)(-1)^{\sum_{j<i} n_j} \delta_{n_i+1, n_i'}\,. \tag{1.4.10}$$

We now calculate

$$\begin{aligned}a_i |\ldots, n_i', \ldots\rangle &= \sum_{n_i} |n_i\rangle \langle n_i| a_i |n_i'\rangle \\ &= \sum_{n_i} |n_i\rangle (1 - n_i)(-1)^{\sum_{j<i} n_j} \delta_{n_i+1, n_i'} \\ &= (2 - n_i')(-1)^{\sum_{j<i} n_j} |\ldots, n_i' - 1, \ldots\rangle\, n_i'\,.\end{aligned} \tag{1.4.11}$$

Here, we have introduced the factor n_i', since, for $n_i' = 0$, the Kronecker delta $\delta_{n_i+1, n_i'} = 0$ always gives zero. The factor n_i' also ensures that the right-hand side cannot become equal to the state $|\ldots, n_i' - 1, \ldots\rangle = |\ldots, -1, \ldots\rangle$.

To summarize, the effects of the creation and annihilation operators are

$$\begin{aligned}a_i^\dagger |\ldots, n_i, \ldots\rangle &= (1 - n_i)(-1)^{\sum_{j<i} n_j} |\ldots, n_i + 1, \ldots\rangle \\ a_i |\ldots, n_i, \ldots\rangle &= n_i(-1)^{\sum_{j<i} n_j} |\ldots, n_i - 1, \ldots\rangle\,.\end{aligned} \tag{1.4.12}$$

It follows from this that

$$a_i a_i^\dagger |\ldots, n_i, \ldots\rangle = (1 - n_i)(-1)^{2\sum_{j<i} n_j}(n_i + 1)|\ldots, n_i, \ldots\rangle$$
$$= (1 - n_i)|\ldots, n_i, \ldots\rangle \qquad (1.4.13a)$$
$$a_i^\dagger a_i |\ldots, n_i, \ldots\rangle = n_i(-1)^{2\sum_{j<i} n_j}(1 - n_i + 1)|\ldots, n_i, \ldots\rangle$$
$$= n_i |\ldots, n_i, \ldots\rangle \,, \qquad (1.4.13b)$$

since for $n_i \in \{0, 1\}$ we have $n_i^2 = n_i$ and $(-1)^{2\sum_{j<i} n_j} = 1$. On account of the property (1.4.13b) one can regard $a_i^\dagger a_i$ as the occupation-number operator for the state $|i\rangle$. By taking the sum of (1.4.13a,b), one obtains the anticommutator

$$[a_i, a_i^\dagger]_+ = 1.$$

In the anticommutator $[a_i, a_j^\dagger]_+$ with $i \neq j$, the phase factor of the two terms is different:

$$[a_i, a_j^\dagger]_+ \propto (1 - n_j)n_i(1 - 1) = 0 \,.$$

Likewise, $[a_i, a_j]_+$ for $i \neq j$, also has different phase factors in the two summands and, since $a_i a_i |\ldots, n_i, \ldots\rangle \propto n_i(n_i - 1) = 0$, one obtains the following *anticommutation rules* for *fermions:*

$$[a_i, a_j]_+ = 0, \quad [a_i^\dagger, a_j^\dagger]_+ = 0, \quad [a_i, a_j^\dagger]_+ = \delta_{ij}. \qquad (1.4.14)$$

1.4.2 Single- and Many-Particle Operators

For fermions, too, the operators can be expressed in terms of creation and annihilation operators. The form is exactly the same as for bosons, (1.3.21) and (1.3.24). Now, however, one has to pay special attention to the order of the creation and annihilation operators.
The important relation

$$\sum_\alpha |i\rangle_\alpha \langle j|_\alpha = a_i^\dagger a_j \,, \qquad (1.4.15)$$

from which, according to (1.3.26), one also obtains two-particle (and many-particle) operators, can be proved as follows: Given the state $S_- |i_1, i_2, \ldots, i_N\rangle$, we assume, without loss of generality, the arrangement to be $i_1 < i_2 < \ldots < i_N$. Application of the left-hand side of (1.4.15) gives

$$\sum_\alpha |i\rangle_\alpha \langle j|_\alpha S_- |i_1, i_2, \ldots, i_N\rangle = S_- \sum_\alpha |i\rangle_\alpha \langle j|_\alpha |i_1, i_2, \ldots, i_N\rangle$$
$$= n_j(1 - n_i)S_- |i_1, i_2, \ldots, i_N\rangle \big|_{j \to i} \,.$$

The symbol $|_{j\to i}$ implies that the state $|j\rangle$ is replaced by $|i\rangle$. In order to bring the i into the right position, one has to carry out $\sum_{k<j} n_k + \sum_{k<i} n_k$ permutations of rows for $i \leq j$ and $\sum_{k<j} n_k + \sum_{k<i} n_k - 1$ permutations for $i > j$.

This yields the same phase factor as does the right-hand side of (1.4.15):

$$a_i^\dagger a_j \,|\ldots, n_i, \ldots, n_j, \ldots\rangle = n_j(-1)^{\Sigma_{k<j}\, n_k}\, a_i^\dagger\, |\ldots, n_i, \ldots, n_j-1, \ldots\rangle$$

$$= n_i(1-n_i)(-1)^{\Sigma_{k<i}\, n_k + \Sigma_{k<j}\, n_k - \delta_{i>j}}\,|\ldots, n_i+1, \ldots, n_j-1, \ldots\rangle\,.$$

In summary, for bosons and fermions, the single- and two-particle operators can be written, respectively, as

$$T = \sum_{i,j} t_{ij} a_i^\dagger a_j \tag{1.4.16a}$$

$$F = \frac{1}{2} \sum_{i,j,k,m} \langle i,j|\, f^{(2)}\,|k,m\rangle\, a_i^\dagger a_j^\dagger a_m a_k, \tag{1.4.16b}$$

where the operators a_i obey the cómmutation relations (1.3.7) for bosons and, for fermions, the anticommutation relations (1.4.14). The Hamiltonian of a many-particle system with kinetic energy T, potential energy U and a two-particle interaction $f^{(2)}$ has the form

$$H = \sum_{i,j}(t_{ij} + U_{ij})a_i^\dagger a_j + \frac{1}{2}\sum_{i,j,k,m} \langle i,j|\, f^{(2)}\,|k,m\rangle\, a_i^\dagger a_j^\dagger a_m a_k\,, \tag{1.4.16c}$$

where the matrix elements are defined in (1.3.21, 1.3.22, 1.3.25) and, for fermions, particular attention must be paid to the order of the two annihilation operators in the two-particle operator.

From this point on, the development of the theory can be presented simultaneously for bosons and fermions.

1.5 Field Operators

1.5.1 Transformations Between Different Basis Systems

Consider two basis systems $\{|i\rangle\}$ and $\{|\lambda\rangle\}$. What is the relationship between the operators a_i and a_λ?
The state $|\lambda\rangle$ can be expanded in the basis $\{|i\rangle\}$:

$$|\lambda\rangle = \sum_i |i\rangle \langle i|\lambda\rangle\,. \tag{1.5.1}$$

The operator $a_i^\dagger$ creates particles in the state $|i\rangle$. Hence, the superposition $\sum_i \langle i|\lambda\rangle\, a_i^\dagger$ yields one particle in the state $|\lambda\rangle$. This leads to the relation

$$a_\lambda^\dagger = \sum_i \langle i|\lambda\rangle\, a_i^\dagger \tag{1.5.2a}$$

with the adjoint

$$a_\lambda = \sum_i \langle \lambda | i \rangle a_i. \tag{1.5.2b}$$

The position eigenstates $|\mathbf{x}\rangle$ represent an important special case

$$\langle \mathbf{x} | i \rangle = \varphi_i(\mathbf{x}), \tag{1.5.3}$$

where $\varphi_i(\mathbf{x})$ is the single-particle wave function in the coordinate representation. The creation and annihilation operators corresponding to the eigenstates of position are called field operators.

1.5.2 Field Operators

The *field operators* are defined by

$$\psi(\mathbf{x}) = \sum_i \varphi_i(\mathbf{x}) a_i \tag{1.5.4a}$$

$$\psi^\dagger(\mathbf{x}) = \sum_i \varphi_i^*(\mathbf{x}) a_i^\dagger . \tag{1.5.4b}$$

The operator $\psi^\dagger(\mathbf{x})$ ($\psi(\mathbf{x})$) creates (annihilates) a particle in the position eigenstate $|\mathbf{x}\rangle$, i.e., at the position $\mathbf{x}$. The field operators obey the following commutation relations:

$$[\psi(\mathbf{x}), \psi(\mathbf{x}')]_\pm = 0 , \tag{1.5.5a}$$

$$[\psi^\dagger(\mathbf{x}), \psi^\dagger(\mathbf{x}')]_\pm = 0 , \tag{1.5.5b}$$

$$[\psi(\mathbf{x}), \psi^\dagger(\mathbf{x}')]_\pm = \sum_{i,j} \varphi_i(\mathbf{x}) \varphi_j^*(\mathbf{x}') [a_i, a_j^\dagger]_\pm \tag{1.5.5c}$$

$$= \sum_{i,j} \varphi_i(\mathbf{x}) \varphi_j^*(\mathbf{x}') \delta_{ij} = \delta^{(3)}(\mathbf{x} - \mathbf{x}') ,$$

where the upper sign applies to fermions and the lower one to bosons.

We shall now express a few important operators in terms of the field operators.

Kinetic energy[7]

$$\sum_{i,j} a_i^\dagger T_{ij} a_j = \sum_{i,j} \int d^3x \, a_i^\dagger \varphi_i^*(\mathbf{x}) \left(-\frac{\hbar^2}{2m} \nabla^2 \right) \varphi_j(\mathbf{x}) a_j$$

$$= \frac{\hbar^2}{2m} \int d^3x \, \nabla \psi^\dagger(\mathbf{x}) \nabla \psi(\mathbf{x}) \tag{1.5.6a}$$

---

[7] The second line in (1.5.6a) holds when the wave function on which the operator acts decreases sufficiently fast at infinity that one can neglect the surface contribution to the partial integration.

Single-particle potential

$$\sum_{i,j} a_i^\dagger U_{ij} a_j = \sum_{i,j} \int d^3x \, a_i^\dagger \varphi_i^*(\mathbf{x}) U(\mathbf{x}) \varphi_j(\mathbf{x}) a_j$$

$$= \int d^3x \, U(\mathbf{x}) \psi^\dagger(\mathbf{x}) \psi(\mathbf{x}) \tag{1.5.6b}$$

Two-particle interaction or any two-particle operator

$$\frac{1}{2} \sum_{i,j,k,m} \int d^3x d^3x' \, \varphi_i^*(\mathbf{x}) \varphi_j^*(\mathbf{x}') V(\mathbf{x},\mathbf{x}') \varphi_k(\mathbf{x}) \varphi_m(\mathbf{x}') a_i^\dagger a_j^\dagger a_m a_k$$

$$= \frac{1}{2} \int d^3x d^3x' \, V(\mathbf{x},\mathbf{x}') \psi^\dagger(\mathbf{x}) \psi^\dagger(\mathbf{x}') \psi(\mathbf{x}') \psi(\mathbf{x}) \tag{1.5.6c}$$

Hamiltonian

$$H = \int d^3x \left(\frac{\hbar^2}{2m} \boldsymbol{\nabla} \psi^\dagger(\mathbf{x}) \boldsymbol{\nabla} \psi(\mathbf{x}) + U(\mathbf{x}) \psi^\dagger(\mathbf{x}) \psi(\mathbf{x}) \right) +$$

$$\frac{1}{2} \int d^3x d^3x' \, \psi^\dagger(\mathbf{x}) \psi^\dagger(\mathbf{x}') V(\mathbf{x},\mathbf{x}') \psi(\mathbf{x}') \psi(\mathbf{x}) \tag{1.5.6d}$$

Particle density (particle-number density)
The particle-density operator is given by

$$n(\mathbf{x}) = \sum_\alpha \delta^{(3)}(\mathbf{x} - \mathbf{x}_\alpha) \, . \tag{1.5.7}$$

Hence its representation in terms of creation and annihilation operators is

$$n(\mathbf{x}) = \sum_{i,j} a_i^\dagger a_j \int d^3y \, \varphi_i^*(\mathbf{y}) \delta^{(3)}(\mathbf{x} - \mathbf{y}) \varphi_j(\mathbf{y})$$

$$= \sum_{i,j} a_i^\dagger a_j \varphi_i^*(\mathbf{x}) \varphi_j(\mathbf{x}). \tag{1.5.8}$$

This representation is valid in any basis and can also be expressed in terms of the field operators

$$n(\mathbf{x}) = \psi^\dagger(\mathbf{x}) \psi(\mathbf{x}). \tag{1.5.9}$$

Total-particle-number operator

$$\hat{N} = \int d^3x \, n(\mathbf{x}) = \int d^3x \, \psi^\dagger(\mathbf{x}) \psi(\mathbf{x}) \, . \tag{1.5.10}$$

Formally, at least, the particle-density operator (1.5.9) of the many-particle system looks like the probability density of a particle in the state $\psi(\mathbf{x})$. However, the analogy is no more than a formal one since the former is an operator and the latter a complex function. This formal correspondence has given rise to the term *second quantization*, since the operators, in the creation and annihilation operator formalism, can be obtained by replacing the wave function $\psi(\mathbf{x})$ in the single-particle densities by the operator $\psi(\mathbf{x})$. This immediately enables one to write down, e.g., the current-density operator (see Problem 1.6)

$$\mathbf{j}(\mathbf{x}) = \frac{\hbar}{2im}[\psi^\dagger(\mathbf{x})\boldsymbol{\nabla}\psi(\mathbf{x}) - (\boldsymbol{\nabla}\psi^\dagger(\mathbf{x}))\psi(\mathbf{x})] . \tag{1.5.11}$$

The kinetic energy (1.5.12) has a formal similarity to the expectation value of the kinetic energy of a single particle, where, however, the wavefunction is replaced by the field operator.

Remark. The representations of the operators in terms of field operators that we found above could also have been obtained directly. For example, for the particle-number density

$$\int d^3\xi \, d^3\xi' \, \psi^\dagger(\boldsymbol{\xi}) \, \langle\boldsymbol{\xi}| \, \delta^{(3)}(\mathbf{x}-\widehat{\boldsymbol{\xi}}) \, |\boldsymbol{\xi}'\rangle \, \psi(\boldsymbol{\xi}') = \psi^\dagger(\mathbf{x})\psi(\mathbf{x}), \tag{1.5.12}$$

where $\widehat{\boldsymbol{\xi}}$ is the position operator of a single particle and where we have made use of the fact that the matrix element within the integral is equal to $\delta^{(3)}(\mathbf{x}-\boldsymbol{\xi})\delta^{(3)}(\boldsymbol{\xi}-\boldsymbol{\xi}')$. In general, for a k-particle operator V_k:

$$\int d^3\xi_1 \ldots d^3\xi_k d^3\xi_1' \ldots d^3\xi_k' \, \psi^\dagger(\boldsymbol{\xi}_1) \ldots \psi^\dagger(\boldsymbol{\xi}_k)$$
$$\langle\boldsymbol{\xi}_1\boldsymbol{\xi}_2 \ldots \boldsymbol{\xi}_k| \, V_k \, |\boldsymbol{\xi}_1'\boldsymbol{\xi}_2' \ldots \boldsymbol{\xi}_k'\rangle \, \psi(\boldsymbol{\xi}_k') \ldots \psi(\boldsymbol{\xi}_1'). \tag{1.5.13}$$

1.5.3 Field Equations

The *equations of motion* of the field operators $\psi(\mathbf{x},t)$ in the *Heisenberg representation*

$$\psi(\mathbf{x},t) = e^{iHt/\hbar} \, \psi(\mathbf{x},0) \, e^{-iHt/\hbar} \tag{1.5.14}$$

read, for the Hamiltonian (1.5.6d),

$$i\hbar\frac{\partial}{\partial t}\psi(\mathbf{x},t) = \left(-\frac{\hbar^2}{2m}\boldsymbol{\nabla}^2 + U(\mathbf{x})\right)\psi(\mathbf{x},t) +$$
$$+ \int d^3x' \, \psi^\dagger(\mathbf{x}',t)V(\mathbf{x},\mathbf{x}')\psi(\mathbf{x}',t)\psi(\mathbf{x},t). \tag{1.5.15}$$

The structure is that of a nonlinear Schrödinger equation, another reason for using the expression "second quantization".

Proof: One starts from the Heisenberg equation of motion

$$i\hbar\frac{\partial}{\partial t}\psi(\mathbf{x},t) = -[H,\psi(\mathbf{x},t)] = -e^{iHt/\hbar}\,[H,\psi(\mathbf{x},0)]\,e^{-iHt/\hbar} . \tag{1.5.16}$$

Using the relation

$$[AB,C]_- = A[B,C]_\pm \mp [A,C]_\pm B \quad \begin{matrix} \text{Fermi} \\ \text{Bose} \end{matrix} , \tag{1.5.17}$$

one obtains for the commutators with the kinetic energy:

$$\int d^3x' \frac{\hbar^2}{2m} [\boldsymbol{\nabla}' \psi^\dagger(\mathbf{x}') \boldsymbol{\nabla}' \psi(\mathbf{x}'), \psi(\mathbf{x})]$$

$$= \int d^3x' \frac{\hbar^2}{2m} (-\boldsymbol{\nabla}' \delta^{(3)}(\mathbf{x}' - \mathbf{x}) \cdot \boldsymbol{\nabla}' \psi(\mathbf{x}')) = \frac{\hbar^2}{2m} \boldsymbol{\nabla}^2 \psi(\mathbf{x}) \ ,$$

the potential energy:

$$\int d^3x' \, U(\mathbf{x}') [\psi^\dagger(\mathbf{x}') \psi(\mathbf{x}'), \psi(\mathbf{x})]$$

$$= \int d^3x' \, U(\mathbf{x}') (-\delta^{(3)}(\mathbf{x}' - \mathbf{x}) \psi(\mathbf{x}')) = -U(\mathbf{x}) \psi(\mathbf{x}) \ ,$$

and the interaction:

$$\frac{1}{2} \left[\int d^3x' d^3x'' \, \psi^\dagger(\mathbf{x}') \psi^\dagger(\mathbf{x}'') V(\mathbf{x}', \mathbf{x}'') \psi(\mathbf{x}'') \psi(\mathbf{x}'), \psi(\mathbf{x}) \right]$$

$$= \frac{1}{2} \int d^3x' \int d^3x'' \, [\psi^\dagger(\mathbf{x}') \psi^\dagger(\mathbf{x}''), \psi(\mathbf{x})] V(\mathbf{x}', \mathbf{x}'') \psi(\mathbf{x}'') \psi(\mathbf{x}')$$

$$= \frac{1}{2} \int d^3x' \int d^3x'' \, \left\{ \pm \delta^{(3)}(\mathbf{x}'' - \mathbf{x}) \psi^\dagger(\mathbf{x}') - \psi^\dagger(\mathbf{x}'') \delta^{(3)}(\mathbf{x}' - \mathbf{x}) \right\}$$

$$\times \, V(\mathbf{x}', \mathbf{x}'') \psi(\mathbf{x}'') \psi(\mathbf{x}')$$

$$= - \int d^3x' \, \psi^\dagger(\mathbf{x}') V(\mathbf{x}, \mathbf{x}') \psi(\mathbf{x}') \psi(\mathbf{x}).$$

In this last equation, (1.5.17) and (1.5.5c) are used to proceed from the second line. Also, after the third line, in addition to $\psi(\mathbf{x}'') \psi(\mathbf{x}') = \mp \psi(\mathbf{x}') \psi(\mathbf{x}'')$, the symmetry $V(\mathbf{x}, \mathbf{x}') = V(\mathbf{x}', \mathbf{x})$ is exploited. Together, these expressions give the equation of motion (1.5.15) of the field operator, which is also known as the *field equation*.

The equation of motion for the adjoint field operator reads:

$$i\hbar \dot{\psi}^\dagger(\mathbf{x}, t) = - \left\{ -\frac{\hbar^2}{2m} \boldsymbol{\nabla}^2 + U(\mathbf{x}) \right\} \psi^\dagger(\mathbf{x}, t)$$

$$- \int d^3x' \, \psi^\dagger(\mathbf{x}, t) \psi^\dagger(\mathbf{x}', t) V(\mathbf{x}, \mathbf{x}') \psi(\mathbf{x}', t), \qquad (1.5.18)$$

where it is assumed that $V(\mathbf{x}, \mathbf{x}')^* = V(\mathbf{x}, \mathbf{x}')$.

If (1.5.15) is multiplied from the left by $\psi^\dagger(\mathbf{x}, t)$ and (1.5.18) from the right by $\psi(\mathbf{x}, t)$, one obtains the equation of motion for the density operator

$$\dot{n}(\mathbf{x}, t) = \left(\psi^\dagger \dot{\psi} + \dot{\psi}^\dagger \psi \right) = \frac{1}{i\hbar} \left(-\frac{\hbar^2}{2m} \right) \left\{ \psi^\dagger \boldsymbol{\nabla}^2 \psi - (\boldsymbol{\nabla}^2 \psi^\dagger) \, \psi \right\} \ ,$$

and thus

$$\dot{n}(\mathbf{x}) = -\boldsymbol{\nabla} \mathbf{j}(\mathbf{x}), \qquad (1.5.19)$$

where $\mathbf{j}(\mathbf{x})$ is the particle current density defined in (1.5.11). Equation (1.5.19) is the continuity equation for the particle-number density.

1.6 Momentum Representation

1.6.1 Momentum Eigenfunctions and the Hamiltonian

The momentum representation is particularly useful in translationally invariant systems. We base our considerations on a rectangular normalization volume of dimensions L_x, L_y and L_z. The momentum eigenfunctions, which are used in place of $\varphi_i(\mathbf{x})$, are normalized to 1 and are given by

$$\varphi_{\mathbf{k}}(\mathbf{x}) = e^{i\mathbf{k}\cdot\mathbf{x}}/\sqrt{V} \qquad (1.6.1)$$

with the volume $V = L_x L_y L_z$. By assuming periodic boundary conditions

$$e^{ik(x+L_x)} = e^{ikx}, \text{etc.}, \qquad (1.6.2a)$$

the allowed values of the wave vector $\mathbf{k}$ are restricted to

$$\mathbf{k} = 2\pi\left(\frac{n_x}{L_x}, \frac{n_y}{L_y}, \frac{n_z}{L_z}\right), n_x = 0, \pm1, \dots, n_y = 0, \pm1, \dots, n_z = 0, \pm1, \dots .$$
$$(1.6.2b)$$

The eigenfunctions (1.6.1) obey the following orthonormality relation:

$$\int d^3x \varphi_{\mathbf{k}}^*(\mathbf{x})\varphi_{\mathbf{k}'}(\mathbf{x}) = \delta_{\mathbf{k},\mathbf{k}'}. \qquad (1.6.3)$$

In order to represent the Hamiltonian in second-quantized form, we need the matrix elements of the operators that it contains. The kinetic energy is proportional to

$$\int \varphi_{\mathbf{k}'}^* \left(-\nabla^2\right) \varphi_{\mathbf{k}} d^3x = \delta_{\mathbf{k},\mathbf{k}'}\mathbf{k}^2 \qquad (1.6.4a)$$

and the matrix element of the single-particle potential is given by the Fourier transform of the latter:

$$\int \varphi_{\mathbf{k}'}^*(\mathbf{x})U(\mathbf{x})\varphi_{\mathbf{k}}(\mathbf{x})d^3x = \frac{1}{V}U_{\mathbf{k}'-\mathbf{k}}. \qquad (1.6.4b)$$

For two-particle potentials $V(\mathbf{x} - \mathbf{x}')$ that depend only on the relative coordinates of the two particles, it is useful to introduce their Fourier transform

$$V_{\mathbf{q}} = \int d^3x e^{-i\mathbf{q}\cdot\mathbf{x}}V(\mathbf{x}), \qquad (1.6.5a)$$

and also its inverse

$$V(\mathbf{x}) = \frac{1}{V}\sum_{\mathbf{q}} V_{\mathbf{q}}e^{i\mathbf{q}\cdot\mathbf{x}}. \qquad (1.6.5b)$$

For the matrix element of the two-particle potential, one then finds

$$\langle \mathbf{p}', \mathbf{k}' | V(\mathbf{x} - \mathbf{x}') | \mathbf{p}, \mathbf{k} \rangle$$

$$= \frac{1}{V^2} \int d^3x \, d^3x' e^{-i\mathbf{p}'\cdot\mathbf{x}} e^{-i\mathbf{k}'\cdot\mathbf{x}'} V(\mathbf{x} - \mathbf{x}') e^{i\mathbf{k}\cdot\mathbf{x}'} e^{i\mathbf{p}\cdot\mathbf{x}}$$

$$= \frac{1}{V^3} \sum_{\mathbf{q}} V_{\mathbf{q}} \int d^3x \int d^3x' e^{-i\mathbf{p}'\cdot\mathbf{x} - i\mathbf{k}'\cdot\mathbf{x}' + i\mathbf{q}\cdot(\mathbf{x}-\mathbf{x}') + i\mathbf{k}\cdot\mathbf{x}' + i\mathbf{p}\cdot\mathbf{x}}$$

$$= \frac{1}{V^3} \sum_{\mathbf{q}} V_{\mathbf{q}} V \delta_{-\mathbf{p}'+\mathbf{q}+\mathbf{p},0} V \delta_{-\mathbf{k}'-\mathbf{q}+\mathbf{k},0}.$$

$$(1.6.5c)$$

Inserting (1.6.5a,b,c) into the general representation (1.4.16c) of the Hamiltonian yields:

$$H = \sum_{\mathbf{k}} \frac{(\hbar\mathbf{k})^2}{2m} a_{\mathbf{k}}^{\dagger} a_{\mathbf{k}} + \frac{1}{V} \sum_{\mathbf{k}',\mathbf{k}} U_{\mathbf{k}'-\mathbf{k}} a_{\mathbf{k}'}^{\dagger} a_{\mathbf{k}} + \frac{1}{2V} \sum_{\mathbf{q},\mathbf{p},\mathbf{k}} V_{\mathbf{q}} a_{\mathbf{p}+\mathbf{q}}^{\dagger} a_{\mathbf{k}-\mathbf{q}}^{\dagger} a_{\mathbf{k}} a_{\mathbf{p}}.$$

$$(1.6.6)$$

The creation operators of a particle with wave vector $\mathbf{k}$ (i.e., in the state $\varphi_{\mathbf{k}}$) are denoted by $a_{\mathbf{k}}^{\dagger}$ and the annihilation operators by $a_{\mathbf{k}}$. Their commutation relations are

$$[a_{\mathbf{k}}, a_{\mathbf{k}'}]_{\pm} = 0, \quad [a_{\mathbf{k}}^{\dagger}, a_{\mathbf{k}'}^{\dagger}]_{\pm} = 0 \quad \text{and} \quad [a_{\mathbf{k}}, a_{\mathbf{k}'}^{\dagger}]_{\pm} = \delta_{\mathbf{k}\mathbf{k}'}. \qquad (1.6.7)$$

The interaction term allows a pictorial interpretation. It causes the annihilation of two particles with wave vectors $\mathbf{k}$ and $\mathbf{p}$ and creates in their place two particles with wave vectors $\mathbf{k} - \mathbf{q}$ and $\mathbf{p} + \mathbf{q}$. This is represented in Fig. 1.1a. The full lines denote the particles and the dotted lines the interaction potential $V_{\mathbf{q}}$. The amplitude for this transition is proportional to $V_{\mathbf{q}}$. This dia-

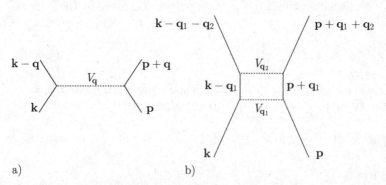

a)   b)

Fig. 1.1. a) Diagrammatic representation of the interaction term in the Hamiltonian (1.6.6) **b)** The diagrammatic representation of the double scattering of two particles

grammatic form is a useful way of representing the perturbation-theoretical description of such processes. The double scattering of two particles can be represented as shown in Fig. 1.1b, where one must sum over all intermediate states.

1.6.2 Fourier Transformation of the Density

The other operators considered in the previous section can also be expressed in the momentum representation. As an important example, we shall look at the density operator. The Fourier transform of the density operator[8] is defined by

$$\hat{n}_{\mathbf{q}} = \int d^3x \, n(\mathbf{x}) e^{-i\mathbf{q}\cdot\mathbf{x}} = \int d^3x \, \psi^\dagger(\mathbf{x})\psi(\mathbf{x}) e^{-i\mathbf{q}\cdot\mathbf{x}} \ . \tag{1.6.8}$$

From (1.5.4a,b) we insert

$$\psi(\mathbf{x}) = \frac{1}{\sqrt{V}} \sum_{\mathbf{p}} e^{i\mathbf{p}\cdot\mathbf{x}} a_{\mathbf{p}}, \ \psi^\dagger(\mathbf{x}) = \frac{1}{\sqrt{V}} \sum_{\mathbf{p}} e^{-i\mathbf{p}\cdot\mathbf{x}} a_{\mathbf{p}}^\dagger \ , \tag{1.6.9}$$

which yields

$$\hat{n}_{\mathbf{q}} = \int d^3x \frac{1}{V} \sum_{\mathbf{p}} \sum_{\mathbf{k}} e^{-i\mathbf{p}\cdot\mathbf{x}} a_{\mathbf{p}}^\dagger e^{i\mathbf{k}\cdot\mathbf{x}} a_{\mathbf{k}} e^{-i\mathbf{q}\cdot\mathbf{x}} \ ,$$

and thus, with (1.6.3), one finally obtains

$$\hat{n}_{\mathbf{q}} = \sum_{\mathbf{p}} a_{\mathbf{p}}^\dagger a_{\mathbf{p}+\mathbf{q}} \ . \tag{1.6.10}$$

We have thus found the Fourier transform of the density operator in the momentum representation.

The occupation-number operator for the state $|\mathbf{p}\rangle$ is also denoted by $\hat{n}_{\mathbf{p}} \equiv a_{\mathbf{p}}^\dagger a_{\mathbf{p}}$. It will always be clear from the context which one of the two meanings is meant. The operator for the total number of particles (1.3.13) in the momentum representation reads

$$\hat{N} = \sum_{\mathbf{p}} a_{\mathbf{p}}^\dagger a_{\mathbf{p}} \ . \tag{1.6.11}$$

1.6.3 The Inclusion of Spin

Up until now, we have not explicitly considered the spin. One can think of it as being included in the previous formulas as part of the spatial degree

---

[8] The hat on the operator, as used here for $\hat{n}_{\mathbf{q}}$ and previously for the occupation-number operator, will only be retained where it is needed to avoid confusion.

of freedom $\mathbf{x}$. If the spin is to be given explicitly, then one has to make the replacements $\psi(\mathbf{x}) \to \psi_\sigma(\mathbf{x})$ and $a_\mathbf{p} \to a_{\mathbf{p}\sigma}$ and, in addition, introduce the sum over σ, the z component of the spin. The particle-number density, for example, then takes the form

$$n(\mathbf{x}) = \sum_\sigma \psi_\sigma^\dagger(\mathbf{x})\psi_\sigma(\mathbf{x})$$

$$\hat{n}_\mathbf{q} = \sum_{\mathbf{p},\sigma} a_{\mathbf{p}\sigma}^\dagger a_{\mathbf{p}+\mathbf{q}\sigma} \, . \tag{1.6.12}$$

The Hamiltonian for the case of a spin-independent interaction reads:

$$H = \sum_\sigma \int d^3x \left(\frac{\hbar^2}{2m} \boldsymbol{\nabla}\psi_\sigma^\dagger \boldsymbol{\nabla}\psi_\sigma + U(\mathbf{x})\psi_\sigma^\dagger(\mathbf{x})\psi_\sigma(\mathbf{x}) \right)$$

$$+ \frac{1}{2} \sum_{\sigma,\sigma'} \int d^3x\, d^3x'\, \psi_\sigma^\dagger(\mathbf{x})\psi_{\sigma'}^\dagger(\mathbf{x}')V(\mathbf{x},\mathbf{x}')\psi_{\sigma'}(\mathbf{x}')\psi_\sigma(\mathbf{x}) \, , \tag{1.6.13}$$

the corresponding form applying in the momentum representation.

For spin-$\frac{1}{2}$ fermions, the two possible spin quantum numbers for the z component of $\mathbf{S}$ are $\pm\frac{\hbar}{2}$. The spin density operator

$$\mathbf{S}(\mathbf{x}) = \sum_{\alpha=1}^N \delta(\mathbf{x} - \mathbf{x}_\alpha)\mathbf{S}_\alpha \tag{1.6.14a}$$

is, in this case,

$$\mathbf{S}(\mathbf{x}) = \frac{\hbar}{2} \sum_{\sigma,\sigma'} \psi_\sigma^\dagger(\mathbf{x})\boldsymbol{\sigma}_{\sigma\sigma'}\psi_{\sigma'}(\mathbf{x}), \tag{1.6.14b}$$

where $\boldsymbol{\sigma}_{\sigma\sigma'}$ are the matrix elements of the Pauli matrices.

The commutation relations of the field operators and the operators in the momentum representation read:

$$[\psi_\sigma(\mathbf{x}), \psi_{\sigma'}(\mathbf{x}')]_\pm = 0 \, , \qquad [\psi_\sigma^\dagger(\mathbf{x}), \psi_{\sigma'}^\dagger(\mathbf{x}')]_\pm = 0$$

$$[\psi_\sigma(\mathbf{x}), \psi_{\sigma'}^\dagger(\mathbf{x}')]_\pm = \delta_{\sigma\sigma'}\delta(\mathbf{x} - \mathbf{x}') \tag{1.6.15}$$

and

$$[a_{\mathbf{k}\sigma}, a_{\mathbf{k}'\sigma'}]_\pm = 0, \quad [a_{\mathbf{k}\sigma}^\dagger, a_{\mathbf{k}'\sigma'}^\dagger]_\pm = 0, \quad [a_{\mathbf{k}\sigma}, a_{\mathbf{k}'\sigma'}^\dagger]_\pm = \delta_{\mathbf{k}\mathbf{k}'}\delta_{\sigma\sigma'} \, . \tag{1.6.16}$$

The equations of motion are given by

$$i\hbar\frac{\partial}{\partial t}\psi_\sigma(\mathbf{x},t) = \left(-\frac{\hbar^2}{2m}\boldsymbol{\nabla}^2 + U(\mathbf{x}) \right) \psi_\sigma(\mathbf{x},t)$$

$$+ \sum_{\sigma'} \int d^3x'\, \psi_{\sigma'}^\dagger(\mathbf{x}',t)V(\mathbf{x},\mathbf{x}')\psi_{\sigma'}(\mathbf{x}',t)\psi_\sigma(\mathbf{x},t) \tag{1.6.17}$$

and

$$i\hbar\dot{a}_{\mathbf{k}\sigma}(t) = \frac{(\hbar\mathbf{k})^2}{2m}a_{\mathbf{k}\sigma}(t) + \frac{1}{V}\sum_{\mathbf{k}'}U_{\mathbf{k}-\mathbf{k}'}a_{\mathbf{k}'\sigma}(t)$$

$$+\frac{1}{V}\sum_{\mathbf{p},\mathbf{q},\sigma'}V_{\mathbf{q}}a^{\dagger}_{\mathbf{p}+\mathbf{q}\sigma'}(t)a_{\mathbf{p}\sigma'}(t)a_{\mathbf{k}+\mathbf{q}\sigma}(t) . \tag{1.6.18}$$

Problems

1.1 Show that the fully symmetrized (antisymmetrized) basis functions

$$S_{\pm}\varphi_{i_1}(x_1)\varphi_{i_2}(x_2)\,...\,\varphi_{i_N}(x_N)$$

are complete in the space of the symmetric (antisymmetric) wave functions $\psi_{s/a}(x_1, x_2, \ldots, x_N)$.

Hint: Assume that the product states $\varphi_{i_1}(x_1)\ldots\varphi_{i_N}(x_N)$, composed of the single-particle wave functions $\varphi_i(x)$, form a complete basis set, and express $\psi_{s/a}$ in this basis. Show that the expansion coefficients $c^{s/a}_{i_1,\ldots,i_N}$ possess the symmetry property $\frac{1}{\sqrt{N!}}S_{\pm}c^{s/a}_{i_1,\ldots,i_N} = c^{s/a}_{i_1,\ldots,i_N}$. The above assertion then follows directly by utilizing the identity $\frac{1}{\sqrt{N!}}S_{\pm}\psi_{s/a} = \psi_{s/a}$ demonstrated in the main text.

1.2 Consider the three-particle state $|\alpha\rangle|\beta\rangle|\gamma\rangle$, where the particle number is determined by its position in the product.
a) Apply the elements of the permutation group S_3. One thereby finds six different states, which can be combined into four invariant subspaces.
b) Consider the following basis, given in (1.1.16c), of one of these subspaces, comprising two states:

$$|\psi_1\rangle = \frac{1}{\sqrt{12}}\bigg(2\,|\alpha\rangle|\beta\rangle|\gamma\rangle +2\,|\beta\rangle|\alpha\rangle|\gamma\rangle -|\alpha\rangle|\gamma\rangle|\beta\rangle -|\gamma\rangle|\beta\rangle|\alpha\rangle$$

$$-|\gamma\rangle|\alpha\rangle|\beta\rangle -|\beta\rangle|\gamma\rangle|\alpha\rangle\bigg) ,$$

$$|\psi_2\rangle = \frac{1}{2}\bigg(0+0 -|\alpha\rangle|\gamma\rangle|\beta\rangle +|\gamma\rangle|\beta\rangle|\alpha\rangle +|\gamma\rangle|\alpha\rangle|\beta\rangle -|\beta\rangle|\gamma\rangle|\alpha\rangle\bigg)$$

and find the corresponding two-dimensional representation of S_3.

1.3 For a simple harmonic oscillator, $[a, a^{\dagger}] = 1$, (or for the equivalent Bose operator) prove the following relations:

$$[a, e^{\alpha a^{\dagger}}] = \alpha e^{\alpha a^{\dagger}} , \qquad e^{-\alpha a^{\dagger}}\,a\,e^{\alpha a^{\dagger}} = a + \alpha ,$$

$$e^{-\alpha a^{\dagger}}\,e^{\beta a}\,e^{\alpha a^{\dagger}} = e^{\beta\alpha}e^{\beta a} , \qquad e^{\alpha a^{\dagger}a}\,a\,e^{-\alpha a^{\dagger}a} = e^{-\alpha}\,a ,$$

where α and β are complex numbers.
Hint:
a) First demonstrate the validity of the following relations

$$\left[a, f(a^{\dagger})\right] = \frac{\partial}{\partial a^{\dagger}}f(a^{\dagger}) , \qquad \left[a^{\dagger}, f(a)\right] = -\frac{\partial}{\partial a}f(a) .$$

b) In some parts of the problem it is useful to consider the left-hand side of the identity as a function of α, to derive a differential equation for these functions, and then to solve the corresponding initial value problem.

c) The Baker–Hausdorff identity

$$e^A B e^{-A} = B + [A, B] + \frac{1}{2!}[A, [A, B]] + \ldots$$

can likewise be used to prove some of the above relations.

1.4 For independent harmonic oscillators (or noninteracting bosons) described by the Hamiltonian

$$H = \sum_i \epsilon_i a_i^\dagger a_i$$

determine the equation of motion for the creation and annihilation operators in the Heisenberg representation,

$$a_i(t) = e^{iHt/\hbar} a_i e^{-iHt/\hbar} .$$

Give the solution of the equation of motion by (i) solving the corresponding initial value problem and (ii) by explicitly carrying out the commutator operations in the expression $a_i(t) = e^{iHt/\hbar} a_i e^{-iHt/\hbar}$.

1.5 Consider a two-particle potential $V(\mathbf{x}', \mathbf{x}'')$ symmetric in $\mathbf{x}'$ and $\mathbf{x}''$. Calculate the commutator

$$\frac{1}{2}\left[\int d^3x' \int d^3x'' \psi^\dagger(\mathbf{x}')\psi^\dagger(\mathbf{x}'')V(\mathbf{x}',\mathbf{x}'')\psi(\mathbf{x}'')\psi(\mathbf{x}'), \psi(\mathbf{x}) \right],$$

for fermionic and bosonic field operators $\psi(\mathbf{x})$.

1.6 (a) Verify, for an N-particle system, the form of the current-density operator,

$$\mathbf{j}(\mathbf{x}) = \frac{1}{2}\sum_{\alpha=1}^{N}\left\{ \frac{\mathbf{p}_\alpha}{m}, \delta(\mathbf{x}-\mathbf{x}_\alpha) \right\}$$

in second quantization. Use a basis consisting of plane waves. Also give the form of the operator in the momentum representation, i.e., evaluate the integral, $\mathbf{j}(\mathbf{q}) = \int d^3x\, e^{-i\mathbf{q}\cdot\mathbf{x}}\mathbf{j}(\mathbf{x})$.

(b) For spin-$\frac{1}{2}$ particles, determine, in the momentum representation, the spin-density operator,

$$\mathbf{S}(\mathbf{x}) = \sum_{\alpha=1}^{N}\delta(\mathbf{x}-\mathbf{x}_\alpha)\mathbf{S}_\alpha ,$$

in second quantization.

1.7 Consider electrons on a lattice with the single-particle wave function localized at the lattice point $\mathbf{R}_i$ given by $\varphi_{i\sigma}(\mathbf{x}) = \chi_\sigma \varphi_i(\mathbf{x})$ with $\varphi_i(\mathbf{x}) = \phi(\mathbf{x} - \mathbf{R}_i)$. A Hamiltonian, $H = T + V$, consisting of a spin-independent single-particle operator $T = \sum_{\alpha=1}^{N} t_\alpha$ and a two-particle operator $V = \frac{1}{2}\sum_{\alpha \neq \beta} V^{(2)}(\mathbf{x}_\alpha, \mathbf{x}_\beta)$ can be represented in the basis $\{\varphi_{i\sigma}\}$ by

$$H = \sum_{i,j} \sum_\sigma t_{ij} a_{i\sigma}^\dagger a_{j\sigma} + \frac{1}{2} \sum_{i,j,k,l} \sum_{\sigma,\sigma'} V_{ijkl} a_{i\sigma}^\dagger a_{j\sigma'}^\dagger a_{l\sigma'} a_{k\sigma},$$

where the matrix elements are given by $t_{ij} = \langle i \mid t \mid j \rangle$ and $V_{ijkl} = \langle ij \mid V^{(2)} \mid kl \rangle$. If one assumes that the overlap of the wave functions $\varphi_i(\mathbf{x})$ at different lattice points is negligible, one can make the following approximations:

$$t_{ij} = \begin{cases} w & \text{for } i = j, \\ t & \text{for } i \text{ and } j \text{ adjacent sites}, \\ 0 & \text{otherwise} \end{cases}$$

$$V_{ijkl} = V_{ij}\delta_{il}\delta_{jk} \quad \text{with} \quad V_{ij} = \int d^3x \int d^3y \mid \varphi_i(\mathbf{x}) \mid^2 V^{(2)}(\mathbf{x},\mathbf{y}) \mid \varphi_j(\mathbf{y}) \mid^2.$$

(a) Determine the matrix elements V_{ij} for a contact potential

$$V = \frac{\lambda}{2} \sum_{\alpha \neq \beta} \delta(\mathbf{x}_\alpha - \mathbf{x}_\beta)$$

between the electrons for the following cases: (i) "on-site" interaction $i = j$, and (ii) nearest-neighbor interaction, i.e., i and j adjacent lattice points. Assume a square lattice with lattice constant a and wave functions that are Gaussians, $\varphi(\mathbf{x}) = \frac{1}{\Delta^{3/2}\pi^{3/4}} \exp\{-\mathbf{x}^2/2\Delta^2\}$.
(b) In the limit $\Delta \ll a$, the "on-site" interaction $U = V_{ii}$ is the dominant contribution. Determine for this limiting case the form of the Hamiltonian in second quantization. The model thereby obtained is known as the *Hubbard model*.

1.8 Show, for bosons and fermions, that the particle-number operator $\hat{N} = \sum_i a_i^\dagger a_i$ commutes with the Hamiltonian

$$H = \sum_{ij} a_i^\dagger \langle i \mid T \mid j \rangle a_j + \frac{1}{2} \sum_{ijkl} a_i^\dagger a_j^\dagger \langle ij \mid V \mid kl \rangle a_l a_k.$$

1.9 Determine, for bosons and fermions, the thermal expectation value of the occupation-number operator $\hat{n}_i$ for the state $|i\rangle$ in the grand canonical ensemble, whose density matrix is given by

$$\rho_G = \frac{1}{Z_G} e^{-\beta(H - \mu \hat{N})}$$

with $Z_G = \text{Tr}\left(e^{-\beta(H - \mu \hat{N})} \right)$.

1.10 (a) Show, by verifying the relation

$$n(\mathbf{x}) \, |\phi\rangle = \delta(\mathbf{x} - \mathbf{x}') \, |\phi\rangle \ ,$$

that the state

$$|\phi\rangle = \psi^\dagger(\mathbf{x}') \, |0\rangle$$

($|0\rangle$ = vacuum state) describes a particle with the position $\mathbf{x}'$.
(b) The operator for the total particle number reads:

$$\hat{N} = \int d^3x \, n(\mathbf{x}) \ .$$

Show that for spinless particles

$$[\psi(\mathbf{x}), \hat{N}] = \psi(\mathbf{x}) \ .$$

2. Spin-1/2 Fermions

In this chapter, we shall apply the second quantization formalism to a number of simple problems. To begin with, we consider a gas of noninteracting spin-$\frac{1}{2}$ fermions for which we will obtain correlation functions and, subsequently, some properties of the electron gas that take into account the Coulomb interaction. Finally, a compact derivation of the Hartree–Fock equations for atoms will be presented.

2.1 Noninteracting Fermions

2.1.1 The Fermi Sphere, Excitations

In the ground state of N free fermions, $|\phi_0\rangle$, all single-particle states lie within the Fermi sphere (Fig. 2.1), i.e., states with wave number up to k_F, the Fermi wave number, are occupied:

$$|\phi_0\rangle = \prod_{\substack{\mathbf{p} \\ |\mathbf{p}| < k_F}} \prod_{\sigma} a_{\mathbf{p}\sigma}^{\dagger} |0\rangle \ . \tag{2.1.1}$$

Fig. 2.1. The Fermi sphere

The expectation value of the particle-number operator in momentum space is

$$n_{\mathbf{p},\sigma} = \langle\phi_0| a_{\mathbf{p}\sigma}^{\dagger} a_{\mathbf{p}\sigma} |\phi_0\rangle = \begin{cases} 1 & |\mathbf{p}| \leq k_F \\ 0 & |\mathbf{p}| > k_F \end{cases} . \tag{2.1.2}$$

For $|\mathbf{p}| > k_F$, we have $a_{\mathbf{p}\sigma} |\phi_0\rangle = \prod_{\substack{\mathbf{p}' \\ |\mathbf{p}'| < k_F}} \prod_{\sigma'} a_{\mathbf{p}'\sigma'}^{\dagger} a_{\mathbf{p}\sigma} |0\rangle = 0$. According to

(2.1.2), the total particle number is related to the Fermi momentum by[1]

---

[1] $\sum_{\mathbf{k}} f(\mathbf{k}) = \sum_{\mathbf{k}} \frac{\Delta}{\left(\frac{2\pi}{L}\right)^3} f(\mathbf{k}) = \left(\frac{L}{2\pi}\right)^3 \int d^3k f(\mathbf{k})$. The volume of $\mathbf{k}$-space per point is $\Delta = \left(\frac{2\pi}{L}\right)^3$, c.f. Eq. (1.6.2b).

$$N = \sum_{\mathbf{p},\sigma} n_{\mathbf{p}\sigma} = 2 \sum_{|\mathbf{p}| \leq k_F} 1 = 2V \int_0^{k_F} \frac{d^3p}{(2\pi)^3} = \frac{V k_F^3}{3\pi^2}, \tag{2.1.3}$$

whence it follows that

$$k_F^3 = \frac{3\pi^2 N}{V} = 3\pi^2 n. \tag{2.1.4}$$

Here, k_F is the Fermi wave vector, $p_F = \hbar k_F$ the Fermi momentum[2], and $n = \frac{N}{V}$ the mean particle density. The Fermi energy is defined by $\epsilon_F = (\hbar k_F)^2/(2m)$.

For the $\mathbf{x}$-dependence of the ground-state expectation value of the particle density, one obtains

$$\langle n(\mathbf{x}) \rangle = \sum_\sigma \langle \phi_0 | \psi_\sigma^\dagger(\mathbf{x}) \psi_\sigma(\mathbf{x}) | \phi_0 \rangle$$

$$= \sum_\sigma \sum_{\mathbf{p},\mathbf{p}'} \frac{e^{-i\mathbf{p}\cdot\mathbf{x}} e^{i\mathbf{p}'\cdot\mathbf{x}}}{V} \langle \phi_0 | a_{\mathbf{p}\sigma}^\dagger a_{\mathbf{p}'\sigma} | \phi_0 \rangle$$

$$= \sum_\sigma \sum_{\mathbf{p},\mathbf{p}'} \frac{e^{-i(\mathbf{p}-\mathbf{p}')\cdot\mathbf{x}}}{V} \delta_{\mathbf{p}\mathbf{p}'} n_{\mathbf{p}\sigma}$$

$$= \frac{1}{V} \sum_{\mathbf{p},\sigma} n_{\mathbf{p}\sigma} = n.$$

As was to be expected, the density is homogeneous.

The simplest excitation of a degenerate electron gas is obtained by promoting an electron from a state within the Fermi sphere to a state outside this sphere (see Fig. 2.2). One also describes this as the creation of an electron–hole pair; its state is written as

$$|\phi\rangle = a_{\mathbf{k}_2\sigma_2}^\dagger a_{\mathbf{k}_1\sigma_1} |\phi_0\rangle. \tag{2.1.5}$$

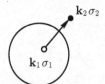

Fig. 2.2. Excited state of a degenerate electron gas; electron–hole pair

The absence of an electron in the state $|\mathbf{k}_1, \sigma_1\rangle$ has an effect similar to that of a positively charged particle (hole). If one defines $b_{\mathbf{k}\sigma} \equiv a_{-\mathbf{k},-\sigma}^\dagger$ and $b_{\mathbf{k}\sigma}^\dagger \equiv a_{-\mathbf{k},-\sigma}$, then the hole annihilation and creation operators b and $b^\dagger$ likewise satisfy anticommutation relations.

---

[2] We denote wave vectors by $\mathbf{p}, \mathbf{q}, \mathbf{k}$ etc. Solely p_F has the dimension of "momentum".

2.1.2 Single-Particle Correlation Function

The correlation function of the field operators in the ground state

$$G_\sigma(\mathbf{x} - \mathbf{x}') = \langle \phi_0 | \psi_\sigma^\dagger(\mathbf{x}) \psi_\sigma(\mathbf{x}') | \phi_0 \rangle \qquad (2.1.6)$$

signifies the probability amplitude that the annihilation of a particle at $\mathbf{x}'$ and the creation of a particle at $\mathbf{x}$ once more yields the initial state. The function $G_\sigma(\mathbf{x} - \mathbf{x}')$ can also be viewed as the probability amplitude for the transition of the state $\psi_\sigma(\mathbf{x}') | \phi_0 \rangle$ (in which one particle at $\mathbf{x}'$ has been removed) into $\psi_\sigma(\mathbf{x}) | \phi_0 \rangle$ (in which one particle at $\mathbf{x}$ has been removed).

$$G_\sigma(\mathbf{x} - \mathbf{x}') = \langle \phi_0 | \sum_{\mathbf{p},\mathbf{p}'} \frac{1}{V} e^{-i\mathbf{p}\cdot\mathbf{x} + i\mathbf{p}'\cdot\mathbf{x}'} a_{\mathbf{p}\sigma}^\dagger a_{\mathbf{p}'\sigma} | \phi_0 \rangle$$

$$= \frac{1}{V} \sum_{\mathbf{p}} e^{-i\mathbf{p}\cdot(\mathbf{x} - \mathbf{x}')} n_{\mathbf{p},\sigma} = \int \frac{d^3 p}{(2\pi)^3} e^{-i\mathbf{p}\cdot(\mathbf{x} - \mathbf{x}')} \Theta(k_F - p)$$

$$= \frac{1}{(2\pi)^2} \int_0^{k_F} dp\, p^2 \int_{-1}^{1} d\eta\, e^{ip|\mathbf{x} - \mathbf{x}'|\eta},$$

$$(2.1.7)$$

where we have used polar coordinates and introduced the abbreviation $\eta = \cos\theta$. The integration over η yields $\frac{e^{ipr} - e^{-ipr}}{ipr}$ with $r = |\mathbf{x} - \mathbf{x}'|$. Thus, we have

$$G_\sigma(\mathbf{x} - \mathbf{x}') = \frac{1}{2\pi^2 r} \int_0^{k_F} dp\, p \sin pr = \frac{1}{2\pi^2 r^3}(\sin k_F r - k_F r \cos k_F r)$$

$$= \frac{3n}{2} \frac{\sin k_F r - k_F r \cos k_F r}{(k_F r)^3}$$

The single-particle correlation function oscillates with a characteristic period of $1/k_F$ under an envelope which falls off to zero (see Fig. 2.3). The values at $r = 0$ and for $r \to \infty$ are $G_\sigma(r = 0) = \frac{n}{2}$, $\lim_{r \to \infty} G_\sigma(r) = 0$; the zeros are determined by $\tan x = x$, i.e., for large x they are at $\frac{n\pi}{2}$.

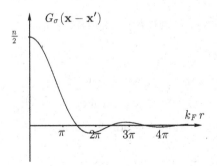

Fig. 2.3. Correlation function $G_\sigma(\mathbf{x} - \mathbf{x}')$ as a function of $k_F r$

Remark. In relation to the first interpretation of $G_\sigma(\mathbf{x})$ given above, it should be noted that the state $\psi_\sigma(\mathbf{x}') |\phi_0\rangle$ is not normalized,

$$\langle\phi_0| \psi_\sigma^\dagger(\mathbf{x}')\psi_\sigma(\mathbf{x}') |\phi_0\rangle = \frac{n}{2}. \tag{2.1.8}$$

The probability amplitude is obtained from the single-particle correlation function by multiplying the latter by the factor $\left(\frac{n}{2}\right)^{-1}$. Now

$$G_\sigma(\mathbf{x} - \mathbf{x}') = \langle\phi_0| \psi_\sigma^\dagger(\mathbf{x})\psi_\sigma(\mathbf{x}') |\phi_0\rangle = \frac{n}{2} \frac{\langle\phi_0| \psi_\sigma^\dagger(\mathbf{x})}{\sqrt{n/2}} \cdot \frac{\psi_\sigma(\mathbf{x}') |\phi_0\rangle}{\sqrt{n/2}}. \tag{2.1.9}$$

The probability amplitude for a transition from the (normalized) state $\frac{\psi_\sigma(\mathbf{x}')|\phi_0\rangle}{\sqrt{n/2}}$ to the (normalized) state $\frac{\psi_\sigma(\mathbf{x})|\phi_0\rangle}{\sqrt{n/2}}$ is equal to the overlap of the two states.

2.1.3 Pair Distribution Function

As a result of the Pauli principle, even noninteracting fermions are corre-lated with one another when they have the same spin. The Pauli principle forbids two fermions with the same spin from possessing the same spatial wave function. Hence, such fermions have a tendency to avoid one another and the probability of their being found close together is relatively small. The Coulomb repulsion enhances this tendency. In the following, however, we will consider only noninteracting fermions.

A measure of the correlations just descibed is the pair distribution func-tion, which can be introduced as follows: Suppose that at point $\mathbf{x}$ a particle is removed from the state $|\phi_0\rangle$ so as to yield the $(N - 1)$-particle state

$$|\phi'(\mathbf{x}, \sigma)\rangle = \psi_\sigma(\mathbf{x}) |\phi_0\rangle. \tag{2.1.10}$$

The density distribution for this state is

$$\begin{aligned}
&\langle\phi'(\mathbf{x}, \sigma)| \psi_{\sigma'}^\dagger(\mathbf{x}')\psi_{\sigma'}(\mathbf{x}') |\phi'(\mathbf{x}, \sigma)\rangle \\
&= \langle\phi_0| \psi_\sigma^\dagger(\mathbf{x})\psi_{\sigma'}^\dagger(\mathbf{x}')\psi_{\sigma'}(\mathbf{x}')\psi_\sigma(\mathbf{x}) |\phi_0\rangle \\
&\equiv \left(\frac{n}{2}\right)^2 g_{\sigma\sigma'}(\mathbf{x} - \mathbf{x}').
\end{aligned} \tag{2.1.11}$$

This expression also defines the *pair distribution function* $g_{\sigma\sigma'}(\mathbf{x} - \mathbf{x}')$. **Note:**

$$\begin{aligned}
\left(\frac{n}{2}\right)^2 g_{\sigma\sigma'}(\mathbf{x} - \mathbf{x}') &= \langle\phi_0| \psi_\sigma^\dagger(\mathbf{x})\psi_\sigma(\mathbf{x})\psi_{\sigma'}^\dagger(\mathbf{x}')\psi_{\sigma'}(\mathbf{x}') |\phi_0\rangle \\
&\quad -\delta_{\sigma\sigma'}\delta(\mathbf{x} - \mathbf{x}') \langle\phi_0| \psi_\sigma^\dagger(\mathbf{x})\psi_{\sigma'}(\mathbf{x}') |\phi_0\rangle \\
&= \langle\phi_0| n(\mathbf{x})n(\mathbf{x}') |\phi_0\rangle - \delta_{\sigma\sigma'}\delta(\mathbf{x} - \mathbf{x}') \langle\phi_0| n(\mathbf{x}) |\phi_0\rangle.
\end{aligned}$$

For the sake of convenience, the pair distribution function is calculated in Fourier space:

$$\left(\frac{n}{2}\right)^2 g_{\sigma\sigma'}(\mathbf{x} - \mathbf{x}') = \frac{1}{V^2} \sum_{\mathbf{k},\mathbf{k}'} \sum_{\mathbf{q},\mathbf{q}'} e^{-i(\mathbf{k}-\mathbf{k}')\cdot\mathbf{x} - i(\mathbf{q}-\mathbf{q}')\cdot\mathbf{x}'}$$

$$\times \langle \phi_0 | a^\dagger_{\mathbf{k}\sigma} a^\dagger_{\mathbf{q}\sigma'} a_{\mathbf{q}'\sigma'} a_{\mathbf{k}'\sigma} | \phi_0 \rangle \,. \tag{2.1.12}$$

We will distinguish two cases:

(i) $\sigma \neq \sigma'$:
For $\sigma \neq \sigma'$, we must have $\mathbf{k} = \mathbf{k}'$ and $\mathbf{q} = \mathbf{q}'$, otherwise the states would be orthogonal to one another:

$$\left(\frac{n}{2}\right)^2 g_{\sigma\sigma'}(\mathbf{x} - \mathbf{x}') = \frac{1}{V^2} \sum_{\mathbf{k},\mathbf{q}} \langle \phi_0 | \hat{n}_{\mathbf{k}\sigma} \hat{n}_{\mathbf{q}\sigma'} | \phi_0 \rangle$$

$$= \frac{1}{V^2} \sum_{\mathbf{k},\mathbf{q}} n_{\mathbf{k}\sigma} n_{\mathbf{q}\sigma'}$$

$$= \frac{1}{V^2} N_\sigma N_{\sigma'} = \frac{1}{V^2} \frac{N}{2} \cdot \frac{N}{2} = \left(\frac{n}{2}\right)^2 \,.$$

Thus, for $\sigma \neq \sigma'$,

$$g_{\sigma\sigma'}(\mathbf{x} - \mathbf{x}') = 1 \tag{2.1.13}$$

independent of the separation. Particles with opposite spin are not affected by the Pauli principle.

(ii) $\sigma = \sigma'$:
For $\sigma = \sigma'$ there are two possibilities: either $\mathbf{k} = \mathbf{k}', \mathbf{q} = \mathbf{q}'$ or $\mathbf{k} = \mathbf{q}', \mathbf{q} = \mathbf{k}'$:

$$\langle \phi_0 | a^\dagger_{\mathbf{k}\sigma} a^\dagger_{\mathbf{q}\sigma} a_{\mathbf{q}'\sigma} a_{\mathbf{k}'\sigma} | \phi_0 \rangle = \delta_{\mathbf{k}\mathbf{k}'} \delta_{\mathbf{q}\mathbf{q}'} \langle \phi_0 | a^\dagger_{\mathbf{k}\sigma} a^\dagger_{\mathbf{q}\sigma} a_{\mathbf{q}\sigma} a_{\mathbf{k}\sigma} | \phi_0 \rangle$$

$$+ \delta_{\mathbf{k}\mathbf{q}'} \delta_{\mathbf{q}\mathbf{k}'} \langle \phi_0 | a^\dagger_{\mathbf{k}\sigma} a^\dagger_{\mathbf{q}\sigma} a_{\mathbf{k}\sigma} a_{\mathbf{q}\sigma} | \phi_0 \rangle$$

$$= (\delta_{\mathbf{k}\mathbf{k}'} \delta_{\mathbf{q}\mathbf{q}'} - \delta_{\mathbf{k}\mathbf{q}'} \delta_{\mathbf{q}\mathbf{k}'}) \langle \phi_0 | a^\dagger_{\mathbf{k}\sigma} a_{\mathbf{k}\sigma} a^\dagger_{\mathbf{q}\sigma} a_{\mathbf{q}\sigma} | \phi_0 \rangle$$

$$= (\delta_{\mathbf{k}\mathbf{k}'} \delta_{\mathbf{q}\mathbf{q}'} - \delta_{\mathbf{k}\mathbf{q}'} \delta_{\mathbf{q}\mathbf{k}'}) n_{\mathbf{k}\sigma} n_{\mathbf{q}\sigma} \,. \tag{2.1.14}$$

Since $(a_{\mathbf{k}\sigma})^2 = 0$, we must have $\mathbf{k} \neq \mathbf{q}$ and thus, by anticommutating – see (1.6.16) – we obtain the expression (2.1.14), and from (2.1.12) one gains:

$$\left(\frac{n}{2}\right)^2 g_{\sigma\sigma}(\mathbf{x} - \mathbf{x}') = \frac{1}{V^2} \sum_{\mathbf{k},\mathbf{q}} \left(1 - e^{-i(\mathbf{k}-\mathbf{q})\cdot(\mathbf{x}-\mathbf{x}')}\right) n_{\mathbf{k}\sigma} n_{\mathbf{q}\sigma}$$

$$= \left(\frac{n}{2}\right)^2 - (G_\sigma(\mathbf{x} - \mathbf{x}'))^2 \,. \tag{2.1.15}$$

With the single-particle correlation function $G_\sigma(\mathbf{x} - \mathbf{x}')$ from (2.1.8) and the abbreviation $x = k_F |\mathbf{x} - \mathbf{x}'|$, we finally obtain

$$g_{\sigma\sigma}(\mathbf{x} - \mathbf{x}') = 1 - \frac{9}{x^6} (\sin x - x \cos x)^2 \,. \tag{2.1.16}$$

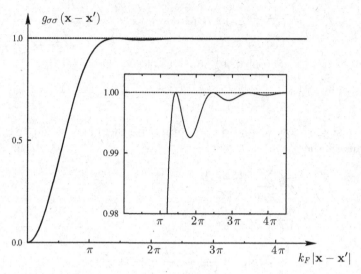

Fig. 2.4. The pair distribution function $g_{\sigma\sigma}(\mathbf{x} - \mathbf{x}')$. The correlation hole and the weak oscillations with wave number k_F should be noted

Let us give a physical interpretation of the pair distribution function (2.1.16) plotted in Fig. 2.4. If a fermion is removed at $\mathbf{x}$, the particle density in the vicinity of this point is strongly reduced. In other words, the probability of finding two fermions with the same spin at separations $\lesssim k_F^{-1}$ is small. The reduction of $g_{\sigma\sigma}(\mathbf{x} - \mathbf{x}')$ at such separations is referred to as an exchange, or correlation hole. It should be emphasized once again that this effective repulsion stems solely from the antisymmetric nature of the state and not from any genuine interaction.

For the noninteracting electron gas at $T = 0$, one has

$$\frac{1}{4} \sum_{\sigma,\sigma'} g_{\sigma\sigma'} = 2^{-1}(1 + g_{\sigma\sigma}(\mathbf{x})) \qquad (2.1.17a)$$

$$\sum_{\sigma,\sigma'} \langle \phi_0 | \psi_\sigma^\dagger(\mathbf{x}) \psi_{\sigma'}^\dagger(\mathbf{0}) \psi_{\sigma'}(\mathbf{0}) \psi_\sigma(\mathbf{x}) | \phi_0 \rangle = \frac{n^2}{4} \sum_{\sigma,\sigma'} g_{\sigma\sigma'}(\mathbf{x})$$

$$= \frac{n^2}{2}(1 + g_{\sigma\sigma}(\mathbf{x}))$$

$$\rightarrow n^2 \text{ for } \mathbf{x} \rightarrow \infty \qquad (2.1.17b)$$

$$\rightarrow \frac{n^2}{2} \text{ for } \mathbf{x} \rightarrow \mathbf{0}.$$

The next section provides a compilation of the definitions of the pair distribution function and other correlation functions. According to this, the spin-dependent pair distribution function

$$g_{\sigma\sigma'}(\mathbf{x}) = \left(\frac{2}{n}\right)^2 \left\langle \psi_\sigma^\dagger(\mathbf{x})\psi_{\sigma'}^\dagger(\mathbf{0})\psi_{\sigma'}(\mathbf{0})\psi_\sigma(\mathbf{x}) \right\rangle$$

is proportional to the probability of finding a particle with spin σ at position $\mathbf{x}$ when it is known with certainty that a particle with spin σ' is located at $\mathbf{0}$. It is equal to the probability that two particles with spins σ and σ' are to be found at a separation $\mathbf{x}$.

*2.1.4 Pair Distribution Function, Density Correlation Functions, and Structure Factor

The definitions and relationships given in this section hold for arbitrary many-body systems and for fermions as well as bosons[3]. The standard definition of the *pair distribution function* of N particles reads:

$$g(\mathbf{x}) = \frac{V}{N(N-1)} \left\langle \sum_{\alpha\neq\beta=1}^{N} \delta(\mathbf{x} - \mathbf{x}_\alpha + \mathbf{x}_\beta) \right\rangle. \tag{2.1.18}$$

Here, $g(\mathbf{x})$ is the probability density that a pair of particles has the separation $\mathbf{x}$; in other words, the probability density that a particle is located at $\mathbf{x}$ when with certainty there is a particle at the position $\mathbf{0}$. As a probability density, $g(\mathbf{x})$ is normalized to 1:

$$\int \frac{d^3x}{V} g(\mathbf{x}) = 1. \tag{2.1.19}$$

The *density–density correlation function* $G(\mathbf{x})$ for translationally invariant systems is given by

$$\begin{aligned} G(\mathbf{x}) &= \langle n(\mathbf{x})n(0)\rangle = \langle n(\mathbf{x}+\mathbf{x}')n(\mathbf{x}')\rangle \\ &= \sum_{\alpha,\beta} \langle \delta(\mathbf{x}+\mathbf{x}'-\mathbf{x}_\alpha)\delta(\mathbf{x}'-\mathbf{x}_\beta)\rangle. \end{aligned} \tag{2.1.20}$$

Due to translational invariance, this is independent of $\mathbf{x}'$ and we may integrate over $\mathbf{x}'$, whence (with $\frac{1}{V}\int d^3x' = 1$) it follows that

$$G(\mathbf{x}) = \frac{1}{V} \sum_{\alpha,\beta} \langle \delta(\mathbf{x} - \mathbf{x}_\alpha + \mathbf{x}_\beta)\rangle .$$

This leads to the relationship

$$\begin{aligned} G(\mathbf{x}) &= \frac{1}{V} \left(\sum_\alpha \delta(\mathbf{x}) + \frac{N(N-1)}{V} g(\mathbf{x}) \right) \\ &= n\delta(\mathbf{x}) + \frac{N(N-1)}{V^2} g(\mathbf{x}) . \end{aligned} \tag{2.1.21}$$

---

[3] The brackets signify an arbitrary expectation value, e.g., a quantum-mechanical expectation value in a particular state or a thermal expectation value.

For interactions of finite range, the densities become independent of each other at large separations:

$$\lim_{\mathbf{x}\to\infty} G(\mathbf{x}) = \langle n(\mathbf{x})\rangle\langle n(0)\rangle = n^2.$$

From this it follows that, for large N,

$$\lim_{\mathbf{x}\to\infty} g(\mathbf{x}) = \frac{V^2}{N(N-1)}n^2 = 1.$$

The *static structure factor* $S(q)$ is defined by

$$S(\mathbf{q}) = \frac{1}{N}\left\langle \sum_{\alpha,\beta} e^{-i\mathbf{q}\cdot(\mathbf{x}_\alpha-\mathbf{x}_\beta)} \right\rangle - N\delta_{\mathbf{q}0}. \qquad (2.1.22)$$

One may also write

$$S(\mathbf{q}) = \frac{1}{N}\sum_{\alpha\neq\beta} \left\langle e^{-i\mathbf{q}\cdot(\mathbf{x}_\alpha-\mathbf{x}_\beta)} \right\rangle + 1 - N\delta_{\mathbf{q}0} \qquad (2.1.23)$$

or

$$S(\mathbf{q}) = \frac{1}{N}\langle \hat{n}_\mathbf{q}\hat{n}_{-\mathbf{q}}\rangle - N\delta_{\mathbf{q}0},$$

where

$$\hat{n}_\mathbf{q} = \int d^3x\, e^{-i\mathbf{q}\cdot\mathbf{x}}n(\mathbf{x}) = \sum_\alpha e^{-i\mathbf{q}\cdot\mathbf{x}_\alpha}.$$

Since $N(N-1) \to N^2$ for large N

$$\int d^3x\, e^{-i\mathbf{q}\cdot\mathbf{x}}g(\mathbf{x}) = \frac{V}{N^2}\int d^3x\, e^{-i\mathbf{q}\cdot\mathbf{x}}\left\langle \sum_{\alpha\neq\beta} \delta(\mathbf{x}-\mathbf{x}_\alpha+\mathbf{x}_\beta) \right\rangle$$

$$= \frac{V}{N^2}\left\langle \sum_{\alpha\neq\beta} e^{-i\mathbf{q}\cdot(\mathbf{x}_\alpha-\mathbf{x}_\beta)} \right\rangle,$$

and it follows that

$$S(\mathbf{q}) = \frac{N}{V}\int d^3x\, e^{-i\mathbf{q}\cdot\mathbf{x}}g(\mathbf{x}) + 1 - N\delta_{\mathbf{q}0}.$$

With

$$N\delta_{\mathbf{q}0} = \frac{N}{V}\int d^3x\, e^{-i\mathbf{q}\cdot\mathbf{x}},$$

one obtains

$$S(\mathbf{q}) - 1 = n \int d^3x e^{-i\mathbf{q} \cdot \mathbf{x}} (g(\mathbf{x}) - 1) \tag{2.1.24a}$$

and the inverse

$$g(\mathbf{x}) - 1 = \frac{1}{n} \int \frac{d^3q}{(2\pi)^3} e^{i\mathbf{q} \cdot \mathbf{x}} (S(\mathbf{q}) - 1) . \tag{2.1.24b}$$

In the classical case,

$$\lim_{q \to 0} S(\mathbf{q}) = nkT\kappa_T , \tag{2.1.25}$$

where κ_T is the isothermal compressibility.

The above definitions yield the following second-quantized representations of the density–density correlation function and the pair-distribution function:

$$G(\mathbf{x} - \mathbf{x}') = \langle \psi^\dagger(\mathbf{x})\psi(\mathbf{x})\psi^\dagger(\mathbf{x}')\psi(\mathbf{x}') \rangle \tag{2.1.26a}$$

$$g(\mathbf{x}) = \frac{V^2}{N^2} \langle \psi^\dagger(\mathbf{x})\psi^\dagger(\mathbf{0})\psi(\mathbf{0})\psi(\mathbf{x}) \rangle. \tag{2.1.26b}$$

The first formula, (2.1.26a), is self-evident; the second follows from the former and (2.1.21) and a permutation of the field operators.

Proof of the last formula based on the definition (2.1.18) and on (1.5.6c):

$$\sum_{\alpha \neq \beta} \delta(\mathbf{x} - \mathbf{x}_\alpha + \mathbf{x}_\beta) \to \int d^3x' d^3x'' \psi^\dagger(\mathbf{x}')\psi^\dagger(\mathbf{x}'')\delta(\mathbf{x} - \mathbf{x}' + \mathbf{x}'')\psi(\mathbf{x}'')\psi(\mathbf{x}')$$

$$= \int d^3x' \psi^\dagger(\mathbf{x}')\psi^\dagger(\mathbf{x}' - \mathbf{x})\psi(\mathbf{x}' - \mathbf{x})\psi(\mathbf{x}')$$

$$\left\langle \sum_{\alpha \neq \beta} \delta(\mathbf{x} - \mathbf{x}_\alpha + \mathbf{x}_\beta) \right\rangle = V \left\langle \psi^\dagger(\mathbf{x}')\psi^\dagger(\mathbf{x}' - \mathbf{x})\psi(\mathbf{x}' - \mathbf{x})\psi(\mathbf{x}') \right\rangle .$$

2.2 Ground State Energy and Elementary Theory of the Electron Gas

2.2.1 Hamiltonian

The Hamiltonian, including the Coulomb repulsion, reads:

$$H = \sum_{\mathbf{k},\sigma} \frac{\hbar^2 k^2}{2m} a^\dagger_{\mathbf{k}\sigma} a_{\mathbf{k}\sigma} + \frac{e^2}{2V} \sum_{\substack{\mathbf{k},\mathbf{k}',\mathbf{q},\sigma,\sigma' \\ q \neq 0}} \frac{4\pi}{q^2} a^\dagger_{\mathbf{k}+\mathbf{q},\sigma} a^\dagger_{\mathbf{k}'-\mathbf{q},\sigma'} a_{\mathbf{k}'\sigma'} a_{\mathbf{k}\sigma} . \tag{2.2.1}$$

The $\mathbf{q} = \mathbf{0}$ contribution, which, because of the long-range nature of the Coulomb interaction, would diverge, is excluded here since it is canceled by the interaction of the electrons with the positive background of ions and by the interaction between the ions. This can be seen from the following.

The interaction energy of the background of positive ions is

$$H_{\text{ion}} = \frac{1}{2} e^2 \int d^3x d^3x' \frac{n(\mathbf{x})n(\mathbf{x}')}{|\mathbf{x}-\mathbf{x}'|} e^{-\mu|\mathbf{x}-\mathbf{x}'|} . \tag{2.2.2a}$$

Here, $n(\mathbf{x}) = \frac{N}{V}$ and we have introduced a cutoff at μ^{-1}. At the end of the calculation we will take $\mu \to 0$

$$H_{\text{ion}} = \frac{1}{2} e^2 \left(\frac{N}{V}\right)^2 V 4\pi \int_0^\infty dr\, r\, e^{-\mu r} = \frac{1}{2} e^2 \frac{N^2}{V} \frac{4\pi}{\mu^2} . \tag{2.2.2a'}$$

The interaction of the electrons with the positive background reads:

$$H_{\text{ion, el}} = -e^2 \sum_{i=1}^{N} \frac{N}{V} \int d^3x \frac{e^{-\mu|\mathbf{x}-\mathbf{x}_i|}}{|\mathbf{x}-\mathbf{x}_i|} = -e^2 \frac{N^2}{V} \frac{4\pi}{\mu^2} . \tag{2.2.2b}$$

Finally, we consider the $\mathbf{q} = 0$ contribution to the electron–electron interaction, where $\frac{4\pi e^2}{q^2} \to \frac{4\pi e^2}{q^2+\mu^2}$,

$$\frac{e^2}{2V} \frac{4\pi}{\mu^2} \sum_{\mathbf{k},\mathbf{k}',\sigma,\sigma'} a_{\mathbf{k}\sigma}^\dagger a_{\mathbf{k}'\sigma'}^\dagger a_{\mathbf{k}'\sigma'} a_{\mathbf{k}\sigma}$$

$$= \frac{e^2}{2V} \frac{4\pi}{\mu^2} \sum_{\mathbf{k},\mathbf{k}',\sigma,\sigma'} \left[a_{\mathbf{k}\sigma}^\dagger a_{\mathbf{k}\sigma} \left(a_{\mathbf{k}'\sigma'}^\dagger a_{\mathbf{k}'\sigma'} - \delta_{\mathbf{k}\mathbf{k}'}\delta_{\sigma\sigma'} \right) \right]$$

$$= \frac{e^2}{2V} \frac{4\pi}{\mu^2} \sum_{\mathbf{k},\mathbf{k}',\sigma,\sigma'} \hat{n}_{\mathbf{k}\sigma} \left(\hat{n}_{\mathbf{k}'\sigma'} - \delta_{\mathbf{k}\mathbf{k}'}\delta_{\sigma\sigma'} \right) \tag{2.2.2c}$$

$$= \frac{e^2}{2V} \frac{4\pi}{\mu^2} (\hat{N}^2 - \hat{N}) = \frac{e^2}{2V} \frac{4\pi}{\mu^2} (N^2 - N).$$

The leading terms, proportional to N^2, in the three evaluated energy contributions cancel one another. The term $-\frac{e^2}{2V} \frac{4\pi}{\mu^2} N$ yields an energy contribution per particle of $\frac{E}{N} \propto \frac{1}{N} \frac{N}{V}$ and vanishes in the thermodynamic limit. The limits are taken in the order $N, V \to \infty$ and then $\mu \to 0$.

2.2.2 Ground State Energy in the Hartree–Fock Approximation

The ground state energy is calculated in perturbation theory by assuming a ground state $|\phi_0\rangle$, in which all single-particle states up to k_F are occupied:

$$|\phi_0\rangle = \prod_{p \le k_F} \prod_\sigma a_{\mathbf{p}\sigma}^\dagger |0\rangle \equiv \left(\prod_{p=0}^{k_F} a_{\mathbf{p}\uparrow}^\dagger \right) \left(\prod_{p=0}^{k_F} a_{\mathbf{p}\downarrow}^\dagger \right) |0\rangle . \tag{2.2.3}$$

The kinetic energy in this state is diagonal:

$$E^{(0)} = \langle \phi_0 | H_{\text{kin}} | \phi_0 \rangle = \frac{\hbar^2}{2m} \sum_{\mathbf{k}, \sigma} k^2 \Theta(k_F - k)$$

$$= \frac{\hbar^2}{2m} 2 \frac{V}{(2\pi)^3} \int d^3 k \, k^2 \Theta(k_F - k)$$

$$= \frac{\hbar^2}{m} \frac{V}{(2\pi)^3} 4\pi \frac{1}{5} k_F^5 = \frac{3\hbar^2 k_F^2}{10m} N = \frac{3}{5} \epsilon_F N = \frac{e^2}{2a_0} \frac{1}{r_s^2} \frac{3}{5} \left(\frac{9\pi}{4} \right)^{2/3} N$$

$$E^{(0)} = \frac{e^2}{2a_0} \frac{2.21}{r_s^2} N \ . \tag{2.2.4}$$

Here, according to (2.1.4), we have used

$$n = \frac{k_F^3}{3\pi^2} = \frac{3}{4\pi r_0^3} = \frac{3}{4\pi a_0^3 r_s^3} \tag{2.2.5}$$

and introduced r_0, the radius of a sphere of volume equal to the volume per particle. The quantity $a_0 = \frac{\hbar^2}{me^2}$ is the Bohr radius and $r_s = \frac{r_0}{a_0}$.
The potential energy in first-order perturbation theory[4] reads:

$$E^{(1)} = \frac{e^2}{2V} \sideset{}{'}\sum_{\mathbf{k}, \mathbf{k}', \mathbf{q}, \sigma, \sigma'} \frac{4\pi}{q^2} \langle \phi_0 | a_{\mathbf{k}+\mathbf{q}, \sigma}^\dagger a_{\mathbf{k}'-\mathbf{q}, \sigma'}^\dagger a_{\mathbf{k}'\sigma'} a_{\mathbf{k}\sigma} | \phi_0 \rangle \ . \tag{2.2.6}$$

The prime on the summation sign indicates that the term $\mathbf{q} = 0$ is excluded. The only contribution for which every annihilation operator is compensated by a creation operator is proportional to $\delta_{\sigma\sigma'} \delta_{\mathbf{k}', \mathbf{k}+\mathbf{q}} a_{\mathbf{k}+\mathbf{q}\sigma}^\dagger a_{\mathbf{k}\sigma'}^\dagger a_{\mathbf{k}+\mathbf{q}\sigma'} a_{\mathbf{k}\sigma}$, thus:

$$E^{(1)} = -\frac{e^2}{2V} \sideset{}{'}\sum_{\mathbf{k}, \mathbf{q}, \sigma} \frac{4\pi}{q^2} n_{\mathbf{k}+\mathbf{q}, \sigma} n_{\mathbf{k}, \sigma}$$

$$= -\frac{e^2}{2V} \sum_{\sigma} \sideset{}{'}\sum_{\mathbf{k}, \mathbf{q}} \frac{4\pi}{q^2} \Theta(k_F - |\mathbf{q} + \mathbf{k}|) \Theta(k_F - k)$$

$$= -\frac{4\pi e^2 V}{(2\pi)^6} \int d^3 k \, \Theta(k_F - k) \int d^3 k' \frac{1}{|\mathbf{k} - \mathbf{k}'|^2} \Theta(k_F - k') \ . \tag{2.2.6'}$$

One then finds

$$-\frac{4\pi e^2}{(2\pi)^3} \int d^3 k' \frac{1}{|\mathbf{k} - \mathbf{k}'|^2} \Theta(k_F - k') = -\frac{2e^2}{\pi} k_F F \left(\frac{k}{k_F} \right),$$

where

$$F(x) = \frac{1}{2} + \frac{1 - x^2}{4x} \log \left| \frac{1 + x}{1 - x} \right| \tag{2.2.6''}$$

---

[4] This first-order perturbation theory can also be considered as the Hartree–Fock theory with the variational state (2.2.3); see also Problem 2.5.

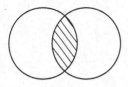

Fig. 2.5. Integration region for $E^{(1)}$ consisting of the region of overlap of two Fermi spheres with relative displacement **q**; see Eq. (2.2.6′)

and

$$
E^{(1)} = -\frac{e^2 k_F V}{\pi} \int\limits_{k<k_F} \frac{d^3 k}{(2\pi)^3} \left[1 + \frac{k_F^2 - k^2}{2kk_F} \log \left| \frac{k_F + k}{k_F - k} \right| \right]
$$

$$
= -N \frac{3}{4} \frac{e^2 k_F}{\pi}
$$

$$
= -\frac{e^2}{2a_0 r_s} \left(\frac{9\pi}{4} \right)^{1/3} \frac{3N}{2\pi} = -\frac{e^2}{2a_0} \frac{0.916}{r_s} N. \tag{2.2.7}
$$

Taking $E^{(0)}$ and $E^{(1)}$ together yields:

$$
\frac{E}{N} = \frac{e^2}{2a_0} \left[\frac{2.21}{r_s^2} - \frac{0.916}{r_s} + \dots \right]_{(r_s \ll 1)}. \tag{2.2.8}
$$

The first term is the kinetic energy, and the second the exchange term. The pressure and the bulk modulus are given by

$$
P = -\left(\frac{\partial E}{\partial V} \right)_N = -\frac{dE}{dr_s} \frac{dr_s}{dV} = \frac{Ne^2}{2a_0} \frac{r_s}{3V} \left[\frac{4.42}{r_s^3} - \frac{0.916}{r_s^2} \right]
$$

and

$$
B = \frac{1}{\kappa} = -V \left(\frac{\partial P}{\partial V} \right) = \frac{Ne^2}{9Va_0} \left[\frac{11.05}{r_s^2} - \frac{1.832}{r_s} \right]. \tag{2.2.9}
$$

For $r_s = 4.83$ the energy takes on its minimum value corresponding to $\frac{E}{N} = -1.29$ eV. This is of the same order of magnitude as in simple metals, e.g., Na $\left(r_s = 3.96, \frac{E}{N} = -1.13 \text{eV} \right)$. However, these values of r_s lie outside the range of validity of the present theory.

Higher order *corrections* to the energy can be obtained in the *random phase approximation* (RPA):

$$
\frac{E}{N \, \text{Ry}} = \left\{ \frac{2.21}{r_s^2} - \frac{0.916}{r_s} + \underbrace{0.062 \ln r_s - 0.096 + A r_s + B r_s \ln r_s + \dots}_{\text{correlation energy}} \right\}
$$

$$
\tag{2.2.10}
$$

where we have made use of the Rydberg, 1 Ry $= \frac{e^2}{2a_0} = \frac{e^4 m}{2\hbar^2} = 13.6$ eV. The RPA yields an energy that contains, in addition to the Hartree–Fock energy, the summation of an infinite series arising from perturbation theory. It is the latter that yields the logarithmic contributions. That perturbation theory should lead to a series in powers of r_s, can be seen from the rescaling in (2.2.23).

Remarks

For $r_s \to \infty$, one expects the electrons to form a *Wigner crystal* [5], i.e., to crystallize. For large r_s one finds the expansion[6]

$$\lim_{r_s \to \infty} \frac{E}{N} = \frac{e^2}{2a_0} \left[-\frac{1.79}{r_s} + \frac{2.64}{r_s^{3/2}} + \ldots \right] , \tag{2.2.11}$$

which, for $r_s \gg 10$, is quantitatively reliable (see Problem 2.7). The Wigner crystal has a lower energy than the fluid. Corrections arising from correlation effects are discussed in other advanced texts[7].

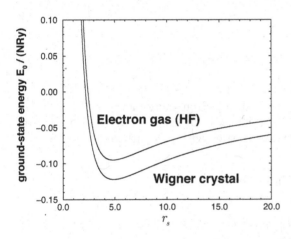

Fig. 2.6. Energies of the electron gas in the Hartree–Fock approximation and of the Wigner crystal, in each case as a function of r_s

To date, Wigner crystallization[5] in three dimensions has not been de- tected experimentally. It is possible that this is due to quantum fluctuations, which destroy (melt) the lattice[6]. On the basis of a Lindemann criterion[8], one finds that the Wigner lattice is stable for $r_s > r_s^c = 0.41 \, \delta^{-4}$, where δ (0.15 < δ < 0.5) is the Lindemann parameter. Even for $\delta = 0.5$, the value of $r_s^c = 6.49$ is already larger than the minimum value of (2.2.11), $r_s = 4.88$. In two dimensions, a triangular lattice structure has been theoretically pre-

---

[5] E.P. Wigner, Phys. Rev. **46**, 1002 (1934) , Trans. Faraday Soc. **34**, 678 (1938)

[6] R.A. Coldwell-Horsfall and A.A. Maradudin, J. Math. Phys. **1**, 395 (1960)

[7] G.D. Mahan, *Many Particle Physics*, Plenum Press, New York, 1990, 2nd edn, Sect. 5.2

[8] See, e.g., J. M. Ziman, *Principles of the Theory of Solids*, 2nd edn, Cambridge University Press, Cambridge, 1972, p.65.

dicted[9] and experimentally observed for electrons on the surface of helium.[10] Its melting curve has also been determined. Figure 2.6 compares the Hartree–Fock energy (2.2.8) with the energy of the Wigner crystal (2.2.11). The minimum of the Hartree–Fock energy as a function of r_s lies at $r_s = 4.83$ and has the value $E/N = -0.095 e^2/2a_0$.

To summarize, the range of validity of the RPA, equation(2.2.10), is restricted to $r_s \ll 1$, whereas (2.2.11) for the Wigner crystal is valid for $r_s \gg 10$; real metals lie between these two regimes: $1.8 \le r_s \le 5.6$.

2.2.3 Modification of Electron Energy Levels due to the Coulomb Interaction

$$H = H_0 + H_{\text{Coul}} , \quad H_0 = \sum_{\mathbf{k},\sigma} \frac{(\hbar k)^2}{2m} a_{\mathbf{k}\sigma}^\dagger a_{\mathbf{k}\sigma}$$

$$H_{\text{Coul}} = \frac{1}{2V} \sum_{\substack{\mathbf{q} \neq 0, \mathbf{p}, \mathbf{k}' \\ \sigma\sigma'}} \frac{4\pi e^2}{q^2} a_{\mathbf{p}+\mathbf{q}\,\sigma}^\dagger a_{\mathbf{k}'-\mathbf{q}\,\sigma'}^\dagger a_{\mathbf{k}'\sigma'} a_{\mathbf{p}\sigma} .$$

The Coulomb interaction modifies the electron energy levels $\epsilon_0(\mathbf{k}) = \frac{(\hbar k)^2}{2m}$. We can calculate this effect approximately by considering the equation of motion of the operator $a_{\mathbf{k}\sigma}(t)$. Let us start with free particles:

$$\dot{a}_{\mathbf{k}\sigma}(t) = \frac{i}{\hbar} \left[\sum_{\mathbf{k}',\sigma'} \epsilon_0(\mathbf{k}') a_{\mathbf{k}'\sigma'}^\dagger a_{\mathbf{k}'\sigma'}, a_{\mathbf{k}\sigma} \right]$$

$$= -\frac{i}{\hbar} \sum_{\mathbf{k}',\sigma'} \epsilon_0(\mathbf{k}') \underbrace{\left[a_{\mathbf{k}'\sigma'}^\dagger , a_{\mathbf{k}\sigma} \right]_+}_{+\delta_{\mathbf{k}\mathbf{k}'}\delta_{\sigma\sigma'}} a_{\mathbf{k}'\sigma'}$$

$$\dot{a}_{\mathbf{k}\sigma}(t) = -\frac{i}{\hbar} \epsilon_0(\mathbf{k}) a_{\mathbf{k}\sigma}(t) . \tag{2.2.12}$$

We now define the correlation function

$$G_{\mathbf{k}\sigma}(t) = \langle \phi_0 | a_{\mathbf{k}\sigma}(t) a_{\mathbf{k}\sigma}^\dagger(0) | \phi_0 \rangle . \tag{2.2.13}$$

Multiplying the equation of motion by $a_{\mathbf{k}\sigma}^\dagger(0)$ yields an equation of motion for $G_{\mathbf{k}\sigma}(t)$:

$$\frac{d}{dt} G_{\mathbf{k}\sigma}(t) = -\frac{i}{\hbar} \epsilon_0(\mathbf{k}) G_{\mathbf{k}\sigma}(t). \tag{2.2.14}$$

---

[9] G. Meissner, H. Namaizawa, and M. Voss, Phys. Rev. **B13**, 1360 (1976); L. Bonsall, and A.A. Maradudin, Phys. Rev. **B15**, 1959 (1977)
[10] C.C. Grimes, and G. Adams, Phys. Rev. Lett. **42**, 795 (1979)

Its solution is

$$G_{\mathbf{k}\sigma}(t) = e^{-\frac{i}{\hbar}\epsilon_0(\mathbf{k})t}(-n_{\mathbf{k}\sigma} + 1), \qquad (2.2.15)$$

since $\langle\phi_0| a_{\mathbf{k}\sigma}(0)a_{\mathbf{k}\sigma}^\dagger(0) |\phi_0\rangle = -n_{\mathbf{k}\sigma} + 1$.

When the Coulomb repulsion is included, the equation of motion for the annihilation operator $a_{\mathbf{k}\sigma}$ reads:

$$\dot{a}_{\mathbf{k}\sigma} = -\frac{i}{\hbar}\left(\epsilon_0(\mathbf{k})a_{\mathbf{k}\sigma} - \frac{1}{V}\sum_{\substack{\mathbf{p},\mathbf{q}\neq 0 \\ \sigma'}}\frac{4\pi e^2}{q^2}a_{\mathbf{p}+\mathbf{q}\,\sigma'}^\dagger a_{\mathbf{k}+\mathbf{q}\,\sigma}\, a_{\mathbf{p}\sigma'}\right), \qquad (2.2.16)$$

as can be immediately seen from the field equation. From this it follows that

$$\frac{d}{dt}G_{\mathbf{k}\sigma}(t) = -\frac{i}{\hbar}\left(\epsilon_0(\mathbf{k})G_{\mathbf{k}\sigma}(t)\right.$$

$$\left. -\frac{1}{V}\sum_{\substack{\mathbf{p},\mathbf{q}\neq 0 \\ \sigma'}}\frac{4\pi e^2}{q^2}\left\langle a_{\mathbf{p}+\mathbf{q}\,\sigma'}^\dagger(t)a_{\mathbf{k}+\mathbf{q}\,\sigma}(t)a_{\mathbf{p}\sigma'}(t)a_{\mathbf{k}\sigma}^\dagger(0)\right\rangle\right). \qquad (2.2.17)$$

On the right-hand side there now appears not only $G_{\mathbf{k}\sigma}(t)$, but also a higher-order correlation function. In a systematic treatment we could derive an equation of motion for this, too. We introduce the following factorization approximation for the expectation value[11]:

$$\left\langle a_{\mathbf{p}+\mathbf{q}\,\sigma'}^\dagger(t)a_{\mathbf{k}+\mathbf{q}\,\sigma}(t)a_{\mathbf{p}\sigma'}(t)a_{\mathbf{k}\sigma}^\dagger(0)\right\rangle$$

$$= \left\langle a_{\mathbf{p}+\mathbf{q}\,\sigma'}^\dagger(t)a_{\mathbf{k}+\mathbf{q}\,\sigma}(t)\right\rangle\left\langle a_{\mathbf{p}\sigma'}(t)a_{\mathbf{k}\sigma}^\dagger(0)\right\rangle \qquad (2.2.18)$$

$$= \delta_{\sigma\sigma'}\delta_{\mathbf{p}\mathbf{k}}\left\langle a_{\mathbf{p}+\mathbf{q}\,\sigma'}^\dagger(t)a_{\mathbf{p}+\mathbf{q}\,\sigma'}(t)\right\rangle\left\langle a_{\mathbf{k}\sigma}(t)a_{\mathbf{k}\sigma}^\dagger(0)\right\rangle.$$

The equation of motion thus reads:

$$\frac{d}{dt}G_{\mathbf{k}\sigma}(t) = -\frac{i}{\hbar}\left(\epsilon_0(\mathbf{k}) - \frac{1}{V}\sum_{\mathbf{q}\neq 0}\frac{4\pi e^2}{q^2}n_{\mathbf{k}+\mathbf{q}\,\sigma}\right)G_{\mathbf{k}\sigma}(t). \qquad (2.2.19)$$

From this, we can read off the energy levels $\epsilon(\mathbf{k})$ as

---

[11] The other possible factorization $\left\langle a_{\mathbf{p}+\mathbf{q}\,\sigma'}^\dagger(t)a_{\mathbf{p}\,\sigma'}(t)\right\rangle\left\langle a_{\mathbf{k}+\mathbf{q}\,\sigma}^\dagger(t)a_{\mathbf{k}\sigma}(0)\right\rangle$ requires $\mathbf{q} = 0$, which is excluded in the summation of Eq. (2.2.17).

$$\epsilon(\mathbf{k}) = \frac{\hbar^2 k^2}{2m} - \frac{1}{V} \sum_{\mathbf{k}'} \frac{4\pi e^2}{|\mathbf{k} - \mathbf{k}'|^2} n_{\mathbf{k}'\sigma} \, . \tag{2.2.20}$$

The second term leads to a change in $\epsilon(\mathbf{k})$,

$$\begin{aligned}
\Delta\epsilon(\mathbf{k}) &= -\int \frac{d^3 k'}{(2\pi)^3} \frac{4\pi e^2}{|\mathbf{k} - \mathbf{k}'|^2} \Theta(k_F - k') \\
&= -\frac{e^2}{\pi} \int_0^{k_F} dk' k'^2 \int_{-1}^{1} d\eta \frac{1}{k^2 + k'^2 - 2kk'\eta} \\
&= -\frac{e^2}{\pi k} \int_0^{k_F} dk' k' \log\left|\frac{k + k'}{k - k'}\right| \\
&= -\frac{2e^2 k_F}{\pi} \underbrace{\left(\frac{1}{2} + \frac{1 - x^2}{4x} \log\left|\frac{1 + x}{1 - x}\right|\right)}_{F(x)} \qquad x = \frac{k}{k_F} \, . \tag{2.2.21}
\end{aligned}$$

Here again the function $F(x)$ of Eq. (2.2.6″) appears. The Hartree–Fock energy levels are reduced in comparison to those of the free electron gas. However, the estimated reduction turns out to be greater than that actually observed. Figure 2.7 shows $F(x)$ and $\epsilon(\mathbf{k})$ in comparison to $\epsilon_0(\mathbf{k}) = \frac{\hbar^2 k^2}{2m}$ for $r_s = 4$.

Notes:

(i) A shorter derivation of the Hartree–Fock energy is obtained by introducing the following approximation in the Hamiltonian

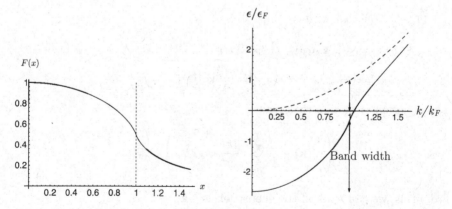

Fig. 2.7. (a) The function $F(x)$, Eq. (2.2.6″), and **(b)** the Hartree–Fock energy levels $\epsilon(\mathbf{k})$ as a function of the wave number for $r_s = 4$, compared with the energy of the free-electron gas $\epsilon_0(\mathbf{k})$ (dashed).

$$\frac{1}{2V} \sum_{\substack{k,k',q \\ \sigma,\sigma'}} \frac{4\pi e^2}{q^2} a^\dagger_{k+q\,\sigma}\, a^\dagger_{k'-q\,\sigma'}\, a_{k'\sigma'} a_{k\sigma} \longrightarrow$$

$$\frac{1}{2V} \sum_{\substack{k,k',q\neq 0 \\ \sigma,\sigma'}} \frac{4\pi e^2}{q^2} \left(\left\langle a^\dagger_{k+q\,\sigma}\, a_{k'\sigma'}\right\rangle a^\dagger_{k'-q\,\sigma'} a_{k\sigma} + a^\dagger_{k+q\,\sigma}\, a_{k'\sigma'} \left\langle a^\dagger_{k'-q\,\sigma'} a_{k\sigma}\right\rangle \right)$$

$$= \frac{2}{2V} \sum_{\substack{k,q \\ \sigma}} \frac{4\pi e^2}{q^2} \left\langle a^\dagger_{k+q\,\sigma}\, a_{k+q\,\sigma}\right\rangle a^\dagger_{k\sigma} a_{k\sigma} .$$

This yields:

$$H = \sum_{k,\sigma} \epsilon(k) a^\dagger_{k\sigma} a_{k\sigma}$$

with

$$\epsilon(k) = \frac{\hbar^2 k^2}{2m} - \frac{1}{V} \sum_q \frac{4\pi e^2}{q^2} \Theta(k_F - |k+q|) .$$

(ii) The perturbation-theoretical expansion in terms of the Coulomb interaction leads to a power series (with logarithmic corrections) in r_s. This structure can be seen from the following scaling of the Hamiltonian:

$$H = \sum_i \frac{p_i^2}{2m} + \frac{1}{2} \sum_{i\neq j} \frac{e^2}{r_{ij}} \qquad (2.2.22)$$

To this end, we carry out a canonical transformation $r' = r/r_0$ $p' = p r_0$. The characteristic length r_0 is defined by $\frac{4\pi}{3} r_0^3 N = V$, i.e.,

$$r_0 = \left(\frac{3V}{4\pi N} \right)^{1/3} .$$

In the new variables the Hamiltonian reads:

$$H = \frac{1}{r_0^2} \left(\sum_i \frac{p_i'^2}{2m} + r_0 \frac{1}{2} \sum_{i\neq j} \frac{e^2}{r'_{ij}} \right) . \qquad (2.2.23)$$

The Coulomb interaction becomes less and less important in comparison to the kinetic energy as r_0 (or r_s) becomes smaller, i.e., as the density of the gas increases.

2.3 Hartree–Fock Equations for Atoms

In this section, we consider atoms (possibly ionized) with N electrons and the nuclear charge number Z. The nucleus is assumed to be fixed and thus the Hamiltonian written in second quantized form is

$$H = \sum_{i,j} a_i^\dagger \langle i| T |j\rangle a_j + \sum_{i,j} a_i^\dagger \langle i| U |j\rangle a_j$$

$$+ \frac{1}{2} \sum_{i,j,k,m} \langle i,j| V |k,m\rangle a_i^\dagger a_j^\dagger a_m a_k \, , \tag{2.3.1}$$

where

$$T = \frac{\mathbf{p}^2}{2m} \tag{2.3.2a}$$

$$U = -\frac{Ze^2}{r} \, , \ r = |\mathbf{x}| \tag{2.3.2b}$$

and

$$V = \frac{e^2}{|\mathbf{x} - \mathbf{x}'|} \tag{2.3.2c}$$

represent the kinetic energy of an electron, the potential felt by an electron due to the nucleus, and the Coulomb repulsion between two electrons, respectively. Although the Hartree and the Hartree–Fock approximations have already been discussed in Sect. 13.3 of QM I[12], we will present here a derivation of the Hartree–Fock equations within the second quantization formalism. This method is easier to follow than that using Slater determinants.

We write the state of the N electrons as

$$|\psi\rangle = a_1^\dagger \ldots a_N^\dagger |0\rangle \, . \tag{2.3.3}$$

Here, $|0\rangle$ is the vacuum state containing no electrons and $a_i^\dagger$ is the creation operator for the state $|i\rangle \equiv |\varphi_i, m_{s_i}\rangle$, $m_{s_i} = \pm\frac{1}{2}$. The states $|i\rangle$ are mutually orthogonal and the $\varphi_i(\mathbf{x})$ are single-particle wave functions which are yet to be determined. We begin by calculating the expectation value for the general Hamiltonian (2.3.1) $\langle\psi| H |\psi\rangle$ without particular reference to the atom. For the single-particle contributions, one immediately finds

$$\sum_{i,j} \langle i| T |j\rangle \langle\psi| a_i^\dagger a_j |\psi\rangle = \sum_{i=1}^{N} \langle i| T |i\rangle \tag{2.3.4a}$$

$$\sum_{i,j} \langle i| U |j\rangle \langle\psi| a_i^\dagger a_j |\psi\rangle = \sum_{i=1}^{N} \langle i| U |i\rangle \, , \tag{2.3.4b}$$

whilst the two-particle contributions are found as

$$\langle\psi| a_i^\dagger a_j^\dagger a_m a_k |\psi\rangle = \langle\psi| (\delta_{im}\delta_{jk} a_m^\dagger a_k^\dagger + \delta_{ik}\delta_{jm} a_k^\dagger a_m^\dagger) a_m a_k |\psi\rangle \tag{2.3.4c}$$

$$= (\delta_{ik}\delta_{jm} - \delta_{im}\delta_{jk}) \Theta(m, k \in 1, \ldots, N) \, .$$

The first factor implies that the expectation value vanishes whenever the creation and annihilation operators fail to compensate one another. The second

---

[12] QM I *op. cit.*

implies that both the operators a_m and a_k must be present in the set $a_1 \ldots a_N$ occurring in the state (2.3.3), otherwise their application to the right on the vacuum state $|0\rangle$ would give zero. Therefore, the total expectation value of H reads:

$$\langle \psi | H | \psi \rangle = \frac{\hbar^2}{2m} \sum_{i=1}^N \int d^3x |\nabla \varphi_i|^2 + \sum_{i=1}^N \int d^3x U(\mathbf{x}) |\varphi_i(\mathbf{x})|^2$$

$$+ \frac{1}{2} \sum_{i,j=1}^N \int d^3x d^3x' V(\mathbf{x} - \mathbf{x}') \left\{ |\varphi_i(\mathbf{x})|^2 |\varphi_j(\mathbf{x}')|^2 \right. \tag{2.3.5}$$

$$\left. - \delta_{m_{s_i}, m_{s_j}} \varphi_i^*(\mathbf{x}) \varphi_j^*(\mathbf{x}') \varphi_i(\mathbf{x}') \varphi_j(\mathbf{x}) \right\} .$$

In the spirit of the Ritz variational principle, the single-particle wave functions $\varphi_i(\mathbf{x})$ are now determined so as to minimize the expectation value of H. As subsidiary conditions, one must take account of the normalizations $\int |\varphi_i|^2 d^3x = 1$; this leads to the additional terms $-\epsilon_i(\int d^3x |\varphi_i(\mathbf{x})|^2 - 1)$ with Lagrange parameters ϵ_i. In all, one thus has to take the functional derivative of $\langle \psi | H | \psi \rangle - \sum_{i=1}^N \epsilon_i \left(\int d^3x |\varphi_i(\mathbf{x})|^2 - 1 \right)$ with respect to $\varphi_i(\mathbf{x})$ and $\varphi_i^*(\mathbf{x})$ and set this equal to zero, where one uses

$$\frac{\delta \varphi_i(\mathbf{x}')}{\delta \varphi_j(\mathbf{x})} = \delta_{ij} \delta(\mathbf{x} - \mathbf{x}') . \tag{2.3.6}$$

The following equations refer once again to atoms, i.e., they take into account (2.3.2a–c). Taking the variational derivative with respect to φ_i^* yields:

$$\left(-\frac{\hbar^2}{2m} \nabla^2 - \frac{Ze^2}{r} \right) \varphi_i(\mathbf{x}) + \sum_{j=1}^N \int d^3x' \frac{e^2}{|\mathbf{x} - \mathbf{x}'|} |\varphi_j(\mathbf{x}')|^2 \varphi_i(\mathbf{x})$$

$$- \sum_{j=1}^N \delta_{m_{s_i} m_{s_j}} \int d^3x' \frac{e^2}{|\mathbf{x} - \mathbf{x}'|} \varphi_j^*(\mathbf{x}') \varphi_i(\mathbf{x}') \cdot \varphi_j(\mathbf{x})$$

$$= \epsilon_i \varphi_i(\mathbf{x}) . \tag{2.3.7}$$

These are the *Hartree–Fock* equations. Compared to the Hartree equations[12], they contain the additional term

$$\int d^3x' \frac{e^2}{|\mathbf{x} - \mathbf{x}'|} |\varphi_i(\mathbf{x}')|^2 \varphi_i(\mathbf{x})$$

$$- \sum_j \delta_{m_{s_i} m_{s_j}} \int d^3x' \frac{e^2}{|\mathbf{x} - \mathbf{x}'|} \varphi_j^*(\mathbf{x}') \varphi_i(\mathbf{x}') \varphi_j(\mathbf{x})$$

$$= - \sum_{j \neq i} \delta_{m_{s_i} m_{s_j}} \int d^3x' \frac{e^2}{|\mathbf{x} - \mathbf{x}'|} \varphi_j^*(\mathbf{x}') \varphi_i(\mathbf{x}') \varphi_j(\mathbf{x}) . \tag{2.3.8}$$

The second term of the interaction on the left-hand side is known as the exchange integral, since it derives from the antisymmetry of the fermion state. The interaction term can also be written in the form

$$\int d^3x' \frac{e^2}{|\mathbf{x}-\mathbf{x}'|} \sum_j \varphi_j^*(\mathbf{x}') \left[\varphi_j(\mathbf{x}')\varphi_i(\mathbf{x}) - \varphi_j(\mathbf{x})\varphi_i(\mathbf{x}')\delta_{m_{s_i} m_{s_j}} \right] .$$

The exchange term is a nonlocal term which only occurs for $m_{s_i} = m_{s_j}$. The term in square brackets is equal to the probability amplitude that i and j are at the positions $\mathbf{x}$ and $\mathbf{x}'$. For further discussion of the Hartree–Fock equations and their physical implications we refer to Sect. 13.3.2 of QM I.

Problems

2.1 Calculate the static structure function for noninteracting fermions

$$S^0(\mathbf{q}) \equiv \frac{1}{N} \langle \phi_0 | \hat{n}_{\mathbf{q}}\hat{n}_{-\mathbf{q}} | \phi_0 \rangle ,$$

where $\hat{n}_{\mathbf{q}} = \sum_{\mathbf{k},\sigma} a_{\mathbf{k}\,\sigma}^\dagger a_{\mathbf{k}+\mathbf{q}\,\sigma}$ is the particle density operator in the momentum representation and $|\phi_0\rangle$ is the ground state. Take the continuum limit $\sum_{\mathbf{k},\sigma} = 2V \int d^3k/(2\pi)^3$ and calculate $S^0(\mathbf{q})$ explicitly.
Hint: Consider the cases $\mathbf{q} = 0$ and $\mathbf{q} \neq 0$ separately.

2.2 Prove the validity of the following relations, which have been used in the evaluation of the energy shift $\Delta\epsilon(\mathbf{k})$ of the electron gas, Eq. (2.2.21):

a) $$-4\pi e^2 \int \frac{d^3k'}{(2\pi)^3} \frac{1}{|\,\mathbf{k}-\mathbf{k}'\,|^2} \Theta(k_F - k') = -\frac{2e^2}{\pi} k_F F(k/k_F) ,$$

with

$$F(x) = \frac{1}{2} + \frac{1-x^2}{4x} \ln \left| \frac{1+x}{1-x} \right| .$$

b) $$E^{(1)} = -\frac{e^2 k_F}{\pi} V \int \frac{d^3k}{(2\pi)^3} \left[1 + \frac{k_F^2 - k^2}{2kk_F} \ln \left| \frac{k_F + k}{k_F - k} \right| \right] \Theta(k_F - k)$$

$$= -\frac{3}{4} \frac{e^2 k_F}{\pi} N = -\frac{e^2}{2a_0 r_s} \left(\frac{9\pi}{4} \right)^{1/3} \frac{3N}{2\pi} ,$$

where r_s is a dimensionless number which characterizes the mean particle separation in units of the Bohr radius $a_0 = \hbar^2/me^2$. Furthermore, $k_F^3 = 3\pi^2 n = 1/(\alpha a_0 r_s)^3$ with $\alpha = (4/9\pi)^{1/3}$.

2.3 Apply the atomic Hartree–Fock equations to the electron gas.
a) Show that the Hartree–Fock equations are solved by plane waves.
b) Replace the nuclei by a uniform positive background charge of the same total charge and show that the Hartree term is canceled by the Coulomb attraction of the positive background and the electrons.

The electronic energy levels are then given by

$$\epsilon(\mathbf{k}) = \frac{(\hbar \mathbf{k})^2}{2m} - \frac{1}{V} \sum_{\mathbf{q}} \frac{4\pi e^2}{|\mathbf{k} - \mathbf{q}|^2} \Theta(k_F - q) .$$

According to Problem 2.2, this can also be written as $\epsilon(\mathbf{k}) = \frac{(\hbar \mathbf{k})^2}{2m} - \frac{2e^2}{\pi} k_F F(k/k_F)$.

2.4 Show that the Hartree–Fock states $|i\rangle \equiv |\varphi_i, m_{s_i}\rangle$ following from (2.3.7) are orthogonal and that the ϵ_i are real.

2.5 Show that, for noninteracting fermions,

$$S^0(\mathbf{q}, \omega) \equiv \frac{1}{N} \int_{-\infty}^{+\infty} dt\, e^{i\omega t} \langle \phi_0 | \hat{n}_{\mathbf{q}}(t) \hat{n}_{-\mathbf{q}}(0) | \phi_0 \rangle$$

$$= \frac{\hbar V}{2\pi^2 N} \int d^3k\, \Theta(k_F - k)\, \Theta(|\mathbf{k} + \mathbf{q}| - k_F)$$

$$\times \delta \left(\hbar\omega - \frac{\hbar^2}{2m}(q^2 + 2\mathbf{k} \cdot \mathbf{q}) \right) .$$

Also, prove the relationship

$$\int_{-\infty}^{+\infty} \frac{d\omega}{2\pi} S^0(\mathbf{q}, \omega) = \begin{cases} N & \text{for } \mathbf{q} = 0 \\ 1 - \frac{1}{N} \sum_{\mathbf{k}, \sigma} n_{\mathbf{k}\sigma} n_{\mathbf{k}+\mathbf{q}\sigma} & \text{for } \mathbf{q} \neq 0 \end{cases} ,$$

where $\hat{n}_{\mathbf{k}\sigma} = a_{\mathbf{k}\sigma}^\dagger a_{\mathbf{k}\sigma}$.

2.6 Derive the following relations for Fermi operators:

a)

$$e^{-\alpha a^\dagger} a e^{\alpha a^\dagger} = a - \alpha^2 a^\dagger + \alpha(a a^\dagger - a^\dagger a)$$

$$e^{-\alpha a} a^\dagger e^{\alpha a} = a^\dagger - \alpha^2 a - \alpha(a a^\dagger - a^\dagger a)$$

b)

$$e^{\alpha a^\dagger a} a e^{-\alpha a^\dagger a} = e^{-\alpha} a$$

$$e^{\alpha a^\dagger a} a^\dagger e^{-\alpha a^\dagger a} = e^{-\alpha} a^\dagger .$$

2.7 According to a prediction made by Wigner[13], at low temperatures and sufficiently low densities, an electron gas should undergo a phase transition to a crystalline structure (bcc). For a qualitative analysis[14], consider the energy of a lattice of electrons embedded in a homogeneous, positively charged background. Assume that the potential in which each electron moves can be approximated by the potential of a uniformly charged sphere of radius $r_0 = r_s a_0$ surrounding each electron.

[13] E.P. Wigner, Phys. Rev. **46**, 1002 (1934)
[14] E.P. Wigner, Trans. Faraday Soc. **34**, 678 (1938)

Here, r_0 is the mean particle separation in the Wigner crystal with electron density n, i.e., $\frac{4\pi}{3}r_0^3 = 1/n$. This leads to a model of independent electrons (Einstein approximation) in an oscillator potential

$$H = \frac{p^2}{2m} + \frac{e^2}{2r_0^3}r^2 - \frac{3e^2}{2r_0} \, .$$

Determine the zero-point energy E_0 of this three-dimensional harmonic oscillator and compare this with the result found in the literature[15]:

$$E_0 = \frac{e^2}{2a_0}\left\{-\frac{1.792}{r_s} + \frac{2.638}{r_s^{3/2}}\right\} \, .$$

By minimizing the zero-point energy, determine the mean separation of the electrons.

---

[15] R.A. Coldwell–Horsfall and A.A. Maradudin, J. Math. Phys. **1**, 395 (1960)

3. Bosons

In this chapter we study the characteristic properties of bosonic many-particle systems. To start with the pair distribution function of noninteracting bosons will be computed in order to investigate correlation effects. Subsequently, the excitations of the weakly interacting Bose gas will be determined and properties of the superfluid phase will be discussed.

3.1 Free Bosons

3.1.1 Pair Distribution Function for Free Bosons

We shall assume that the bosons are noninteracting and that they carry zero spin. Hence, their only quantum number is their momentum. We consider a given state of an N-particle system

$$|\phi\rangle = |n_{\mathbf{p}_0}, n_{\mathbf{p}_1}, \ldots\rangle, \tag{3.1.1}$$

where the occupation numbers can take the values $0, 1, 2, \ldots$ etc. The expectation value of the particle density is

$$\langle\phi| \psi^\dagger(\mathbf{x})\psi(\mathbf{x}) |\phi\rangle = \frac{1}{V} \sum_{\mathbf{k},\mathbf{k}'} e^{-i\mathbf{k}\cdot\mathbf{x}+i\mathbf{k}'\cdot\mathbf{x}} \langle\phi| a_{\mathbf{k}}^\dagger a_{\mathbf{k}'} |\phi\rangle$$

$$= \frac{1}{V} \sum_{\mathbf{k}} n_{\mathbf{k}} = \frac{N}{V} = n . \tag{3.1.2}$$

The density in the state (3.1.1) is independent of position.
The pair distribution function is given by

$$n^2 g(\mathbf{x} - \mathbf{x}') = \langle\phi| \psi^\dagger(\mathbf{x})\psi^\dagger(\mathbf{x}')\psi(\mathbf{x}')\psi(\mathbf{x}) |\phi\rangle$$

$$= \frac{1}{V^2} \sum_{\mathbf{k},\mathbf{k}',\mathbf{q},\mathbf{q}'} e^{-i\mathbf{k}\cdot\mathbf{x}-i\mathbf{q}\cdot\mathbf{x}'+i\mathbf{q}'\cdot\mathbf{x}'+i\mathbf{k}'\cdot\mathbf{x}} \langle\phi| a_{\mathbf{k}}^\dagger a_{\mathbf{q}}^\dagger a_{\mathbf{q}'} a_{\mathbf{k}'} |\phi\rangle . \tag{3.1.3}$$

The expectation value $\langle\phi| a_{\mathbf{k}}^\dagger a_{\mathbf{q}}^\dagger a_{\mathbf{q}'} a_{\mathbf{k}'} |\phi\rangle$ differs from zero only if $\mathbf{k} = \mathbf{k}'$ and $\mathbf{q} = \mathbf{q}'$, or $\mathbf{k} = \mathbf{q}'$ and $\mathbf{q} = \mathbf{k}'$. The case $\mathbf{k} = \mathbf{q}$, which, in contrast to fermions, is possible for bosons, has to be treated separately. Hence, it follows that

$$\langle\phi|\, a_{\mathbf{k}}^{\dagger} a_{\mathbf{q}}^{\dagger} a_{\mathbf{q}'} a_{\mathbf{k}'}\, |\phi\rangle$$

$$= (1 - \delta_{\mathbf{kq}}) \left(\delta_{\mathbf{kk}'}\delta_{\mathbf{qq}'} \langle\phi|\, a_{\mathbf{k}}^{\dagger} a_{\mathbf{q}}^{\dagger} a_{\mathbf{q}} a_{\mathbf{k}}\, |\phi\rangle + \delta_{\mathbf{kq}'}\delta_{\mathbf{qk}'} \langle\phi|\, a_{\mathbf{k}}^{\dagger} a_{\mathbf{q}}^{\dagger} a_{\mathbf{k}} a_{\mathbf{q}}\, |\phi\rangle \right)$$

$$+ \delta_{\mathbf{kq}}\delta_{\mathbf{kk}'}\delta_{\mathbf{qq}'} \langle\phi|\, a_{\mathbf{k}}^{\dagger} a_{\mathbf{k}}^{\dagger} a_{\mathbf{k}} a_{\mathbf{k}}\, |\phi\rangle$$

$$= (1 - \delta_{\mathbf{kq}})(\delta_{\mathbf{kk}'}\delta_{\mathbf{qq}'} + \delta_{\mathbf{kq}'}\delta_{\mathbf{qk}'}) n_{\mathbf{k}} n_{\mathbf{q}} + \delta_{\mathbf{kq}}\delta_{\mathbf{kk}'}\delta_{\mathbf{qq}'} n_{\mathbf{k}}(n_{\mathbf{k}} - 1) \tag{3.1.4}$$

and

$$\langle\phi|\, \psi^{\dagger}(\mathbf{x})\psi^{\dagger}(\mathbf{x}')\psi(\mathbf{x}')\psi(\mathbf{x})\, |\phi\rangle \tag{3.1.5}$$

$$= \frac{1}{V^2}\left\{ \sum_{\mathbf{k},\mathbf{q}}(1 - \delta_{\mathbf{kq}})(1 + e^{-i(\mathbf{k}-\mathbf{q})\cdot(\mathbf{x}-\mathbf{x}')}) n_{\mathbf{k}} n_{\mathbf{q}} + \sum_{\mathbf{k}} n_{\mathbf{k}}(n_{\mathbf{k}} - 1) \right\}$$

$$= \frac{1}{V^2}\left\{ \sum_{\mathbf{k},\mathbf{q}} n_{\mathbf{k}} n_{\mathbf{q}} - \sum_{\mathbf{k}} n_{\mathbf{k}}^2 + \left| \sum_{\mathbf{k}} e^{-i\mathbf{k}\cdot(\mathbf{x}-\mathbf{x}')} n_{\mathbf{k}} \right|^2 - \sum_{\mathbf{k}} n_{\mathbf{k}}^2 \right.$$

$$\left. + \sum_{\mathbf{k}} n_{\mathbf{k}}^2 - \sum_{\mathbf{k}} n_{\mathbf{k}} \right\}$$

$$= n^2 + \left| \frac{1}{V} \sum_{\mathbf{k}} e^{-i\mathbf{k}\cdot(\mathbf{x}-\mathbf{x}')} n_{\mathbf{k}} \right|^2 - \frac{1}{V^2}\sum_{\mathbf{k}} n_{\mathbf{k}}(n_{\mathbf{k}} + 1)\,.$$

In contrast to fermions, the second term here is positive due to the permutation symmetry of the wave function. For fermions, there is no multiple occupancy so the last term does not arise.

We now consider two examples. When all the bosons occupy the same state $\mathbf{p}_0$, then (3.1.5) yields:

$$n^2 g(\mathbf{x} - \mathbf{x}') = n^2 + n^2 - \frac{1}{V^2}N(N + 1) = \frac{N(N-1)}{V^2}\,. \tag{3.1.6}$$

In this case, the pair distribution function is position independent; there are no correlations. The right-hand side signifies that the probability of detecting the first particle is N/V, and that of the second particle $(N - 1)/V$.

If, on the other hand, the particles are distributed over many different momentum values and the distribution is described, e.g., by a Gaussian

$$n_{\mathbf{k}} = \frac{(2\pi)^3 n}{(\sqrt{\pi}\Delta)^3} e^{-(\mathbf{k}-\mathbf{k}_0)^2/\Delta^2} \tag{3.1.7}$$

with the normalization

$$\int \frac{d^3p}{(2\pi)^3} n_{\mathbf{p}} = n\,,$$

it then follows that

$$\int \frac{d^3k}{(2\pi)^3} e^{-i\mathbf{k}\cdot(\mathbf{x}-\mathbf{x}')} n_{\mathbf{k}} = n e^{-\frac{\Delta^2}{4}(\mathbf{x}-\mathbf{x}')^2} e^{-i\mathbf{k}_0\cdot(\mathbf{x}-\mathbf{x}')}$$

and

$$\frac{1}{V}\int \frac{d^3k}{(2\pi)^3} n_{\mathbf{k}}^2 = \frac{1}{V}\left[\frac{(2\pi)^3 n}{(\sqrt{\pi}\Delta)^3}\right]^2 \int \frac{d^3k}{(2\pi)^3} e^{-2(\mathbf{k}-\mathbf{k}_0)^2/\Delta^2} \sim \frac{n^2}{V\Delta^3} .$$

If the density and the width Δ of the momentum distribution are held fixed, then, in the limit of large volume V, the third term in (3.1.5) disappears. The pair distribution function is then given by

$$n^2 g(\mathbf{x}-\mathbf{x}') = n^2 \left(1 + e^{-\frac{\Delta^2}{2}(\mathbf{x}-\mathbf{x}')^2}\right) . \tag{3.1.8}$$

As can be seen from Fig. 3.1, for bosons the probability density of finding two particles at a small separation, i.e., $r < \Delta^{-1}$, is increased. Due to the symmetry of the wave function, bosons have a tendency to "cluster together". From Fig. 3.1, one sees that the probability density of finding two bosons at exactly the same place is twice that at large separations.

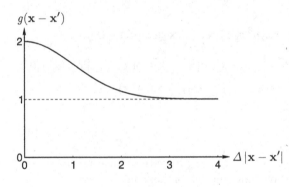

Fig. 3.1. Pair distribution function for bosons

*3.1.2 Two-Particle States of Bosons

In order to investigate the consequences of Bose–Einstein statistics further, we now turn to boson interference and fluctuation processes. Such interference can already be found in two-particle states. The general two-particle state is

$$|2\rangle = \int d^3x_1 d^3x_2 \varphi(\mathbf{x}_1, \mathbf{x}_2)\psi^\dagger(\mathbf{x}_1)\psi^\dagger(\mathbf{x}_2)|0\rangle , \tag{3.1.9}$$

with the normalization $\langle 2|2\rangle = 1$ leading to

$$\langle 2|2\rangle = \int d^3x_1 d^3x_2 \varphi^*(\mathbf{x}_1, \mathbf{x}_2)(\varphi(\mathbf{x}_1, \mathbf{x}_2) + \varphi(\mathbf{x}_2, \mathbf{x}_1)) = 1 . \tag{3.1.10}$$

We could have restricted ourselves from the outset to symmetric $\varphi(\mathbf{x}_1, \mathbf{x}_2)$ since $[\psi^\dagger(\mathbf{x}_1), \psi^\dagger(\mathbf{x}_2)] = 0$ and thus the odd part of $\varphi(\mathbf{x}_1, \mathbf{x}_2)$ makes no contribution.

In the following, we shall consider functions $\varphi(x_1, x_2)$ of the form

$$\varphi(\mathbf{x}_1, \mathbf{x}_2) \propto \varphi_1(\mathbf{x}_1)\varphi_2(\mathbf{x}_2) . \tag{3.1.11}$$

Had the particles been distinguishable, for such a wave function, they would have been completely independent. Furthermore, we assume

$$\int d^3x |\varphi_i(\mathbf{x})|^2 = 1 , \tag{3.1.12}$$

and then the normalization condition (3.1.10) yields:

$$\varphi(\mathbf{x}_1, \mathbf{x}_2) = \frac{\varphi_1(\mathbf{x}_1)\varphi_2(\mathbf{x}_2)}{(1 + |(\varphi_1, \varphi_2)|^2)^{1/2}} \tag{3.1.13}$$

with $(\varphi_i, \varphi_j) \equiv \int d^3x \varphi_i^*(\mathbf{x})\varphi_j(\mathbf{x})$. For the two-particle state (3.1.9) with (3.1.13)[1], the expectation value of the density is

$$\begin{aligned}
\langle 2| n(\mathbf{x}) |2\rangle &= \int d^3x_1 \, d^3x_2 d^3x_1' d^3x_2' \varphi_1^*(\mathbf{x}_1)\varphi_2^*(\mathbf{x}_2)\varphi_1(\mathbf{x}_1')\varphi_2(\mathbf{x}_2') \\
&\quad \times [1 + |(\varphi_1, \varphi_2)|^2]^{-1} \langle 0| \psi(\mathbf{x}_2)\psi(\mathbf{x}_1)\psi^\dagger(\mathbf{x})\psi(\mathbf{x})\psi^\dagger(\mathbf{x}_1')\psi^\dagger(\mathbf{x}_2') |0\rangle \\
&= [|\varphi_1(\mathbf{x})|^2 + |\varphi_2(\mathbf{x})|^2 + (\varphi_1, \varphi_2)\varphi_2^*(\mathbf{x})\varphi_1(\mathbf{x}) + \text{c.c.}] \\
&\quad \times [1 + |(\varphi_1, \varphi_2)|^2]^{-1} .
\end{aligned} \tag{3.1.14}$$

In (3.1.14), in addition to $|\varphi_1(\mathbf{x})|^2 + |\varphi_2(\mathbf{x})|^2$, an interference term occurs. When the two single-particle wave functions are orthogonal, i.e., $(\varphi_1, \varphi_2) = 0$, the density

$$\langle 2| n(\mathbf{x}) |2\rangle = |\varphi_1(\mathbf{x})|^2 + |\varphi_2(\mathbf{x})|^2 \tag{3.1.15}$$

equals the sum of the single-particle densities, as would be the case for independent particles. For two overlapping Gaussians, with separation $2a$, it is easy to calculate the clustering effect for bosons. Let

$$\varphi_1(x) = \frac{1}{\pi^{1/4}}e^{-\frac{1}{2}(x-a)^2} , \quad \varphi_2(x) = \frac{1}{\pi^{1/4}}e^{-\frac{1}{2}(x+a)^2} \tag{3.1.16}$$

---

[1] The Schrödinger two-particle wave function in the coordinate representation corresponding to (3.1.9) with (3.1.13) reads $\frac{\varphi_1(x_1)\varphi_2(x_2)+\varphi_2(x_1)\varphi_1(x_2)}{\sqrt{2}(1+|(\varphi_1,\varphi_2)|^2)^{1/2}}$.

with the properties $(\varphi_i, \varphi_i) = 1$ and $(\varphi_1, \varphi_2) = \frac{1}{\sqrt{\pi}} \int dx e^{-x^2 - a^2} = e^{-a^2}$; for these states the density expectation value (3.1.14) is

$$\langle 2| n(\mathbf{x}) |2\rangle = \frac{1}{\sqrt{\pi}(1 + e^{-2a^2})} \left\{ e^{-(x-a)^2} + e^{-(x+a)^2} + 2e^{-2a^2} e^{-x^2} \right\} .$$

$$(3.1.17)$$

The integrated density

$$\int d^3x \, \langle 2| n(\mathbf{x}) |2\rangle = 2$$

is equal to the number of particles. Figure 3.2 shows $\langle 2| n(\mathbf{x}) |2\rangle$ for the separations $a = 3$ and $a = 1$. For the smaller separation the wave functions overlap and, for small x, the particle density is greater than it would be for independent particles.

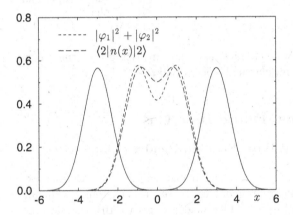

Fig. 3.2. Densities for two-boson states. The full line is the case $a = 3$. Since there is no overlap here, $|\varphi_1|^2 + |\varphi_2|^2$ and $\langle n(\mathbf{x}) \rangle$ are indistinguishable from one another. The dashed lines are for $a = 1$: in this case $\langle 2| n(\mathbf{x}) |2\rangle$ is increased at small separations in comparison to $|\varphi_1|^2 + |\varphi_2|^2$

Photon Correlations

Photons represent the ideal example of noninteracting particles. In photon correlation experiments it has actually been possible to observe the predicted tendency of bosons to cluster together.[2] These correlation effects can be understood theoretically with the help of pair correlations of the form (3.1.8).[3] Since the classical electromagnetic waves of Maxwell's theory are coherent

---

[2] R. Hanbury Brown and R.G. Twiss, Nature **177**, 27 (1956); **178**, 1447 (1956)
[3] E.M. Purcell, Nature **178**, 1449 (1956)

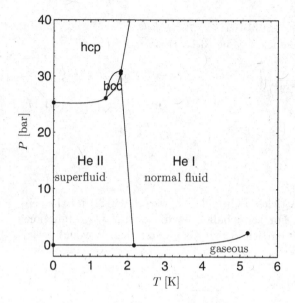

Fig. 3.3. The phase diagram of He^4. The solid phases are: hcp (hexagonal close packed) and bcc (body centered cubic). The fluid region is divided into a normal (He I) and a superfluid (He II) phase

states of photons in quantum mechanics, it is not surprising that these correlation effects also follow from classical electrodynamics.[4]

3.2 Weakly Interacting, Dilute Bose Gas

3.2.1 Quantum Fluids and Bose–Einstein Condensation

The most important Bose fluid is He^4, which has spin $S = 0$. Another example is spin-polarized atomic hydrogen; this, however, is extremely difficult to produce for long enough periods at sufficient density. All other atomic bosons are heavier and more strongly interacting, causing them to crystallize at temperatures far above any possible superfluid transition. At normal pressures, He^4 remains fluid down to $T = 0$ and at the lambda point $T_\lambda = 2.18\,\mathrm{K}$ it enters the superfluid state (Fig. 3.3). The normal and the superfluid phases are also known as He I and He II. In order for He^4 to crystallize, it must be subjected to a pressure of at least 25 bar. Although they are rare in comparison to Fermi fluids, which are realized in He^3 and by every metal, Bose fluids are a rewarding topic of study due to their fascinating properties. Corresponding to superfluidity there is the superconducting phase in fermion systems. He^3, electrons in metals, and electrons in a number of oxidic high-T_c perovskites can form pairs of fermions that obey Bose statistics. Real helium is only mimicked by an ideal Bose gas since, in additions to quantum effects,

---

[4] Discussions of the Hanbury-Brown and Twiss experiments can be found in C. Kittel, *Elementary Statistical Physics*, p. 123, J. Wiley, New York, 1958 and G. Baym, *Lectures on Quantum Mechanics*, p. 431, W.A. Benjamin, London, 1973

it is also plagued by the difficulties associated with a dense fluid. In an *ideal* (i.e., noninteracting) *Bose gas* at temperatures below $T_c(v) = \frac{2\pi\hbar^2/m}{[v\cdot 2.61]^{2/3}}$ (for the mass and density of He^4 this gives 3.14 K) Bose–Einstein condensation occurs[5]. The single-particle ground state becomes macroscopically occupied in conjunction with the disappearance of the chemical potential $\mu \to 0$.

In reality, He^4 atoms have approximately a Lennard–Jones potential,

$$V(r) = 4\epsilon\left[\left(\frac{\sigma}{r}\right)^{12} - \left(\frac{\sigma}{r}\right)^{6}\right] \qquad (3.2.1)$$

$$\epsilon = 1.411 \times 10^{-15}\text{erg}$$

$$\sigma = 2.556\text{Å} .$$

It consists of a repulsive (hard-core) part and an attractive component. At small separations the potential (3.2.1) is equivalent to the potential of an almost ideal hard sphere of diameter 2 Å. For fcc close packing of spheres, this would correspond to a molar volume of 12 cm³, whereas the actually observed molar volume at $P = 30$ bar is 26 cm³. The reason for this higher value lies in the large amplitude of the quantum-mechanical zero-point oscillations. In the fluid phase $V_M = 27$ cm³. The various phases of He^4 and He^3 are also known as quantum fluids or quantum crystals.

Note: Recently, Bose–Einstein condensation has been observed, in a gas of about 2000 spin-polarized ^{87}Rb atoms confined in a quadrupole trap[6][7], some 70 years after its original prediction by Einstein[8], on the basis of statistical considerations of Bose. The transition temperature is 170×10^{-9}K. One might expect that at low temperatures alkali atoms would form a solid; however, even at temperatures in the nano-Kelvin regime, it is possible to maintain a metastable gaseous state.

A similar experiment has been carried out with a gas of 2×10^5 spin-polarized ^{7}Li atoms.[9] In this case, the condensation temperature is $T_c \approx 400 \times 10^{-9}$K. In ^{87}Rb the s-wave scattering length is positive, whereas in ^{7}Li it is negative. Despite this, the gaseous phase of ^{7}Li does not collapse into the fluid or the solid phase, not, at least, in the spatially inhomogeneous case.[9] Bose–Einstein condensation has also been observed in sodium in a sample of 5×10^5 atoms at a density of 10^{14}cm^{-3} and temperatures below 2μK.[10] Finally, also in atomic hydrogen, a condensate of more than 10^8 atoms, with a transition temperature of roughly 50μK, could be maintained for up to 5 seconds.[11]

---

[5] See for instance F. Schwabl, *Statistical Mechanics*, Springer, 2nd ed., Berlin Heidelberg, 2006, Sect. 4.4; in subsequent citations this book will be referred to as SM.

[6] M.H. Andersen, J.R. Enscher, M.R. Matthews, C.E. Wieman, and E. A. Cornell, Science **269**, 198 (1995)

[7] See also G.P. Collins, Physics Today, August 1995, 17

[8] A. Einstein, Sitzber. Kgl. Preuss. Akad. Wiss. **1924**, 261 (1924), ibid. **1925**, 3 (1925); S. Bose, Z. Phys. **26**, 178 (1924)

[9] C.C. Bradley, C.A. Sackett, J.J. Tollett, and R.G. Hulet, Phys. Rev. Lett. **75**, 1687 (1995)

[10] K. B. Davis, M.-O. Mewes, M.R. Andrews, N.J. van Druten, D.S. Durfee, D.M. Kurn, and W. Ketterle, Phys. Rev. Lett. **75**, 2969 (1995)

[11] D. Kleppner, Th. Greytak et al., Phys. Rev. Lett. **81**, 3811 (1998)

These experimental discoveries triggered an avalanche of new physics. See literature on Bose–Einstein condensation, in the bibliography at the end of Part I.

3.2.2 Bogoliubov Theory of the Weakly Interacting Bose Gas

In the momentum representation, the Hamiltonian reads:

$$H = \sum_{\mathbf{k}} \frac{k^2}{2m} a_{\mathbf{k}}^{\dagger} a_{\mathbf{k}} + \frac{1}{2V} \sum_{\mathbf{k},\mathbf{p},\mathbf{q}} V_{\mathbf{q}} a_{\mathbf{k}+\mathbf{q}}^{\dagger} a_{\mathbf{p}-\mathbf{q}}^{\dagger} a_{\mathbf{p}} a_{\mathbf{k}}, \tag{3.2.2}$$

where we have set $\hbar = 1$. This Hamiltonian is still completely general, but in the following we will introduce approximations which restrict the validity of the theory to dilute, weakly interacting Bose gases. The creation and annihilation operators $a_{\mathbf{k}}^{\dagger}$ and $a_{\mathbf{k}}$ satisfy the Bose commutation relations and $V_{\mathbf{q}}$ is the Fourier transform of the two-particle interaction

$$V_{\mathbf{q}} = \int d^3x \, e^{-i\mathbf{q}\cdot\mathbf{x}} V(\mathbf{x}). \tag{3.2.3}$$

At low temperatures, a Bose–Einstein condensation takes place in the $\mathbf{k} = 0$ mode, i.e., in analogy to the ideal Bose gas it is expected that in the ground state[12] $|0\rangle$ the single-particle state with $\mathbf{k} = 0$ is macroscopically occupied,

$$N_0 = \langle 0| a_0^{\dagger} a_0 |0\rangle \lesssim N , \tag{3.2.4a}$$

and thus the number of excited particles is

$$N - N_0 \ll N_0 \lesssim N . \tag{3.2.4b}$$

Hence, we can neglect the interaction of the excited particles with one another and restrict ourselves to the interaction of the excited particles with the condensed particles:

$$H = \sum_{\mathbf{k}} \frac{k^2}{2m} a_{\mathbf{k}}^{\dagger} a_{\mathbf{k}} + \frac{1}{2V} V_0 a_0^{\dagger} a_0^{\dagger} a_0 a_0 + \frac{1}{V} {\sum_{\mathbf{k}}}' (V_0 + V_{\mathbf{k}}) a_0^{\dagger} a_0 a_{\mathbf{k}}^{\dagger} a_{\mathbf{k}}$$

$$+ \frac{1}{2V} {\sum_{\mathbf{k}}}' V_{\mathbf{k}} (a_{\mathbf{k}}^{\dagger} a_{-\mathbf{k}}^{\dagger} a_0 a_0 + a_0^{\dagger} a_0^{\dagger} a_{\mathbf{k}} a_{-\mathbf{k}}) + \mathcal{O}(a_{\mathbf{k}}^3) . \tag{3.2.5}$$

The prime on the sum indicates that the value $\mathbf{k} = 0$ is excluded. Due to momentum conservation, there is no term containing $a_{\mathbf{k}\neq 0}$ and three operators with $\mathbf{k} = 0$.

The effect of a_0 and $a_0^{\dagger}$ on the state with N_0 particles in the condensate is

$$a_0 |N_0, \ldots\rangle = \sqrt{N_0} |N_0 - 1, \ldots\rangle$$

$$a_0^{\dagger} |N_0, \ldots\rangle = \sqrt{N_0 + 1} |N_0 + 1, \ldots\rangle . \tag{3.2.6}$$

---

[12] Here, $|0\rangle$ is the ground state of the N bosons and not the vacuum state with respect to the $a_{\mathbf{k}}$, which would contain no bosons at all. It will emerge that $|0\rangle$ is the vacuum state for the operators $\alpha_{\mathbf{k}}$ to be introduced below.

Since N_0 is such a huge number, $N_0 \approx 10^{23}$, both of these correspond to multiplication by $\sqrt{N_0}$. Furthermore, it is physically obvious that the removal or addition of one particle from or to the condensate will make no difference to the physical properties of the system. In comparison to N_0, the effect of the commutator

$$a_0 a_0^\dagger - a_0^\dagger a_0 = 1$$

is negligible, i.e., the operators

$$a_0 = a_0^\dagger = \sqrt{N_0} \tag{3.2.7}$$

can be approximated by a c-number. The Hamiltonian then becomes

$$
\begin{aligned}
H = & \sum_{\mathbf{k}}' \frac{k^2}{2m} a_{\mathbf{k}}^\dagger a_{\mathbf{k}} + \frac{1}{2V} N_0^2 V_0 \\
& + \frac{N_0}{V} \sum_{\mathbf{k}}' \left[(V_0 + V_{\mathbf{k}}) a_{\mathbf{k}}^\dagger a_{\mathbf{k}} + \frac{1}{2} V_{\mathbf{k}} (a_{\mathbf{k}}^\dagger a_{-\mathbf{k}}^\dagger + a_{\mathbf{k}} a_{-\mathbf{k}}) \right] + \cdots .
\end{aligned}
\tag{3.2.8}
$$

The value of N_0 is unknown at the present stage. It is determined by the density (or the particle number for a given volume) and by the interaction. We express N_0 in terms of the total particle number N and the number of particles in the excited state

$$N = N_0 + \sum_{\mathbf{k}}' a_{\mathbf{k}}^\dagger a_{\mathbf{k}} . \tag{3.2.9}$$

We then have, for example,

$$\frac{V_0}{2V} N_0^2 = \frac{V_0}{2V} N^2 - \frac{NV_0}{V} \sum_{\mathbf{k}}' a_{\mathbf{k}}^\dagger a_{\mathbf{k}} + \frac{V_0}{2V} \sum_{\mathbf{k},\mathbf{k}'}' a_{\mathbf{k}}^\dagger a_{\mathbf{k}} a_{\mathbf{k}'}^\dagger a_{\mathbf{k}'} . \tag{3.2.10}$$

The Hamiltonian follows as

$$
\begin{aligned}
H = & \sum_{\mathbf{k}}' \frac{k^2}{2m} a_{\mathbf{k}}^\dagger a_{\mathbf{k}} + \frac{N}{V} \sum_{\mathbf{k}}' V_{\mathbf{k}} a_{\mathbf{k}}^\dagger a_{\mathbf{k}} + \frac{N^2}{2V} V_0 \\
& + \frac{N}{2V} \sum_{\mathbf{k}}' V_{\mathbf{k}} \left(a_{\mathbf{k}}^\dagger a_{-\mathbf{k}}^\dagger + a_{\mathbf{k}} a_{-\mathbf{k}} \right) + H' .
\end{aligned}
\tag{3.2.11}
$$

The operator H' contains terms with four creation or annihilation operators, and these are of order n'^2, where $n' = \frac{N - N_0}{V}$ is the density of the particles that are not part of the condensate. The Bogoliubov approximation, which amounts to neglecting these anharmonic terms, is a good approximation when $n' \ll n$. We shall see later, when we calculate n', that exactly this condition is fulfilled by the dilute, weakly interacting Bose gas.

If H' is neglected, we have a quadratic form, which still has to be diagonalized. The transformation proceeds in analogy to the theory of antiferromagnetic magnons. We introduce the ansatz[13]

$$a_{\mathbf{k}} = u_{\mathbf{k}}\alpha_{\mathbf{k}} + v_{\mathbf{k}}\alpha^\dagger_{-\mathbf{k}}$$
$$a^\dagger_{\mathbf{k}} = u_{\mathbf{k}}\alpha^\dagger_{\mathbf{k}} + v_{\mathbf{k}}\alpha_{-\mathbf{k}}$$

$(3.2.12)$

with real symmetric coefficients, and demand that the operators α also satisfy Bose commutation relations

$$[\alpha_{\mathbf{k}}, \alpha_{\mathbf{k}'}] = [\alpha^\dagger_{\mathbf{k}}, \alpha^\dagger_{\mathbf{k}'}] = 0, [\alpha_{\mathbf{k}}, \alpha^\dagger_{\mathbf{k}'}] = \delta_{\mathbf{k}\mathbf{k}'} \ .$$

$(3.2.13)$

This is the case when

$$u^2_{\mathbf{k}} - v^2_{\mathbf{k}} = 1.$$

$(3.2.14)$

Proof:

$$[a_{\mathbf{k}}, a_{\mathbf{k}'}] = u_{\mathbf{k}}v_{\mathbf{k}'}\delta_{\mathbf{k},-\mathbf{k}'} + v_{\mathbf{k}}u_{\mathbf{k}'}(-\delta_{\mathbf{k},-\mathbf{k}'}) = 0$$
$$\left[a_{\mathbf{k}}, a^\dagger_{\mathbf{k}'}\right] = u_{\mathbf{k}}u_{\mathbf{k}'}\delta_{\mathbf{k}\mathbf{k}'} + v_{\mathbf{k}}v_{\mathbf{k}'}(-\delta_{\mathbf{k}\mathbf{k}'}) = (u^2_{\mathbf{k}} - v^2_{\mathbf{k}})\delta_{\mathbf{k}\mathbf{k}'} \ .$$

The inverse of the transformation $(3.2.12)$ reads (see Problem 3.3):

$$\alpha_{\mathbf{k}} = u_{\mathbf{k}}a_{\mathbf{k}} - v_{\mathbf{k}}a^\dagger_{-\mathbf{k}}$$
$$\alpha^\dagger_{\mathbf{k}} = u_{\mathbf{k}}a^\dagger_{\mathbf{k}} - v_{\mathbf{k}}a_{-\mathbf{k}} \ .$$

$(3.2.15)$

With the additional calculational step

$$a^\dagger_{\mathbf{k}}a_{\mathbf{k}} = u^2_{\mathbf{k}}\alpha^\dagger_{\mathbf{k}}\alpha_{\mathbf{k}} + v^2_{\mathbf{k}}\alpha_{-\mathbf{k}}\alpha^\dagger_{-\mathbf{k}} + u_{\mathbf{k}}v_{\mathbf{k}}(\alpha^\dagger_{\mathbf{k}}\alpha^\dagger_{-\mathbf{k}} + \alpha_{\mathbf{k}}\alpha_{-\mathbf{k}})$$
$$a^\dagger_{\mathbf{k}}a^\dagger_{-\mathbf{k}} = u^2_{\mathbf{k}}\alpha^\dagger_{\mathbf{k}}\alpha^\dagger_{-\mathbf{k}} + v^2_{\mathbf{k}}\alpha_{\mathbf{k}}\alpha_{-\mathbf{k}} + u_{\mathbf{k}}v_{\mathbf{k}}(\alpha^\dagger_{\mathbf{k}}\alpha_{\mathbf{k}} + \alpha_{-\mathbf{k}}\alpha^\dagger_{-\mathbf{k}})$$
$$a_{\mathbf{k}}a_{-\mathbf{k}} = u^2_{\mathbf{k}}\alpha_{\mathbf{k}}\alpha_{-\mathbf{k}} + v^2_{\mathbf{k}}\alpha^\dagger_{\mathbf{k}}\alpha^\dagger_{-\mathbf{k}} + u_{\mathbf{k}}v_{\mathbf{k}}(\alpha^\dagger_{-\mathbf{k}}\alpha_{-\mathbf{k}} + \alpha_{\mathbf{k}}\alpha^\dagger_{\mathbf{k}}) \ ,$$

one obtains for the Hamiltonian

$$H = \frac{1}{2V}N^2V_0 +$$
$$+ \sum_{\mathbf{k}}{}' \left(\frac{k^2}{2m} + nV_{\mathbf{k}}\right) \left[u^2_{\mathbf{k}}\alpha^\dagger_{\mathbf{k}}\alpha_{\mathbf{k}} + v^2_{\mathbf{k}}\alpha_{\mathbf{k}}\alpha^\dagger_{\mathbf{k}} + u_{\mathbf{k}}v_{\mathbf{k}}\left(\alpha^\dagger_{\mathbf{k}}\alpha^\dagger_{-\mathbf{k}} + \alpha_{\mathbf{k}}\alpha_{-\mathbf{k}}\right)\right]$$
$$+ \frac{N}{2V}\sum_{\mathbf{k}}{}' V_{\mathbf{k}} \left[(u^2_{\mathbf{k}} + v^2_{\mathbf{k}})\left(\alpha^\dagger_{\mathbf{k}}\alpha^\dagger_{-\mathbf{k}} + \alpha_{\mathbf{k}}\alpha_{-\mathbf{k}}\right) + 2u_{\mathbf{k}}v_{\mathbf{k}}\left(\alpha^\dagger_{\mathbf{k}}\alpha_{\mathbf{k}} + \alpha_{\mathbf{k}}\alpha^\dagger_{\mathbf{k}}\right)\right] \ .$$

$(3.2.16)$

In order for the nondiagonal terms to disappear, we require

$$\left(\frac{k^2}{2m} + nV_{\mathbf{k}}\right)u_{\mathbf{k}}v_{\mathbf{k}} + \frac{n}{2}V_{\mathbf{k}}(u^2_{\mathbf{k}} + v^2_{\mathbf{k}}) = 0 \ .$$

$(3.2.17)$

---

[13] The transformation is known as the Bogoliubov transformation. This diagonalization method was originally introduced by T. Holstein and H. Primakoff (Phys. Rev. **58**, 1098 (1940)) for complicated spin-wave Hamiltonians and was rediscovered by N.N. Bogoliubov (J. Phys. (U.S.S.R.) **11**, 23 (1947)).

Together with $u_{\mathbf{k}}^2 - v_{\mathbf{k}}^2 = 1$ from (3.2.14), one now has a system of equations that allow the calculation of $u_{\mathbf{k}}^2$ and $v_{\mathbf{k}}^2$. It is convenient to introduce the definition

$$\omega_{\mathbf{k}} \equiv \left[\left(\frac{k^2}{2m} + nV_{\mathbf{k}} \right)^2 - (nV_{\mathbf{k}})^2 \right]^{1/2} = \left[\left(\frac{k^2}{2m} \right)^2 + \frac{nk^2V_{\mathbf{k}}}{m} \right]^{1/2} . \quad (3.2.18)$$

From (3.2.14) and (3.2.17), one finds $u_{\mathbf{k}}^2$ and $v_{\mathbf{k}}^2$ to be (Problem 3.4)

$$u_{\mathbf{k}}^2 = \frac{\omega_{\mathbf{k}} + \left(\frac{k^2}{2m} + nV_{\mathbf{k}} \right)}{2\omega_{\mathbf{k}}} ,$$

$$v_{\mathbf{k}}^2 = \frac{-\omega_{\mathbf{k}} + \left(\frac{k^2}{2m} + nV_{\mathbf{k}} \right)}{2\omega_{\mathbf{k}}} , \quad (3.2.19)$$

$$u_{\mathbf{k}}v_{\mathbf{k}} = -\frac{nV_{\mathbf{k}}}{2\omega_{\mathbf{k}}} , \quad v_{\mathbf{k}}^2 = \frac{(nV_{\mathbf{k}})^2}{2\omega_{\mathbf{k}}(\omega_{\mathbf{k}} + \frac{k^2}{2m} + nV_{\mathbf{k}})} .$$

Inserting (3.2.19) into the Hamiltonian yields:

$$H = \underbrace{\frac{N^2}{2V}V_0 - \frac{1}{2}\sideset{}{'}\sum_{\mathbf{k}} \left(\frac{k^2}{2m} + nV_{\mathbf{k}} - \omega_{\mathbf{k}} \right)}_{\text{ground-state energy } E_0} + \underbrace{\sideset{}{'}\sum_{\mathbf{k}} \omega_{\mathbf{k}} \alpha_{\mathbf{k}}^\dagger \alpha_{\mathbf{k}}}_{\substack{\text{sum of oscillators} \\ \sim \text{ quasiparticles}}} \quad (3.2.20)$$

The Hamiltonian consists of the ground-state energy and a sum of oscillators of energy $\omega_{\mathbf{k}}$. The excitations that are created by the $\alpha_{\mathbf{k}}^\dagger$ are called *quasiparticles*.

The *ground state* of the system $|0\rangle$ is fixed by the condition that no quasiparticles are excited,

$$\alpha_{\mathbf{k}} |0\rangle = 0 \quad \text{for all } \mathbf{k}. \quad (3.2.21)$$

We can now calculate the number of particles (not quasiparticles) outside the condensate

$$N' = \langle 0| \sideset{}{'}\sum_{\mathbf{k}} a_{\mathbf{k}}^\dagger a_{\mathbf{k}} |0\rangle = \langle 0| \sideset{}{'}\sum_{\mathbf{k}} v_{\mathbf{k}}^2 \alpha_{\mathbf{k}} \alpha_{\mathbf{k}}^\dagger |0\rangle = \sideset{}{'}\sum_{\mathbf{k}} v_{\mathbf{k}}^2 . \quad (3.2.22)$$

For a contact potential $V(\mathbf{x}) = \lambda \delta(\mathbf{x})$, it follows by using (3.2.18) and (3.2.19) that

$$n' \equiv \frac{N'}{V} = \frac{m^{3/2}}{3\pi^2} (\lambda n)^{3/2} . \quad (3.2.23)$$

The expansion parameter is λn, i.e., the strength of the potential times the density. If this expansion parameter is small, consistent with the assumptions made, the density of particles outside the condensate is low. The dependence on λn is nonanalytic and thus cannot be expanded about $\lambda n = 0$. Hence, these results for condensed Bose systems cannot be obtained using straight-forward perturbation theory for the initial Hamiltonian (3.2.2). The number of particles in the condensate is $N_0 = N - n'V$. Its temperature dependence $N_0(T)$ is studied in Problem 3.5.

The ground-state energy (3.2.20) is composed of a term that would be the interaction energy if all particles were in the condensate, and a further negative term. Through the occupation of $\mathbf{k} \neq 0$ Bose states in the ground state (see (3.2.22)), the kinetic energy is increased, whereas the potential energy is reduced.

Excited states of the system are obtained by applying $\alpha_{\mathbf{k}}^{\dagger}$ to the ground state $|0\rangle$. Their energy is $\omega_{\mathbf{k}}$. For small $\mathbf{k}$ one finds from (3.2.18)

$$\omega_{\mathbf{k}} = ck \quad \text{with} \quad c = \sqrt{\frac{nV_0}{m}}. \tag{3.2.24}$$

Thus, the long wavelength excitations are phonons with linear dispersion.

This value for the sound velocity also follows from the compressibility $\kappa = -\frac{1}{V}\frac{\partial V}{\partial P}$:

$$c = \frac{1}{\sqrt{\rho\kappa}} = \sqrt{\frac{\partial P}{\partial \rho}}. \tag{3.2.25}$$

Here, $\rho = mn$ is the mass density and the pressure at zero temperature is given by

$$P = -\frac{\partial E_0}{\partial V}. \tag{3.2.26}$$

For large k, one obtains from (3.2.18)

$$\omega_{\mathbf{k}} = \frac{k^2}{2m} + nV_{\mathbf{k}}. \tag{3.2.27}$$

This corresponds to the dispersion relation for free particles whose energy is shifted by a mean potential of $nV_{\mathbf{k}}$ (see Fig. 3.4). A comparison with the experimental excitation spectrum of He^4 is not justified on account of the restriction to weak interaction and low density; in particular, one cannot attempt to explain the roton minimum (see Sect. 3.2.3) in terms of the k dependence of the potential, since this would require potential strengths outside the domain of validity of this theory (see Problem 3.6).

When $\alpha_{\mathbf{k}}^{\dagger}$ is applied to a state, one speaks of the creation of a quasiparticle with the wave vector $\mathbf{k}$. We shall show furthermore that, for small $\mathbf{k}$, the excitation of a quasiparticle corresponds to a density wave. To this end, we consider the operator for the particle number density

$$n_{\mathbf{k}} = \sum_{\mathbf{p}} a_{\mathbf{p}}^{\dagger} a_{\mathbf{p}+\mathbf{k}} \approx \sqrt{N_0}(a_{-\mathbf{k}}^{\dagger} + a_{\mathbf{k}}) \tag{3.2.28}$$

under the assumption of a macroscopic occupation of the $\mathbf{k} = 0$ state.

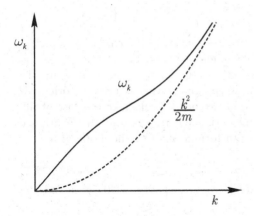

Fig. 3.4. Excitations of the weakly interacting Bose gas

From Eq. (3.2.12) it follows that

$$a_{\mathbf{k}} + a^\dagger_{-\mathbf{k}} = (u_{\mathbf{k}} + v_{\mathbf{k}})(\alpha_{\mathbf{k}} + \alpha^\dagger_{-\mathbf{k}})$$

and therefore

$$n_{\mathbf{k}} = A_{\mathbf{k}} \left(\alpha_{\mathbf{k}} + \alpha^\dagger_{-\mathbf{k}} \right) .$$ (3.2.29)

From Eq. (3.2.19) the amplitude $A_{\mathbf{k}}$ takes the form

$$A_{\mathbf{k}} \equiv \sqrt{N_0}(u_{\mathbf{k}} + v_{\mathbf{k}}) = k\sqrt{\frac{N_0}{2m\omega_{\mathbf{k}}}} .$$

From Eq. (3.2.29) one obtains the density operator

$$\rho(\mathbf{x}) = \tilde{\rho}(\mathbf{x}) + \tilde{\rho}^\dagger(\mathbf{x})$$ (3.2.30a)

in which $\tilde{\rho}(\mathbf{x}) = \sum_{\mathbf{k}} A_{\mathbf{k}} e^{i\mathbf{k}\cdot\mathbf{x}} \alpha_{\mathbf{k}}$, from which it follows that

$$\tilde{\rho}(\mathbf{x}) \left(\alpha^\dagger_{\mathbf{k}} |0\rangle \right) = \sum_{\mathbf{k}'} e^{i\mathbf{k}'\cdot\mathbf{x}} A_{\mathbf{k}'} \alpha_{\mathbf{k}'} \alpha^\dagger_{\mathbf{k}} |0\rangle = e^{i\mathbf{k}\cdot\mathbf{x}} A_{\mathbf{k}} |0\rangle .$$ (3.2.30b)

For a coherent state $|c_{\mathbf{k}}\rangle$ built out of quasi-particle excitations with wave vector $\mathbf{k}$

$$|c_{\mathbf{k}}\rangle = e^{-|c_{\mathbf{k}}|^2/2} \sum_{n=0}^\infty \frac{\left(c_{\mathbf{k}} \alpha^\dagger_{\mathbf{k}} \right)^n}{n!} |0\rangle$$ (3.2.31a)

and in which $c_{\mathbf{k}} = |c_{\mathbf{k}}| e^{-i\delta_{\mathbf{k}}}$ one gains $\tilde{\rho}(\mathbf{x}) |c_{\mathbf{k}}\rangle = A_{\mathbf{k}} c_{\mathbf{k}} e^{i\mathbf{k}\cdot\mathbf{x}} |c_{\mathbf{k}}\rangle$. From this it follows that the expectation value of the density

$$\langle c_{\mathbf{k}}| n(\mathbf{x}) |c_{\mathbf{k}}\rangle = 2A_{\mathbf{k}} |c_{\mathbf{k}}| \cos(\mathbf{k}\mathbf{x} - \delta_{\mathbf{k}}) ,$$ (3.2.31b)

so that a coherent state of this type represents a density wave.

Notes:

(i) Second-order phase transitions are associated with a broken symmetry. In the well known case of a Heisenberg ferromagnet, this symmetry is the invariance of the Hamiltonian with respect to the rotation of all spins. In the ferromagnetic phase, where a finite magnetization is present, oriented, e.g., in the z direction, the rotational invariance is broken. In the case of the Bose–Einstein condensate the gauge invariance is broken, i.e., the invariance of the Hamiltonian with respect to transformation of the field operator

$$\psi(\mathbf{x}) \rightarrow \psi'(\mathbf{x}) = \psi(\mathbf{x})e^{i\alpha} \tag{3.2.32}$$

with a phase α. In the ground state $|0\rangle$, one has $\langle 0|\psi(\mathbf{x})|0\rangle \neq 0$ and the phase is fixed arbitrarily at $\alpha = 0$.

(ii) For finite-ranged potentials, e.g., the spherical well of Problem 3.6, the Fourier transform falls off with increasing wave vector k, leading to a finite ground-state energy in (3.2.20). For the δ-function potential the Fourier transform is a constant, which leads to a divergence at the upper integration limit. To ensure that the ground-state energy E_0 also remains finite for an effective contact potential, the potential strength λ must be replaced by the (finite) scattering length a. In second-order Born approximation, the scattering length is given in terms of λ by

$$a = \frac{m}{4\pi\hbar^2}\lambda\left\{1 - \frac{\lambda}{V}{\sum_{\mathbf{k}}}'\frac{m}{k^2} + \mathcal{O}(\lambda^2)\right\}$$

or, inversely,

$$\lambda = \frac{4\pi\hbar^2 a}{m}\left\{1 + \frac{4\pi\hbar^2 a}{V}{\sum_{\mathbf{k}}}'\frac{1}{k^2} + \mathcal{O}(a^2)\right\} \tag{3.2.33}$$

(see Problem 3.8). Inserting this into (3.2.16) shows that V_0 and $V_{\mathbf{k}}$ must be replaced, here and in all subsequent formulas, by

$$V_0 \rightarrow \frac{4\pi\hbar^2 a}{m}\left\{1 + \frac{4\pi\hbar^2 a}{V}{\sum_{\mathbf{k}}}'\frac{1}{k^2}\right\} \tag{3.2.34a}$$

and

$$V_{\mathbf{k}} \rightarrow \frac{4\pi\hbar^2 a}{m}. \tag{3.2.34b}$$

For the interaction of the excited particles it is sufficient to retain only terms up to first order in a. The value of the ground-state energy is then

$$E_0 = \frac{2\pi\hbar^2}{m}\frac{aN^2}{V}\left\{1 + \frac{128}{15\sqrt{\pi}}\left(\frac{a^3 N}{V}\right)^{1/2}\right\}. \tag{3.2.35}$$

*3.2.3 Superfluidity

Superfluidity refers to a state in which the fluid can flow past objects without exerting a drag and where objects can move through the fluid without slowing down. This property holds only up to a certain critical velocity which we will now relate to the quasiparticle spectrum. The excitation spectrum of real helium, as derived from neutron scattering measurements, displays, according to Fig. 3.5, the following characteristics. For small p, the excitation energy varies linearly with the momentum

$$\epsilon_{\mathbf{p}} = cp .$$ (3.2.36a)

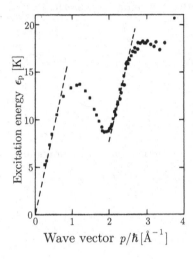

Fig. 3.5. The quasiparticle excitations in superfluid He^4. Phonons and rotons according to Henshaw and Woods[14]

The excitations in this region are called phonons; their velocity – the sound velocity – is $c = 238 \, \text{m/s}$. The second characteristic feature of the excitation spectrum is the minimum at $p_0 = 1.91 \, \text{Å}^{-1} \hbar$. Here, the excitations are referred to as rotons and can be described by the dispersion relation

$$\epsilon_{\mathbf{p}} = \Delta + \frac{(|\mathbf{p}| - p_0)^2}{2\mu}$$ (3.2.36b)

with the effective mass $\mu = 0.16 \, m_{\text{He}}$ and the energy gap $\Delta/k = 8.6 \, \text{K}$. The condensation of helium and the resulting quasiparticle dispersion relation ((3.2.36a,b), Fig. 3.5) has essential consequences for the dynamical behavior of He^4 in the He-II phase. It leads to superfluidity and to the two-fluid model. To see this, we consider the flow of helium through a tube in two different inertial frames. In frame K, the tube is at rest and the fluid moves with a velocity $-\mathbf{v}$. In frame K$_0$, the helium is at rest and the tube moves with a velocity $\mathbf{v}$ (see Fig. 3.6).

---
[14] D.G. Henshaw and A.D. Woods, Phys. Rev. **121**, 1266 (1961)

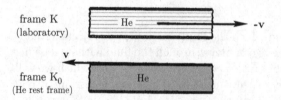

Fig. 3.6. Superfluid helium in the rest frame of the tube (laboratory frame, K) and in the rest frame of the fluid, K_0

The total energies (E, E_0) and the total momenta $(\mathbf{P}, \mathbf{P}_0)$ in the two frames (K, K_0) are related to one another through a Galilei transformation:

$$\mathbf{P} = \mathbf{P}_0 - M\mathbf{v} \qquad (3.2.37\text{a})$$

$$E = E_0 - \mathbf{P}_0 \cdot \mathbf{v} + \frac{M\mathbf{v}^2}{2} , \qquad (3.2.37\text{b})$$

where we have introduced

$$\sum_i \mathbf{p}_i = \mathbf{P} , \qquad \sum_i \mathbf{p}_{i0} = \mathbf{P}_0 , \qquad \sum_i m = M .$$

One can derive (3.2.37a,b) by using the Galilei transformation of the individual particles

$$\mathbf{x}_i = \mathbf{x}_{i0} - \mathbf{v}t$$
$$\mathbf{p}_i = \mathbf{p}_{i0} - m\mathbf{v} .$$

Thus,

$$\mathbf{P} = \sum \mathbf{p}_i = \sum (\mathbf{p}_{i0} - m\mathbf{v}) = \mathbf{P}_0 - M\mathbf{v} .$$

The energy transforms as follows:

$$E = \sum_i \frac{1}{2m}\mathbf{p}_i^2 + \sum_{\langle i,j \rangle} V(\mathbf{x}_i - \mathbf{x}_j)$$

$$= \sum_i \frac{m}{2} \left(\frac{\mathbf{p}_{i0}}{m} - \mathbf{v} \right)^2 + \sum_{\langle i,j \rangle} V(\mathbf{x}_{i0} - \mathbf{x}_{j0})$$

$$= \sum_i \frac{\mathbf{p}_{i0}^2}{2m} - \mathbf{P}_0 \cdot \mathbf{v} + \frac{M}{2}\mathbf{v}^2 + \sum_{\langle i,j \rangle} V(\mathbf{x}_{i0} - \mathbf{x}_{j0})$$

$$= E_0 - \mathbf{P}_0 \cdot \mathbf{v} + \frac{M}{2}\mathbf{v}^2 .$$

In a normal fluid, any flow that might initially be present will be degraded by frictional losses. When viewed in the frame K_0, this means that, in the fluid, excitations are created which move with the wall of the tube, such that more and more fluid is pulled along with the moving tube. Seen from the tube frame K, the same process can be interpreted as a deceleration of the fluid flow. In order that such excitations actually occur, the energy of the fluid must simultaneously decrease. We now have to examine whether, for the particular excitation spectrum of He-II, Fig. 3.5, the moving fluid can reduce its energy through the creation of excitations.

Is it energetically favorable for quasiparticles to be excited? We first consider helium at the temperature $T = 0$, i.e., in the ground state. In the ground state the energy and momentum in the frame K_0 are given by

$$E_0^g \quad \text{and} \quad \mathbf{P}_0 = 0 .$$
(3.2.38a)

Thus, in K, these quantities are

$$E^g = E_0^g + \frac{M\mathbf{v}^2}{2} \quad \text{and} \quad \mathbf{P} = -M\mathbf{v} .$$
(3.2.38b)

If a quasiparticle with momentum $\mathbf{p} = \hbar\mathbf{k}$ and energy $\epsilon(\mathbf{p}) = \hbar\omega_{\mathbf{k}}$ is created, the energy and momentum in the frame K_0 have the values

$$E_0 = E_0^g + \epsilon(\mathbf{p}) \quad \text{and} \quad \mathbf{P}_0 = \mathbf{p} ,$$
(3.2.38c)

whence, from (3.2.37a,b), the energy in K follows as

$$E = E_0^g + \epsilon(\mathbf{p}) - \mathbf{p} \cdot \mathbf{v} + \frac{M\mathbf{v}^2}{2} \quad \text{and} \quad \mathbf{P} = \mathbf{p} - M\mathbf{v} .$$
(3.2.38d)

The excitation energy in K (the tube frame) is thus

$$\Delta E = \epsilon(\mathbf{p}) - \mathbf{p} \cdot \mathbf{v} .$$
(3.2.39)

Here, ΔE is the energy change in the fluid due to the creation of an excitation in the tube frame K. Only when $\Delta E < 0$ does the flowing fluid lose energy. Since $\epsilon - \mathbf{pv}$ has its smallest value when $\mathbf{p}$ is parallel to $\mathbf{v}$, the inequality

$$v > \frac{\epsilon}{p}$$
(3.2.40a)

must be satisfied in order for an excitation to occur. From (3.2.40a) and the experimental excitation spectrum, one obtains the critical velocity (Fig. 3.7)

$$v_c = \left(\frac{\epsilon}{p}\right)_{\min} \approx 60\,\text{m/s} .$$
(3.2.40b)

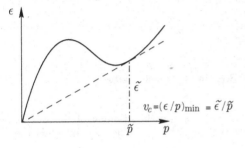

Fig. 3.7. Quasiparticles and critical velocity

If the flow velocity is smaller than v_c, no quasiparticles are excited and the fluid flows unimpeded and loss-free through the tube. This is the phenomenon of superfluidity. The existence of a finite critical velocity is closely related to the form of the excitation spectrum, which has a finite group velocity at $\mathbf{p} = 0$ and is everywhere greater than zero (Fig. 3.7).

The value (3.2.40b) of the critical velocity is observed experimentally for the motion of ions in He-II. The critical velocity for flows in capillaries is much smaller than v_c, since vortices already occur at lower velocities. Such excitations have not been considered here.

Problems

3.1 Consider the following two-particle boson state

$$|2\rangle = \int d^3x_1 \int d^3x_2 \varphi(\mathbf{x}_1, \mathbf{x}_2) \psi^\dagger(\mathbf{x}_1)\psi^\dagger(\mathbf{x}_2)|0\rangle .$$

a) Confirm the normalization condition (3.1.10).
b) Verify the result (3.1.14) for the expectation value $\langle 2| n(\mathbf{x})|2\rangle$ on the assumption that $\varphi(\mathbf{x}_1, \mathbf{x}_2) \propto \varphi_1(\mathbf{x}_1)\varphi_2(\mathbf{x}_2)$.

3.2 The Heisenberg model of a ferromagnet is defined by the Hamiltonian

$$H = -\frac{1}{2}\sum_{\mathbf{l},\mathbf{l}'} J(|\mathbf{l} - \mathbf{l}'|)\mathbf{S}_\mathbf{l} \cdot \mathbf{S}_{\mathbf{l}'} ,$$

where $\mathbf{l}$ and $\mathbf{l}'$ are nearest neighbor sites on a cubic lattice. By means of the *Holstein–Primakoff transformation*

$$S_i^+ = \sqrt{2S}\varphi(\hat{n}_i)a_i ,$$
$$S_i^- = \sqrt{2S}a_i^\dagger\varphi(\hat{n}_i) ,$$
$$S_i^z = S - \hat{n}_i ,$$

with $S_i^\pm = S_i^x \pm S_i^y$, $\varphi(\hat{n}_i) = \sqrt{1 - \hat{n}_i/2S}$, $\hat{n}_i = a_i^\dagger a_i$ and $[a_i, a_j^\dagger] = \delta_{ij}$, $[a_i, a_j] = 0$ one can express the Hamiltonian in terms of Bose operators a_i.
a) Show that the commutation relations for the spin operators are satisfied.
b) Write down the Hamiltonian to second order (harmonic approximation) in terms of the Bose operators a_i by regarding the square-roots in the above transformation as a shorthand for the series expansion.
c) Diagonalize H (by means of a Fourier transformation) and determine the dispersion relation of the spin waves (magnons).

3.3 Confirm the inverse (3.2.15) of the Bogoliubov transformation.

3.4 By means of the Bogoliubov transformation, the Hamiltonian of the weakly interacting Bose gas can be brought into diagonal form. One thereby finds the condition (3.2.17):

$$\left(\frac{k^2}{2m} + nV_k\right) u_k v_k + \frac{n}{2} V_k \left(u_k^2 + v_k^2\right) = 0 .$$

Confirm the results (3.2.18) and (3.2.19).

3.5 Determine the temperature dependence of the number of particles in the condensate, $N_0(T)$, for a contact potential $V_k = \lambda$.
a) Proceed by first calculating the thermodynamic expectation value of the particle number operator $\hat{N} = \sum_k a_k^\dagger a_k$ by rewriting it in terms of the quasiparticle operators α_k. One finds (in the continuum limit: $\frac{1}{V}\sum_k \to \int \frac{d^3 k}{(2\pi)^3}$)

$$N = N_0(T) + 2\frac{(mn\lambda)^{3/2}}{\pi^2}\left(\frac{1}{6} + U_1(\gamma)\right) ,$$

where $\gamma = \beta \frac{k_0^2}{2m}$, $k_0^2 = 4mn\lambda$, $\beta = 1/kT$, k the Boltzmann constant, and

$$U_n(\gamma) = \int_0^\infty dy \frac{x^n}{e^{\gamma y} - 1} , \quad \text{with } y = x\sqrt{x^2 + 1} .$$

b) Show that, for low temperatures, the depletion of the condensate increases quadratically with temperature

$$\frac{N_0(T)}{V} = \frac{N_0(0)}{V} - \frac{m}{12c}(kT)^2 ,$$

where $c = \sqrt{\frac{n\lambda}{m}}$. Also, discuss the limiting case of high temperatures and compare the result obtained with the results from the theory of the Bose–Einstein condensation of noninteracting bosons below the transition temperature.
    Lit.: R.A. Ferrell, N. Menyhárd, H. Schmidt, F. Schwabl and P. Szépfalusy, Ann. Phys. (N.Y.) **47**, 565 (1968); Phys. Rev. Lett. **18**, 891 (1967); Phys. Letters **24A**, 493 (1967); K. Kehr, Z. Phys. **221**, 291 (1969).

3.6 a) Determine the excitation spectrum ω_k of the weakly interacting Bose gas for the spherical well potential $V(\mathbf{x}) = \lambda'\Theta(R - |\mathbf{x}|)$. Analyze the limiting case $R \to 0$ and compare the result with the excitation spectrum for the contact potential $V_k = \lambda$. The comparison yields $\lambda = \frac{4\pi}{3}\lambda' R^3$.
b) Determine the range of the parameter $k_0 R$, where $k_0^2 = 4mn\lambda$, in which the excitation spectrum displays a "roton minimum". Discuss the extent to which this parameter range lies within or outside the range of validity of the Bogoliubov theory of weakly interacting bosons.
    Hints: Rewrite the spectrum in terms of the dimensionless quantities $x = k/k_0$ and $y = k_0 R$ and consider the derivative of the spectrum with respect to x. The condition for the derivative to vanish should be investigated graphically.

3.7 Show that (3.2.11) yields the Hamiltonian (3.2.16), which in turn leads to (3.2.20).

3.8 Ground-state energy for bosons with contact interaction

Consider a system of N identical bosons of mass m interacting with one another via a two-particle contact potential,

$$H = \sum_{\alpha=1}^{N} \frac{p_i^2}{2m} + \lambda \sum_{i<j} \delta(\mathbf{x}_i - \mathbf{x}_j).$$

In the limit of weak interaction, the Bogoliubov transformation can be used to express the Hamiltonian in the form

$$H = \frac{N^2}{2V}\lambda - \frac{1}{2}{\sum_k}' \left(\frac{k^2}{2m} + n\lambda - \omega_\mathbf{k} \right) + {\sum_k}' \omega_\mathbf{k} a_\mathbf{k}^\dagger a_\mathbf{k}.$$

The ground-state energy

$$E_0 = \frac{N^2\lambda}{2V} - {\sum_k}' \left(\frac{k^2}{2m} + n\lambda - \omega_\mathbf{k} \right)$$

diverges at the upper integration limit (ultraviolet divergence). The reason for this is the unphysical form of the contact potential. The divergence is removed by introducing the physical scattering length, which describes the s-wave scattering by a short-range potential (L.D. Landau and E.M. Lifshitz, *Course of Theoretical Physics*, Vol. 9, E.M. Lifshitz and E.P. Pitaevskii, *Statistical Physics 2*, Pergamon Press, New York, 1981, §6). Show that the scattering amplitude f of particles in the condensate is given, in first-order perturbation theory, by

$$a := -f(\mathbf{k}_1 = \mathbf{k}_2 = \mathbf{k}_3 = \mathbf{k}_4 = 0) = \frac{m}{4\pi\hbar^2}\lambda \left\{ 1 - \frac{\lambda}{V}{\sum_k}' \frac{m}{k^2} + \mathcal{O}(\lambda^2) \right\}.$$

Eliminate λ from the expression for the ground-state energy by introducing a. For small values of a/r_0, where $r_0 = (N/V)^{-1/3}$ is the mean separation of particles, show that the ground-state energy is given by

$$E_0 = N\frac{2\pi\hbar^2 an}{m} \left\{ 1 + \frac{128}{15\sqrt{\pi}} \left(\frac{a}{r_0} \right)^{3/2} \right\}.$$

Calculate from this the chemical potential $\mu = \frac{\partial E_0}{\partial N}$ and the sound velocity c

$$c = \sqrt{\frac{\partial P}{\partial \rho}}, \quad \rho = mn, \quad P = -\frac{\partial E_0}{\partial V}.$$

4. Correlation Functions, Scattering, and Response

In the following, we investigate the dynamical properties of many-particle systems on a microscopic, quantum-mechanical basis. We begin by expressing experimentally relevant quantities such as the inelastic scattering cross-section and the dynamical susceptibility (which describes the response of the system to time-dependent fields) in terms of microscopic entities such as the dynamical correlation functions. To this end the time-evolution operator will be determined perturbatively using the interaction representation. General properties of these correlation functions and their interrelations are then derived using the symmetry properties of the system, causality, and the specific definitions in terms of equilibrium expectation values. Finally, we calculate correlation functions for a few physically relevant models.

4.1 Scattering and Response

Before entering into the details, let us make some remarks about the physical motivation behind the subject of this chapter. If a time-dependent field $E e^{i(\mathbf{k}\cdot\mathbf{x}-\omega t)}$ is applied to a many-particle system (solid, liquid, or gas), this induces a "polarization" (Fig. 4.1)

$$P(\mathbf{k}, \omega) e^{i(\mathbf{k}\cdot\mathbf{x}-\omega t)} + P(2\mathbf{k}, 2\omega) e^{i2(\mathbf{k}\cdot\mathbf{x}-\omega t)} + \dots . \tag{4.1.1}$$

The first term has the same periodicity as the applied field; in addition, nonlinear effects give rise to higher harmonics. The linear susceptibility is

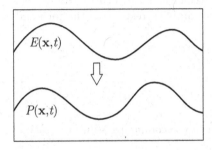

$E(\mathbf{x}, t)$

$P(\mathbf{x}, t)$

Fig. 4.1. An external field $E(\mathbf{x}, t)$ induces a polarization $P(\mathbf{x}, t)$

defined by

$$\chi(\mathbf{k}, \omega) = \lim_{E \to 0} \frac{P(\mathbf{k}, \omega)}{E} \, , \tag{4.1.2}$$

which is a property of the unperturbed sample and must be expressible solely in terms of quantities that characterize the many-particle system. In this chapter we will derive general microscopic expressions for this type of susceptibility.

Another possibility for obtaining information about a many-particle system is to carry out *scattering experiments* with particles, e.g., neutrons, electrons, or photons. The wavelength of these particles must be comparable with the scale of the structures that one wants to resolve, and their energy must be of the same order of magnitude as the excitation energies of the quasiparticles that are to be measured. An important tool is *neutron scattering*, since thermal neutrons, as available from nuclear reactors, ideally satisfy these conditions for experiments on solids.[1] Since neutrons are neutral, their interaction with the nuclei is short-ranged; in contrast to electrons, they penetrate deep into the solid. Furthermore, due to their magnetic moment and the associated dipole interaction with magnetic moments of the solid, neutrons can also be used to investigate magnetic properties.

We begin by considering a completely general scattering process and will specialize later to the case of neutron scattering. The calculation of the *inelastic scattering cross-section* proceeds as follows. We consider a many-particle system, such as a solid or a liquid, that is described by the Hamiltonian H_0. The constituents (atoms, ions) of this substance are described by coordinates $\mathbf{x}_\alpha$ which, in addition to the spatial coordinates, may also represent other degrees of freedom. The incident particles, e.g., neutrons or electrons, which are scattered by this sample, have mass m, spatial coordinate $\mathbf{x}$, and spin m_s.

The total Hamiltonian then reads:

$$H = H_0 + \frac{\mathbf{p}^2}{2m} + W(\{\mathbf{x}_\alpha\}, \mathbf{x}) \, . \tag{4.1.3}$$

This comprises the Hamiltonian of the target, H_0, the kinetic energy of the incident particle, and the interaction between this projectile particle and the target, $W(\{\mathbf{x}_\alpha\}, \mathbf{x})$. In second quantization with respect to the incident particle, the Hamiltonian reads:

---

[1] The neutron wavelength depends on the energy according to $\lambda(\text{nm}) = \frac{0.0286}{\sqrt{E(\text{eV})}}$ and thus $\lambda = 0.18\text{nm}$ for $E = 25\text{meV} \cong 290\text{K}$.

$$H = H_0 + \frac{\mathbf{p}^2}{2m} + \sum_{\mathbf{k}'\mathbf{k}''\sigma'\sigma''} a_{\mathbf{k}'\sigma'}^\dagger a_{\mathbf{k}''\sigma''}$$

$$\times \frac{1}{V} \int d^3x \, e^{-i(\mathbf{k}'-\mathbf{k}'')\cdot\mathbf{x}} W^{\sigma'\sigma''}(\{\mathbf{x}_\alpha\}, \mathbf{x})$$

$$= H_0 + \frac{\mathbf{p}^2}{2m} + \sum_{\mathbf{k}'\mathbf{k}''\sigma'\sigma''} a_{\mathbf{k}'\sigma'}^\dagger a_{\mathbf{k}''\sigma''} W_{\mathbf{k}'-\mathbf{k}''}^{\sigma'\sigma''}(\{\mathbf{x}_\alpha\}), \qquad (4.1.4)$$

where $a_{\mathbf{k}'\sigma'}^\dagger$ ($a_{\mathbf{k}''\sigma''}$) creates (annihilates) a projectile particle. We write the eigenstates of H_0 as $|n\rangle$, i.e.,

$$H_0 |n\rangle = E_n |n\rangle . \qquad (4.1.5)$$

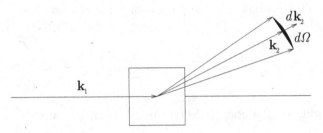

Fig. 4.2. Inelastic scattering with momentum transfer $\mathbf{k} = \mathbf{k}_1 - \mathbf{k}_2$ and energy transfer $\hbar\omega = \frac{\hbar^2}{2m}(k_1^2 - k_2^2)$

In the scattering setup sketched in Fig. 4.2, a particle with wave vector $\mathbf{k}_1$ and spin m_{s1} is incident on a substance initially in the state $|n_1\rangle$. Thus, the initial state of the system as a whole is $|\mathbf{k}_1, m_{s1}, n_1\rangle$. The corresponding final state is $|\mathbf{k}_2, m_{s2}, n_2\rangle$. Due to its interaction with the target, the incident particle is deflected, i.e., the direction of its momentum is changed and, for inelastic scattering, so is the length of its wave vector (momentum). If the interaction is spin dependent, the spin quantum number may also be changed.

The transition probability per unit time can be obtained from Fermi's golden rule[2]

$$\Gamma(\mathbf{k}_1, m_{s1}, n_1 \to \mathbf{k}_2, m_{s2}, n_2)$$
$$= \frac{2\pi}{\hbar} |\langle \mathbf{k}_2, m_{s2}, n_2| W |\mathbf{k}_1, m_{s1}, n_1\rangle|^2 \, \delta(E_{n_1} - E_{n_2} + \hbar\omega) . \qquad (4.1.6)$$

Here,

$$\hbar\omega = \frac{\hbar^2}{2m}(k_1^2 - k_2^2) \qquad (4.1.7a)$$

---

[2] See, e.g., QM I, Eq. (16.40)

and

$$\mathbf{k} = \mathbf{k}_1 - \mathbf{k}_2 \tag{4.1.7b}$$

are the energy and momentum transfer of the projectile to the target, and

$$\epsilon = \frac{\hbar^2 \mathbf{k}_2^2}{2m} \tag{4.1.7c}$$

the final energy of the particle. The matrix element in the golden rule becomes

$$\langle \mathbf{k}_2, m_{s2}, n_2 | \, W \, | \mathbf{k}_1, m_{s1}, n_1 \rangle = W^{\sigma'\sigma''}_{\mathbf{k}_2 - \mathbf{k}_1}(\{\mathbf{x}_\alpha\}) \,. \tag{4.1.8}$$

We take the distribution of initial states to be $p(n_1)$ with $\sum_{n_1} p(n_1) = 1$ and the distribution of the spin states of the particle to be $p_s(m_{s1})$ with $\sum_{m_{s1}} p_s(m_{s1}) = 1$. If only $\mathbf{k}_2$ is measured, and the spin is not analyzed, the transition probability of interest is

$$\Gamma(\mathbf{k}_1 \to \mathbf{k}_2) = \sum_{n_2 n_1} \sum_{m_{s1} m_{s2}} p(n_1) p_s(m_{s1}) \Gamma(\mathbf{k}_1, m_{s1}, n_1 \to \mathbf{k}_2, m_{s2}, n_2) \,. \tag{4.1.9}$$

The differential scattering cross-section (effective target area) per atom is defined by

$$\frac{d^2\sigma}{d\Omega d\epsilon} d\Omega d\epsilon = \frac{\text{probablility of transition into } d\Omega d\epsilon/s}{\text{number of scatterers} \times \text{flux of incident particles}} \,. \tag{4.1.10}$$

Here, $d\Omega$ is an element of solid angle and the flux of incident particles is equal to the magnitude of their current density. The number of scatterers is N and the normalization volume is L^3. The states of the incident particles are

$$\psi_{\mathbf{k}_1}(\mathbf{x}) = \frac{1}{L^{3/2}} e^{i\mathbf{k}_1 \cdot \mathbf{x}} \,. \tag{4.1.11}$$

The current density follows as:

$$\mathbf{j}(\mathbf{x}) = \frac{-i\hbar}{2m}(\psi^* \nabla \psi - (\nabla \psi^*)\psi) = \frac{\hbar \mathbf{k}_1}{mL^3} \,, \tag{4.1.12}$$

and for the differential scattering cross-section one obtains

$$\frac{d^2\sigma}{d\Omega d\epsilon} d\Omega d\epsilon = \frac{1}{N} \frac{mL^3}{\hbar k_1} \Gamma(\mathbf{k}_1 \to \mathbf{k}_2) \left(\frac{L}{2\pi}\right)^3 d^3 k_2 \,, \tag{4.1.13}$$

since the number of final states, i.e., the number of $\mathbf{k}_2$ values in the interval $d^3 k_2$ is $\left(\frac{L}{2\pi}\right)^3 d^3 k_2$. With $\epsilon = \frac{\hbar^2 k_2^2}{2m}$, it follows that $d\epsilon = \hbar^2 k_2 \, dk_2/m$ and $d^3 k_2 = \frac{m}{\hbar^2} k_2 \, d\epsilon \, d\Omega$.

We thus find

$$
\frac{d^2\sigma}{d\Omega d\epsilon} = \left(\frac{m}{2\pi\hbar^2}\right)^2 \frac{k_2}{k_1}\frac{L^6}{N}
\tag{4.1.14}
$$
$$
\times \sum_{\substack{n_1,n_2\\ m_{s1},m_{s2}}} p(n_1)p_s(m_{s1})\,|\langle \mathbf{k}_1, m_{s1}, n_1|\,W\,|\mathbf{k}_2, m_{s2}, n_2\rangle|^2\,\delta(E_{n_1} - E_{n_2} + \hbar\omega)\,.
$$

We now consider the particular case of neutron scattering and investigate the scattering of neutrons by nuclei. The range of the nuclear force is $R \approx 10^{-12}$cm and thus $k_1 R \approx 10^{-4} \ll 1$ and, therefore, for thermal neutrons one has only s-wave scattering. In this case, the interaction can be represented by an effective pseudopotential

$$
W(\{\mathbf{x}_\alpha\},\mathbf{x}) = \frac{2\pi\hbar^2}{m}\sum_{\alpha=1}^{N} a_\alpha \delta(\mathbf{x}_\alpha - \mathbf{x})\,,
\tag{4.1.15}
$$

to be used within the Born approximation, where a_α are the scattering lengths of the nuclei. This yields:

$$
\frac{d^2\sigma}{d\Omega d\epsilon} = \frac{k_2}{k_1}\frac{1}{N}\sum_{n_1 n_2} p(n_1)\left|\sum_{\alpha=1}^{N} a_\alpha \langle n_1|\,e^{-i\mathbf{k}\cdot\mathbf{x}_\alpha}\,|n_2\rangle\right|^2 \delta(E_{n_1} - E_{n_2} + \hbar\omega)\,.
\tag{4.1.16}
$$

Here, we have used

$$
\langle \mathbf{k}_1|\,W\,|\mathbf{k}_2\rangle = \frac{2\pi\hbar^2}{mL^3}\int d^3x\, e^{-i\mathbf{k}_1\cdot\mathbf{x}}\sum_\alpha a_\alpha \delta(\mathbf{x} - \mathbf{x}_\alpha)e^{i\mathbf{k}_2\cdot\mathbf{x}}
$$
$$
= \frac{2\pi\hbar^2}{mL^3}\sum_\alpha a_\alpha e^{-i(\mathbf{k}_1-\mathbf{k}_2)\cdot\mathbf{x}_\alpha}
\tag{4.1.17}
$$

and the fact that the interaction is independent of spin. Written out explicitly, the expression (4.1.16) assumes the form

$$
\sum_{\alpha\beta} a_\alpha a_\beta \langle \ldots e^{-i\mathbf{k}\cdot\mathbf{x}_\alpha}\ldots\rangle\langle\ldots e^{i\mathbf{k}\cdot\mathbf{x}_\beta}\ldots\rangle\delta(E_{n_1} - E_{n_2} + \hbar\omega)
\tag{4.1.16'}
$$

and still has to be averaged over the various isotopes with different scattering lengths. One assumes that their distribution is random, i.e., spatially uncorrelated:

$$
\bar{a} = \frac{1}{N}\sum_{\alpha=1}^{N} a_\alpha\,,\quad \overline{a^2} = \frac{1}{N}\sum_{\alpha=1}^{N} a_\alpha^2
$$
$$
\overline{a_\alpha a_\beta} = \begin{cases} \bar{a}^2 & \text{for } \alpha \neq \beta \\ \overline{a^2} & \text{for } \alpha = \beta \end{cases} = \bar{a}^2 + \delta_{\alpha\beta}(\overline{a^2} - \bar{a}^2)\,.
\tag{4.1.18}
$$

This gives rise to a decomposition of the scattering cross-section into a coherent and an incoherent part[3]:

$$\frac{d^2\sigma}{d\Omega d\epsilon} = A_{\text{coh}} S_{\text{coh}}(\mathbf{k}, \omega) + A_{\text{inc}} S_{\text{inc}}(\mathbf{k}, \omega) . \tag{4.1.19a}$$

Here, the various terms signify

$$A_{\text{coh}} = \overline{a}^2 \frac{k_2}{k_1} \quad , \quad A_{\text{inc}} = (\overline{a^2} - \overline{a}^2)\frac{k_2}{k_1} \tag{4.1.19b}$$

and

$$
\begin{aligned}
S_{\text{coh}}(\mathbf{k}, \omega) &= \frac{1}{N} \sum_{\alpha\beta} \sum_{n_1 n_2} p(n_1) \langle n_1 | e^{-i\mathbf{k}\cdot\mathbf{x}_\alpha} | n_2 \rangle \langle n_2 | e^{i\mathbf{k}\cdot\mathbf{x}_\beta} | n_1 \rangle \\
&\quad \times \delta(E_{n_1} - E_{n_2} + \hbar\omega) , \\
S_{\text{inc}}(\mathbf{k}, \omega) &= \frac{1}{N} \sum_{\alpha} \sum_{n_1 n_2} p(n_1) \left| \langle n_1 | e^{-i\mathbf{k}\cdot\mathbf{x}_\alpha} | n_2 \rangle \right|^2 \\
&\quad \times \delta(E_{n_1} - E_{n_2} + \hbar\omega) ,
\end{aligned}
\tag{4.1.20}
$$

the suffices standing for coherent and incoherent.

In the coherent part, the amplitudes stemming from the different atoms are superposed. This gives rise to interference terms which contain information about the correlation between different atoms. In the incoherent scattering cross-section, it is the intensities rather than the amplitudes of the scattering from different atoms that are added. It contains no interference terms and the information which it yields relates to the autocorrelation, i.e., the correlation of each atom with itself. For later use we note here that, for systems in equilibrium,

$$p(n_1) = \frac{e^{-\beta E_{n_1}}}{Z} , \tag{4.1.21a}$$

which corresponds to the density matrix of the canonical ensemble[4]

$$\rho = e^{-\beta H_0}/Z , \quad Z = \text{Tr} \, e^{-\beta H_0} . \tag{4.1.21b}$$

We shall also make use of the following representation of the delta function

$$\delta(\omega) = \int \frac{dt}{2\pi} e^{i\omega t} . \tag{4.1.22}$$

The coherent scattering cross-section contains the factor

$$
\begin{aligned}
&\frac{1}{\hbar} \int \frac{dt}{2\pi} e^{i(E_{n_1} - E_{n_2} + \hbar\omega)t/\hbar} \langle n_1 | e^{-i\mathbf{k}\cdot\mathbf{x}_\alpha} | n_2 \rangle \\
&= \frac{1}{2\pi\hbar} \int dt \, e^{i\omega t} \langle n_1 | e^{iH_0 t/\hbar} e^{-i\mathbf{k}\cdot\mathbf{x}_\alpha} e^{-iH_0 t/\hbar} | n_2 \rangle \tag{4.1.23} \\
&= \frac{1}{2\pi\hbar} \int dt \, e^{i\omega t} \langle n_1 | e^{-i\mathbf{k}\cdot\mathbf{x}_\alpha(t)} | n_2 \rangle . \tag{4.1.24}
\end{aligned}
$$

---

[3] See, e.g., L. van Hove, Phys. Rev. **95**, 249 (1954)

[4] See, e.g., SM, sect. 2.6. $\beta = 1/kT$ in terms of Boltzmann's constant k and the temperature T.

Hence, by making use of the completeness relation

$$\sum_{n_2} |n_2\rangle \langle n_2| = 1 \,,$$

one finds

$$S_{\substack{\text{coh} \\ \text{inc}}}(\mathbf{k}, \omega) = \int \frac{dt}{2\pi\hbar} e^{i\omega t} \frac{1}{N} \sum_{\alpha\beta} \left\langle e^{-i\mathbf{k}\cdot\mathbf{x}_\alpha(t)} e^{i\mathbf{k}\cdot\mathbf{x}_\beta(0)} \right\rangle \left(\begin{matrix} 1 \\ \delta_{\alpha\beta} \end{matrix} \right) . \qquad (4.1.25)$$

The correlation functions in (4.1.25) are evaluated using the density matrix of the many-particle system (4.1.21), the thermal average of an operator O being defined by

$$\langle O \rangle = \sum_n \frac{e^{-\beta H_0}}{Z} \langle n| O |n\rangle = \text{Tr}\,(\rho O) \,. \qquad (4.1.26)$$

One refers to $S_{\text{coh(inc)}}(\mathbf{k}, \omega)$ as the coherent (incoherent) dynamical structure function. Both contain an elastic ($\omega = 0$) and an inelastic ($\omega \neq 0$) component. Using the density operator

$$\rho(\mathbf{x}, t) = \sum_{\alpha=1}^N \delta(\mathbf{x} - \mathbf{x}_\alpha(t)) \qquad (4.1.27)$$

and its Fourier transform

$$\rho_{\mathbf{k}}(t) = \frac{1}{\sqrt{V}} \int d^3 x e^{-i\mathbf{k}\cdot\mathbf{x}} \rho(\mathbf{x}, t) = \frac{1}{\sqrt{V}} \sum_{\alpha=1}^N e^{-i\mathbf{k}\cdot\mathbf{x}_\alpha(t)} \,, \qquad (4.1.28)$$

it follows from (4.1.25) that

$$S_{\text{coh}}(\mathbf{k}, \omega) = \int \frac{dt}{2\pi\hbar} e^{i\omega t} \frac{V}{N} \langle \rho_{\mathbf{k}}(t) \rho_{-\mathbf{k}}(0) \rangle \,. \qquad (4.1.29)$$

Thus, the coherent scattering cross-section can be represented by the Fourier transform of the density–density correlation function, where $\hbar\mathbf{k}$ is the momentum transfer and $\hbar\omega$ the energy transfer from the neutron to the target system. An important application is the scattering from solids to determine the lattice dynamics. The one-phonon scattering yields, as a function of frequency ω, resonances at the values $\pm\omega_{t_1}(\mathbf{k})$, $\pm\omega_{t_2}(\mathbf{k})$, and $\pm\omega_l(\mathbf{k})$, the frequencies of the two transverse, and the longitudinal phonons. The width of the resonances is determined by the lifetime of the phonons. The background intensity is due to multiphonon scattering (see Sect. 4.7(i) and Problem 4.5). The intensity of the single-phonon lines also depends on the scattering geometry via the scalar product of $\mathbf{k}$ with the polarization vector of the phonons and via the Debye–Waller factor. As a schematic example of the shape of

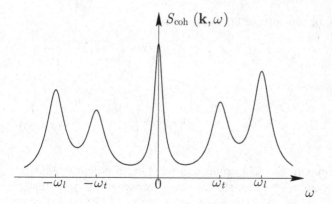

Fig. 4.3. Coherent scattering cross-section as a function of ω for fixed momentum transfer $\mathbf{k}$. Resonances (peaks) are seen at the transverse ($\pm\omega_t(\mathbf{k})$) and longitudinal ($\pm\omega_l(\mathbf{k})$) phonon frequencies, as well as at $\omega = 0$

the scattering cross section, we show in Fig. 4.3 S_{coh} for fixed $\mathbf{k}$ as a function of the frequency ω. The resonances at finite frequencies are due to a transverse and a longitudinal acoustic phonon, and, furthermore, one sees a quasi-elastic peak at $\omega = 0$. Quasi-elastic peaks may result from disorder and from relaxation and diffusion processes (Sect. 4.7(ii)).

The coherent scattering cross-section is a source of direct information about density excitations, such as phonons in solids and fluids. The incoherent component is a sum of the scattering intensities of the individual scatterers. It contains information about the autocorrelations.

For other scattering experiments (e.g., with photons, electrons, or atoms) one can likewise represent the scattering cross-section in terms of correlation functions of the many-particle system. We shall pursue the detailed properties of the differential scattering cross-section here no further. These preliminary remarks are intended mainly as additional motivation for the sections that are to follow, where we will see that the correlation functions and the susceptibility are related to one another. Causality will allow us to derive dispersion relations. Time-reversal invariance and translational invariance will yield symmetry relations, and from the static limit and the commutation relations we will derive sum rules.

4.2 Density Matrix, Correlation Functions

The Hamiltonian of the many-particle system will be denoted by H_0 and is assumed to be time independent. The formal solution of the Schrödinger equation

$$i\hbar\frac{\partial}{\partial t}\left|\psi,t\right\rangle = H_0\left|\psi,t\right\rangle \tag{4.2.1}$$

is then

$$|\psi, t\rangle = U_0(t, t_0) |\psi, t_0\rangle \quad . \tag{4.2.2}$$

Due to the time independence of H_0, the unitary operator $U_0(t, t_0)$ (with $U_0(t_0, t_0) = 1$) is given by

$$U_0(t, t_0) = e^{-iH_0(t-t_0)/\hbar} \quad . \tag{4.2.3}$$

The Heisenberg state

$$|\psi_H\rangle = U_0^\dagger(t, t_0) |\psi, t\rangle = |\psi, t_0\rangle \tag{4.2.4}$$

is time independent and the Heisenberg operators

$$A(t) = U_0^\dagger(t, t_0) A U_0(t, t_0) = e^{iH_0(t-t_0)/\hbar} A e^{-iH_0(t-t_0)/\hbar} \quad , \tag{4.2.5}$$

corresponding to the Schrödinger operators $A, B, ..$, satisfy the equation of motion (Heisenberg equation of motion)

$$\frac{d}{dt} A(t) = \frac{i}{\hbar} [H_0, A(t)] \quad . \tag{4.2.6}$$

The density matrix of the canonical ensemble is

$$\rho = \frac{e^{-\beta H_0}}{Z} \quad .$$

with the canonical partition function

$$Z = \mathrm{Tr}\, e^{-\beta H_0} \quad , \tag{4.2.7}$$

and for the grand canonical ensemble

$$\rho = \frac{e^{-\beta(H_0 - \mu N)}}{Z_G} \tag{4.2.8}$$

with the grand canonical partition function

$$Z_G = \mathrm{Tr}\, e^{-\beta(H_0 - \mu N)}$$

$$= \sum_N \sum_m e^{-\beta(E_m(N) - \mu N)} \left[\equiv \sum_n e^{-\beta(E_n - N_n \mu)} \right] \quad .$$

Since H_0 is a constant of motion, these density matrices are time independent, as indeed must be the case for equilibrium density matrices. The mean values in these ensembles are defined by

$$\langle O \rangle = \mathrm{Tr}(\rho O) \quad . \tag{4.2.9}$$

In particular, we now wish to investigate the correlation functions

$$C(t, t') = \langle A(t)B(t')\rangle$$
$$= \text{Tr}(\rho\, e^{iH_0 t/\hbar} A e^{-iH_0 t/\hbar} e^{iH_0 t'/\hbar} B e^{-iH_0 t'/\hbar})$$
$$= \text{Tr}(\rho\, e^{iH_0(t-t')/\hbar} A e^{-iH_0(t-t')/\hbar} B)$$
$$= C(t - t', 0) \quad . \tag{4.2.10}$$

Without loss of generality, we have set $t_0 = 0$ and used the cyclic invariance of the trace and also $[\rho, H_0] = 0$. The correlation functions depend only on the time difference; equation (4.2.10) expresses *temporal translational invariance*.

The following *definitions* will prove to be useful:

$$G^>_{AB}(t) = \langle A(t)B(0)\rangle , \tag{4.2.11a}$$

$$G^<_{AB}(t) = \langle B(0)A(t)\rangle . \tag{4.2.11b}$$

Their Fourier transforms are defined by

$$G^{\gtrless}_{AB}(\omega) = \int dt\, e^{i\omega t} G^{\gtrless}_{AB}(t) . \tag{4.2.12}$$

By inserting (4.2.5) into (4.2.12), taking energy eigenstates as a basis, and introducing intermediate states by means of the closure relation $\mathbb{1} = \sum_m |m\rangle\langle m|$, we obtain the following *spectral representation* for $G^>_{AB}(\omega)$:

$$G^>_{AB}(\omega) = \frac{2\pi}{Z} \sum_{n,m} e^{-\beta(E_n - \mu N_n)} \langle n| A |m\rangle \langle m| B |n\rangle$$
$$\times \delta\left(\frac{E_n - E_m}{\hbar} + \omega\right) \tag{4.2.13a}$$

$$G^<_{AB}(\omega) = \frac{2\pi}{Z} \sum_{n,m} e^{-\beta(E_n - \mu N_n)} \langle n| B |m\rangle \langle m| A |n\rangle$$
$$\times \delta\left(\frac{E_m - E_n}{\hbar} + \omega\right) . \tag{4.2.13b}$$

From this, it is immediately obvious that

$$G^>_{AB}(-\omega) = G^<_{BA}(\omega) \tag{4.2.14a}$$

$$G^<_{AB}(\omega) = G^>_{AB}(\omega)e^{-\beta\hbar\omega} . \tag{4.2.14b}$$

To derive the first relation, one compares $G^>_{AB}(-\omega)$ with (4.2.13a). The second follows if one exchanges n with m in (4.2.13b) and uses the δ-function. The latter relation is always applicable in the canonical ensemble and is valid in the grand canonical ensemble when the operators A and B do not change the number of particles. If, however, B increases the particle number by Δn_B, then $N_m - N_n = \Delta n_B$ and (4.2.14b) must be replaced by

$$G^<_{AB}(\omega) = G^>_{AB}(\omega)e^{-\beta(\hbar\omega - \mu \Delta n_B)} \ . \tag{4.2.14b'}$$

Inserting $A = \rho_{\mathbf{k}}$ and $B = \rho_{-\mathbf{k}}$ into (4.2.14a,b) yields the following relationship for the density–density correlation function:

$$S_{\text{coh}}(\mathbf{k}, -\omega) = e^{-\beta\hbar\omega}S_{\text{coh}}(-\mathbf{k}, \omega) \ . \tag{4.2.15}$$

For systems possessing inversion symmetry $S(\mathbf{k}, \omega) = S(-\mathbf{k}, \omega)$, and hence

$$S_{\text{coh}}(\mathbf{k}, -\omega) = e^{-\beta\hbar\omega}S_{\text{coh}}(\mathbf{k}, \omega) \ . \tag{4.2.16}$$

This relation implies that, apart from a factor $\frac{k_2}{k_1}$ in (4.1.19), the *anti-Stokes lines* (energy loss by the sample) are weaker by a factor $e^{-\beta\hbar\omega}$ than the *Stokes lines* (energy gain)[5]. For $T \to 0$ we have $S_{\text{coh}}(\mathbf{k}, \omega < 0) \to 0$, since the system is then in the ground state and cannot transfer any energy to the scattered particle. The above relationship expresses what is known as detailed balance (Fig. 4.4):

$$W_{n \to n'}P^e_n = W_{n' \to n}P^e_{n'} \quad \text{or}$$
$$W_{n \to n'} = W_{n' \to n}e^{-\beta(E_{n'} - E_n)} \ . \tag{4.2.17}$$

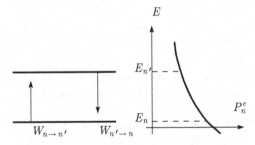

Fig. 4.4. Illustration concerning detailed balance

Here, $W_{n \to n'}$ and $W_{n' \to n}$ are the transition probabilities from the level n to the level n' and vice versa, and P^e_n and $P^e_{n'}$ are the equilibrium occupation probabilities. Detailed balance implies that these quantities are related to one another in such a way that the occupation probabilities do not change as a result of the transition processes.

4.3 Dynamical Susceptibility

We now wish to derive a microscopic expression for the dynamical susceptibility. To this end, we assume that the system is influenced by an external

---

[5] From the measurement of the ratio of the Stokes and anti-Stokes lines in Raman scattering the temperature of a system may be determined.

force $F(t)$ which couples to the operator B.[6] The Hamiltonian then has the form

$$H = H_0 + H'(t) \tag{4.3.1a}$$

$$H'(t) = -F(t)B . \tag{4.3.1b}$$

For $t \leq t_0$, we assume that $F(t) = 0$, and that the system is in equilibrium. We are interested in the response to the perturbation (4.3.1b). The mean value of A at time t is given by

$$
\begin{aligned}
\langle A(t) \rangle &= \text{Tr}\,(\rho_S(t)A) = \text{Tr}\,(U(t,t_0)\,\rho_S(t_0)\,U^\dagger(t,t_0)\,A) \tag{4.3.2} \\
&= \text{Tr}\,(\rho_S(t_0)\,U^\dagger(t,t_0)\,A\,U(t,t_0)) \\
&= \text{Tr}\,(\frac{e^{-\beta H_0}}{Z}\,U^\dagger(t,t_0)\,A\,U(t,t_0)) ,
\end{aligned}
$$

where the notation $\langle A(t) \rangle$ is to be understood as the mean value of the Heisenberg operator (4.2.5). Here we have introduced the time-evolution operator $U(t,t_0)$ for the entire Hamiltonian H and inserted the solution

$$\rho_S(t) = U(t,t_0)\rho_S(t_0)U^\dagger(t,t_0)$$

of the von Neumann equation

$$\dot{\rho}_S = -\frac{i}{\hbar}[H, \rho_S] .$$

Then, using the cyclic invariance of the trace and assuming a canonical equilibrium density matrix at time t_0, we end up with the mean value of the operator A in the Heisenberg representation.

The time-evolution operator $U(t,t_0)$ can be determined perturbation theoretically in the *interaction representation*. For this, we need the equation of motion for $U(t,t_0)$. From

$$i\hbar \frac{d}{dt} |\psi,t\rangle = H|\psi,t\rangle ,$$

it follows that

$$i\hbar \frac{d}{dt} U(t,t_0)|\psi_0\rangle = HU(t,t_0)|\psi_0\rangle$$

and, thus,

$$\left(i\hbar \frac{d}{dt} U(t,t_0) - HU(t,t_0)\right)|\psi_0\rangle = 0$$

---

[6] Physical forces are real and observables, e.g., the density $\rho(\mathbf{x})$, are represented by hermitian operators. Nonetheless, we shall also consider the correlation functions for nonhermitian operators such as $\rho_\mathbf{k}$ ($\rho_\mathbf{k}^\dagger = \rho_{-\mathbf{k}}$), since we may also be interested in the properties of individual Fourier components. $F(t)$ is a c-number.

for every $|\psi_0\rangle$, which yields the equation of motion

$$i\hbar \frac{d}{dt} U(t, t_0) = H U(t, t_0) \quad . \tag{4.3.3}$$

We now make the ansatz

$$U(t, t_0) = e^{-iH_0(t-t_0)/\hbar} U'(t, t_0) . \tag{4.3.4}$$

This gives

$$i\hbar \frac{d}{dt} U' = e^{iH_0(t-t_0)/\hbar}(-H_0 + H)U ,$$

and thus

$$i\hbar \frac{d}{dt} U'(t, t_0) = H'_I(t) U'(t, t_0) \quad , \tag{4.3.5}$$

where the interaction representation of H'

$$H'_I(t) = e^{iH_0(t-t_0)/\hbar} H'(t) e^{-iH_0(t-t_0)/\hbar} \tag{4.3.6}$$

has been introduced. The integration of (4.3.5) yields for the time evolution operator in the interaction representation $U'(t, t_0)$ the integral equation

$$U'(t, t_0) = 1 + \frac{1}{i\hbar} \int_{t_0}^{t} dt' H'_I(t') U'(t', t_0) \tag{4.3.7}$$

and its iteration

$$U'(t, t_0) = 1 + \frac{1}{i\hbar} \int_{t_0}^{t} dt' H'_I(t')$$

$$+ \frac{1}{(i\hbar)^2} \int_{t_0}^{t} dt' \int_{t_0}^{t'} dt'' H'_I(t') H'_I(t'') + \dots \tag{4.3.8}$$

$$= T \exp\left(-\frac{i}{\hbar} \int_{t_0}^{t} dt' H'_I(t')\right) .$$

Here, T is the time-ordering operator. The second representation of (4.3.8) is not required at present, but will be discussed in more detail in Part III.

For the linear response, we need only the first two terms in (4.3.8). Inserting these into (4.3.2), we obtain, to first order in $F(t)$,

$$\langle A(t) \rangle = \langle A \rangle_0 + \frac{1}{i\hbar} \int_{t_0}^{t} dt' \left\langle \left[e^{iH_0(t-t_0)/\hbar} A e^{-iH_0(t-t_0)/\hbar}, H'_I(t') \right] \right\rangle_0$$

$$= \langle A \rangle_0 - \frac{1}{i\hbar} \int_{t_0}^{t} dt' \langle [A(t), B(t')] \rangle_0 F(t') \quad . \tag{4.3.9}$$

The subscript 0 indicates that the expectation value is calculated with the density matrix $e^{-\beta H_0}/Z$ of the unperturbed system. In the first term we have exploited the cyclic invariance of the trace

$$\langle A(t)\rangle_0 = \mathrm{Tr}\left(\frac{e^{-\beta H_0}}{Z}e^{-iH_0(t-t_0)/\hbar}Ae^{iH_0(t-t_0)/\hbar}\right) = \langle A\rangle_0 \ .$$

We now assume that the initial time at which the system is in equilibrium, with density matrix $e^{-\beta H_0}/Z$, lies in the distant past. In other words, we take the limit $t_0 \to -\infty$, which, however, does not prevent us from switching on the force $F(t')$ at a later instant. For the change in the expectation value due to the perturbation, we obtain

$$\Delta\langle A(t)\rangle = \langle A(t)\rangle - \langle A\rangle_0 = \int_{-\infty}^{\infty} dt' \chi_{AB}(t-t')F(t') \quad . \tag{4.3.10}$$

Here we have introduced the *dynamical susceptibility*, or *linear response function*

$$\chi_{AB}(t-t') = \frac{i}{\hbar}\Theta(t-t')\langle[A(t),B(t')]\rangle_0 \ , \tag{4.3.11}$$

which is given by the expectation value of the commutator of the two Heisenberg operators $A(t)$ and $B(t')$ (with respect to the Hamiltonian H_0). The step function arises from the upper integration boundary in Eq. (4.3.9) and expresses causality. Within the equilibrium expectation value we can make the replacements

$$A(t) \to e^{iH_0t/\hbar}Ae^{-iH_0t/\hbar} \text{ and } B(t) \to e^{iH_0t/\hbar}Be^{-iH_0t/\hbar} \ .$$

Equation (4.3.10) determines, to first order, the effect on the observable A of a force that couples to B.

We also define the *Fourier transform* of the *dynamical susceptibility*

$$\chi_{AB}(z) = \int_{-\infty}^{\infty} dt \, e^{izt}\chi_{AB}(t) \quad , \tag{4.3.12}$$

where z may be complex (see Sect. 4.4). In order to find its physical significance, we consider a periodic perturbation which is switched on very slowly ($\epsilon \to 0, \epsilon > 0$):

$$H' = -\left(BF_\omega e^{-i\omega t'} + B^\dagger F_\omega^* e^{i\omega t'}\right)e^{\epsilon t'} \quad . \tag{4.3.13}$$

For this perturbation, it follows from (4.3.10) and (4.3.12) that

$$\Delta\langle A(t)\rangle = \int_{-\infty}^{\infty} dt' \left(\chi_{AB}(t-t')F_\omega e^{-i\omega t'} + \chi_{AB^\dagger}(t-t')F_\omega^* e^{i\omega t'}\right)e^{\epsilon t'}$$

$$= \chi_{AB}(\omega)F_\omega e^{-i\omega t} + \chi_{AB^\dagger}(-\omega)F_\omega^* e^{i\omega t} \quad . \tag{4.3.14}$$

The factor $e^{\epsilon t}$ that appears in the intermediate step can be put equal to 1 since $\epsilon \to 0$. The effect of the periodic perturbation (4.3.13) on $\Delta\langle A(t)\rangle$ is thus proportional to the force (including its periodicity) and to the Fourier transform of the susceptibility. Resonances in the susceptibility express themselves as a strong reaction to forces of the corresponding frequency.

4.4 Dispersion Relations

The *causality principle* demands that the response of a system can only be induced by a perturbation occurring at an earlier time. This is the source of the step function in (4.3.11), i.e.,

$$\chi_{AB}(t) = 0 \quad \text{for } t < 0 \ . \tag{4.4.1}$$

This leads to the *theorem*: $\chi_{AB}(z)$ is analytical in the upper half plane.
Proof. χ_{AB} is only nonzero for $t > 0$, where it is finite. Thus, the factor $e^{-\text{Im} \, zt}$ guarantees the convergence of the Fourier integral (4.3.12).

For z in the upper half plane, the analyticity of $\chi_{AB}(z)$ allows us to use Cauchy's integral theorem to write

$$\chi_{AB}(z) = \frac{1}{2\pi i} \int_C dz' \frac{\chi_{AB}(z')}{z' - z} \ . \tag{4.4.2}$$

Here, C is a closed loop in the analytic region. We choose the path shown in Fig. 4.5; along the real axis, and around a semicircle in the upper half plane, with both parts allowed to expand to infinity.

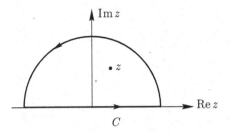

Fig. 4.5. Integration path C for deriving the dispersion relation

We now assume that $\chi_{AB}(z)$ becomes sufficiently small at infinity that the semicircular part of the integration path contributes nothing. We then have

$$\chi_{AB}(z) = \frac{1}{2\pi i} \int_{-\infty}^{\infty} dx' \frac{\chi_{AB}(x')}{x' - z} \ . \tag{4.4.3}$$

For real z it follows from (4.4.3) that

$$\chi_{AB}(x) = \lim_{\epsilon \to 0} \chi_{AB}(x + i\epsilon) = \lim_{\epsilon \to 0} \int \frac{dx'}{2\pi i} \frac{\chi_{AB}(x')}{x' - x - i\epsilon}$$

$$= \int \frac{dx'}{2\pi i} \left[P \frac{1}{x' - x} + i\pi \delta(x' - x) \right] \chi_{AB}(x') ,$$

i.e.,

$$\chi_{AB}(x) = \frac{1}{\pi i} P \int dx' \frac{\chi_{AB}(x')}{x' - x} . \tag{4.4.4}$$

We encounter here the Cauchy principal value

$$P \int dx' \frac{f(x')}{x' - x} \equiv \lim_{\epsilon \to 0} \left(\int_{-\infty}^{x - \epsilon} dx' + \int_{x + \epsilon}^{\infty} dx' \right) \frac{f(x')}{x' - x} .$$

We then arrive at the *dispersion relations* (also known as Kramers–Kronig relations)

$$\text{Re } \chi_{AB}(\omega) = \frac{1}{\pi} P \int d\omega' \frac{\text{Im } \chi_{AB}(\omega')}{\omega' - \omega} \tag{4.4.5a}$$

and

$$\text{Im } \chi_{AB}(\omega) = -\frac{1}{\pi} P \int d\omega' \frac{\text{Re } \chi_{AB}(\omega')}{\omega' - \omega} . \tag{4.4.5b}$$

These relationships between the real and imaginary parts of the susceptibility are a consequence of causality.

4.5 Spectral Representation

We define[7] the dissipative response

$$\chi''_{AB}(t) = \frac{1}{2\hbar} \langle [A(t), B(0)] \rangle \tag{4.5.1a}$$

and

$$\chi''_{AB}(\omega) = \int_{-\infty}^{\infty} dt \, e^{i\omega t} \chi''_{AB}(t) . \tag{4.5.1b}$$

Given the Fourier representation of the step function

---

[7] Here, and below we omit the index 0 from the expectation value. The notation $\langle \, \rangle$ represents the expectation value with respect to the Hamiltonian H_0 of the entire system without external perturbation.

$$\Theta(t) = \lim_{\epsilon \to 0^+} \int_{-\infty}^{\infty} \frac{d\omega}{2\pi} e^{-i\omega t} \frac{i}{\omega + i\epsilon} , \tag{4.5.2}$$

we find

$$\begin{aligned}
\chi_{AB}(\omega) &= \int dt \, e^{i\omega t} \, \Theta(t) \, 2i \, \chi_{AB}''(t) \\
&= \frac{1}{\pi} \int_{-\infty}^{\infty} d\omega' \frac{\chi_{AB}''(\omega')}{\omega' - \omega - i\epsilon} \\
&= \frac{1}{\pi} P \int d\omega' \frac{\chi_{AB}''(\omega')}{\omega' - \omega} + i\chi_{AB}''(\omega) ,
\end{aligned} \tag{4.5.3}$$

where, in expressions such as the second line of (4.5.3), it should always be understood that the limit $\epsilon \to 0^+$ is taken. This yields the following decomposition of $\chi_{AB}(\omega)$:

$$\chi_{AB}(\omega) = \chi_{AB}'(\omega) + i\chi_{AB}''(\omega) , \tag{4.5.4}$$

with

$$\chi_{AB}'(\omega) = \frac{1}{\pi} P \int d\omega' \frac{\chi_{AB}''(\omega')}{\omega' - \omega} . \tag{4.5.5}$$

When $\chi_{AB}''(\omega)$ is real, then, according to (4.5.5), so is $\chi_{AB}'(\omega)$ and (4.5.4) represents the separation into real and imaginary parts. The relation (4.5.5) is then identical to the dispersion relation (4.4.5a). The question as to the reality of $\chi_{AB}''(\omega)$ will be dealt with in Sect. 4.8.

4.6 Fluctuation–Dissipation Theorem

With the definitions (4.5.1b) and (4.2.11) we find

$$\chi_{AB}''(\omega) = \frac{1}{2\hbar} \left(G_{AB}^{>}(\omega) - G_{AB}^{<}(\omega) \right) , \tag{4.6.1a}$$

which, together with (4.2.14b), yields

$$\chi_{AB}''(\omega) = \frac{1}{2\hbar} G_{AB}^{>}(\omega) \left(1 - e^{-\beta\hbar\omega} \right) . \tag{4.6.1b}$$

These relations between $G^{>}$ and χ'' are known as the *fluctuation–dissipation theorem*. Together with the relation (4.5.3), one obtains for the dynamical susceptibility

$$\chi_{AB}(\omega) = \frac{1}{2\pi\hbar} \int_{-\infty}^{\infty} d\omega' \frac{G_{AB}^{>}(\omega')(1 - e^{-\beta\hbar\omega'})}{\omega' - \omega - i\epsilon} . \tag{4.6.2}$$

Classical limit $\beta\hbar\omega \ll 1$: The classical limit refers to the frequency and temperature region for which $\beta\hbar\omega \ll 1$. The fluctuation–dissipation theorem (4.6.1) then simplifies to

$$\chi''_{AB}(\omega) = \frac{\beta\omega}{2} G^>_{AB}(\omega) \quad .$$ (4.6.3)

In the classical limit (i. e. $G^>_{AB}(\omega') \neq 0$ only for $\beta\hbar|\omega'| \ll 1$) one obtains from (4.6.2)

$$\chi_{AB}(0) = \beta \int \frac{d\omega'}{2\pi} G^>_{AB}(\omega') = \beta G^>_{AB}(t = 0) \quad .$$ (4.6.4)

Hence the static susceptibility ($\omega = 0$) is given in the classical limit by the equal-time correlation function of A and B divided by kT.

The name fluctuation–dissipation theorem for (4.6.1) is appropriate since $G_{AB}(\omega)$ is a measure of the correlation between fluctuations of A and B, whilst χ''_{AB} describes the dissipation.

That χ'' has to do with *dissipation* can be seen as follows: For a perturbation of the form

$$H' = \Theta(t) \left(A^\dagger F e^{-i\omega t} + A F^* e^{i\omega t} \right) \quad ,$$ (4.6.5)

where F is a complex number, the golden rule gives a transition rate per unit time from the state n into the state m of

$$\begin{aligned}
\Gamma_{n \to m} = \frac{2\pi}{\hbar} &\left\{ \delta(E_m - E_n - \hbar\omega)|\langle m| A^\dagger F |n\rangle|^2 \right. \\
&\left. + \delta(E_m - E_n + \hbar\omega)|\langle m| A F^* |n\rangle|^2 \right\} .
\end{aligned}$$ (4.6.6)

The power of the external force (= the energy absorbed per unit time), with the help of (4.6.1a) and (4.2.13a), is found to be

$$\begin{aligned}
W &= \sum_{n,m} \frac{e^{-\beta E_n}}{Z} \Gamma_{n \to m} (E_m - E_n) \\
&= \frac{2\pi}{\hbar} \hbar\omega \frac{1}{2\pi\hbar} \left(G^>_{AA^\dagger}(\omega) - G^<_{AA^\dagger}(\omega) \right) |F|^2 \\
&= 2\omega \chi''_{AA^\dagger}(\omega)|F|^2 \quad ,
\end{aligned}$$ (4.6.7)

where a canonical distribution has been assumed for the initial states. We have thus shown that $\chi''_{AA^\dagger}(\omega)$ determines the energy absorption and, therefore, the strength of the dissipation. For frequencies at which $\chi''_{AA^\dagger}(\omega)$ is large, i.e., in the vicinity of resonances, the absorption per unit time is large as well.

Remark. If the expectation values of the operators A and B are finite, in some of the relations of Chap. 4, it can be expedient to use the operators $\hat{A}(t) = A(t) - \langle A \rangle$ and $\hat{B}(t) = B(t) - \langle B \rangle$, in order to avoid contributions proportional to $\delta(\omega)$, e. g. $\varrho(\mathbf{x}, t)$ or $\varrho_{\mathbf{k}=0}(t)$. Since the commutator remains unchanged $\chi_{AB}(t) = \chi_{\hat{A}\hat{B}}(t)$, $\chi''_{AB}(t) = \chi_{\hat{A}\hat{B}}(t)$ etc. hold.

4.7 Examples of Applications

To gain familiarity with some characteristic forms of response and correlation functions, we will give these for three typical examples: for a harmonic crystal, for diffusive dynamics, and for a damped harmonic oscillator.

(i) **Harmonic crystal.** As a first, quantum-mechanical example, we calculate the susceptibility for the displacements in a harmonic crystal. For the sake of simplicity we consider a Bravais lattice, i.e., a lattice with one atom per unit cell. We first recall a few basic facts from solid state physics concerning lattice dynamics[8]. The atoms and lattice points are labeled by vectors $\mathbf{n} = (n_1, n_2, n_3)$ of natural numbers $n_i = 1, \ldots, N_i$, where $N = N_1 N_2 N_3$ is the number of lattice points. The cartesian coordinates are characterized by the indices $i = 1, 2, 3$. We denote the equilibrium positions of the atoms (i.e., the lattice points) by $\mathbf{a_n}$, so that the actual position of the atom $\mathbf{n}$ is $\mathbf{x_n} = \mathbf{a_n} + \mathbf{u_n}$, where $\mathbf{u_n}$ is the displacement from the equilibrium position. The latter can be represented by normal coordinates $Q_{\mathbf{k}, \lambda}$

$$u_{\mathbf{n}}^i(t) = \frac{1}{\sqrt{NM}} \sum_{\mathbf{k}, \lambda} e^{i\mathbf{k} \cdot \mathbf{a_n}} \epsilon^i(\mathbf{k}, \lambda) Q_{\mathbf{k}, \lambda}(t) , \qquad (4.7.1)$$

where M the mass of an atom, $\mathbf{k}$ the wave vector, and $\epsilon^i(\mathbf{k}, \lambda)$ the components of the three polarization vectors, $\lambda = 1, 2, 3$. The normal coordinates can be expressed in terms of creation and annihilation operators $a_{\mathbf{k}, \lambda}^\dagger$ and $a_{\mathbf{k}, \lambda}$ for *phonons* with wave vector $\mathbf{k}$ and polarization λ

$$Q_{\mathbf{k}, \lambda}(t) = \sqrt{\frac{\hbar}{2\omega_{\mathbf{k}, \lambda}}} \left(a_{\mathbf{k}, \lambda}(t) + a_{-\mathbf{k}, \lambda}^\dagger(t) \right) \qquad (4.7.2)$$

with the three acoustic phonon frequencies $\omega_{\mathbf{k}, \lambda}$. Here, we use the Heisenberg representation

$$a_{\mathbf{k}, \lambda}(t) = e^{-i\omega_{\mathbf{k}, \lambda} t} a_{\mathbf{k}, \lambda}(0) . \qquad (4.7.3)$$

After this transformation, the Hamiltonian takes the form

$$H = \sum_{\mathbf{k}, \lambda} \hbar \omega_{\mathbf{k}, \lambda} \left(a_{\mathbf{k}, \lambda}^\dagger a_{\mathbf{k}, \lambda} + \frac{1}{2} \right) \qquad (4.7.4)$$

($a_{\mathbf{k}, \lambda} \equiv a_{\mathbf{k}, \lambda}(0)$). From the commutation relations of the $\mathbf{x_n}$ and their adjoint momenta one obtains for the creation and annihilation operators the standard commutator form,

---

[8] See, e.g., C. Kittel, *Quantum Theory of Solids*, 2nd revised printing, J. Wiley, New York, 1987

$$\left[a_{\mathbf{k},\lambda}, a_{\mathbf{k}',\lambda'}^{\dagger}\right] = \delta_{\lambda\lambda'}\delta_{\mathbf{kk}'}$$

$$\left[a_{\mathbf{k},\lambda}, a_{\mathbf{k}',\lambda'}\right] = \left[a_{\mathbf{k},\lambda}^{\dagger}, a_{\mathbf{k}',\lambda'}^{\dagger}\right] = 0 .$$

(4.7.5)

The dynamical susceptibility for the displacements is defined by

$$\chi^{ij}(\mathbf{n} - \mathbf{n}', t) = \frac{i}{\hbar}\Theta(t)\left\langle\left[u_{\mathbf{n}}^{i}(t), u_{\mathbf{n}'}^{j}(0)\right]\right\rangle$$

(4.7.6)

and can be expressed in terms of

$$\chi''^{ij}(\mathbf{n} - \mathbf{n}', t) = \frac{1}{2\hbar}\left\langle\left[u_{\mathbf{n}}^{i}(t), u_{\mathbf{n}'}^{j}(0)\right]\right\rangle$$

(4.7.7)

as

$$\chi^{ij}(\mathbf{n} - \mathbf{n}', t) = 2i\Theta(t)\chi''^{ij}(\mathbf{n} - \mathbf{n}', t) .$$

(4.7.8)

The phonon correlation function is defined by

$$D^{ij}(\mathbf{n} - \mathbf{n}', t) = \left\langle u_{\mathbf{n}}^{i}(t)u_{\mathbf{n}'}^{j}(0)\right\rangle .$$

(4.7.9)

For all of these quantities it has been assumed that the system is translationally invariant, i.e., one considers either an infinitely large crystal or a finite crystal with periodic boundary conditions. For the physical quantities of interest, this idealization is of no consequence. The translational invariance means that (4.7.6) and (4.7.7) depend only on the difference $\mathbf{n} - \mathbf{n}'$. The calculation of $\chi''^{ij}(\mathbf{n} - \mathbf{n}', t)$ leads, with the utilization of (4.7.1), (4.7.2), (4.7.3), and (4.7.5), to

$$\chi''^{ij}(\mathbf{n} - \mathbf{n}', t) = \frac{1}{2\hbar}\frac{1}{NM}\sum_{\substack{\mathbf{k},\lambda \\ \mathbf{k}',\lambda'}} e^{i\mathbf{k}\cdot\mathbf{a_n}+i\mathbf{k}'\cdot\mathbf{a_{n'}}}\epsilon^{i}(\mathbf{k}, \lambda)\epsilon^{j}(\mathbf{k}', \lambda')$$

$$\times \frac{\hbar}{\sqrt{4\omega_{\mathbf{k},\lambda}\omega_{\mathbf{k}',\lambda'}}}\left\langle\left[\left(a_{\mathbf{k},\lambda}e^{-i\omega_{\mathbf{k},\lambda}t} + a_{-\mathbf{k},\lambda}^{\dagger}e^{i\omega_{\mathbf{k},\lambda}t}\right), \left(a_{\mathbf{k}',\lambda'} + a_{-\mathbf{k}',\lambda'}^{\dagger}\right)\right]\right\rangle$$

$$= \frac{1}{4NM}\sum_{\mathbf{k},\lambda} e^{i\mathbf{k}\cdot(\mathbf{a_n}-\mathbf{a_{n'}})}\epsilon^{i}(\mathbf{k}, \lambda)\epsilon^{*j}(\mathbf{k}, \lambda)\frac{1}{\omega_{\mathbf{k},\lambda}}\left(e^{-i\omega_{\mathbf{k},\lambda}t} - e^{i\omega_{\mathbf{k},\lambda}t}\right) .$$

(4.7.10)

In the following, we shall make use of the fact that the polarization vectors for Bravais lattices are real.[9] For (4.7.6), this yields:

---

[9] In non-Bravais lattices the unit cell contains $r \geq 2$ atoms (ions). The number of phonon branches is $3r$, i.e., $\lambda = 1, \ldots, 3r$. Furthermore, the polarization vectors $\epsilon(\mathbf{k}, \lambda)$ are in general complex and in the results (4.7.11) to (4.7.18) the second factor $\epsilon^{j}(\ldots, \lambda)$ must be replaced by $\epsilon^{j*}(\ldots, \lambda)$.

$$\chi^{ij}(\mathbf{n}-\mathbf{n}',t) = \frac{1}{NM} \sum_{\mathbf{k},\lambda} e^{i\mathbf{k}\cdot(\mathbf{a_n}-\mathbf{a_{n'}})} \frac{\epsilon^i(\mathbf{k},\lambda)\epsilon^j(\mathbf{k},\lambda)}{\omega_{\mathbf{k},\lambda}} \sin\omega_{\mathbf{k},\lambda}t\,\Theta(t)$$

(4.7.11)

and for the temporal Fourier transform

$$\chi^{ij}(\mathbf{n}-\mathbf{n}',\omega) = \frac{1}{NM} \sum_{\mathbf{k},\lambda} e^{i\mathbf{k}\cdot(\mathbf{a_n}-\mathbf{a_{n'}})} \frac{\epsilon^i(\mathbf{k},\lambda)\epsilon^j(\mathbf{k},\lambda)}{\omega_{\mathbf{k},\lambda}} \int_0^\infty dt\, e^{i\omega t} \sin\omega_{\mathbf{k},\lambda}t\,.$$

(4.7.12)

Using the equations (A.22), (A.23), and (A.24) from QM I,

$$\int_0^\infty ds e^{isz} = 2\pi\delta_+(z) = \left[\pi\delta(z) + iP\left(\frac{1}{z}\right)\right] = i\lim_{\epsilon\to 0}\frac{1}{z+i\epsilon}$$

(4.7.13)

$$\int_0^\infty ds e^{-isz} = 2\pi\delta_-(z) = \left[\pi\delta(z) - iP\left(\frac{1}{z}\right)\right] = -i\lim_{\epsilon\to 0}\frac{1}{z-i\epsilon}\,,$$

one obtains, for real z,

$$\chi^{ij}(\mathbf{n}-\mathbf{n}',\omega) = \lim_{\epsilon\to 0}\frac{1}{2NM} \sum_{\mathbf{k},\lambda} e^{i\mathbf{k}\cdot(\mathbf{a_n}-\mathbf{a_{n'}})} \frac{\epsilon^i(\mathbf{k},\lambda)\epsilon^j(\mathbf{k},\lambda)}{\omega_{\mathbf{k},\lambda}}$$

(4.7.14a)

$$\times \left\{ \frac{1}{\omega+\omega_{\mathbf{k},\lambda}+i\epsilon} - \frac{1}{\omega-\omega_{\mathbf{k},\lambda}+i\epsilon} \right\}$$

and for the spatial Fourier transform

$$\chi^{ij}(\mathbf{q},\omega) = \sum_{\mathbf{n}} e^{-i\mathbf{q}\cdot\mathbf{a_n}} \chi^{ij}(\mathbf{n},\omega) = \frac{1}{2M} \sum_{\lambda} \frac{\epsilon^i(\mathbf{q},\lambda)\epsilon^j(\mathbf{q},\lambda)}{\omega_{\mathbf{q},\lambda}}$$

(4.7.14b)

$$\times \left\{ \frac{1}{\omega+\omega_{\mathbf{q},\lambda}+i\epsilon} - \frac{1}{\omega-\omega_{\mathbf{q},\lambda}+i\epsilon} \right\}\,.$$

For the decompositions

$$\chi^{ij}(\mathbf{n}-\mathbf{n}',\omega) = \chi'^{ij}(\mathbf{n}-\mathbf{n}',\omega) + i\chi''^{ij}(\mathbf{n}-\mathbf{n}',\omega)$$

(4.7.15a)

and

$$\chi^{ij}(\mathbf{q},\omega) = \chi'^{ij}(\mathbf{q},\omega) + i\chi''^{ij}(\mathbf{q},\omega)$$

(4.7.15b)

this leads to

$$\chi'^{ij}(\mathbf{n}-\mathbf{n}',\omega) = \frac{1}{2NM}\sum_{\mathbf{k},\lambda} e^{i\mathbf{k}\cdot(\mathbf{a_n}-\mathbf{a_{n'}})}\frac{\epsilon^i(\mathbf{k},\lambda)\epsilon^j(\mathbf{k},\lambda)}{\omega_{\mathbf{k},\lambda}}$$

$$\times \left\{ \mathrm{P}\left(\frac{1}{\omega+\omega_{\mathbf{k},\lambda}}\right) - \mathrm{P}\left(\frac{1}{\omega-\omega_{\mathbf{k},\lambda}}\right)\right\} \qquad (4.7.16a)$$

$$\chi'^{ij}(\mathbf{q},\omega) = \sum_{\mathbf{n}} e^{-i\mathbf{q}\cdot\mathbf{a_n}}\chi^{ij}(\mathbf{n},\omega)$$

$$= \frac{1}{2M}\sum_{\lambda}\frac{\epsilon^i(\mathbf{q},\lambda)\epsilon^j(\mathbf{q},\lambda)}{\omega_{\mathbf{q},\lambda}}$$

$$\times \left\{ \mathrm{P}\left(\frac{1}{\omega+\omega_{\mathbf{q},\lambda}}\right) - \mathrm{P}\left(\frac{1}{\omega-\omega_{\mathbf{q},\lambda}}\right)\right\} \qquad (4.7.16b)$$

$$\chi''^{ij}(\mathbf{n}-\mathbf{n}',\omega) = \frac{\pi}{2NM}\sum_{\mathbf{k},\lambda} e^{i\mathbf{k}\cdot(\mathbf{a_n}-\mathbf{a_{n'}})}\frac{\epsilon^i(\mathbf{k},\lambda)\epsilon^j(\mathbf{k},\lambda)}{\omega_{\mathbf{k},\lambda}}$$

$$\times [\delta(\omega-\omega_{\mathbf{k},\lambda}) - \delta(\omega+\omega_{\mathbf{k},\lambda})] \qquad (4.7.17a)$$

$$\chi''^{ij}(\mathbf{q},\omega) = \sum_{\mathbf{n}} e^{-i\mathbf{q}\cdot\mathbf{a_n}}\chi^{ij}(\mathbf{n},\omega)$$

$$= \frac{\pi}{2M}\sum_{\lambda}\frac{\epsilon^i(\mathbf{q},\lambda)\epsilon^j(\mathbf{q},\lambda)}{\omega_{\mathbf{q},\lambda}}$$

$$\times [\delta(\omega-\omega_{\mathbf{q},\lambda}) - \delta(\omega+\omega_{\mathbf{q},\lambda})] \ . \qquad (4.7.17b)$$

The phonon correlation function (4.7.9) can be either calculated directly, or determined with the help of the fluctuation–dissipation theorem from $\chi''^{ij}(\mathbf{n}-\mathbf{n}',\omega)$:

$$D^{ij}(\mathbf{n}-\mathbf{n}',\omega) = 2\hbar\frac{e^{\beta\hbar\omega}}{e^{\beta\hbar\omega}-1}\chi''^{ij}(\mathbf{n}-\mathbf{n}',\omega)$$

$$= 2\hbar\left[1+n(\omega)\right]\chi''^{ij}(\mathbf{n}-\mathbf{n}',\omega)$$

$$= \frac{\pi\hbar}{NM}\sum_{\mathbf{k},\lambda} e^{i\mathbf{k}\cdot(\mathbf{a_n}-\mathbf{a_{n'}})}\frac{\epsilon^i(\mathbf{k},\lambda)\epsilon^j(\mathbf{k},\lambda)}{\omega_{\mathbf{k},\lambda}} \qquad (4.7.18a)$$

$$\times \left\{(1+n_{\mathbf{k},\lambda})\delta(\omega-\omega_{\mathbf{k},\lambda}) - n_{\mathbf{k},\lambda}\delta(\omega+\omega_{\mathbf{k},\lambda})\right\} \ ;$$

analogously, it also follows that

$$D^{ij}(\mathbf{q},\omega) = 2\hbar\left[1+n(\omega)\right]\chi''^{ij}(\mathbf{q},\omega)$$

$$= \frac{\pi\hbar}{M}\sum_{\lambda}\frac{\epsilon^i(\mathbf{q},\lambda)\epsilon^j(\mathbf{q},\lambda)}{\omega_{\mathbf{q},\lambda}} \qquad (4.7.18b)$$

$$\times \left\{(1+n_{\mathbf{q},\lambda})\delta(\omega-\omega_{\mathbf{q},\lambda}) - n_{\mathbf{q},\lambda}\delta(\omega+\omega_{\mathbf{q},\lambda})\right\} \ .$$

Here,

$$n_{\mathbf{q},\lambda} = \left\langle a_{\mathbf{q},\lambda}^{\dagger} a_{\mathbf{q},\lambda} \right\rangle = \frac{1}{e^{\beta \hbar \omega_{\mathbf{q},\lambda}} - 1} \tag{4.7.19}$$

is the average thermal occupation number for phonons of wave vector $\mathbf{q}$ and polarization λ. The phonon resonances in $D^{ij}(\mathbf{q},\omega)$ for a particular $\mathbf{q}$ are sharp δ-function-like peaks at the positions $\pm\omega_{\mathbf{q},\lambda}$. The expansion of the density–density correlation function, which determines the inelastic neutron scattering cross-section, has as one of its contributions the phonon correlation function (4.7.18b). The excitations of the many-particle system (in this case the phonons) express themselves as resonances in the scattering cross-section. In reality, the phonons interact with one another and also with other excitations of the system, e.g, with the electrons in a metal. This leads to damping of the phonons. The essential effect of this is captured by replacing the quantity ϵ by a finite damping constant $\gamma(\mathbf{q},\lambda)$. The phonon resonances in (4.7.18) then acquire a finite width. See Fig. 4.3 and Problem 4.5.

(ii) **Diffusion.** The diffusion equation for $M(\mathbf{x},t)$ reads:

$$\dot{M}(\mathbf{x},t) = D\nabla^2 M(\mathbf{x},t) \quad , \tag{4.7.20}$$

where D is the diffusion constant and $M(\mathbf{x},t)$ can represent, for example, the magnetization density of a paramagnet. From (4.7.20) one readily finds[10,11]

$$\chi(\mathbf{q},\omega) = \chi(\mathbf{q})\frac{iDq^2}{\omega + iDq^2}$$

$$\chi'(\mathbf{q},\omega) = \chi(\mathbf{q})\frac{(Dq^2)^2}{\omega^2 + (Dq^2)^2}$$

$$\chi''(\mathbf{q},\omega) = \chi(\mathbf{q})\frac{Dq^2\omega}{\omega^2 + (Dq^2)^2} \tag{4.7.21}$$

$$G^{>}(\mathbf{q},\omega) = \chi(\mathbf{q})\frac{2\hbar\omega}{1 - e^{-\beta\hbar\omega}}\frac{Dq^2}{\omega^2 + (Dq^2)^2} \quad .$$

Figure 4.6 shows $\chi'(\mathbf{q},\omega)$, $\chi''(\mathbf{q},\omega)$, and $G^{>}(\mathbf{q},\omega)$. One sees that $\chi'(\mathbf{q},\omega)$ is symmetric in ω, whereas $\chi''(\mathbf{q},\omega)$ is antisymmetric. The form of $G^{>}(\mathbf{q},\omega)$ also depends on the value of $\beta\hbar Dq^2$, which in Fig. 4.6c is taken to be $\beta\hbar Dq^2 = 0.1$. In order to emphasize the different weights of the Stokes and anti-Stokes components, Fig. 4.6d is drawn for the value $\beta\hbar Dq^2 = 1$. However, it should be stressed that, for diffusive dynamics, this is unrealistic since, in the hydrodynamic regime, the frequencies are always smaller than kT.

---

[10] $M(\mathbf{x},t)$ is a macroscopic quantity; from the knowledge of its dynamics the dynamical susceptibility can be deduced (Problem 4.1). The same is true for the oscillator Q (see Problem 4.2),

[11] Here, we have also used $\chi' = \mathrm{Re}\,\chi$, $\chi'' = \mathrm{Im}\,\chi$, which, according to Sect. 4.8, holds for $Q^{\dagger} = Q$ and $M_{-\mathbf{q}} = M_{\mathbf{q}}^{\dagger}$.

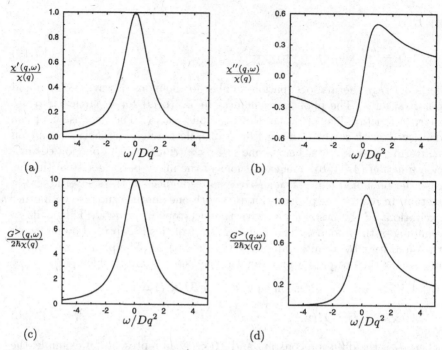

Fig. 4.6. Diffusive dynamics: (a) Real part and (b) imaginary part of the dynamical susceptibility (4.7.21). The curves in (c) and (d) show $G^>$ divided by the static susceptibility as a function of $\frac{\omega}{Dq^2}$; (c) for $\beta\hbar Dq^2 = 0.1$ and (d) for $\beta\hbar Dq^2 = 1$

(iii) **Damped oscillator.** We now consider a damped *harmonic oscillator*

$$m\left[\frac{d^2}{dt^2} + \gamma\frac{d}{dt} + \omega_0^2\right]Q = 0 \tag{4.7.22}$$

with mass m, frequency ω_0, and damping constant γ. If, on the right-hand side of the equation of motion (4.7.22), one adds an external force F, then, in the static limit, one obtains $\frac{Q}{F} = 1/m\omega_0^2$. Since this relationship defines the static susceptibility, the eigenfrequency of the oscillator depends on its mass and the static susceptibility χ, according to $\omega_0^2 = \frac{1}{m\chi}$. From the equation of motion (4.7.22) with a periodic frequency-dependent external force one finds for the dynamical susceptibility [10,11] $\chi(\omega)$ and for $G^>(\omega)$

$$\chi(\omega) = \frac{1/m}{-\omega^2 + \omega_0^2 - i\omega\gamma}$$

$$\chi'(\omega) = \frac{1}{m} \frac{-\omega^2 + \omega_0^2}{(-\omega^2 + \omega_0^2)^2 + \omega^2\gamma^2}$$

$$\chi''(\omega) = \frac{1}{m} \frac{\omega\gamma}{(-\omega^2 + \omega_0^2)^2 + \omega^2\gamma^2}$$

$$G^>(\omega) = \frac{2\hbar\omega}{m(1 - e^{-\beta\hbar\omega})} \frac{\gamma}{(-\omega^2 + \omega_0^2)^2 + \omega^2\gamma^2} .$$

(4.7.23)

These quantities, each divided by $\chi = 1/m\omega_0^2$, are shown in Fig. 4.7 as functions of ω/ω_0. Here, the ratio of the damping constant to the oscillator frequency has been taken as $\gamma/\omega_0 = 0.4$. One sees that χ' and χ'' are symmetric and antisymmetric, respectively. Figure 4.7c shows $G^>(\omega)$ at $\beta\hbar\omega_0 = 0.1$, whereas Fig. 4.7d is for $\beta\hbar\omega_0 = 1$. As in Fig. 4.6c,d, the asymmetry becomes apparent when the temperature is lowered. The differences between the intensities of the Stokes and anti-Stokes lines can be used, for example, to determine the temperature of a sample by Raman scattering.

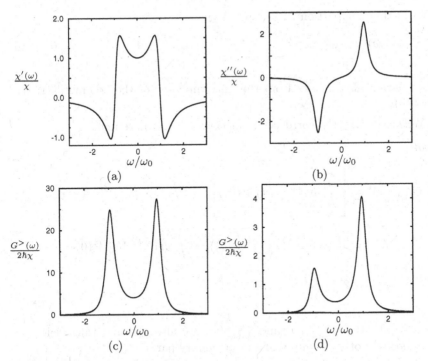

Fig. 4.7. $\chi'(\omega), \chi''(\omega)$ and $G^>(\omega)$ for the harmonic oscillator $\frac{\gamma}{\omega_0} = 0.4$. The two plots of $G^>(\omega)$ are for different values of $\beta\hbar\omega_0$, namely in (c) $\beta\hbar\omega_0 = 0.1$ and in (d) $\beta\hbar\omega_0 = 1.0$

*4.8 Symmetry Properties

4.8.1 General Symmetry Relations

In the two previous figures we have seen that $\chi'(\omega)$ is symmetric and $\chi''(\omega)$ antisymmetric, and that in $G^>(\omega)$ the Stokes line is stronger than the anti-Stokes line. We will now undertake a general investigation of the conditions under which these symmetry properties hold. The symmetry properties that will be discussed here are either of a purely mathematical nature and a direct consequence of the definitions and of the usual properties of commutators together with the dispersion relations and the relationships (4.2.14a,b), or they are of a physical nature and follow from the symmetry properties of the Hamiltonian, such as translational invariance, rotational invariance, inversion symmetry, or time-reversal invariance. It follows from (4.6.1b) and (4.2.14b) that

$$\chi''_{AB}(-\omega) = \frac{1}{2\hbar}G^>_{AB}(-\omega)\left[1 - e^{\beta\hbar\omega}\right] = \frac{1}{2\hbar}e^{-\beta\hbar\omega}G^>_{BA}(\omega)\left[1 - e^{\beta\hbar\omega}\right]$$

$$(4.8.1a)$$

and a further comparison with (4.6.1b) yields:

$$\chi''_{AB}(-\omega) = -\chi''_{BA}(\omega) .$$

$$(4.8.1b)$$

This relation also follows from the antisymmetry of the commutator; see (4.8.12b).

When $B = A^\dagger$, the correlation functions $G^{\gtrless}_{AA^\dagger}(\omega)$ are real.

Proof:

$$G^>_{AA^\dagger}(\omega)^* = \left[\int_{-\infty}^{\infty} dt\, e^{i\omega t}\langle A(t)A^\dagger(0)\rangle\right]^* = \int_{-\infty}^{\infty} dt\, e^{-i\omega t}\langle A(0)A^\dagger(t)\rangle$$

$$= \int_{-\infty}^{\infty} dt\, e^{-i\omega t}\langle A(-t)A^\dagger(0)\rangle = \int_{-\infty}^{\infty} dt\, e^{i\omega t}\langle A(t)A^\dagger(0)\rangle$$

$$= G^>_{AA^\dagger}(\omega) .$$

$$(4.8.2)$$

For $B = A^\dagger$, then $\chi'_{AA^\dagger}(\omega)$ and $\chi''_{AA^\dagger}(\omega)$ are also real and thus yield the decomposition of $\chi_{AA^\dagger}$ into real and imaginary parts:

$$\operatorname{Im} \chi_{AA^\dagger} = \chi''_{AA^\dagger} \quad , \quad \operatorname{Re} \chi_{AA^\dagger} = \chi'_{AA^\dagger} .$$

$$(4.8.3)$$

These properties are satisfied by the density–density correlation function. The *definitions* of *density correlation* and *density-response* functions read:

$$S(\mathbf{k}, \omega) = \int\limits_{-\infty}^{\infty} dt\, e^{i\omega t} S(\mathbf{k}, t) = \int dt\, d^3x\, e^{-i(\mathbf{k}\cdot\mathbf{x}-\omega t)} S(\mathbf{x}, t) \;, \tag{4.8.4a}$$

where

$$S(\mathbf{x}, t) = \langle \rho(\mathbf{x}, t)\rho(0, 0)\rangle \tag{4.8.4b}$$

denotes the correlation of the density operator (4.1.27). It follows with (4.1.28) that

$$S(\mathbf{k}, t) = \langle \rho_{\mathbf{k}}(t)\rho_{-\mathbf{k}}(0)\rangle \;. \tag{4.8.4c}$$

The susceptibility or response function is defined correspondingly through $\chi(\mathbf{k}, \omega)$ or

$$\chi''(\mathbf{k}, \omega) = \int\limits_{-\infty}^{\infty} dt\, e^{i\omega t} \frac{1}{2\hbar}\langle[\rho_{\mathbf{k}}(t), \rho_{-\mathbf{k}}(0)]\rangle \;. \tag{4.8.5}$$

The relationship between the density correlation function and $S_{\text{coh}}(\mathbf{k}, \omega)$ reads:

$$S(\mathbf{k}, \omega) = \frac{N}{V} 2\pi\hbar S_{\text{coh}}(\mathbf{k}, \omega) \quad . \tag{4.8.6}$$

Further symmetry properties result in the presence of space *inversion symmetry*. Since we then have $\chi''(-\mathbf{k}, \omega) = \chi''(\mathbf{k}, \omega)$, it follows from (4.8.1b) that

$$\chi''(\mathbf{k}, -\omega) = -\chi''(\mathbf{k}, \omega) \quad . \tag{4.8.7a}$$

Thus χ'' is an odd function of ω and, due to (4.8.3), is also real. Correspondingly, $\chi'(\mathbf{k}, \omega)$ is even:

$$\chi'(\mathbf{k}, -\omega) = \chi'(\mathbf{k}, \omega) \quad . \tag{4.8.7b}$$

This can be seen by means of the dispersion relation, since

$$\chi'(\mathbf{k}, -\omega) = P\int\limits_{-\infty}^{\infty} \frac{d\omega'}{\pi} \frac{\chi''(\mathbf{k}, \omega')}{\omega' + \omega} = -P\int\limits_{-\infty}^{\infty} \frac{d\omega'}{\pi} \frac{\chi''(\mathbf{k}, -\omega')}{\omega' + \omega}$$

$$= P\int\limits_{-\infty}^{\infty} \frac{d\omega'}{\pi} \frac{\chi''(\mathbf{k}, \omega')}{\omega' - \omega} = \chi'(\mathbf{k}, \omega) \;. \tag{4.8.8}$$

For systems with inversion symmetry the density susceptibility can, according to (4.6.1a) and (4.2.14a), be represented in the form

$$\chi''(\mathbf{k},\omega) = \frac{1}{2\hbar}\left(S(\mathbf{k},\omega) - S(\mathbf{k},-\omega)\right) . \tag{4.8.9}$$

Inserting this into the dispersion relation, one finds

$$\chi'(\mathbf{k},\omega) = \frac{1}{2\hbar\pi}P\int\limits_{-\infty}^{\infty} d\omega' S(\mathbf{k},\omega')\left[\frac{1}{\omega'-\omega} - \frac{1}{-\omega'-\omega}\right]$$

$$= \frac{1}{\hbar\pi}P\int\limits_{-\infty}^{\infty} d\omega'\frac{\omega' S(\mathbf{k},\omega')}{\omega'^2 - \omega^2} . \tag{4.8.10}$$

From this one obtains the asymptotic behavior

$$\lim_{\omega\to0}\chi'(\mathbf{k},\omega) = \frac{1}{\hbar\pi}P\int\limits_{-\infty}^{\infty} d\omega'\frac{S(\mathbf{k},\omega')}{\omega'} \tag{4.8.11a}$$

$$\lim_{\omega\to\infty}\omega^2\chi'(\mathbf{k},\omega) = -\frac{1}{\hbar\pi}\int\limits_{-\infty}^{\infty} d\omega'\omega' S(\mathbf{k},\omega') . \tag{4.8.11b}$$

4.8.2 Symmetry Properties of the Response Function for Hermitian Operators

4.8.2.1 Hermitian Operators

Examples of hermitian operators are the density $\rho(\mathbf{x},t)$ and the momentum density $\mathbf{P}(\mathbf{x},t)$. For arbitrary, and in particular also for hermitian, operators A and B, one has the following symmetry relations:

$$\chi''_{AB}(t-t') = -\chi''_{BA}(t'-t) \tag{4.8.12a}$$

$$\chi''_{AB}(\omega) = -\chi''_{BA}(-\omega) . \tag{4.8.12b}$$

This follows from the antisymmetry of the commutator. The relation for the Fourier transform is identical to (4.8.1b). Likewise, from the definition (4.5.1a), one can conclude directly

$$\chi''_{AB}(t-t')^* = -\chi''_{AB}(t-t') \tag{4.8.13a}$$

i.e., $\chi''_{AB}(t-t')$ is imaginary (the commutator of two hermitian operators is antihermitian) and

$$\chi''_{AB}(\omega)^* = -\chi''_{AB}(-\omega) . \tag{4.8.13b}$$

Taken together, (4.8.12) and (4.8.13) yield:

$$\chi''_{AB}(t-t')^* = +\chi''_{BA}(t'-t) \tag{4.8.14a}$$

and

$$\chi''_{AB}(\omega)^* = \chi''_{BA}(\omega) \quad . \tag{4.8.14b}$$

Remark: For both the correlation function and the susceptibility, the translational invariance implies

$$G^{\gtrless}_{A(\mathbf{x})B(\mathbf{x}')} = G^{\gtrless}_{AB}(\mathbf{x}-\mathbf{x}',\dots) \tag{4.8.15a}$$

and the rotational invariance

$$G^{\gtrless}_{A(\mathbf{x})B(\mathbf{x}')} = G^{\gtrless}_{AB}(|\mathbf{x}-\mathbf{x}'|,\dots) \quad . \tag{4.8.15b}$$

It thus follows from (4.8.14b) for systems with spatial translational and rotational invariance that

$$\chi''_{A(\mathbf{x})A(\mathbf{x}')}(\omega) = \chi''_{AA}(|\mathbf{x}-\mathbf{x}'|,\omega) \tag{4.8.16}$$

is real and antisymmetric in ω.
For different operators, it is the behavior under the *time-reversal transformation* that determines whether or not χ'' is real.

4.8.2.2 Time Reversal, Spatial and Temporal Translations

Time-reversal invariance

Under the time-reversal operation (Sect. 11.4.2.3), an operator $A(\mathbf{x},t)$ transforms as follows:

$$A(\mathbf{x},t) \to A'(\mathbf{x},t) = \mathcal{T} A(\mathbf{x},t)\mathcal{T}^{-1} = \epsilon_A A(\mathbf{x},-t) . \tag{4.8.17}$$

ϵ_A is known as the signature and can take the following values:

$\epsilon_A = 1$ (e.g., for position and for electric field)

$\epsilon_A = -1$ (e.g., for velocity, angular momentum, and magnetic field).

For the expectation value of an operator B one finds

$$\langle \alpha | B | \alpha \rangle = \langle \mathcal{T} B \alpha | \mathcal{T} \alpha \rangle = \langle \mathcal{T} B \mathcal{T}^{-1} \mathcal{T} \alpha | \mathcal{T} \alpha \rangle$$
$$= \langle \mathcal{T} \alpha | (\mathcal{T} B \mathcal{T}^{-1})^\dagger | \mathcal{T} \alpha \rangle . \tag{4.8.18a}$$

Making use of (4.8.17), one obtains

$$(\mathcal{T}[A(\mathbf{x},t),B(\mathbf{x}',t')]\mathcal{T}^{-1})^\dagger = \epsilon_A \epsilon_B [A(\mathbf{x},-t),B(\mathbf{x}',-t')]^\dagger$$
$$= -\epsilon_A \epsilon_B [A(\mathbf{x},-t),B(\mathbf{x}',-t')] . \tag{4.8.18b}$$

For time-reversal-invariant Hamiltonians, this yields:

$$\chi''_{AB}(t-t') = -\epsilon_A\epsilon_B\chi''_{AB}(t'-t) \qquad (4.8.19a)$$

and

$$\chi''_{AB}(\omega) = -\epsilon_A\epsilon_B\chi''_{AB}(-\omega) = \epsilon_A\epsilon_B\chi''_{BA}(\omega) \quad . \qquad (4.8.19b)$$

When $\epsilon_A = \epsilon_B$, then $\chi''_{AB}(\omega)$ is symmetric under exchange of A and B, odd in ω, and real. When $\epsilon_A = -\epsilon_B$, then $\chi''_{AB}(\omega)$ is antisymmetric under the exchange of A and B, even in ω, and imaginary. If a magnetic field is present, then its direction is reversed under a time-reversal transformation

$$\chi''_{AB}(\omega;\mathbf{B}) = \epsilon_A\epsilon_B\chi''_{BA}(\omega;-\mathbf{B})$$

$$= -\epsilon_A\epsilon_B\chi''_{AB}(-\omega;-\mathbf{B}). \qquad (4.8.20)$$

Finally, we remark that, from (4.8.13b) and (4.5.3),

$$\chi^*_{AB}(\omega) = \chi_{AB}(-\omega) . \qquad (4.8.21)$$

This relation guarantees that the response (4.3.14) is real.

Translational invariance of the correlation function

$$\begin{aligned}
f(\mathbf{x},t;\mathbf{x}',t') &\equiv \langle A(\mathbf{x},t)B(\mathbf{x}',t')\rangle \\
&= \langle T_\mathbf{a}^{-1}T_\mathbf{a}A(\mathbf{x},t)T_\mathbf{a}^{-1}T_\mathbf{a}B(\mathbf{x}',t')T_\mathbf{a}^{-1}T_\mathbf{a}\rangle \\
&= \langle T_\mathbf{a}^{-1}A(\mathbf{x}+\mathbf{a},t)B(\mathbf{x}'+\mathbf{a},t')T_\mathbf{a}\rangle .
\end{aligned}$$

If the density matrix ρ commutes with $T_\mathbf{a}$, i.e., $[T_\mathbf{a},\rho] = 0$, then, due to the cyclic invariance of the trace, it follows that

$$\langle A(\mathbf{x},t)B(\mathbf{x}',t')\rangle = \langle A(\mathbf{x}+\mathbf{a},t)B(\mathbf{x}'+\mathbf{a},t')\rangle \qquad (4.8.22)$$

$$= f(\mathbf{x}-\mathbf{x}',t;0,t') \quad ,$$

where in the last step we have set $\mathbf{a} = -\mathbf{x}'$. Thus, spatial and temporal translational invariance together yield:

$$f(\mathbf{x},t;\mathbf{x}',t') = f(\mathbf{x}-\mathbf{x}',t-t') . \qquad (4.8.23)$$

Rotational invariance

A system can be translationally invariant without being rotationally invariant. When rotational invariance holds, then (for any rotation matrix R)

$$f(\mathbf{x}-\mathbf{x}',t-t') = f(R(\mathbf{x}-\mathbf{x}'),t-t') = f(|\mathbf{x}-\mathbf{x}'|,t-t') , \qquad (4.8.24)$$

independent of the direction.

Fourier transformation for *translationally invariant* systems yields:

$$\tilde{f}(\mathbf{k}, t; \mathbf{k}', t') = \int d^3x\, d^3x'\, e^{-i\mathbf{k}\cdot\mathbf{x} - i\mathbf{k}'\cdot\mathbf{x}'} f(\mathbf{x}, t; \mathbf{x}', t')$$

$$= \int d^3x\, d^3x'\, e^{-i\mathbf{k}\cdot\mathbf{x} - i\mathbf{k}'\cdot\mathbf{x}'} f(\mathbf{x} - \mathbf{x}', t - t') .$$

Substituting $\mathbf{y} = \mathbf{x} - \mathbf{x}'$ leads to

$$\tilde{f}(\mathbf{k}, t; \mathbf{k}', t') = \int d^3x' \int d^3y\, e^{-i\mathbf{k}\cdot(\mathbf{y}+\mathbf{x}') - i\mathbf{k}'\cdot\mathbf{x}'} f(\mathbf{y}, t - t')$$

$$= (2\pi)^3 \delta^{(3)}(\mathbf{k} + \mathbf{k}') \tilde{f}(\mathbf{k}, t - t') .$$

If rotational invariance holds, then

$$\tilde{f}(\mathbf{k}, t - t') = \tilde{f}(|\mathbf{k}|, t - t') . \tag{4.8.25}$$

4.8.2.3 The Classical Limit

We have already seen (Eqs. (4.6.3),(4.6.4)) that, in the classical limit ($\hbar\omega \ll kT$):

$$\chi''_{AB}(\omega) = \frac{\beta\omega}{2} G^>_{AB}(\omega) \quad \text{and} \tag{4.8.26a}$$

$$\chi_{AB}(0) = \beta G^>_{AB}(t = 0) . \tag{4.8.26b}$$

From the time-reversal relation for $\chi''_{AB}(\omega)$, Eq. (4.8.19b), it follows that

$$G^>_{AB}(-\omega) = \epsilon_A \epsilon_B G^>_{AB}(\omega) . \tag{4.8.27}$$

When $\epsilon_A = \epsilon_B$, then $G^>_{AB}(\omega)$ is symmetric in ω, real, and symmetric upon interchange of A and B. (The latter follows from the fluctuation–dissipation theorem and from the symmetry of $\chi''_{AB}(\omega)$). When $\epsilon_A = -\epsilon_B$, then $G^>_{AB}$ is odd in ω, antisymmetric upon interchange of A and B, and imaginary. For $\epsilon_A = \epsilon_B$, equation (4.8.26a) is equivalent to

$$\text{Im } \chi_{AB}(\omega) = \frac{\beta\omega}{2} G^>_{AB}(\omega) . \tag{4.8.28}$$

The half-range Fourier transform of $G^>_{AB}(t)$, i.e. the Fourier transform of $\Theta(t) G^>_{AB}(t)$ satisfies in the classical limit

$$G_{AB}^H(\omega) \equiv \int_0^\infty dt\, e^{i\omega t} G_{AB}^>(t)$$

$$= \int_0^\infty dt\, e^{i\omega t} \int_{-\infty}^\infty \frac{d\omega'}{2\pi} e^{-i\omega' t} G_{AB}^>(\omega')$$

$$= \int_{-\infty}^\infty dt\, e^{i\omega t} \int_{-\infty}^\infty \frac{d\omega''}{2\pi} \frac{ie^{-i\omega'' t}}{\omega'' + i\epsilon} \int_{-\infty}^\infty \frac{d\omega'}{2\pi} e^{-i\omega' t} G_{AB}^>(\omega')$$

$$= -\frac{i}{2\pi} \int_{-\infty}^\infty d\omega' \frac{G_{AB}^>(\omega')}{\omega' - \omega - i\epsilon}$$

$$= -\frac{i}{2\pi} \frac{2}{\beta} \int_{-\infty}^\infty d\omega' \frac{\chi_{AB}''(\omega')}{\omega'(\omega' - \omega - i\epsilon)}$$

$$= -\frac{i}{\pi\beta} \int_{-\infty}^\infty d\omega' \chi_{AB}''(\omega') \left(\frac{1}{\omega'} - \frac{1}{\omega' - \omega - i\epsilon}\right) \frac{1}{-\omega - i\epsilon}$$

$$= \frac{i}{\beta\omega} (\chi_{AB}(0) - \chi_{AB}(\omega)) . \tag{4.8.29}$$

4.8.2.4 Kubo Relaxation Function

The Kubo relaxation function is particularly useful for the description of the relaxation of the deviation $\delta\langle A(t)\rangle$ after the external force has been switched off (see SM, Appendix H).

The *Kubo relaxation function* of two operators A and B is defined by

$$\phi_{AB}(t) = \frac{i}{\hbar} \int_t^\infty dt'\, \langle [A(t'), B(0)]\rangle e^{-\epsilon t'} \tag{4.8.30}$$

and its half-range Fourier transform is given by

$$\phi_{AB}(\omega) = \int_0^\infty dt\, e^{i\omega t} \phi_{AB}(t) . \tag{4.8.31}$$

It is related to the dynamical susceptibility via

$$\phi_{AB}(t = 0) = \chi_{AB}(\omega = 0) \tag{4.8.32a}$$

and

$$\phi_{AB}(\omega) = \frac{1}{i\omega}(\chi_{AB}(\omega) - \chi_{AB}(0)) . \tag{4.8.32b}$$

The first relation follows from a comparison of (4.3.12) with (4.8.31) and the second from a short calculation (Problem 4.6). Equation (4.8.29) thus implies that, in the classical limit,

$$\phi_{AB}(\omega) = \beta G_{AB}^H(\omega) . \tag{4.8.33}$$

4.9 Sum Rules

4.9.1 General Structure of Sum Rules

We start from the definitions (4.5.1a,b)

$$\frac{1}{\hbar}\langle[A(t),B(0)]\rangle = \int \frac{d\omega}{\pi} e^{-i\omega t}\chi''_{AB}(\omega)$$ (4.9.1)

and differentiate this n times with respect to time:

$$\frac{1}{\hbar}\left\langle\left[\frac{d^n}{dt^n}A(t),B(0)\right]\right\rangle = \int \frac{d\omega}{\pi}(-i\omega)^n e^{-i\omega t}\chi''_{AB}(\omega)\ .$$

Repeated substitution of the Heisenberg equation yields, for $t = 0$,

$$\int \frac{d\omega}{\pi}\omega^n \chi''_{AB}(\omega) = \frac{i^n}{\hbar}\left\langle\left[\frac{d^n}{dt^n}A(t)\Big|_{t=0}, B(0)\right]\right\rangle$$

$$= \frac{1}{\hbar^{n+1}}\langle[[\ldots[A,H_0],\ldots,H_0],B]\rangle.$$ (4.9.2)

The right-hand side contains an n-fold commutator of A with H_0. If these commutators lead to simple expressions, then (4.9.2) provides information about moments of the dissipative part of the susceptibility. Such relations are known as sum rules.

The f-sum rule: An important example is the f-sum rule for the density–density susceptibility, which, with the help of (4.8.9), can be represented as a sum rule for the correlation function

$$\int \frac{d\omega}{2\pi}\omega\chi''(\mathbf{k},\omega) = \int \frac{d\omega}{2\pi\hbar}\omega S(\mathbf{k},\omega) = \frac{i}{2\hbar}\langle[\dot{\rho}_{\mathbf{k}}(t),\rho_{-\mathbf{k}}(t)]\rangle\ .$$

The commutator on the right-hand side can be calculated with $\dot{\rho}_{\mathbf{k}} = i\mathbf{k}\cdot\mathbf{j}_{\mathbf{k}}$, which yields, for purely coordinate-dependent potentials, the standard form of the f-sum rule

$$\int \frac{d\omega}{2\pi}\frac{\omega}{\hbar}S(\mathbf{k},\omega) = \frac{k^2}{2m}n\ ,$$ (4.9.3)

where $n = \frac{N}{V}$ is the particle number density.

There are also sum rules that result from the fact that, in many cases,[12] in the limit $\mathbf{k} \to 0$ and $\omega \to 0$ the dynamical susceptibility must transform into a susceptibility known from equilibrium statistical mechanics.

Compressibility sum rule: As an example, we will use (4.8.11a) to give the compressibility sum rule for the density response function:

---

[12] P.C. Kwok and T.D. Schultz, J. Phys. **C2**, 1196 (1969)

$$\lim_{k \to 0} P \int \frac{d\omega}{\pi} \frac{1}{\hbar} \frac{S(\mathbf{k}, \omega)}{\omega} = n \left(\frac{\partial n}{\partial P} \right)_T = n^2 \kappa_T \quad . \tag{4.9.4}$$

Here we have made use of the relationship

$$\chi'(0, 0) = \frac{1}{V} \left(\frac{\partial N}{\partial \mu} \right)_{T, V} = -\frac{N^2}{V^3} \left(\frac{\partial V}{\partial P} \right)_{T, N}$$

$$= -n^3 \left(\frac{\partial \frac{1}{n}}{\partial P} \right)_{T, N} = n \left(\frac{\partial n}{\partial P} \right)_{T, N} \quad ,$$

which derives from (4.8.26b) and from thermodynamics[13].
   The *static form factor* is defined by

$$S(\mathbf{k}) = \langle \rho_{\mathbf{k}} \rho_{-\mathbf{k}} \rangle \quad . \tag{4.9.5}$$

This determines the elastic scattering and is related to $S(\mathbf{k}, \omega)$ via

$$\int \frac{d\omega}{2\pi} S(\mathbf{k}, \omega) = S(\mathbf{k}) \quad . \tag{4.9.6}$$

The static form factor $S(\mathbf{k})$ can be deduced from x-ray scattering.
   Equations (4.9.3), (4.9.4), and (4.9.6) provide us with three sum rules for the density correlation function. The sum rules give precise relationships between $S(\mathbf{k}, \omega)$ and static quantities. When these static quantities are known from theory or experiment and one has some idea of the form of $S(\mathbf{k}, \omega)$, it is then possible to use the sum rules to determine the parameters involved in $S(\mathbf{k}, \omega)$. We shall elucidate this for the example of excitations in superfluid helium, discussed in Sect. 3.2.3.

4.9.2 Application to the Excitations in He II

Let us return to the excitations in superfluid helium of Sect. 3.2.3. We approximate $S(\mathbf{q}, \omega)$ by an infinitely sharp density resonance (phonon, roton) and assume $T = 0$ so that only the Stokes component is present:

$$S(\mathbf{q}, \omega) = Z_{\mathbf{q}} \delta(\omega - \epsilon_{\mathbf{q}}/\hbar) \quad . \tag{4.9.7}$$

Inserting this into the *f*-sum rule (4.9.3) and the form factor (4.9.6) yields:

$$\epsilon_{\mathbf{q}} = \frac{\hbar^2 n q^2}{2m S(\mathbf{q})} \quad . \tag{4.9.8}$$

---

[13] See, e.g., L.D. Landau and E.M. Lifschitz, *Course of Theoretical Physics, Vol. 5. Statistical Physics* 3rd edn. Part 1, E.M. Lifshitz, L.P. Pitajevski, Pergamon, Oxford, 1980; F. Schwabl, *Statistical Mechanics*, 2nd ed., Springer, Berlin Heidelberg, 2006, Eq. (3.2.10)

The f-sum rule (4.9.3) and compressibility sum rule (4.9.4) give, in the limit $q \to 0$,

$$\epsilon_{\mathbf{q}} = \hbar s_T q = \hbar \sqrt{\frac{1}{m} \left(\frac{\partial P}{\partial n} \right)_T} q \ , \ Z_{\mathbf{q}} = \frac{\pi \hbar n q}{m s_T} , \ S(\mathbf{q}) = \frac{\hbar n q}{2 m s_T} , \qquad (4.9.9)$$

where we have introduced the isothermal sound velocity $s_T = \sqrt{\left(\frac{\partial P}{\partial n} \right)_T / m}$. The relationship (4.9.8) between the energy of the excitations and the static form factor was first derived by Feynman[14]. Figure 4.8 shows experimental results for these two quantities. For small q, it is seen that $S(\mathbf{q})$ increases linearly with q, yielding the linear dispersion relation in the phonon regime. The maximum of $S(\mathbf{q})$ at $q \approx 2\text{\AA}^{-1}$ leads to the roton minimum.

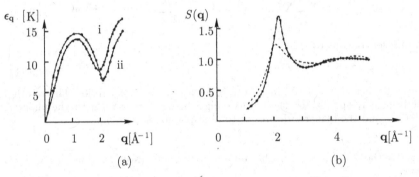

Fig. 4.8. (a) The excitations of He II at low temperatures: (i) under vapor pressure, (ii) at 25.3 atm. (b) The static form factor [15]

Problems

4.1 Confirm the validity of Eq. (4.7.21) by adding an external magnetic field $H(\mathbf{x}, t)$ to the diffusion equation (4.7.20).

4.2 For the classical damped harmonic oscillator

$$\left(\frac{d^2}{dt^2} + \gamma \frac{d}{dt} + \omega_0^2 \right) Q(t) = F(t)/m$$

determine the following functions: $\chi(\omega)$, $\chi'(\omega)$ $\chi''(\omega)$, and $G^{>}(\omega)$.

---

[14] R. Feynman, Phys. Rev. B **94**, 262 (1954)
[15] D.G. Henshaw, Phys. Rev. **119**, 9 (1960); D.G. Henshaw and A.D.B. Woods, Phys. Rev. **121**, 1266 (1961)

Hint: Solve the equation of motion in Fourier space and determine the dynamical susceptibility from $\chi(\omega) = \frac{dQ(\omega)}{dF(\omega)}$.

4.3 Prove the *f*-sum rule,

$$\int \frac{d\omega}{2\pi}\omega\chi''(\mathbf{k},\omega) = \int \frac{d\omega}{2\pi\hbar}\omega S(\mathbf{k},\omega) = \frac{k^2}{2m}n$$

for the density–density correlation function.
Hint: Calculate $\frac{i}{2\hbar}\langle[\dot{\rho}_{\mathbf{k}},\rho_{-\mathbf{k}}]\rangle$.

4.4 Show, for $B = A^\dagger$, that $G^>_{AB}(\omega)$, $G^<_{AB}(\omega)$, $\chi'_{AB}(\omega)$, and $\chi''_{AB}(\omega)$ are real.

4.5 Show that the coherent neutron scattering cross-section for harmonic phonons, Eqs. (4.1.29) and (4.7.1) ff., can be written as

$$S_{\mathrm{coh}}(\mathbf{k},\omega) = e^{-2W}\frac{1}{N}\sum_{n,m}e^{-i(\mathbf{a_n}-\mathbf{a_m})\cdot\mathbf{k}}\int_{-\infty}^{\infty}\frac{dt}{2\pi\hbar}e^{i\omega t}e^{\langle\mathbf{k}\cdot\mathbf{u_n}(t)\mathbf{k}\cdot\mathbf{u_m}(0)\rangle} \quad (4.9.10)$$

with the Debye–Waller factor

$$e^{-2W} = e^{-\langle(\mathbf{k}\cdot\mathbf{u_n}(0))^2\rangle}. \quad (4.9.11)$$

Expand the last exponential function in $S_{\mathrm{coh}}(\mathbf{k},\omega)$ as a Taylor series. The zeroth-order term corresponds to elastic scattering, the first-order term to one-phonon scattering, and the higher-order terms to multiphonon scattering.

4.6 Derive the relation (4.8.32b) by suitable partial integration, and using $\phi_{AB}(t = \infty) = 0$.

Bibliography for Part I

A.A. Abrikosov, L.P. Gorkov, and I.E. Dzyaloshinski, *Methods of Quantum Field Theory in Statistical Physics* (Prentice Hall, Englewood Cliffs 1963)

L.E. Ballentine, *Quantum Mechanics* (Prentice Hall, Englewood Cliffs 1990)

G. Baym, *Lectures on Quantum Mechanics* (Benjamin, London 1973)

A.L. Fetter and J.D. Walecka, *Quantum Theory of Many-Particle Systems* (McGraw-Hill, New York 1971)

A. Griffin, *Excitations in a Bose-condensed Liquid*, Cambridge University Press, Cambridge, 1993

L.D. Landau and E.M. Lifshitz, *Course of Theoretical Physics*, Volume 9, E.M. Lifshitz and E.P. Pitaevskii, *Statistical Physics 2*, Pergamon Press, New York, 1981

G.D. Mahan, *Many-Particle Physics* (3rd edition, Kluwer Academic/Plenum Publishers, New York 2000)

P.C. Martin, *Measurements and Correlation Functions* (Gordon and Breach, New York 1968)

P. Nozières and D. Pines, *The Theory of Quantum Liquids*, Volume I, Normal Fermi Liquids (Benjamin, New York 1966)

P. Nozières and D. Pines, *The Theory of Quantum Liquids*, Volume II, Superfluid Bose Liquids (Addison-Wesley, New York 1990)

C.J. Pethick and H. Smith, *Bose–Einstein Condensation in Diluted Gases* (University Press, Cambridge 2002)

L. Pitaevskii and S. Stringari, *Bose–Einstein Condensation* (Clarendon Press, Oxford 2003)

J.J. Sakurai, *Modern Quantum Mechanics* (Addison-Wesley, Redwood City 1985)

E.P. Wigner, *Group Theory* and its applications to the Quantum Mechanics of Atomic Spectra (Academic, New York 1959)

J.M. Ziman, *Elements of Advanced Quantum Theory* (Cambridge University Press, Cambridge 1969)

Part II

Relativistic Wave Equations

5. Relativistic Wave Equations and their Derivation

In this chapter, the relativistic wave equations for particles with spin 0 and 1/2 will be determined on the basis of the correspondence principle and further general considerations.

5.1 Introduction

Quantum theory is based on the following axioms[1]:

1. The state of a system is described by a state vector $|\psi\rangle$ in a linear space.
2. The observables are represented by hermitian operators $A...$, and functions of observables by the corresponding functions of the operators.
3. The mean (expectation) value of an observable in the state $|\psi\rangle$ is given by $\langle A \rangle = \langle \psi | A | \psi \rangle$.
4. The time evolution is determined by the Schrödinger equation involving the Hamiltonian H

$$i\hbar \frac{\partial |\psi\rangle}{\partial t} = H |\psi\rangle \ . \tag{5.1.1}$$

5. If, in a measurement of the observable A, the value a_n is found, then the original state changes to the corresponding eigenstate $|n\rangle$ of A.

We consider the Schrödinger equation for a free particle in the coordinate representation

$$i\hbar \frac{\partial \psi}{\partial t} = -\frac{\hbar^2}{2m} \boldsymbol{\nabla}^2 \psi \ . \tag{5.1.2}$$

It is evident from the differing orders of the time and the space derivatives that this equation is not Lorentz covariant, i.e., that it changes its structure under a transition from one inertial system to another.

Efforts to formulate a relativistic quantum mechanics began with attempts to use the correspondence principle in order to derive a relativistic wave equation intended to replace the Schrödinger equation. The first such equation was due to Schrödinger (1926)[2], Gordon (1926)[3], and Klein

---

[1] See QM I, Sect. 8.3.
[2] E. Schrödinger, Ann. Physik **81**, 109 (1926)
[3] W. Gordon, Z. Physik **40**, 117 (1926)

(1927)[4]. This scalar wave equation of second order, which is now known as the Klein–Gordon equation, was initially dismissed, since it led to negative probability densities. The year 1928 saw the publication of the Dirac equation[5]. This equation pertains to particles with spin 1/2 and is able to describe many of the single-particle properties of fermions. The Dirac equation, like the Klein–Gordon equation, possesses solutions with negative energy, which, in the framework of wave mechanics, leads to difficulties (see below). To prevent transitions of an electron into lower lying states of negative energy, in 1930[6] Dirac postulated that the states of negative energy should all be occupied. Missing particles in these otherwise occupied states represent particles with opposite charge (antiparticles). This necessarily leads to a many-particle theory, or to a quantum field theory. By reinterpreting the Klein–Gordon equation as the basis of a field theory, Pauli and Weisskopf[7] showed that this could describe mesons with spin zero, e.g., π mesons. The field theories based upon the Dirac and Klein–Gordon equations correspond to the Maxwell equations for the electromagnetic field, and the d'Alembert equation for the four-potential.

The Schrödinger equation, as well as the other axioms of quantum theory, remain unchanged. Only the Hamiltonian is changed and now represents a quantized field. The elementary particles are excitations of the fields (mesons, electrons, photons, etc.).

It will be instructive to now follow the historical development rather than begin immediately with quantum field theory. For one thing, it is conceptually easier to investigate the properties of the Dirac equation in its interpretation as a single-particle wave equation. Furthermore, it is exactly these single-particle solutions that are needed as basis states for expanding the field operators. At low energies one can neglect decay processes and thus, here, the quantum field theory gives the same physical predictions as the elementary single-particle theory.

5.2 The Klein–Gordon Equation

5.2.1 Derivation by Means of the Correspondence Principle

In order to derive relativistic wave equations, we first recall the correspondence principle[8]. When classical quantities were replaced by the operators

$$\text{energy} \quad E \longrightarrow i\hbar \frac{\partial}{\partial t}$$

[4] O. Klein, Z. Physik **41**, 407 (1927)
[5] P.A.M. Dirac, Proc. Roy. Soc. (London) **A117**, 610 (1928); ibid. **A118**, 351 (1928)
[6] P.A.M. Dirac, Proc. Roy. Soc. (London) **A126**, 360 (1930)
[7] W. Pauli and V. Weisskopf, Helv. Phys. Acta **7**, 709 (1934)
[8] See, e.g., QM I, Sect. 2.5.1

and

$$\text{momentum} \qquad \mathbf{p} \longrightarrow \frac{\hbar}{\mathrm{i}}\boldsymbol{\nabla}\,, \qquad\qquad (5.2.1)$$

we obtained from the nonrelativistic energy of a free particle

$$E = \frac{\mathbf{p}^2}{2m}\,, \qquad\qquad (5.2.2)$$

the free time-dependent *Schrödinger equation*

$$\mathrm{i}\hbar\frac{\partial}{\partial t}\psi = -\frac{\hbar^2\boldsymbol{\nabla}^2}{2m}\psi\,. \qquad\qquad (5.2.3)$$

This equation is obviously not Lorentz covariant due to the different orders of the time and space derivatives.

We now recall some relevant features of the special theory of relativity.[9] We will use the following conventions: The components of the space–time four-vectors will be denoted by Greek indices, and the components of spatial three-vectors by Latin indices or the cartesian coordinates x, y, z. In addition, we will use Einstein's summation convention: Greek indices that appear twice, one contravariant and one covariant, are summed over, the same applying to corresponding Latin indices.

Starting from $x^\mu(s) = (ct, \mathbf{x})$, the contravariant four-vector representation of the world line as a function of the proper time s, one obtains the four-velocity $\dot{x}^\mu(s)$. The differential of the proper time is related to dx^0 via $ds = \sqrt{1-(v/c)^2}\,dx^0$, where

$$\mathbf{v} = c\,(d\mathbf{x}/dx^0) \qquad\qquad (5.2.4a)$$

is the velocity. For the four-momentum this yields:

$$p^\mu = mc\dot{x}^\mu(s) = \frac{1}{\sqrt{1-(v/c)^2}}\begin{pmatrix} mc \\ m\mathbf{v} \end{pmatrix} = \text{four-momentum} = \begin{pmatrix} E/c \\ \mathbf{p} \end{pmatrix}. $$
$$(5.2.4b)$$

In the last expression we have used the fact that, according to relativistic dynamics, $p^0 = mc/\sqrt{1-(v/c)^2}$ represents the kinetic energy of the particle. Therefore, according to the special theory of relativity, the energy E and the momentum p_x, p_y, p_z transform as the components of a contravariant four-vector

$$p^\mu = (p^0, p^1, p^2, p^3) = \left(\frac{E}{c}, p_x, p_y, p_z\right). \qquad\qquad (5.2.5a)$$

---

[9] The most important properties of the Lorentz group will be summarized in Sect. 6.1.

The metric tensor

$$g_{\mu\nu} = \begin{pmatrix} 1 & 0 & 0 & 0 \\ 0 & -1 & 0 & 0 \\ 0 & 0 & -1 & 0 \\ 0 & 0 & 0 & -1 \end{pmatrix} \tag{5.2.6}$$

yields the covariant components

$$p_\mu = g_{\mu\nu}p^\nu = \left(\frac{E}{c}, -\mathbf{p}\right) . \tag{5.2.5b}$$

According to Eq. (5.2.4b), the invariant scalar product of the four-momentum is given by

$$p_\mu p^\mu = \frac{E^2}{c^2} - \mathbf{p}^2 = m^2 c^2 , \tag{5.2.7}$$

with the rest mass m and the velocity of light c.

From the energy–momentum relation following from (5.2.7),

$$E = \sqrt{\mathbf{p}^2 c^2 + m^2 c^4} , \tag{5.2.8}$$

one would, according to the *correspondence principle* (5.2.1), initially arrive at the following *wave equation*:

$$i\hbar \frac{\partial}{\partial t}\psi = \sqrt{-\hbar^2 c^2 \mathbf{\nabla}^2 + m^2 c^4}\ \psi . \tag{5.2.9}$$

An obvious difficulty with this equation lies in the square root of the spatial derivative; its Taylor expansion leads to infinitely high derivatives. Time and space do not occur symmetrically.

Instead, we start from the squared relation:

$$E^2 = \mathbf{p}^2 c^2 + m^2 c^4 \tag{5.2.10}$$

and obtain

$$-\hbar^2 \frac{\partial^2}{\partial t^2}\psi = (-\hbar^2 c^2 \mathbf{\nabla}^2 + m^2 c^4)\psi . \tag{5.2.11}$$

This equation can be written in the even more compact and clearly Lorentz-covariant form

$$\left(\partial_\mu \partial^\mu + \left(\frac{mc}{\hbar}\right)^2\right)\psi = 0 . \tag{5.2.11'}$$

Here x^μ is the space–time position vector

$$x^\mu = (x^0 = ct, \mathbf{x})$$

and the covariant vector

$$\partial_\mu = \frac{\partial}{\partial x^\mu}$$

is the four-dimensional generalization of the gradient vector. As is known from electrodynamics, the d'Alembert operator $\Box \equiv \partial_\mu \partial^\mu$ is invariant under Lorentz transformations. Also appearing here is the Compton wavelength $\hbar/mc$ of a particle with mass m. Equation (5.2.11′) is known as the *Klein–Gordon* equation. It was originally introduced and studied by Schrödinger, and by Gordon and Klein.

We will now investigate the most important properties of the Klein–Gordon equation.

5.2.2 The Continuity Equation

To derive a continuity equation one takes ψ^* times (5.2.11′)

$$\psi^* \left(\partial_\mu \partial^\mu + \left(\frac{mc}{\hbar} \right)^2 \right) \psi = 0$$

and subtracts the complex conjugate of this equation

$$\psi \left(\partial_\mu \partial^\mu + \left(\frac{mc}{\hbar} \right)^2 \right) \psi^* = 0 \,.$$

This yields

$$\psi^* \partial_\mu \partial^\mu \psi - \psi \partial_\mu \partial^\mu \psi^* = 0$$
$$\partial_\mu (\psi^* \partial^\mu \psi - \psi \partial^\mu \psi^*) = 0 \,.$$

Multiplying by $\frac{\hbar}{2mi}$, so that the current density is equal to that in the non-relativistic case, one obtains

$$\frac{\partial}{\partial t} \left(\frac{i\hbar}{2mc^2} \left(\psi^* \frac{\partial \psi}{\partial t} - \psi \frac{\partial \psi^*}{\partial t} \right) \right) + \boldsymbol{\nabla} \cdot \frac{\hbar}{2mi} \left[\psi^* \boldsymbol{\nabla} \psi - \psi \boldsymbol{\nabla} \psi^* \right] = 0 \,.$$

$$(5.2.12)$$

This has the form of a *continuity equation*

$$\dot\rho + \operatorname{div} \mathbf{j} = 0 \,, \tag{5.2.12′}$$

with density

$$\rho = \frac{i\hbar}{2mc^2} \left(\psi^* \frac{\partial \psi}{\partial t} - \psi \frac{\partial \psi^*}{\partial t} \right) \tag{5.2.13a}$$

and current density

$$\mathbf{j} = \frac{\hbar}{2mi} \left(\psi^* \boldsymbol{\nabla} \psi - \psi \boldsymbol{\nabla} \psi^* \right) \,. \tag{5.2.13b}$$

Here, ρ is not positive definite and thus cannot be directly interpreted as a probability density, although $e\rho(\mathbf{x}, t)$ can possibly be conceived as the corresponding charge density. The Klein–Gordon equation is a second-order differential equation in t and thus the initial values of ψ and $\frac{\partial \psi}{\partial t}$ can be chosen independently, so that ρ as a function of $\mathbf{x}$ can be both positive and negative.

5.2.3 Free Solutions of the Klein–Gordon Equation

Equation (5.2.11) is known as the free Klein–Gordon equation in order to distinguish it from generalizations that additionally contain external potentials or electromagnetic fields (see Sect. 5.3.5). There are two free solutions in the form of plane waves:

$$\psi(\mathbf{x}, t) = \mathrm{e}^{\mathrm{i}(Et - \mathbf{p} \cdot \mathbf{x})/\hbar} \tag{5.2.14}$$

with

$$E = \pm\sqrt{\mathbf{p}^2 c^2 + m^2 c^4} \ .$$

Both positive and negative energies occur here and the energy is not bounded from below. This scalar theory does not contain spin and could only describe particles with zero spin.

Hence, the Klein–Gordon equation was rejected initially because the primary aim was a theory for the electron. Dirac[5] had instead introduced a first-order differential equation with positive density, as already mentioned at the beginning of this chapter. It will later emerge that this, too, has solutions with negative energies. The unoccupied states of negative energy describe antiparticles. As a quantized field theory, the Klein–Gordon equation describes mesons[7]. The hermitian scalar Klein–Gordon field describes neutral mesons with spin 0. The nonhermitian pseudoscalar Klein–Gordon field describes charged mesons with spin 0 and their antiparticles.

We shall therefore proceed by constructing a wave equation for spin-1/2 fermions and only return to the Klein–Gordon equation in connection with motion in a Coulomb potential (π^--mesons).

5.3 Dirac Equation

5.3.1 Derivation of the Dirac Equation

We will now attempt to find a wave equation of the form

$$\mathrm{i}\hbar\frac{\partial \psi}{\partial t} = \left(\frac{\hbar c}{\mathrm{i}}\alpha^k \partial_k + \beta mc^2\right)\psi \equiv H\psi \ . \tag{5.3.1}$$

Spatial components will be denoted by Latin indices, where repeated indices are to be summed over. The second derivative $\frac{\partial^2}{\partial t^2}$ in the Klein–Gordon equation leads to a density $\rho = \left(\psi^* \frac{\partial}{\partial t} \psi - \text{c.c.} \right)$. In order that the density be positive, we postulate a differential equation of first order. The requirement of relativistic covariance demands that the spatial derivatives may only be of first order, too. The Dirac Hamiltonian H is linear in the momentum operator and in the rest energy. The coefficients in (5.3.1) cannot simply be numbers: if they were, the equation would not even be form invariant (having the same coefficients) with respect to spatial rotations. α^k and β must be hermitian matrices in order for H to be hermitian, which is in turn necessary for a positive, conserved probability density to exist. Thus α^k and β are $N \times N$ matrices and

$$\psi = \begin{pmatrix} \psi_1 \\ \vdots \\ \psi_N \end{pmatrix} \quad \text{an } N\text{-component column vector .}$$

We shall impose the following requirements on equation (5.3.1):

(i) The components of ψ must satisfy the Klein–Gordon equation so that plane waves fulfil the relativistic energy–momentum relation $E^2 = p^2 c^2 + m^2 c^4$.

(ii) There exists a conserved four-current whose zeroth component is a positive density.

(iii) The equation must be Lorentz covariant. This means that it has the same form in all reference frames that are connencted by a Poincaré transformation.

The resulting equation (5.3.1) is named, after its discoverer, the *Dirac equation*. We must now look at the consequences that arise from the conditions (i)–(iii). Let us first consider condition (i). The two-fold application of H yields

$$-\hbar^2 \frac{\partial^2}{\partial t^2} \psi = -\hbar^2 c^2 \sum_{ij} \frac{1}{2} \left(\alpha^i \alpha^j + \alpha^j \alpha^i \right) \partial_i \partial_j \psi$$

$$+ \frac{\hbar m c^3}{i} \sum_{i=1}^{3} \left(\alpha^i \beta + \beta \alpha^i \right) \partial_i \psi + \beta^2 m^2 c^4 \psi . \tag{5.3.2}$$

Here, we have made use of $\partial_i \partial_j = \partial_j \partial_i$ to symmetrize the first term on the right-hand side. *Comparison with the Klein–Gordon equation* (5.2.11′) leads to the three conditions

$$\alpha^i \alpha^j + \alpha^j \alpha^i = 2\delta^{ij} \, \mathbb{1} , \tag{5.3.3a}$$

$$\alpha^i \beta + \beta \alpha^i = 0 , \tag{5.3.3b}$$

$$\alpha^{i\,2} = \beta^2 = \mathbb{1} . \tag{5.3.3c}$$

5.3.2 The Continuity Equation

The row vectors adjoint to ψ are defined by

$$\psi^\dagger = (\psi_1^*, \ldots, \psi_N^*) \ .$$

Multiplying the Dirac equation from the left by $\psi^\dagger$, we obtain

$$i\hbar\psi^\dagger \frac{\partial\psi}{\partial t} = \frac{\hbar c}{i}\psi^\dagger \alpha^i \partial_i \psi + mc^2\psi^\dagger \beta\psi \ . \tag{5.3.4a}$$

The complex conjugate relation reads:

$$-i\hbar \frac{\partial\psi^\dagger}{\partial t}\psi = -\frac{\hbar c}{i}\left(\partial_i\psi^\dagger\right)\alpha^{i\dagger}\psi + mc^2\psi^\dagger \beta^\dagger\psi \ . \tag{5.3.4b}$$

The difference of these two equations yields:

$$\frac{\partial}{\partial t}\left(\psi^\dagger\psi\right) = -c\left(\left(\partial_i\psi^\dagger\right)\alpha^{i\dagger}\psi + \psi^\dagger\alpha^i\partial_i\psi\right) + \frac{imc^2}{\hbar}\left(\psi^\dagger\beta^\dagger\psi - \psi^\dagger\beta\psi\right) \ . \tag{5.3.5}$$

In order for this to take the form of a continuity equation, the matrices α and β must be hermitian, i.e.,

$$\alpha^{i\dagger} = \alpha^i \ , \quad \beta^\dagger = \beta \ . \tag{5.3.6}$$

Then the *density*

$$\rho \equiv \psi^\dagger\psi = \sum_{\alpha=1}^{N} \psi_\alpha^* \psi_\alpha \tag{5.3.7a}$$

and the *current density*

$$j^k \equiv c\psi^\dagger\alpha^k\psi \tag{5.3.7b}$$

satisfy the *continuity equation*

$$\frac{\partial}{\partial t}\rho + \operatorname{div}\mathbf{j} = 0 \ . \tag{5.3.8}$$

With the zeroth component of j^μ,

$$j^0 \equiv c\rho \ , \tag{5.3.9}$$

we may define a four-current-density

$$j^\mu \equiv (j^0, j^k) \tag{5.3.9'}$$

and write the continuity equation in the form

$$\partial_\mu j^\mu = \frac{1}{c}\frac{\partial}{\partial t}j^0 + \frac{\partial}{\partial x^k}j^k = 0 \ . \tag{5.3.10}$$

The density defined in (5.3.7a) is positive definite and, within the framework of the single particle theory, can be given the preliminary interpretation of a probability density.

5.3.3 Properties of the Dirac Matrices

The matrices α^k, β anticommute and their square is equal to 1; see Eq. (5.3.3a–c). From $(\alpha^k)^2 = \beta^2 = \mathbb{1}$, it follows that the matrices α^k and β possess only the eigenvalues ± 1.

We may now write (5.3.3b) in the form

$$\alpha^k = -\beta \alpha^k \beta \ .$$

Using the cyclic invariance of the trace, we obtain

$$\mathrm{Tr}\,\alpha^k = -\mathrm{Tr}\,\beta \alpha^k \beta = -\mathrm{Tr}\,\alpha^k \beta^2 = -\mathrm{Tr}\,\alpha^k \ .$$

From this, and from an equivalent calculation for β, one obtains

$$\mathrm{Tr}\,\alpha^k = \mathrm{Tr}\,\beta = 0 \ . \tag{5.3.11}$$

Hence, the number of positive and negative eigenvalues must be equal and, therefore, N is even. $N = 2$ is not sufficient since the 2×2 matrices $\mathbb{1}$, σ_x, σ_y, σ_z contain only 3 mutually anticommuting matrices. $N = 4$ is the smallest dimension in which it is possible to realize the algebraic structure (5.3.3a–c).

A particular representation of the matrices is

$$\alpha^i = \begin{pmatrix} 0 & \sigma^i \\ \sigma^i & 0 \end{pmatrix} \ , \quad \beta = \begin{pmatrix} \mathbb{1} & 0 \\ 0 & -\mathbb{1} \end{pmatrix} \ , \tag{5.3.12}$$

where the 4×4 matrices are constructed from the Pauli matrices

$$\sigma^1 = \begin{pmatrix} 0 & 1 \\ 1 & 0 \end{pmatrix} \ , \quad \sigma^2 = \begin{pmatrix} 0 & -i \\ i & 0 \end{pmatrix} \ , \quad \sigma^3 = \begin{pmatrix} 1 & 0 \\ 0 & -1 \end{pmatrix} \tag{5.3.13}$$

and the two-dimensional unit matrix. It is easy to see that the matrices (5.3.12) satisfy the conditions (5.3.3a–c):

$$\text{e.g.,} \quad \alpha^i \beta + \beta \alpha^i = \begin{pmatrix} 0 & -\sigma^i \\ \sigma^i & 0 \end{pmatrix} + \begin{pmatrix} 0 & \sigma^i \\ -\sigma^i & 0 \end{pmatrix} = 0 \ .$$

The Dirac equation (5.3.1), in combination with the matrices (5.3.12), is referred to as the *"standard representation"* of the Dirac equation. One calls ψ a four-spinor or spinor for short (or sometimes a bispinor, in particular when ψ is represented by two two-component spinors). $\psi^\dagger$ is called the *hermitian adjoint* spinor. It will be shown in Sect. 6.2.1 that under Lorentz transformations spinors possess specific transformation properties.

5.3.4 The Dirac Equation in Covariant Form

In order to ensure that time and space derivatives are multiplied by matrices with similar algebraic properties, we multiply the Dirac equation (5.3.1) by β/c to obtain

$$-i\hbar\beta\partial_0\psi - i\hbar\beta\alpha^i\partial_i\psi + mc\psi = 0 \ . \tag{5.3.14}$$

We now define new Dirac matrices

$$\gamma^0 \equiv \beta$$
$$\gamma^i \equiv \beta\alpha^i \ . \tag{5.3.15}$$

These possess the following properties:

γ^0 is hermitian and $(\gamma^0)^2 = \mathbb{1}$. However, γ^k is antihermitian.

$(\gamma^k)^\dagger = -\gamma^k$ and $(\gamma^k)^2 = -\mathbb{1}$.

Proof:

$$\left(\gamma^k\right)^\dagger = \alpha^k\beta = -\beta\alpha^k = -\gamma^k \ ,$$

$$\left(\gamma^k\right)^2 = \beta\alpha^k\beta\alpha^k = -\mathbb{1} \ .$$

These relations, together with

$$\gamma^0\gamma^k + \gamma^k\gamma^0 = \beta\beta\alpha^k + \beta\alpha^k\beta = 0 \qquad \text{and}$$

$$\gamma^k\gamma^l + \gamma^l\gamma^k = \beta\alpha^k\beta\alpha^l + \beta\alpha^l\beta\alpha^k = 0 \quad \text{for } k \neq l$$

lead to the fundamental algebraic structure of the Dirac matrices

$$\gamma^\mu\gamma^\nu + \gamma^\nu\gamma^\mu = 2g^{\mu\nu}\mathbb{1} \ . \tag{5.3.16}$$

The *Dirac equation* (5.3.14) now assumes the form

$$\left(-i\gamma^\mu\partial_\mu + \frac{mc}{\hbar}\right)\psi = 0 \ . \tag{5.3.17}$$

It will be convenient to use the shorthand notation originally introduced by Feynman:

$$\slashed{v} \equiv \gamma \cdot v \equiv \gamma^\mu v_\mu = \gamma_\mu v^\mu = \gamma^0 v^0 - \boldsymbol{\gamma}\mathbf{v} \ . \tag{5.3.18}$$

Here, v^μ stands for any vector. The Feynman slash implies scalar multiplication by γ_μ. In the fourth term we have introduced the covariant components of the γ matrices

$$\gamma_\mu = g_{\mu\nu}\gamma^\nu \ . \tag{5.3.19}$$

In this notation the Dirac equation may be written in the compact form

$$\left(-i\slashed{\partial} + \frac{mc}{\hbar}\right)\psi = 0 \ . \tag{5.3.20}$$

Finally, we also give the γ matrices in the particular representation (5.3.12). From (5.3.12) and (5.3.15) it follows that

$$\gamma^0 = \begin{pmatrix} \mathbb{1} & 0 \\ 0 & -\mathbb{1} \end{pmatrix}, \quad \gamma^i = \begin{pmatrix} 0 & \sigma^i \\ -\sigma^i & 0 \end{pmatrix}. \tag{5.3.21}$$

Remark. A representation of the γ matrices that is equivalent to (5.3.21) and which also satisfies the algebraic relations (5.3.16) is obtained by replacing

$$\gamma \rightarrow M\gamma M^{-1},$$

where M is an arbitrary nonsingular matrix. Other frequently encountered representations are the Majorana representation and the chiral representation (see Sect. 11.3, Remark (ii) and Eq. (11.6.12a–c)).

5.3.5 Nonrelativistic Limit and Coupling to the Electromagnetic Field

5.3.5.1 Particles at Rest

The form (5.3.1) is a particularly suitable starting point when dealing with the nonrelativistic limit. We first consider a free particle *at rest*, i.e., with wave vector $\mathbf{k} = 0$. The spatial derivatives in the Dirac equation then vanish and the equation then simplifies to

$$i\hbar \frac{\partial \psi}{\partial t} = \beta mc^2 \psi. \tag{5.3.17'}$$

This equation possesses the following four solutions

$$\psi_1^{(+)} = e^{-\frac{imc^2}{\hbar}t} \begin{pmatrix} 1 \\ 0 \\ 0 \\ 0 \end{pmatrix}, \quad \psi_2^{(+)} = e^{-\frac{imc^2}{\hbar}t} \begin{pmatrix} 0 \\ 1 \\ 0 \\ 0 \end{pmatrix},$$

$$\tag{5.3.22}$$

$$\psi_1^{(-)} = e^{\frac{imc^2}{\hbar}t} \begin{pmatrix} 0 \\ 0 \\ 1 \\ 0 \end{pmatrix}, \quad \psi_2^{(-)} = e^{\frac{imc^2}{\hbar}t} \begin{pmatrix} 0 \\ 0 \\ 0 \\ 1 \end{pmatrix}.$$

The $\psi_1^{(+)}$, $\psi_2^{(+)}$ and $\psi_1^{(-)}$, $\psi_2^{(-)}$ correspond to positive- and negative-energy solutions, respectively. The interpretation of the negative-energy solutions must be postponed until later. For the moment we will confine ourselves to the positive-energy solutions.

5.3.5.2 Coupling to the Electromagnetic Field

We shall immediately proceed one step further and consider the coupling to an *electromagnetic field*, which will allow us to derive the Pauli equation.

In analogy with the nonrelativistic theory, the canonical momentum $\mathbf{p}$ is replaced by the kinetic momentum $(\mathbf{p} - \frac{e}{c}\mathbf{A})$, and the rest energy in the Dirac Hamiltonian is augmented by the scalar electrical potential $e\Phi$,

$$i\hbar \frac{\partial \psi}{\partial t} = \left(c\boldsymbol{\alpha} \cdot \left(\mathbf{p} - \frac{e}{c}\mathbf{A} \right) + \beta mc^2 + e\Phi \right) \psi . \tag{5.3.23}$$

Here, e is the charge of the particle, i.e., $e = -e_0$ for the electron. At the end of this section we will arrive at (5.3.23), starting from (5.3.17).

5.3.5.3 Nonrelativistic Limit. The Pauli Equation

In order to discuss the *nonrelativistic limit*, we use the explicit representation (5.3.12) of the Dirac matrices and decompose the four-spinors into two two-component column vectors $\tilde{\varphi}$ and $\tilde{\chi}$

$$\psi \equiv \begin{pmatrix} \tilde{\varphi} \\ \tilde{\chi} \end{pmatrix} , \tag{5.3.24}$$

with

$$i\hbar \frac{\partial}{\partial t} \begin{pmatrix} \tilde{\varphi} \\ \tilde{\chi} \end{pmatrix} = c \begin{pmatrix} \boldsymbol{\sigma} \cdot \boldsymbol{\pi} \, \tilde{\chi} \\ \boldsymbol{\sigma} \cdot \boldsymbol{\pi} \, \tilde{\varphi} \end{pmatrix} + e\Phi \begin{pmatrix} \tilde{\varphi} \\ \tilde{\chi} \end{pmatrix} + mc^2 \begin{pmatrix} \tilde{\varphi} \\ -\tilde{\chi} \end{pmatrix} , \tag{5.3.25}$$

where

$$\boldsymbol{\pi} = \mathbf{p} - \frac{e}{c}\mathbf{A} \tag{5.3.26}$$

is the operator of the kinetic momentum.

In the nonrelativistic limit, the rest energy mc^2 is the largest energy involved. Thus, to find solutions with positive energy, we write

$$\begin{pmatrix} \tilde{\varphi} \\ \tilde{\chi} \end{pmatrix} = e^{-\frac{imc^2}{\hbar}t} \begin{pmatrix} \varphi \\ \chi \end{pmatrix} , \tag{5.3.27}$$

where $\begin{pmatrix} \varphi \\ \chi \end{pmatrix}$ are considered to vary slowly with time and satisfy the equation

$$i\hbar \frac{\partial}{\partial t} \begin{pmatrix} \varphi \\ \chi \end{pmatrix} = c \begin{pmatrix} \boldsymbol{\sigma} \cdot \boldsymbol{\pi} \, \chi \\ \boldsymbol{\sigma} \cdot \boldsymbol{\pi} \, \varphi \end{pmatrix} + e\Phi \begin{pmatrix} \varphi \\ \chi \end{pmatrix} - 2mc^2 \begin{pmatrix} 0 \\ \chi \end{pmatrix} . \tag{5.3.25'}$$

In the second equation, $\hbar\dot{\chi}$ and $e\Phi\chi$ may be neglected in comparison to $2mc^2\chi$, and the latter then solved approximately as

$$\chi = \frac{\boldsymbol{\sigma} \cdot \boldsymbol{\pi}}{2mc} \varphi . \tag{5.3.28}$$

From this one sees that, in the nonrelativistic limit, χ is a factor of order $\sim v/c$ smaller than φ. One thus refers to φ as the large, and χ as the small, component of the spinor.

Inserting (5.3.28) into the first of the two equations (5.3.25') yields

$$i\hbar\frac{\partial\varphi}{\partial t} = \left(\frac{1}{2m}(\boldsymbol{\sigma}\cdot\boldsymbol{\pi})(\boldsymbol{\sigma}\cdot\boldsymbol{\pi}) + e\Phi\right)\varphi \ . \tag{5.3.29}$$

To proceed further we use the identity

$$\boldsymbol{\sigma}\cdot\mathbf{a}\,\boldsymbol{\sigma}\cdot\mathbf{b} = \mathbf{a}\cdot\mathbf{b} + i\boldsymbol{\sigma}\cdot(\mathbf{a}\times\mathbf{b}) \ ,$$

which follows from[10,11] $\sigma^i\sigma^j = \delta_{ij} + i\epsilon^{ijk}\sigma^k$, which in turn yields:

$$\boldsymbol{\sigma}\cdot\boldsymbol{\pi}\,\boldsymbol{\sigma}\cdot\boldsymbol{\pi} = \boldsymbol{\pi}^2 + i\boldsymbol{\sigma}\cdot\boldsymbol{\pi}\times\boldsymbol{\pi} = \boldsymbol{\pi}^2 - \frac{e\hbar}{c}\boldsymbol{\sigma}\cdot\mathbf{B} \ .$$

Here, we have used[12]

$$(\boldsymbol{\pi}\times\boldsymbol{\pi})^i\varphi = -i\hbar\left(\frac{-e}{c}\right)\varepsilon^{ijk}\left(\partial_j A^k - A^k\partial_j\right)\varphi$$

$$= i\frac{\hbar e}{c}\varepsilon^{ijk}\left(\partial_j A^k\right)\varphi = i\frac{\hbar e}{c}B^i\varphi$$

with $B^i = \varepsilon^{ijk}\partial_j A^k$. This rearrangement can also be very easily carried out by application of the expression

$$\boldsymbol{\nabla}\times\mathbf{A}\varphi + \mathbf{A}\times\boldsymbol{\nabla}\varphi = \boldsymbol{\nabla}\times\mathbf{A}\,\varphi - \boldsymbol{\nabla}\varphi\times\mathbf{A} = (\boldsymbol{\nabla}\times\mathbf{A})\,\varphi \ .$$

We thus finally obtain

$$i\hbar\frac{\partial\varphi}{\partial t} = \left[\frac{1}{2m}\left(\mathbf{p} - \frac{e}{c}\mathbf{A}\right)^2 - \frac{e\hbar}{2mc}\boldsymbol{\sigma}\cdot\mathbf{B} + e\Phi\right]\varphi \ . \tag{5.3.29'}$$

This result is identical to the *Pauli equation* for the Pauli spinor φ, as is known from nonrelativistic quantum mechanics[13]. The two components of φ describe the spin of the electron. In addition, one automatically obtains the correct gyromagnetic ratio $g = 2$ for the electron. In order to see this, we simply need to repeat the steps familiar to us from nonrelativistic wave mechanics. We assume a homogeneous magnetic field $\mathbf{B}$ that can be represented by the vector potential $\mathbf{A}$:

---

[10] Here, ε^{ijk} is the totally antisymmetric tensor of third rank

$$\varepsilon^{ijk} = \begin{cases} 1 \text{ for even permutations of (123)} \\ -1 \text{ for odd permutations of (123)} \\ 0 \text{ otherwise} \end{cases}$$

[11] QM I, Eq.(9.18a)
[12] Vectors such as $\mathbf{E}$, $\mathbf{B}$ and vector products that are only defined as three-vectors are always written in component form with upper indices; likewise the ε tensor. Here, too, we sum over repeated indices.
[13] See, e.g., QM I, Chap. 9.

$$\mathbf{B} = \text{curl}\,\mathbf{A}\,, \quad \mathbf{A} = \frac{1}{2}\mathbf{B} \times \mathbf{x}\,. \tag{5.3.30a}$$

Introducing the orbital angular momentum $\mathbf{L}$ and the spin $\mathbf{S}$ as

$$\mathbf{L} = \mathbf{x} \times \mathbf{p}\,, \quad \mathbf{S} = \frac{1}{2}\hbar\boldsymbol{\sigma}\,, \tag{5.3.30b}$$

then, for (5.3.30a), it follows[14,15] that

$$i\hbar\frac{\partial\varphi}{\partial t} = \left(\frac{\mathbf{p}^2}{2m} - \frac{e}{2mc}(\mathbf{L} + 2\mathbf{S}) \cdot \mathbf{B} + \frac{e^2}{2mc^2}\mathbf{A}^2 + e\varPhi\right)\varphi\,. \tag{5.3.31}$$

The eigenvalues of the projection of the spin operator $\mathbf{S}\hat{\mathbf{e}}$ onto an arbitrary unit vector $\hat{\mathbf{e}}$ are $\pm\hbar/2$. According to (5.3.31), the interaction with the electromagnetic field is of the form

$$H_{\text{int}} = -\boldsymbol{\mu} \cdot \mathbf{B} + \frac{e^2}{2mc^2}\mathbf{A}^2 + e\varPhi\,, \tag{5.3.32}$$

in which the magnetic moment

$$\boldsymbol{\mu} = \boldsymbol{\mu}_{\text{orbit}} + \boldsymbol{\mu}_{\text{spin}} = \frac{e}{2mc}(\mathbf{L} + 2\mathbf{S}) \tag{5.3.33}$$

is a combination of orbital and spin contributions. The spin moment is of magnitude

$$\boldsymbol{\mu}_{\text{spin}} = g\frac{e}{2mc}\mathbf{S}\,, \tag{5.3.34}$$

with the gyromagnetic ratio (or Landé factor)

$$g = 2\,. \tag{5.3.35}$$

For the electron, $\frac{e}{2mc} = -\frac{\mu_B}{\hbar}$ can be expressed in terms of the Bohr magneton $\mu_B = \frac{e_0\hbar}{2mc} = 0.927 \times 10^{-20}\text{erg/G}$.

We are now in a position to justify the approximations made in this section. The solution φ of (5.3.31) has a time behavior that is characterized by the Larmor frequency or, for $e\varPhi = \frac{-Ze_0^2}{r}$, by the Rydberg energy (Ry $\propto mc^2\alpha^2$, with the fine structure constant $\alpha = e_0^2/\hbar c$). For the hydrogen and other nonrelativistic atoms (small atomic numbers Z), mc^2 is very much larger than either of these two energies, thus justifying for such atoms the approximation introduced previously in the equation of motion for χ.

---

[14] See, e.g., QM I, Chap. 9.
[15] One finds $-\mathbf{p} \cdot \mathbf{A} - \mathbf{A} \cdot \mathbf{p} = -2\mathbf{A} \cdot \mathbf{p} = -2\frac{1}{2}(\mathbf{B} \times \mathbf{x}) \cdot \mathbf{p} = -(\mathbf{x} \times \mathbf{p}) \cdot \mathbf{B} = -\mathbf{L} \cdot \mathbf{B}$, since $(\mathbf{p} \cdot \mathbf{A}) = \frac{\hbar}{i}(\boldsymbol{\nabla} \cdot \mathbf{A}) = 0$.

5.3.5.4 Supplement Concerning Coupling to an Electromagnetic Field

We wish now to use a different approach to derive the Dirac equation in an external field and, to facilitate this, we begin with a few remarks on relativistic notation. The *momentum operator* in covariant and contravariant form reads:

$$p_\mu = i\hbar\partial_\mu \quad \text{and} \quad p^\mu = i\hbar\partial^\mu \ . \tag{5.3.36}$$

Here, $\partial_\mu = \frac{\partial}{\partial x^\mu}$ and $\partial^\mu = \frac{\partial}{\partial x_\mu}$. For the time and space components, this implies

$$p^0 = p_0 = i\hbar\frac{\partial}{\partial ct} \ , \quad p^1 = -p_1 = i\hbar\frac{\partial}{\partial x_1} = \frac{\hbar}{i}\frac{\partial}{\partial x^1} \ . \tag{5.3.37}$$

The coupling to the electromagnetic field is achieved by making the replacement

$$p_\mu \to p_\mu - \frac{e}{c}A_\mu \ , \tag{5.3.38}$$

where $A^\mu = (\varPhi, \mathbf{A})$ is the four-potential. The structure which arises here is well known from electrodynamics and, since its generalization to other gauge theories, is termed *minimal coupling*.

This implies

$$i\hbar\frac{\partial}{\partial x^\mu} \to i\hbar\frac{\partial}{\partial x^\mu} - \frac{e}{c}A_\mu \tag{5.3.39}$$

which explicitly written in components reads:

$$\begin{cases} i\hbar\dfrac{\partial}{\partial t} \to i\hbar\dfrac{\partial}{\partial t} - e\varPhi \\[2mm] \dfrac{\hbar}{i}\dfrac{\partial}{\partial x^i} \to \dfrac{\hbar}{i}\dfrac{\partial}{\partial x^i} + \dfrac{e}{c}A_i = \dfrac{\hbar}{i}\dfrac{\partial}{\partial x^i} - \dfrac{e}{c}A^i \ . \end{cases} \tag{5.3.39'}$$

For the spatial components this is identical to the replacement $\frac{\hbar}{i}\boldsymbol{\nabla} \to \frac{\hbar}{i}\boldsymbol{\nabla} - \frac{e}{c}\mathbf{A}$ or $\mathbf{p} \to \mathbf{p} - \frac{e}{c}\mathbf{A}$. In the noncovariant representation of the Dirac equation, the substitution (5.3.39') immediately leads once again to (5.3.23).

If one inserts (5.3.39) into the Dirac equation (5.3.17), one obtains

$$\left(-\gamma^\mu\left(i\hbar\partial_\mu - \frac{e}{c}A_\mu\right) + mc\right)\psi = 0 \ , \tag{5.3.40}$$

which is the Dirac equation in relativistic covariant form in the presence of an electromagnetic field.

Remarks:

(i) Equation (5.3.23) follows directly when one multiplies (5.3.40), i.e.

$$\gamma^0 \left(i\hbar\partial_0 - \frac{e}{c}A_0 \right)\psi = -\gamma^i \left(i\hbar\partial_i - \frac{e}{c}A_i \right)\psi + mc\psi$$

by γ^0:

$$i\hbar\partial_0\psi = \alpha^i \left(-i\hbar\partial_i - \frac{e}{c}A^i \right)\psi + \frac{e}{c}A_0\psi + mc\beta\psi$$

$$i\hbar\frac{\partial}{\partial t}\psi = c\boldsymbol{\alpha}\cdot\left(\mathbf{p} - \frac{e}{c}\mathbf{A} \right)\psi + e\Phi\psi + mc^2\beta\psi \ .$$

(ii) The minimal coupling, i.e., the replacement of derivatives by derivatives minus four-potentials, has as a consequence the invariance of the Dirac equation (5.3.40) with respect to gauge transformations (of the first kind):

$$\psi(x) \rightarrow e^{-i\frac{e}{\hbar c}\alpha(x)}\psi(x) \ , \qquad A_\mu(x) \rightarrow A_\mu(x) + \partial_\mu\alpha(x) \ .$$

(iii) For electrons, $m = m_e$, and the characteristic length in the Dirac equation equals the Compton wavelength of the electron

$$\lambda_c = \frac{\hbar}{m_e c} = 3.8 \times 10^{-11}\text{cm} \ .$$

Problems

5.1 Show that the matrices (5.3.12) obey the algebraic relations (5.3.3a–c).

5.2 Show that the representation (5.3.21) follows from (5.3.12).

5.3 Particles in a homogeneous magnetic field.
Determine the energy levels that result from the Dirac equation for a (relativistic) particle of mass m and charge e in a homogeneous magnetic field **B**. Use the gauge $A^0 = A^1 = A^3 = 0$, $A^2 = Bx$.

6. Lorentz Transformations and Covariance of the Dirac Equation

In this chapter, we shall investigate how the Lorentz covariance of the Dirac equation determines the transformation properties of spinors under Lorentz transformations. We begin by summarizing a few properties of Lorentz transformations, with which the reader is assumed to be familiar. The reader who is principally interested in the solution of specific problems may wish to omit the next sections and proceed directly to Sect. 6.3 and the subsequent chapters.

6.1 Lorentz Transformations

The contravariant and covariant components of the position vector read:

$$
\begin{array}{llllll}
x^\mu & : & x^0 = ct \,, & x^1 = x \,, & x^2 = y \,, & x^3 = z & \text{contravariant} \\
x_\mu & : & x_0 = ct \,, & x_1 = -x \,, & x_2 = -y \,, & x_3 = -z & \text{covariant}\,.
\end{array}
\tag{6.1.1}
$$

The metric tensor is defined by

$$
g = (g_{\mu\nu}) = (g^{\mu\nu}) = \begin{pmatrix} 1 & 0 & 0 & 0 \\ 0 & -1 & 0 & 0 \\ 0 & 0 & -1 & 0 \\ 0 & 0 & 0 & -1 \end{pmatrix}
\tag{6.1.2a}
$$

and relates covariant and contravariant components

$$
x_\mu = g_{\mu\nu} x^\nu \,, \qquad x^\mu = g^{\mu\nu} x_\nu \,.
\tag{6.1.3}
$$

Furthermore, we note that

$$
g^\mu{}_\nu = g^{\mu\sigma} g_{\sigma\nu} \equiv \delta^\mu{}_\nu \,,
\tag{6.1.2b}
$$

i.e.,

$$
(g^\mu{}_\nu) = (\delta^\mu{}_\nu) = \begin{pmatrix} 1 & 0 & 0 & 0 \\ 0 & 1 & 0 & 0 \\ 0 & 0 & 1 & 0 \\ 0 & 0 & 0 & 1 \end{pmatrix} \,.
$$

The d'Alembert operator is defined by

$$\Box = \frac{1}{c^2}\frac{\partial^2}{\partial t^2} - \sum_{i=1}^{3}\frac{\partial^2}{\partial x^{i\,2}} = \partial_\mu \partial^\mu = g_{\mu\nu}\partial^\mu \partial^\nu \ . \tag{6.1.4}$$

Inertial frames are frames of reference in which, in the absence of forces, particles move uniformly. Lorentz transformations tell us how the coordinates of two inertial frames transform into one another.

The coordinates of two reference systems in uniform motion must be related to one another by a linear transformation. Thus, the inhomogeneous Lorentz transformations (also known as Poincaré transformations) possess the form

$$x'^\mu = \Lambda^\mu{}_\nu x^\nu + a^\mu \ , \tag{6.1.5}$$

where $\Lambda^\mu{}_\nu$ and a^μ are real.

Remarks:

(i) *On the linearity of the Lorentz transformation*:
   Suppose that x' and x are the coordinates of an event in the inertial frames I' and I, respectively. For the transformation one could write

   $$x' = f(x) \ .$$

   In the absence of forces, particles in I and I' move uniformly, i.e., their world lines are straight lines (this is actually the definition of an inertial frame). Transformations under which straight lines are mapped onto straight lines are affinities, and thus of the form (6.1.5). The parametric representation of the equation of a straight line $x^\mu = e^\mu s + d^\mu$ is mapped by such an affine transformation onto another equation for a straight line.

(ii) *Principle of relativity*: The laws of nature are the same in all inertial frames. There is no such thing as an "absolute" frame of reference. The *requirement that the d'Alembert operator be invariant* (6.1.4) yields

   $$\Lambda^\lambda{}_\mu g^{\mu\nu} \Lambda^\rho{}_\nu = g^{\lambda\rho} \ , \tag{6.1.6a}$$

   or, in matrix form,

   $$\Lambda g \Lambda^T = g \ . \tag{6.1.6b}$$

   Proof: $\partial_\mu \equiv \dfrac{\partial}{\partial x^\mu} = \dfrac{\partial x'^\lambda}{\partial x^\mu}\dfrac{\partial}{\partial x'^\lambda} = \Lambda^\lambda{}_\mu \partial'_\lambda$

   $$\partial_\mu g^{\mu\nu}\partial_\nu = \Lambda^\lambda{}_\mu \partial'_\lambda g^{\mu\nu} \Lambda^\rho{}_\nu \partial'_\rho \overset{!}{=} \partial'_\lambda g^{\lambda\rho}\partial'_\rho$$
   $$\Rightarrow \Lambda^\lambda{}_\mu g^{\mu\nu} \Lambda^\rho{}_\nu = g^{\lambda\rho} \ .$$

The relations (6.1.6a,b) define the Lorentz transformations.

Definition: Poincaré group $\equiv$ {inhomogeneous Lorentz transformation, $a^\mu \neq 0$}

The group of homogeneous Lorentz transformations contains all elements with $a^\mu = 0$.

A homogeneous Lorentz transformation can be denoted by the shorthand form (Λ, a), e.g.,

translation group $(1, a)$

rotation group $(D, 0)$.

From the defining equation (6.1.6a,b) follow two important characteristics of Lorentz transformations:

(i) From the definition (6.1.6a), it follows that $(\det \Lambda)^2 = 1$, thus

$$\det \Lambda = \pm 1 .\tag{6.1.7}$$

(ii) Consider now the matrix element $\lambda = 0$, $\rho = 0$ of the defining equation (6.1.6a)

$$\Lambda^0{}_\mu g^{\mu\nu} \Lambda^0{}_\nu = 1 = (\Lambda^0{}_0)^2 - \sum_k (\Lambda^0{}_k)^2 = 1 .$$

This leads to

$$\Lambda^0{}_0 \geq 1 \quad \text{or} \quad \Lambda^0{}_0 \leq -1 .\tag{6.1.8}$$

The sign of the determinant of Λ and the sign of $\Lambda^0{}_0$ can be used to classify the elements of the Lorentz group (Table 6.1). The Lorentz transformations can be combined as follows into the Lorentz group $\mathcal{L}$, and its subgroups or subsets (e.g., $\mathcal{L}_+^\downarrow$ means the set of all elements $L_+^\downarrow$):

Table 6.1. Classification of the elements of the Lorentz group

		sgn $\Lambda^0{}_0$	det Λ
proper orthochronous	$L_+^\uparrow$	1	1
improper orthochronous*	$L_-^\uparrow$	1	-1
time-reflection type**	$L_-^\downarrow$	-1	-1
space–time inversion type***	$L_+^\downarrow$	-1	1

* spatial reflection   ** time reflection   *** space–time inversion

$$P = \begin{pmatrix} 1 & 0 & 0 & 0 \\ 0 & -1 & 0 & 0 \\ 0 & 0 & -1 & 0 \\ 0 & 0 & 0 & -1 \end{pmatrix} \quad T = \begin{pmatrix} -1 & 0 & 0 & 0 \\ 0 & 1 & 0 & 0 \\ 0 & 0 & 1 & 0 \\ 0 & 0 & 0 & 1 \end{pmatrix} \quad PT = \begin{pmatrix} -1 & 0 & 0 & 0 \\ 0 & -1 & 0 & 0 \\ 0 & 0 & -1 & 0 \\ 0 & 0 & 0 & -1 \end{pmatrix} \tag{6.1.9}$$

$\mathcal{L}$ Lorentz group (L.G.)

$\mathcal{L}_+^\uparrow$ restricted L.G. (is an invariant subgroup)

$\mathcal{L}^\uparrow = \mathcal{L}_+^\uparrow \cup \mathcal{L}_-^\uparrow$ orthochronous L.G.

$\mathcal{L}_+ = \mathcal{L}_+^\uparrow \cup \mathcal{L}_+^\downarrow$ proper L.G.

$\mathcal{L}_0 = \mathcal{L}_+^\uparrow \cup \mathcal{L}_-^\downarrow$ orthochronous L.G.

$\mathcal{L}_-^\uparrow = P \cdot \mathcal{L}_+^\uparrow$

$\mathcal{L}_-^\downarrow = T \cdot \mathcal{L}_+^\uparrow$

$\mathcal{L}_+^\downarrow = P \cdot T \cdot \mathcal{L}_+^\uparrow$

The last three subsets of $\mathcal{L}$ do not constitute subgroups.

$$\mathcal{L} = \mathcal{L}^\uparrow \cup T\mathcal{L}^\uparrow = \mathcal{L}_+^\uparrow \cup P\mathcal{L}_+^\uparrow \cup T\mathcal{L}_+^\uparrow \cup PT\mathcal{L}_+^\uparrow \tag{6.1.10}$$

$\mathcal{L}^\uparrow$ is an invariant subgroup of $\mathcal{L}$; $T\mathcal{L}^\uparrow$ is a coset to $\mathcal{L}^\uparrow$.
$\mathcal{L}_+^\uparrow$ is an invariant subgroup of $\mathcal{L}$; $P\mathcal{L}_+^\uparrow$, $T\mathcal{L}_+^\uparrow$, $PT\mathcal{L}_+^\uparrow$ are cosets of $\mathcal{L}$ with respect to $\mathcal{L}_+^\uparrow$. Furthermore, $\mathcal{L}^\uparrow$, $\mathcal{L}_+$, and $\mathcal{L}_0$ are invariant subgroups of $\mathcal{L}$ with the factor groups (E, P), (E, P, T, PT), and (E, T).

Every Lorentz transformation is either proper and orthochronous or can be written as the product of an element of the proper-orthochronous Lorentz group with one of the discrete transformations P, T, or PT.

$\mathcal{L}_+^\uparrow$, the *restricted Lorentz group* = the *proper orthochronous L.G.* consists of all elements with $\det \Lambda = 1$ and $\Lambda^0{}_0 \geq 1$; this includes:

(a) Rotations

(b) Pure Lorentz transformations (= transformations under which space and time are transformed). The prototype is a Lorentz transformation in the x^1 direction

$$L_1(\eta) = \begin{pmatrix} L^0{}_0 & L^0{}_1 & 0 & 0 \\ L^1{}_0 & L^1{}_1 & 0 & 0 \\ 0 & 0 & 1 & 0 \\ 0 & 0 & 0 & 1 \end{pmatrix} = \begin{pmatrix} \cosh\eta & -\sinh\eta & 0 & 0 \\ -\sinh\eta & \cosh\eta & 0 & 0 \\ 0 & 0 & 1 & 0 \\ 0 & 0 & 0 & 1 \end{pmatrix}$$

$$= \begin{pmatrix} \dfrac{1}{\sqrt{1-\beta^2}} & -\dfrac{\beta}{\sqrt{1-\beta^2}} & 0 & 0 \\ -\dfrac{\beta}{\sqrt{1-\beta^2}} & \dfrac{1}{\sqrt{1-\beta^2}} & 0 & 0 \\ 0 & 0 & 1 & 0 \\ 0 & 0 & 0 & 1 \end{pmatrix}, \tag{6.1.11}$$

with $\tanh\eta = \beta$. For this Lorentz transformation the inertial frame I' moves with respect to I with a velocity $v = c\beta$ in the x^1 direction.

6.2 Lorentz Covariance of the Dirac Equation

6.2.1 Lorentz Covariance and Transformation of Spinors

The *principle of relativity* states that the laws of nature are identical in every inertial reference frame.

We consider two·inertial frames I and I' with the space–time coordinates x and x'. Let the wave function of a particle in these two frames be ψ and ψ', respectively. We write the Poincaré transformation between I and I' as

$$x' = \Lambda x + a \ . \tag{6.2.1}$$

It must be possible to construct the wave function ψ' from ψ. This means that there must be a local relationship between ψ' and ψ:

$$\psi'(x') = F(\psi(x)) = F(\psi(\Lambda^{-1}(x' - a))) \ . \tag{6.2.2}$$

The principle of relativity together with the functional relation (6.2.2) necessarily leads to the requirement of *Lorentz covariance*: The Dirac equation in I is transformed by (6.2.1) and (6.2.2) into a Dirac equation in I'. (The Dirac equation is form invariant with respect to Poincaré transformations.) In order that both ψ and ψ' may satisfy the linear Dirac equation, their functional relationship must be linear, i.e.,

$$\psi'(x') = S(\Lambda)\psi(x) = S(\Lambda)\psi(\Lambda^{-1}(x' - a)) \ . \tag{6.2.3}$$

Here, $S(\Lambda)$ is a 4×4 matrix, with which the spinor ψ is to be multiplied. We will determine $S(\Lambda)$ below. In components, the transformation reads:

$$\psi'_\alpha(x') = \sum_{\beta=1}^{4} S_{\alpha\beta}(\Lambda)\psi_\beta(\Lambda^{-1}(x' - a)) \ . \tag{6.2.3'}$$

The Lorentz covariance of the Dirac equation requires that ψ' obey the equation

$$(-i\gamma^\mu \partial'_\mu + m)\psi'(x') = 0 \ , \qquad (c = 1, \ \hbar = 1) \tag{6.2.4}$$

where

$$\partial'_\mu = \frac{\partial}{\partial x'^\mu} \ .$$

The γ matrices are unchanged under the Lorentz transformation. In order to determine S, we need to convert the Dirac equation in the primed and unprimed coordinate systems into one another. The Dirac equation in the unprimed coordinate system

$$(-i\gamma^\mu \partial_\mu + m)\psi(x) = 0 \tag{6.2.5}$$

can, by means of the relation

$$\frac{\partial}{\partial x^\mu} = \frac{\partial x'^\nu}{\partial x^\mu}\frac{\partial}{\partial x'^\nu} = \Lambda^\nu{}_\mu \partial'_\nu$$

and

$$S^{-1}\psi'(x') = \psi(x) \,,$$

be brought into the form

$$(-\mathrm{i}\gamma^\mu \Lambda^\nu{}_\mu \partial'_\nu + m)S^{-1}(\Lambda)\psi'(x') = 0 \,. \tag{6.2.6}$$

After multiplying from the left by S, one obtains[1]

$$-\mathrm{i}S\Lambda^\nu{}_\mu \gamma^\mu S^{-1}\partial'_\nu \psi'(x') + m\psi'(x') = 0 \,. \tag{6.2.6'}$$

From a comparison of (6.2.6') with (6.2.4), it follows that the Dirac equation is form invariant under Lorentz transformations, provided $S(\Lambda)$ satisfies the following condition:

$$S(\Lambda)^{-1}\gamma^\nu S(\Lambda) = \Lambda^\nu{}_\mu \gamma^\mu \,. \tag{6.2.7}$$

It is possible to show (see next section) that this equation has nonsingular solutions for $S(\Lambda)$.[2] A wave function that transforms under a Lorentz transformation according to $\psi' = S\psi$ is known as a *four-component Lorentz spinor*.

6.2.2 Determination of the Representation $S(\Lambda)$

6.2.2.1 Infinitesimal Lorentz Transformations

We first consider *infinitesimal (proper, orthochronous) Lorentz transformations*

$$\Lambda^\nu{}_\mu = g^\nu{}_\mu + \Delta\omega^\nu{}_\mu \tag{6.2.8a}$$

with infinitesimal and antisymmetric $\Delta\omega^{\nu\mu}$

$$\Delta\omega^{\nu\mu} = -\Delta\omega^{\mu\nu} \,. \tag{6.2.8b}$$

This equation implies that $\Delta\omega^{\nu\mu}$ can have only 6 independent nonvanishing elements.

---

[1] We recall here that the $\Lambda^\nu{}_\mu$ are matrix elements that, of course, commute with the γ matrices.

[2] The existence of such an $S(\Lambda)$ follows from the fact that the matrices $\Lambda^\mu_\nu \gamma^\nu$ obey the same anticommutation rules (5.3.16) as the γ^μ by virtue of (6.1.6a), and from Pauli's fundamental theorem (property 7 on page 146). These transformations will be determined explicitly below.

These transformations satisfy the defining relation for Lorentz transformations

$$\Lambda^\lambda{}_\mu g^{\mu\nu} \Lambda^\rho{}_\nu = g^{\lambda\rho} , \qquad (6.1.6a)$$

as can be seen by inserting (6.2.8) into this equation:

$$g^\lambda{}_\mu g^{\mu\nu} g^\rho{}_\nu + \Delta\omega^{\lambda\rho} + \Delta\omega^{\rho\lambda} + O\left((\Delta\omega)^2\right) = g^{\lambda\rho} . \qquad (6.2.9)$$

Each of the 6 independent elements of $\Delta\omega^{\mu\nu}$ generates an infinitesimal Lorentz transformation. We consider some typical special cases:

$$\Delta\omega^0{}_1 = -\Delta\omega^{01} = -\Delta\beta : \begin{array}{l} \text{Transformation onto a coordinate} \\ \text{system moving with velocity } c\Delta\beta \\ \cdot \text{ in the } x \text{ direction} \end{array} \qquad (6.2.10)$$

$$\Delta\omega^1{}_2 = -\Delta\omega^{12} = \quad \Delta\varphi : \begin{array}{l} \text{Transformation onto a coordinate} \\ \text{system that is rotated by an angle} \\ \Delta\varphi \text{ about the } z \text{ axis. (See Fig. 6.1)} \end{array} \qquad (6.2.11)$$

The spatial components are transformed under this *passive* transformation as follows:

$$\begin{array}{l} x'^1 = x^1 + \Delta\varphi x^2 \\ x'^2 = -\Delta\varphi x^1 + x^2 \\ x'^3 = x^3 \end{array} \quad \text{or} \quad \mathbf{x}' = \mathbf{x} + \begin{pmatrix} 0 \\ 0 \\ -\Delta\varphi \end{pmatrix} \times \mathbf{x} = \mathbf{x} + \begin{vmatrix} \mathbf{e}_1 & \mathbf{e}_2 & \mathbf{e}_3 \\ 0 & 0 & -\Delta\varphi \\ x^1 & x^2 & x^3 \end{vmatrix}$$

$$\qquad (6.2.12)$$

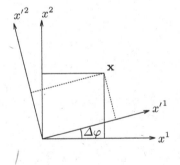

Fig. 6.1. Infinitesimal rotation, passive transformation

It must be possible to expand S as a power series in $\Delta\omega^{\nu\mu}$. We write

$$S = \mathbb{1} + \tau , \quad S^{-1} = \mathbb{1} - \tau , \qquad (6.2.13)$$

where τ is likewise infinitesimal i.e. of order $O(\Delta\omega^{\nu\mu})$. We insert (6.2.13) into the equation for S, namely $S^{-1}\gamma^\mu S = \Lambda^\mu{}_\nu \gamma^\nu$, and get

$$(\mathbb{1} - \tau)\gamma^\mu(\mathbb{1} + \tau) = \gamma^\mu + \gamma^\mu\tau - \tau\gamma^\mu = \gamma^\mu + \Delta\omega^\mu{}_\nu \gamma^\nu , \qquad (6.2.14)$$

from which the equation determining τ follows as

$$\gamma^\mu \tau - \tau \gamma^\mu = \Delta\omega^\mu{}_\nu \gamma^\nu \ . \tag{6.2.14'}$$

To within an additive multiple of $\mathbb{1}$, this unambiguously determines τ. Given two solutions of (6.2.14'), the difference between them has to commute with all γ^μ and thus is proportional to $\mathbb{1}$ (see Sect. 6.2.5, Property 6). The normalization condition $\det S = 1$ removes this ambiguity, since it implies to first order in $\Delta\omega^{\mu\nu}$ that

$$\det S = \det(\mathbb{1} + \tau) = \det \mathbb{1} + \mathrm{Tr}\,\tau = 1 + \mathrm{Tr}\,\tau = 1 \ . \tag{6.2.15}$$

It thus follows that

$$\mathrm{Tr}\,\tau = 0 \ . \tag{6.2.16}$$

Equations (6.2.14') and (6.2.16) have the solution

$$\tau = \frac{1}{8}\Delta\omega^{\mu\nu}(\gamma_\mu\gamma_\nu - \gamma_\nu\gamma_\mu) = -\frac{\mathrm{i}}{4}\Delta\omega^{\mu\nu}\sigma_{\mu\nu} \ , \tag{6.2.17}$$

where we have introduced the definition

$$\sigma_{\mu\nu} = \frac{\mathrm{i}}{2}[\gamma_\mu, \gamma_\nu] \ . \tag{6.2.18}$$

Equation (6.2.17) can be derived by calculating the commutator of τ with γ^μ; the vanishing of the trace is guaranteed by the general properties of the γ matrices (Property 3, Sect. 6.2.5).

6.2.2.2 Rotation About the z Axis

We first consider the rotation R_3 about the z axis as given by (6.2.11). According to (6.2.11) and (6.2.17),

$$\tau(R_3) = \frac{\mathrm{i}}{2}\Delta\varphi\sigma_{12} \ ,$$

and with

$$\sigma^{12} = \sigma_{12} = \frac{\mathrm{i}}{2}[\gamma_1, \gamma_2] = \mathrm{i}\gamma_1\gamma_2 = \mathrm{i}\begin{pmatrix} 0 & \sigma^1 \\ -\sigma^1 & 0 \end{pmatrix}\begin{pmatrix} 0 & \sigma^2 \\ -\sigma^2 & 0 \end{pmatrix} = \begin{pmatrix} \sigma^3 & 0 \\ 0 & \sigma^3 \end{pmatrix} \tag{6.2.19}$$

it follows that

$$S = 1 + \frac{\mathrm{i}}{2}\Delta\varphi\sigma^{12} = 1 + \frac{\mathrm{i}}{2}\Delta\varphi\begin{pmatrix} \sigma^3 & 0 \\ 0 & \sigma^3 \end{pmatrix} \ . \tag{6.2.20}$$

By a succession of infinitesimal rotations we can construct the transformation matrix S for a *finite rotation* through an angle ϑ. This is achieved by decomposing the finite rotation into a sequence of N steps ϑ/N

$$\psi'(x') = S\psi(x) = \lim_{N \to \infty} \left(1 + \frac{i}{2N}\vartheta\sigma^{12}\right)^N \psi(x)$$

$$= e^{\frac{i}{2}\vartheta\sigma^{12}}\psi$$

$$= \left(\cos\frac{\vartheta}{2} + i\sigma^{12}\sin\frac{\vartheta}{2}\right)\psi(x) . \tag{6.2.21}$$

For the coordinates and other four-vectors, this succession of transformations implies that

$$x' = \lim_{N \to \infty} \left(\mathbb{1} + \frac{\vartheta}{N}\begin{pmatrix} 0 & 0 & 0 & 0 \\ 0 & 0 & 1 & 0 \\ 0 & -1 & 0 & 0 \\ 0 & 0 & 0 & 0 \end{pmatrix}\right) \cdots \left(\mathbb{1} + \frac{\vartheta}{N}\begin{pmatrix} 0 & 0 & 0 & 0 \\ 0 & 0 & 1 & 0 \\ 0 & -1 & 0 & 0 \\ 0 & 0 & 0 & 0 \end{pmatrix}\right) x$$

$$= \exp\left\{\vartheta\begin{pmatrix} 0 & 0 & 0 & 0 \\ 0 & 0 & 1 & 0 \\ 0 & -1 & 0 & 0 \\ 0 & 0 & 0 & 0 \end{pmatrix}\right\} x = \begin{pmatrix} 1 & 0 & 0 & 0 \\ 0 & \cos\vartheta & \sin\vartheta & 0 \\ 0 & -\sin\vartheta & \cos\vartheta & 0 \\ 0 & 0 & 0 & 1 \end{pmatrix} x , \tag{6.2.22}$$

and is thus identical to the usual rotation matrix for rotation through an angle ϑ. The transformation S for rotations (6.2.21) is *unitary* ($S^{-1} = S^\dagger$). From (6.2.21), one sees that

$$S(2\pi) = -\mathbb{1} \tag{6.2.23a}$$
$$S(4\pi) = \mathbb{1} . \tag{6.2.23b}$$

This means that spinors do not regain their initial value after a rotation through 2π, but only after a rotation through 4π, a fact that is also confirmed by neutron scattering experiments[3]. We draw attention here to the analogy with the transformation of Pauli spinors with respect to rotations:

$$\varphi'(x') = e^{\frac{i}{2}\boldsymbol{\omega}\cdot\boldsymbol{\sigma}}\varphi(x) . \tag{6.2.24}$$

6.2.2.3 Lorentz Transformation Along the x^1 Direction

According to (6.2.10),

$$\Delta\omega^{01} = \Delta\beta \tag{6.2.25}$$

and (6.2.17) becomes

$$\tau(L_1) = \frac{1}{2}\Delta\beta\gamma_0\gamma_1 = \frac{1}{2}\Delta\beta\alpha_1 . \tag{6.2.26}$$

We may now determine S for a finite Lorentz transformation along the x^1 axis. For the velocity $\frac{v}{c}$, we have $\tanh\eta = \frac{v}{c}$.

---

[3] H. Rauch et al., Phys. Lett. **54A**, 425 (1975); S.A. Werner et al., Phys. Rev. Lett. **35**, 1053 (1975); also described in J.J. Sakurai, *Modern Quantum Mechanics*, p.162, Addison-Wesley, Red Wood City (1985).

The decomposition of η into N steps of $\frac{\eta}{N}$ leads to the following transformation of the coordinates and other four-vectors:

$$x'^{\mu} = \lim_{N\to\infty} \left(g + \frac{\eta}{N}I\right)^{\mu}{}_{\nu_1} \left(g + \frac{\eta}{N}I\right)^{\nu_1}{}_{\nu_2} \cdots \left(g + \frac{\eta}{N}I\right)^{\nu_{N-1}}{}_{\nu} x^{\nu}$$

$$g^{\mu}{}_{\nu} = \delta^{\mu}{}_{\nu},$$

$$I^{\nu}{}_{\mu} = \begin{pmatrix} 0 & -1 & 0 & 0 \\ -1 & 0 & 0 & 0 \\ 0 & 0 & 0 & 0 \\ 0 & 0 & 0 & 0 \end{pmatrix}, \quad I^2 = \begin{pmatrix} 1 & 0 & 0 & 0 \\ 0 & 1 & 0 & 0 \\ 0 & 0 & 0 & 0 \\ 0 & 0 & 0 & 0 \end{pmatrix}, \quad I^3 = I$$

$$x' = e^{\eta I} x = \left(1 + \eta I + \frac{1}{2!}\eta^2 I^2 + \frac{1}{3!}\eta^3 I + \frac{1}{4!}I^2 \cdots\right) x$$

$$x'^{\mu} = \left(1 - I^2 + I^2 \cosh\eta + I \sinh\eta\right)^{\mu}{}_{\nu} x^{\nu}$$

$$= \begin{pmatrix} \cosh\eta & -\sinh\eta & 0 & 0 \\ -\sinh\eta & \cosh\eta & 0 & 0 \\ 0 & 0 & 1 & 0 \\ 0 & 0 & 0 & 1 \end{pmatrix} \begin{pmatrix} x^0 \\ x^1 \\ x^2 \\ x^3 \end{pmatrix}. \tag{6.2.27}$$

The N-fold application of the infinitesimal Lorentz transformation

$$L_1\left(\frac{\eta}{N}\right) = \mathbb{1} + \frac{\eta}{N}I$$

then leads, in the limit of large N, to the Lorentz transformation (6.1.11)

$$L_1(\eta) = e^{\eta I} = \begin{pmatrix} \cosh\eta & -\sinh\eta & 0 & 0 \\ -\sinh\eta & \cosh\eta & 0 & 0 \\ 0 & 0 & 1 & 0 \\ 0 & 0 & 0 & 1 \end{pmatrix}. \tag{6.2.27'}$$

We note that the N infinitesimal steps of $\frac{\eta}{N}$ add up to η. However, this does not imply a simple addition of velocities.

We now calculate the corresponding spinor transformation

$$S(L_1) = \lim_{N\to\infty} \left(1 + \frac{1}{2}\frac{\eta}{N}\alpha_1\right)^N = e^{\frac{\eta}{2}\alpha_1}$$

$$= \mathbb{1}\cosh\frac{\eta}{2} + \alpha_1 \sinh\frac{\eta}{2}. \tag{6.2.28}$$

For homogenous restricted Lorentz transformations, S is *hermitian* ($S(L_1)^{\dagger} = S(L_1)$).

For *general infinitesimal transformations*, characterized by infinitesimal antisymmetric $\Delta\omega^{\mu\nu}$, equation (6.2.17) implies that

$$S(\Lambda) = \mathbb{1} - \frac{i}{4}\sigma_{\mu\nu}\Delta\omega^{\mu\nu}. \tag{6.2.29a}$$

This yields the finite transformation

$$S(\Lambda) = e^{-\frac{i}{4}\sigma_{\mu\nu}\omega^{\mu\nu}} \tag{6.2.29b}$$

with $\omega^{\mu\nu} = -\omega^{\nu\mu}$ and the Lorentz transformation reads $\Lambda = e^{\omega}$, where the matrix elements of ω are equal to $\omega^{\mu}_{\ \nu}$. For example, one can represent a rotation through an angle ϑ about an arbitrary axis $\hat{n}$ as

$$S = e^{\frac{i}{2}\vartheta\hat{n}\cdot\Sigma} , \tag{6.2.29c}$$

where

$$\Sigma = \begin{pmatrix} \sigma & 0 \\ 0 & \sigma \end{pmatrix} . \tag{6.2.29d}$$

6.2.2.4 Spatial Reflection, Parity

The Lorentz transformation corresponding to a spatial reflection is represented by

$$\Lambda^{\mu}_{\ \nu} = \begin{pmatrix} 1 & 0 & 0 & 0 \\ 0 & -1 & 0 & 0 \\ 0 & 0 & -1 & 0 \\ 0 & 0 & 0 & -1 \end{pmatrix} . \tag{6.2.30}$$

The associated S is determined, according to (6.2.7), from

$$S^{-1}\gamma^{\mu}S = \Lambda^{\mu}_{\ \nu}\gamma^{\nu} = \sum_{\nu=1}^{4} g^{\mu\nu}\gamma^{\nu} = g^{\mu\mu}\gamma^{\mu} , \tag{6.2.31}$$

where no summation over μ is implied. One immediately sees that the solution of (6.2.31), which we shall denote in this case by P, is given by

$$S = P \equiv e^{i\varphi}\gamma^{0} . \tag{6.2.32}$$

Here, $e^{i\varphi}$ is an unobservable phase factor. This is conventionally taken to have one of the four values $\pm 1, \pm i$; four reflections then yield the identity $\mathbb{1}$. The spinors transform under a spatial reflection according to

$$\psi'(x') \equiv \psi'(\mathbf{x}', t) = \psi'(-\mathbf{x}, t) = e^{i\varphi}\gamma^{0}\psi(x) = e^{i\varphi}\gamma^{0}\psi(-\mathbf{x}', t) . \tag{6.2.33}$$

The complete spatial reflection (parity) transformation for spinors is denoted by

$$\mathcal{P} = e^{i\varphi}\gamma^{0}\mathcal{P}^{(0)} , \tag{6.2.33'}$$

where $\mathcal{P}^{(0)}$ causes the spatial reflection $\mathbf{x} \to -\mathbf{x}$.

From the relationship $\gamma^{0} \equiv \beta = \begin{pmatrix} \mathbb{1} & 0 \\ 0 & -\mathbb{1} \end{pmatrix}$ one sees in the rest frame of the particle, spinors of positive and negative energy (Eq. (5.3.22)) that are eigenstates of P – with opposite eigenvalues, i.e., opposite parity. *This means that the intrinsic parities of particles and antiparticles are opposite.*

6.2.3 Further Properties of S

For the calculation of the transformation of bilinear forms such as $j^\mu(x)$, we need to establish a relationship between the adjoint transformations $S^\dagger$ and S^{-1}.

Assertion:

$$S^\dagger \gamma^0 = b \gamma^0 S^{-1} \quad , \tag{6.2.34a}$$

where

$$b = \pm 1 \quad \text{for} \quad \Lambda^{00} \begin{cases} \geq +1 \\ \leq -1 \end{cases} . \tag{6.2.34b}$$

Proof: We take as our starting point Eq. (6.2.7)

$$S^{-1}\gamma^\mu S = \Lambda^\mu{}_\nu \gamma^\nu , \qquad \Lambda^\mu{}_\nu \text{ real}, \tag{6.2.35}$$

and write down the adjoint relation

$$(\Lambda^\mu{}_\nu \gamma^\nu)^\dagger = S^\dagger \gamma^{\mu\dagger} S^{\dagger-1} . \tag{6.2.36}$$

The hermitian adjoint matrix can be expressed most concisely as

$$\gamma^{\mu\dagger} = \gamma^0 \gamma^\mu \gamma^0 . \tag{6.2.37}$$

By means of the anticommutation relations, one easily checks that (6.2.37) is in accord with $\gamma^{0\dagger} = \gamma^0$, $\gamma^{k\dagger} = -\gamma^k$. We insert this into the left- and the right-hand sides of (6.2.36) and then multiply by γ^0 from the left- and right-hand side to gain

$$\gamma^0 \Lambda^\mu{}_\nu \gamma^0 \gamma^\nu \gamma^0 \gamma^0 \gamma^0 = \gamma^0 S^\dagger \gamma^0 \gamma^\mu \gamma^0 S^{\dagger-1} \gamma^0$$

$$\Lambda^\mu{}_\nu \gamma^\nu = S^{-1} \gamma^\mu S = \gamma^0 S^\dagger \gamma^0 \gamma^\mu (\gamma^0 S^\dagger \gamma^0)^{-1} ,$$

since $(\gamma^0)^{-1} = \gamma^0$. Furthermore, on the left-hand side we have made the substitution $\Lambda^\mu{}_\nu \gamma^\nu = S^{-1}\gamma^\mu S$. We now multiply by S and S^{-1}:

$$\gamma^\mu = S\gamma^0 S^\dagger \gamma^0 \gamma^\mu (\gamma^0 S^\dagger \gamma^0)^{-1} S^{-1} \equiv (S\gamma^0 S^\dagger \gamma^0)\gamma^\mu (S\gamma^0 S^\dagger \gamma^0)^{-1} .$$

Thus, $S\gamma^0 S^\dagger \gamma^0$ commutes with all γ^μ and is therefore a multiple of the unit matrix

$$S\gamma^0 S^\dagger \gamma^0 = b\,\mathbb{1} \quad , \tag{6.2.38}$$

which also implies that

$$S\gamma^0 S^\dagger = b\gamma^0 \tag{6.2.39}$$

and yields the relation we are seeking[4]

$$S^\dagger \gamma^0 = b(S\gamma^0)^{-1} = b\gamma^0 S^{-1} . \qquad (6.2.34a)$$

Since $(\gamma^0)^\dagger = \gamma^0$ and $S\gamma^0 S^\dagger$ are hermitian, by taking the adjoint of (6.2.39) one obtains $S\gamma^0 S^\dagger = b^*\gamma^0$, from which it follows that

$$b^* = b \qquad (6.2.40)$$

and thus b is real. Making use of the fact that the normalization of S is fixed by $\det S = 1$, on calculating the determinant of (6.2.39), one obtains $b^4 = 1$. This, together with (6.2.40), yields:

$$b = \pm 1 . \qquad (6.2.41)$$

The significance of the sign in (6.2.41) becomes apparent when one considers

$$S^\dagger S = S^\dagger \gamma^0 \gamma^0 S = b\gamma^0 S^{-1} \gamma^0 S = b\gamma^0 \Lambda^0{}_\nu \gamma^\nu$$

$$= b\Lambda^0{}_0 \mathbb{1} + \sum_{k=1}^{3} b\Lambda^0{}_k \underbrace{\gamma^0 \gamma^k}_{\alpha^k} . \qquad (6.2.42)$$

$S^\dagger S$ has positive definite eigenvalues, as can be seen from the following. Firstly, $\det S^\dagger S = 1$ is equal to the product of all the eigenvalues, and these must therefore all be nonzero. Furthermore, $S^\dagger S$ is hermitian and its eigenfunctions satisfy $S^\dagger S\psi_a = a\psi_a$, whence

$$a\psi_a^\dagger \psi_a = \psi_a^\dagger S^\dagger S\psi_a = (S\psi_a)^\dagger S\psi_a > 0$$

and thus $a > 0$. Since the trace of $S^\dagger S$ is equal to the sum of all the eigenvalues, we have, in view of (6.2.42) and using $\operatorname{Tr}\alpha^k = 0$,

$$0 < \operatorname{Tr}(S^\dagger S) = 4b\Lambda^0{}_0 .$$

Thus, $b\Lambda^0{}_0 > 0$. Hence, we have the following relationship between the signs of Λ^{00} and b:

$$\begin{aligned} \Lambda^{00} &\geq 1 \quad \text{for} \quad b = 1 \\ \Lambda^{00} &\leq -1 \quad \text{for} \quad b = -1 . \end{aligned} \qquad (6.2.34b)$$

For Lorentz transformations that do not change the direction of time, we have $b = 1$; while those that do cause time reversal have $b = -1$.

---

[4] Note: For the Lorentz transformation $L_+^\uparrow$ (restricted L.T. and rotations) and for spatial reflections, one can derive this relation with $b = 1$ from the explicit representations.

6.2.4 Transformation of Bilinear Forms

The *adjoint* spinor is defined by

$$\bar{\psi} = \psi^\dagger \gamma^0 \ . \tag{6.2.43}$$

We recall that $\psi^\dagger$ is referred to as a hermitian adjoint spinor. The additional introduction of $\bar{\psi}$ is useful because it allows quantities such as the current density to be written in a concise form. We obtain the following transformation behavior under a Lorentz transformation:

$$\psi' = S\psi \implies \psi'^\dagger = \psi^\dagger S^\dagger_{,} \implies \bar{\psi}' = \psi^\dagger S^\dagger \gamma^0 = b\,\psi^\dagger \gamma^0 S^{-1} \ ,$$

thus,

$$\bar{\psi}' = b\,\bar{\psi}S^{-1} \ . \tag{6.2.44}$$

Given the above definition, the current density (5.3.7) reads:

$$j^\mu = c\,\psi^\dagger \gamma^0 \gamma^\mu \psi = c\,\bar{\psi}\gamma^\mu \psi \tag{6.2.45}$$

and thus transforms as

$$j^{\mu'} = cb\,\bar{\psi}S^{-1}\gamma^\mu S\psi = \Lambda^\mu{}_\nu cb\,\bar{\psi}\gamma^\nu \psi = b\Lambda^\mu{}_\nu j^\nu \ . \tag{6.2.46}$$

Hence, j^μ transforms in the same way as a vector for Lorentz transformations without time reflection. In the same way one immediately sees, using (6.2.3) and (6.2.44), that $\bar{\psi}(x)\psi(x)$ transforms as a scalar:

$$\begin{aligned}
\bar{\psi}'(x')\psi'(x') &= b\bar{\psi}(x')S^{-1}S\psi(x') \\
&= b\,\bar{\psi}(x)\psi(x) \ .
\end{aligned} \tag{6.2.47a}$$

We now summarize the transformation behavior of the most important bilinear quantities under *orthochronous Lorentz transformations* , i.e., transformations that *do not reverse the direction of time*:

$$\bar{\psi}'(x')\psi'(x') = \bar{\psi}(x)\psi(x) \qquad\qquad\quad \text{scalar} \qquad\quad \tag{6.2.47a}$$

$$\bar{\psi}'(x')\gamma^\mu\psi'(x') = \Lambda^\mu{}_\nu\bar{\psi}(x)\gamma^\nu\psi(x) \qquad\quad \text{vector} \qquad\quad \tag{6.2.47b}$$

$$\bar{\psi}'(x')\sigma^{\mu\nu}\psi'(x') = \Lambda^\mu{}_\rho\Lambda^\nu{}_\sigma\bar{\psi}(x)\sigma^{\rho\sigma}\psi(x) \quad \text{antisymmetric tensor}$$
$$\tag{6.2.47c}$$

$$\bar{\psi}'(x')\gamma_5\gamma^\mu\psi'(x') = (\det\Lambda)\Lambda^\mu{}_\nu\bar{\psi}(x)\gamma_5\gamma^\nu\psi(x) \quad \text{pseudovector} \quad \tag{6.2.47d}$$

$$\bar{\psi}'(x')\gamma_5\psi'(x') = (\det\Lambda)\bar{\psi}(x)\gamma_5\psi(x) \qquad\qquad \text{pseudoscalar,} \quad \tag{6.2.47e}$$

where $\gamma_5 = \mathrm{i}\gamma^0\gamma^1\gamma^2\gamma^3$. We recall that $\det\Lambda = \pm 1$; for spatial reflections the sign is -1.

6.2.5 Properties of the γ Matrices

We remind the reader of the definition of γ^5 from the previous section:

$$\gamma_5 \equiv \gamma^5 \equiv i\gamma^0\gamma^1\gamma^2\gamma^3 \tag{6.2.48}$$

and draw the reader's attention to the fact that somewhat different definitions may also be encountered in the literature. In the standard representation (5.3.21) of the Dirac matrices, γ^5 has the form

$$\gamma^5 = \begin{pmatrix} 0 & \mathbb{1} \\ \mathbb{1} & 0 \end{pmatrix}. \tag{6.2.48'}$$

The matrix γ^5 satisfies the relations

$$\{\gamma^5, \gamma^\mu\} = 0 \tag{6.2.49a}$$

and

$$(\gamma^5)^2 = \mathbb{1}. \tag{6.2.49b}$$

By forming products of γ^μ, one can construct 16 linearly independent 4×4 matrices. These are

$$\Gamma^S = \mathbb{1} \tag{6.2.50a}$$
$$\Gamma^V_\mu = \gamma_\mu \tag{6.2.50b}$$
$$\Gamma^T_{\mu\nu} = \sigma_{\mu\nu} = \frac{i}{2}[\gamma_\mu, \gamma_\nu] \tag{6.2.50c}$$
$$\Gamma^A_\mu = \gamma_5\gamma_\mu \tag{6.2.50d}$$
$$\Gamma^P = \gamma_5. \tag{6.2.50e}$$

The upper indices indicate scalar, vector, tensor, axial vector (= pseudovector), and pseudoscalar. These matrices have the following *properties*[5]:

1. $(\Gamma^a)^2 = \pm\mathbb{1}$ (6.2.51a)

2. For every Γ^a except $\Gamma^S \equiv \mathbb{1}$, there exists a Γ^b, such that

$$\Gamma^a\Gamma^b = -\Gamma^b\Gamma^a. \tag{6.2.51b}$$

3. For $a \neq S$ we have $\text{Tr}\,\Gamma^a = 0.$ (6.2.51c)
   Proof: $\text{Tr}\,\Gamma^a(\Gamma^b)^2 = -\text{Tr}\,\Gamma^b\Gamma^a\Gamma^b = -\text{Tr}\,\Gamma^a(\Gamma^b)^2$
            Since $(\Gamma^b)^2 = \pm 1$, it follows that $\text{Tr}\,\Gamma^a = -\text{Tr}\,\Gamma^a$, thus proving the assertion.

---

[5] Only some of these properties will be proved here; other proofs are included as problems.

4. For every pair Γ^a, Γ^b $a \neq b$ there is a $\Gamma^c \neq \mathbb{1}$, such that $\Gamma_a\Gamma_b = \beta\Gamma_c$, $\beta = \pm 1, \pm i$.
   Proof follows by considering the Γ.
5. The matrices Γ^a are linearly independent.
   Suppose that $\sum_a x_a\Gamma^a = 0$ with complex coefficients x_a. From property 3 above one then has

   $$\text{Tr} \sum_a x_a\Gamma^a = x_S = 0 .$$

   Multiplication by Γ_a and use of the properties 1 and 4 shows that subsequent formation of the trace leads to $x_a = 0$.
6. If a 4×4 matrix X commutes with every γ^μ, then $X \propto \mathbb{1}$.
7. Given two sets of γ matrices, γ and γ', both of which satisfy

   $$\{\gamma^\mu, \gamma^\nu\} = 2g^{\mu\nu} ,$$

   there must exist a nonsingular M

   $$\gamma'^\mu = M\gamma^\mu M^{-1} . \tag{6.2.51d}$$

   This M is unique to within a constant factor (Pauli's fundamental theorem).

6.3 Solutions of the Dirac Equation for Free Particles

6.3.1 Spinors with Finite Momentum

We now seek solutions of the free Dirac equation (5.3.1) or (5.3.17)

$$(-i\partial\!\!\!/ + m)\psi(x) = 0 . \tag{6.3.1}$$

Here, and below, we will set $\hbar = c = 1$.
   For particles at rest, these solutions [see (5.3.22)] read:

$$
\begin{aligned}
\psi^{(+)}(x) &= u_r(m, \mathbf{0})\, e^{-imt} & r &= 1, 2 \\
\psi^{(-)}(x) &= v_r(m, \mathbf{0})\, e^{imt} ,
\end{aligned}
\tag{6.3.2}
$$

for the positive and negative energy solutions respectively, with

$$
u_1(m, \mathbf{0}) = \begin{pmatrix} 1 \\ 0 \\ 0 \\ 0 \end{pmatrix} , \quad
u_2(m, \mathbf{0}) = \begin{pmatrix} 0 \\ 1 \\ 0 \\ 0 \end{pmatrix} ,
$$

$$
v_1(m, \mathbf{0}) = \begin{pmatrix} 0 \\ 0 \\ 1 \\ 0 \end{pmatrix} , \quad
v_2(m, \mathbf{0}) = \begin{pmatrix} 0 \\ 0 \\ 0 \\ 1 \end{pmatrix} ,
\tag{6.3.3}
$$

and are normalized to unity. These solutions of the Dirac equation are eigenfunctions of the Dirac Hamiltonian H with eigenvalues $\pm m$, and also of the operator (the matrix already introduced in (6.2.19))

$$\sigma^{12} = \frac{i}{2}[\gamma^1, \gamma^2] = \begin{pmatrix} \sigma^3 & 0 \\ 0 & \sigma^3 \end{pmatrix} \tag{6.3.4}$$

with eigenvalues $+1$ (for $r = 1$) and -1 (for $r = 2$). Later we will show that σ^{12} is related to the spin.

We now seek solutions of the Dirac equation for finite momentum in the form[6]

$$\psi^{(+)}(x) = u_r(k)\,e^{-ik\cdot x} \qquad \text{positive energy} \tag{6.3.5a}$$
$$\psi^{(-)}(x) = v_r(k)\,e^{ik\cdot x} \qquad \text{negative energy} \tag{6.3.5b}$$

with $k^0 > 0$. Since (6.3.5a,b) must also satisfy the Klein–Gordon equation, we know from (5.2.14) that

$$k_\mu k^\mu = m^2 \,, \tag{6.3.6}$$

or

$$E \equiv k^0 = \left(\mathbf{k}^2 + m^2\right)^{1/2} \,, \tag{6.3.7}$$

where k^0 is also written as E; i.e., k is the four-momentum of a particle with mass m.

The spinors $u_r(k)$ and $v_r(k)$ can be found by Lorentz transformation of the spinors (6.3.3) for particles at rest: We transform into a coordinate system that is moving with velocity $-\mathbf{v}$ with respect to the rest frame and then, from the rest-state solutions, we obtain the free wave functions for electrons with velocity $\mathbf{v}$. However, a more straightforward approach is to determine the solutions directly from the Dirac equation. Inserting (6.3.5a,b) into the Dirac equation (6.3.1) yields:

$$(\not{k} - m)u_r(k) = 0 \quad \text{and} \quad (\not{k} + m)v_r(k) = 0 \,. \tag{6.3.8}$$

Furthermore, we have

$$\not{k}\not{k} = k_\mu \gamma^\mu k_\nu \gamma^\nu = k_\mu k_\nu \frac{1}{2}\{\gamma^\mu, \gamma^\nu\} = k_\mu k_\nu g^{\mu\nu} \,. \tag{6.3.9}$$

Thus, from (6.3.6), one obtains

$$(\not{k} - m)(\not{k} + m) = k^2 - m^2 = 0 \,. \tag{6.3.10}$$

Hence one simply needs to apply $(\not{k} + m)$ to the $u_r(m, \mathbf{0})$ and $(\not{k} - m)$ to the $v_r(m, \mathbf{0})$ in order to obtain the solutions $u_r(k)$ and $v_r(k)$ of (6.3.8). The

---

[6] We write the four-momentum as k, the four-coordinates as x, and their scalar product as $k \cdot x$.

normalization remains as yet unspecified; it must be chosen such that it is compatible with the solution (6.3.3), and such that $\bar{\psi}\psi$ transforms as a scalar (Eq. (6.2.47a)). As we will see below, this is achieved by means of the factor $1/\sqrt{2m(m+E)}$:

$$u_r(k) = \frac{\not{k}+m}{\sqrt{2m(m+E)}} u_r(m,\mathbf{0}) = \begin{pmatrix} \left(\dfrac{E+m}{2m}\right)^{1/2} \chi_r \\[2mm] \dfrac{\boldsymbol{\sigma}\cdot\mathbf{k}}{(2m(m+E))^{1/2}} \chi_r \end{pmatrix} \qquad (6.3.11a)$$

$$v_r(k) = \frac{-\not{k}+m}{\sqrt{2m(m+E)}} v_r(m,\mathbf{0}) = \begin{pmatrix} \dfrac{\boldsymbol{\sigma}\cdot\mathbf{k}}{(2m(m+E))^{1/2}} \chi_r \\[2mm] \left(\dfrac{E+m}{2m}\right)^{1/2} \chi_r \end{pmatrix}. \qquad (6.3.11b)$$

Here, the solutions are represented by $u_r(m,\mathbf{0}) = \binom{\chi_r}{0}$ and $v_r(m,\mathbf{0}) = \binom{0}{\chi_r}$ with $\chi_1 = \binom{1}{0}$ and $\chi_2 = \binom{0}{1}$.

In this calclculation we have made use of

$$\not{k}\begin{pmatrix}\chi_r\\0\end{pmatrix} = \left[k^0\begin{pmatrix}\mathbb{1}&0\\0&-\mathbb{1}\end{pmatrix} - k^i\begin{pmatrix}0&\sigma^i\\-\sigma^i&0\end{pmatrix}\right]\begin{pmatrix}\chi_r\\0\end{pmatrix}$$

$$= \begin{pmatrix}k^0\chi_r\\0\end{pmatrix} + \begin{pmatrix}0\\k^i\sigma^i\chi_r\end{pmatrix} = \begin{pmatrix}E\chi_r\\\mathbf{k}\cdot\boldsymbol{\sigma}\chi_r\end{pmatrix}$$

and

$$-\not{k}\begin{pmatrix}0\\\chi_r\end{pmatrix} = \begin{pmatrix}0\\k^0\chi_r\end{pmatrix} + \begin{pmatrix}k^i\sigma^i\chi_r\\0\end{pmatrix}, \quad r = 1,2 .$$

From (6.3.11a,b) one finds for the adjoint spinors defined in (6.2.43)

$$\bar{u}_r(k) = \bar{u}_r(m,\mathbf{0}) \frac{\not{k}+m}{\sqrt{2m(m+E)}} \qquad (6.3.12a)$$

$$\bar{v}_r(k) = \bar{v}_r(m,\mathbf{0}) \frac{-\not{k}+m}{\sqrt{2m(m+E)}} . \qquad (6.3.12b)$$

Proof: $\bar{u}_r(k) = u_r^\dagger(k)\gamma^0 = u_r^\dagger(m,\mathbf{0})\dfrac{(\gamma^{\mu\dagger}k_\mu+m)\gamma^0}{\sqrt{2m(m+E)}} = u_r^\dagger(m,\mathbf{0})\dfrac{\gamma^0(\gamma^\mu k_\mu+m)}{\sqrt{2m(m+E)}}$,

since $\gamma^{\mu\dagger} = \gamma^0\gamma^\mu\gamma^0$ and $(\gamma^0)^2 = \mathbb{1}$

Furthermore, the adjoint spinors satisfy the equations

$$\bar{u}_r(k)(\not{k}-m) = 0 \qquad (6.3.13a)$$

and

$$\bar{v}_r(k)\,(\not{k} + m) = 0 \, , \tag{6.3.13b}$$

as can be seen from (6.3.10) and (6.3.12a,b) or (6.3.8).

6.3.2 Orthogonality Relations and Density

We shall need to know a number of formal properties of the solutions found above for later use. From (6.3.11) and (6.2.37) it follows that:

$$\bar{u}_r(k)u_s(k) = \bar{u}_r(m,\mathbf{0})\frac{(\not{k} + m)^2}{2m(m + E)}u_s(m,\mathbf{0}) \, . \tag{6.3.14a}$$

With

$$\begin{aligned}
\bar{u}_r(m,\mathbf{0})(\not{k} + m)^2 u_s(m,\mathbf{0}) &= \bar{u}_r(m,\mathbf{0})(\not{k}^2 + 2m\not{k} + m^2)u_s(m,\mathbf{0}) \\
&= \bar{u}_r(m,\mathbf{0})(2m^2 + 2m\not{k})u_s(m,\mathbf{0}) \\
&= \bar{u}_r(m,\mathbf{0})(2m^2 + 2mk^0\gamma^0)u_s(m,\mathbf{0}) \\
&= 2m(m + E)\bar{u}_r(m,\mathbf{0})u_s(m,\mathbf{0}) \\
&= 2m(m + E)\delta_{rs} \, ,
\end{aligned} \tag{6.3.14b}$$

$$\begin{aligned}
\bar{u}_r(k)v_s(k) &= \bar{u}_r(m,\mathbf{0})\frac{\not{k}^2 - m^2}{2m(m + E)}v_s(m,\mathbf{0}) \\
&= \bar{u}_r(m,\mathbf{0})\,0\,v_s(m,\mathbf{0}) = 0
\end{aligned} \tag{6.3.14c}$$

and a similar calculation for $v_r(k)$, equations (6.3.14a,b) yield the

orthogonality relations

$$\begin{array}{llll}
\bar{u}_r(k)\,u_s(k) &= \delta_{rs} & \bar{u}_r(k)\,v_s(k) &= 0 \\
\bar{v}_r(k)\,v_s(k) &= -\delta_{rs} & \bar{v}_r(k)\,u_s(k) &= 0.
\end{array} \tag{6.3.15}$$

Remarks:

(i) This normalization remains invariant under orthochronous Lorentz transformations:

$$\bar{u}'_r\,u'_s = u_r^\dagger\,S^\dagger\,\gamma^0\,S\,u_s = u_r^\dagger\,\gamma^0\,S^{-1}\,S\,u_s = \bar{u}_r\,u_s = \delta_{rs} \, . \tag{6.3.16}$$

(ii) For these spinors, $\bar{\psi}(x)\psi(x)$ is a scalar,

$$\bar{\psi}^{(+)}(x)\psi^{(+)}(x) = e^{ik \cdot x}\bar{u}_r(k)u_r(k)e^{-ikx} \doteq 1 \, , \tag{6.3.17}$$

is independent of k, and thus independent of the reference frame. In general, for a superposition of positive energy solutions, i.e., for

$$\psi^{(+)}(x) = \sum_{r=1}^{2} c_r u_r \,, \text{ with } \sum_{r=1}^{2} |c_r|^2 = 1 \,, \tag{6.3.18a}$$

one has the relation

$$\bar{\psi}^{(+)}(x)\psi^{(+)}(x) = \sum_{r,s} \bar{u}_r(k)u_s(k)c_r^* c_s = \sum_{r=1}^{2} |c_r|^2 = 1 \,. \tag{6.3.18b}$$

Analogous relationships hold for $\psi^{(-)}$.

(iii) If one determines $u_r(k)$ through a Lorentz transformation corresponding to $-\mathbf{v}$, this yields exactly the above spinors. Viewed as an active transformation, this amounts to transforming $u_r(m, \mathbf{0})$ to the velocity $\mathbf{v}$. Such a transformation is known as a "boost".

The *density* for a plane wave ($c = 1$) is $\rho = j^0 = \bar{\psi}\gamma^0\psi$. This is not a Lorentz-invariant quantity since it is the zero-component of a four-vector:

$$\bar{\psi}_r^{(+)}(x)\gamma^0 \, \psi_s^{(+)}(x) = \bar{u}_r(k)\gamma^0 \, u_s(k)$$

$$= \bar{u}_r(k)\frac{\{\not{k}, \gamma^0\}}{2m} u_s(k) = \frac{E}{m} \delta_{rs} \tag{6.3.19a}$$

$$\bar{\psi}_r^{(-)}(x)\gamma^0 \, \psi_s^{(-)}(x) = \bar{v}_r(k)\gamma^0 \, v_s(k)$$

$$= -\bar{v}_r(k)\frac{\{\not{k}, \gamma^0\}}{2m} v_s(k) = \frac{E}{m} \delta_{rs} \,. \tag{6.3.19b}$$

In the intermediate steps here, we have used $u_s(k) = (\not{k}/m)u_s(k)$, $\bar{u}_s(k) = \bar{u}_s(k)(\not{k}/m)$ (Eqs. (6.3.8) and (6.3.13)) etc.

Note. The spinors are normalized such that the density in the rest frame is unity. Under a Lorentz transformation, the density times the volume must remain constant. The volume is reduced by a factor $\sqrt{1 - \beta^2}$ and thus the density must increase by the reciprocal factor $\frac{1}{\sqrt{1-\beta^2}} = \frac{E}{m}$.

We now extend the sequence of equations (6.3.19).

For $\quad \psi_r^{(+)}(x) = e^{-i(k^0 x^0 - \mathbf{k}\cdot\mathbf{x})}u_r(k)$

and $\quad \psi_s^{(-)}(x) = e^{i(k^0 x^0 + \mathbf{k}\cdot\mathbf{x})}v_s(\tilde{k})$ $\qquad$ (6.3.20)

with the four-momentum $\tilde{k} = (k^0, -\mathbf{k})$, one obtains

$$\bar{\psi}_r^{(-)}(x)\gamma^0 \, \psi_s^{(+)}(x) = e^{-2ik^0 x^0} \bar{v}_r(\tilde{k})\gamma^0 \, u_s(k)$$

$$= \frac{1}{2}e^{-2ik^0 x^0} \bar{v}_r(\tilde{k})\left(-\frac{\tilde{\not{k}}}{m}\gamma^0 + \gamma^0 \frac{\not{k}}{m} \right) u_s(k) \quad (6.3.19c)$$

$$= 0$$

since the terms proportional to k_0 cancel and since $\{k_i\gamma^i, \gamma^0\} = 0$. In this sense, positive and negative energy states are orthogonal for opposite energies and equal momenta.

6.3.3 Projection Operators

The operators

$$\Lambda_\pm(k) = \frac{\pm \not{k} + m}{2m} \tag{6.3.21}$$

project onto the spinors of positive and negative energy, respectively:

$$\begin{array}{ll} \Lambda_+ u_r(k) = u_r(k) & \Lambda_- v_r(k) = v_r(k) \\ \Lambda_+ v_r(k) = 0 & \Lambda_- u_r(k) = 0 \end{array} . \tag{6.3.22}$$

Thus, the projection operators $\Lambda_\pm(k)$ can also be represented in the form

$$\Lambda_+(k) = \sum_{r=1,2} u_r(k) \otimes \bar{u}_r(k)$$
$$\Lambda_-(k) = -\sum_{r=1,2} v_r(k) \otimes \bar{v}_r(k) . \tag{6.3.23}$$

The tensor product $\otimes$ is defined by

$$(a \otimes \bar{b})_{\alpha\beta} = a_\alpha \bar{b}_\beta . \tag{6.3.24}$$

In matrix form, the tensor product of a spinor a and an adjoint spinor $\bar{b}$ reads:

$$\begin{pmatrix} a_1 \\ a_2 \\ a_3 \\ a_4 \end{pmatrix} (\bar{b}_1, \bar{b}_2, \bar{b}_3, \bar{b}_4) = \begin{pmatrix} a_1\bar{b}_1 & a_1\bar{b}_2 & a_1\bar{b}_3 & a_1\bar{b}_4 \\ a_2\bar{b}_1 & a_2\bar{b}_2 & a_2\bar{b}_3 & a_2\bar{b}_4 \\ a_3\bar{b}_1 & a_3\bar{b}_2 & a_3\bar{b}_3 & a_3\bar{b}_4 \\ a_4\bar{b}_1 & a_4\bar{b}_2 & a_4\bar{b}_3 & a_4\bar{b}_4 \end{pmatrix} .$$

The projection operators have the following properties:

$$\Lambda_\pm^2(k) = \Lambda_\pm(k) \tag{6.3.25a}$$
$$\mathrm{Tr}\, \Lambda_\pm(k) = 2 \tag{6.3.25b}$$
$$\Lambda_+(k) + \Lambda_-(k) = 1 . \tag{6.3.25c}$$

Proof:

$$\Lambda_\pm(k)^2 = \frac{(\pm \not{k} + m)^2}{4m^2} = \frac{\not{k}^2 \pm 2\not{k}m + m^2}{4m^2} = \frac{m^2 \pm 2\not{k}m + m^2}{4m^2}$$
$$= \frac{2m(\pm \not{k} + m)}{4m^2} = \Lambda_\pm(k)$$

$$\mathrm{Tr}\, \Lambda_\pm(k) = \frac{4m}{2m} = 2$$

The validity of the assertion that $\Lambda_\pm$ projects onto positive and negative energy states can be seen in both of the representations, (6.3.21) and (6.3.22), by applying them to the states $u_r(k)$ and $v_r(k)$. A further important projection operator, $P(n)$, which, in the rest frame projects onto the spin orientation n, will be discussed in Problem 6.15.

Problems

6.1 Prove the group property of the Poincaré group.

6.2 Show, by using the transformation properties of x_μ, that $\partial^\mu \equiv \partial/\partial x_\mu$ ($\partial_\mu \equiv \partial/\partial x^\mu$) transforms as a contravariant (covariant) vector.

6.3 Show that the N-fold application of the infinitesimal rotation in Minkowski space (Eq. (6.2.22))

$$\Lambda = 1 + \frac{\vartheta}{N} \begin{pmatrix} 0 & 0 & 0 & 0 \\ 0 & 0 & 1 & 0 \\ 0 & -1 & 0 & 0 \\ 0 & 0 & 0 & 0 \end{pmatrix}$$

leads, in the limit $N \to \infty$, to a rotation about the z axis through an angle ϑ (the last step in (6.2.22)).

6.4 Derive the quadratic form of the Dirac equation

$$\left[\left(i\hbar\partial - \frac{e}{c}A \right)^2 - \frac{i\hbar e}{c}(\alpha E + i\Sigma B) - m^2 c^2 \right]\psi = 0$$

for the case of external electromagnetic fields. Write the result using the electromagnetic field tensor $F_{\mu\nu} = A_{\mu,\nu} - A_{\nu,\mu}$, and also in a form explicitly dependent on E and B.

Hint: Multiply the Dirac equation from the left by $\gamma^\nu \left(i\hbar\partial_\nu - \frac{e}{c}A_\nu \right) + mc$ and, by using the commutation relations for the γ matrices, bring the expression obtained into quadratic form in terms of the field tensor

$$\left[\left(i\hbar\partial - \frac{e}{c}A \right)^2 - \frac{\hbar e}{2c}\sigma^{\mu\nu}F_{\mu\nu} - m^2 c^2 \right]\psi = 0.$$

The assertion follows by evaluating the expression $\sigma^{\mu\nu}F_{\mu\nu}$ using the explicit form of the field tensor as a function of the fields E and B.

6.5 Consider the quadratic form of the Dirac equation from Problem 6.4 with the fields $E = E_0\,(1,0,0)$ and $B = B\,(0,0,1)$, where it is assumed that $E_0/Bc \le 1$. Choose the gauge $A = B\,(0,x,0)$ and solve the equation with the ansatz

$$\psi(x) = e^{-iEt/\hbar} e^{i(k_y y + k_z z)} \varphi(x)\Phi\,,$$

where Φ is a four-spinor that is independent of time and space coordinates. Calculate the energy eigenvalues for an electron. Show that the solution agrees with that obtained from Problem 5.3 when one considers the nonrelativistic limit, i.e., $E_0/Bc \ll 1$.

Hint: Given the above ansatz for ψ, one obtains the following form for the quadratic Dirac equation:

$$[K(x,\partial_x)\mathbb{1} + M]\,\varphi(x)\Phi = 0\,,$$

where $K(x, \partial_x)$ is an operator that contains constant contributions, ∂_x and x. The matrix M is independent of ∂_x and x; it has the property $M^2 \propto \mathbb{1}$. This suggests that the bispinor Φ has the form $\Phi = (\mathbb{1} + \lambda M)\Phi_0$. Determine λ and the eigenvalues of M. With these eigenvalues, the matrix differential equation reverts into an ordinary differential equation of the oscillator type.

6.6 Show that equation (6.2.14′)

$$[\gamma^\mu, \tau] = \Delta\omega^{\mu\nu}\gamma_\nu$$

is satisfied by

$$\tau = \frac{1}{8}\Delta\omega^{\mu\nu}(\gamma_\mu\gamma_\nu - \gamma_\nu\gamma_\mu) \ .$$

6.7 Prove that $\gamma^{\mu\dagger} = \gamma^0\gamma^\mu\gamma^0$.

6.8 Show that the relation

$$S^\dagger\gamma^0 = b\gamma^0 S^{-1}$$

is satisfied with $b = 1$ by the explicit representations of the elements of the Poincaré group found in the main text (rotation, pure Lorentz transformation, spatial reflection).

6.9 Show that $\bar{\psi}(x)\gamma_5\psi(x)$ is a pseudoscalar, $\bar{\psi}(x)\gamma_5\gamma^\mu\psi(x)$ a pseudovector, and $\bar{\psi}(x)\sigma^{\mu\nu}\psi(x)$ a tensor.

6.10 Properties of the matrices Γ^a.
Taking as your starting point the definitions (6.2.50a–e), derive the following properties of these matrices:
(i) For every Γ^a (except Γ^S) there exists a Γ^b such that $\Gamma^a\Gamma^b = -\Gamma^b\Gamma^a$.
(ii) For every pair Γ^a, Γ^b, $(a \neq b)$ there exists a $\Gamma^c \neq \mathbb{1}$ such that $\Gamma^a\Gamma^b = \beta\Gamma^c$ with $\beta = \pm 1, \pm i$.

6.11 Show that if a 4×4 matrix X commutes with all γ^μ, then this matrix X is proportional to the unit matrix.
Hint: Every 4×4 matrix can, according to Problem 6.1, be written as a linear combination of the 16 matrices Γ^a (basis!).

6.12 Prove Pauli's fundamental theorem for Dirac matrices: For any two four-dimensional representations γ_μ and γ'_μ of the Dirac algebra both of which satisfy the relation

$$\{\gamma_\mu, \gamma_\nu\} = 2g_{\mu\nu}$$

there exists a nonsingular transformation M such that

$$\gamma'_\mu = M\gamma_\mu M^{-1} \ .$$

M is uniquely determined to within a constant prefactor.

6.13 From the solution of the field-free Dirac equation in the rest frame, determine the four-spinors $\psi^\pm(x)$ of a particle moving with the velocity $\mathbf{v}$. Do this by applying a Lorentz transformation (into a coordinate system moving with the velocity $-\mathbf{v}$) to the solutions in the rest frame.

6.14 Starting from

$$\Lambda_+(k) = \sum_{r=1,2} u_r(k) \otimes \bar{u}_r(k) , \qquad \Lambda_-(k) = -\sum_{r=1,2} v_r(k) \otimes \bar{v}_r(k) ,$$

prove the validity of the representations for $\Lambda_\pm(k)$ given in (6.3.22).

6.15 (i) Given the definition $P(n) = \frac{1}{2}(1+\gamma_5 \slashed{n})$, show that, under the assumptions $n^2 = -1$ and $n_\mu k^\mu = 0$, the following relations are satisfied

(a) $[\Lambda_\pm(k), P(n)] = 0$,

(b) $\Lambda_+(k)P(n) + \Lambda_-(k)P(n) + \Lambda_+(k)P(-n) + \Lambda_-(k)P(-n) = 1$,

(c) $\mathrm{Tr}\,[\Lambda_\pm(k)P(\pm n)] = 1$,

(d) $P(n)^2 = P(n)$

(ii) Consider the special case $n = (0, \hat{e}_z)$ where $P(n) = \frac{1}{2}\begin{pmatrix} 1+\sigma^3 & 0 \\ 0 & 1-\sigma^3 \end{pmatrix}$.

7. Orbital Angular Momentum and Spin

We have seen that, in nonrelativistic quantum mechanics, the angular momentum operator is the generator of rotations and commutes with the Hamiltonians of rotationally invariant (i.e., spherically symmetric) systems[1]. It thus plays a special role for such systems. For this reason, as a preliminary to the next topic – the Coulomb potential – we present here a detailed investigation of angular momentum in relativistic quantum mechanics.

7.1 Passive and Active Transformations

For positive energy states, in the non-relativistic limit we derived the Pauli equation with the Landé factor $g = 2$ (Sect. 5.3.5). From this, we concluded that the Dirac equation describes particles with spin $S = 1/2$. Following on from the transformation behavior of spinors, we shall now investigate angular momentum in general.

In order to give the reader useful background information, we will start with some remarks concerning active and passive transformations. Consider a given state Z, which in the reference frame I is described by the spinor $\psi(x)$. When regarded from the reference frame I', which results from I through the Lorentz transformation

$$x' = \Lambda x , \tag{7.1.1}$$

the spinor takes the form,

$$\psi'(x') = S\psi(\Lambda^{-1}x') , \qquad \text{passive with } \Lambda . \tag{7.1.2a}$$

A transformation of this type is known as a *passive transformation*. One and the same state is viewed from two different coordinate systems, which is indicated in Fig. 7.1 by $\psi(x) \hat{=} \psi'(x')$.

On the other hand, one can also transform the state and then view the resulting state Z' exactly as the starting state Z from one and the same reference frame I. In this case one speaks of an *active transformation*. For vectors and scalars, it is clear what is meant by their active transformation

---

[1] See QM I, Sect. 5.1

(rotation, Lorentz transformation). The active transformation of a vector by the transformation Λ corresponds to the passive transformation of the coordinate system by Λ^{-1}. For spinors, the active transformation is defined in exactly this way (see Fig. 7.1).

The state Z', which arises through the transformation Λ^{-1}, appears in I exactly as Z in I', i.e.,

$$\psi'(x) = S\psi(\Lambda^{-1}x) \qquad \text{active with } \Lambda^{-1} \tag{7.1.2b}$$

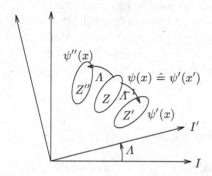

Fig. 7.1. Schematic representation of the passive and active transformation of a spinor; the enclosed area is intended to indicate the region in which the spinor is finite

The state Z'', which results from Z through the active transformation Λ, by definition appears the same in I' as does Z in I, i.e., it takes the form $\psi(x')$. Since I is obtained from I' by the Lorentz transformation Λ^{-1}, in I the spinor Z'' has the form

$$\psi''(x) = S^{-1}\psi(\Lambda x) , \qquad \text{active with } \Lambda . \tag{7.1.2c}$$

7.2 Rotations and Angular Momentum

Under the infinitesimal Lorentz transformation

$$\Lambda^\mu{}_\nu = g^\mu{}_\nu + \Delta\omega^\mu{}_\nu , \quad (\Lambda^{-1})^\mu{}_\nu = g^\mu{}_\nu - \Delta\omega^\mu{}_\nu , \tag{7.2.1}$$

a spinor $\psi(x)$ transforms as

$$\psi'(x') = S\psi(\Lambda^{-1}x') \qquad \text{passive with } \Lambda \tag{7.2.2a}$$

or

$$\psi'(x) = S\psi(\Lambda^{-1}x) \qquad \text{active with } \Lambda^{-1} . \tag{7.2.2b}$$

We now use the results gained in Sect. 6.2.2.1 (Eqs. (6.2.8) and (6.2.13)) to obtain

$$\psi'(x) = (1 - \frac{i}{4} \Delta\omega^{\mu\nu}\sigma_{\mu\nu})\psi(x^\rho - \Delta\omega^\rho_{\ \nu}x^\nu) \,. \tag{7.2.3}$$

Taylor expansion of the spinor yields $(1 - \Delta\omega^\mu_{\ \nu}x^\nu\partial_\mu)\,\psi(x)$, so that

$$\psi'(x) = (1 + \Delta\omega^{\mu\nu}(-\frac{i}{4}\sigma_{\mu\nu} + x_\mu\partial_\nu))\,\psi(x) \,. \tag{7.2.3'}$$

We now consider the special case of rotations through $\Delta\varphi$, which are represented by

$$\Delta\omega^{ij} = -\epsilon^{ijk}\Delta\varphi^k \tag{7.2.4}$$

(the direction of $\Delta\varphi$ specifies the rotation axis and $|\Delta\varphi|$ the angle of rotation). If one also uses

$$\sigma^{ij} = \sigma_{ij} = \epsilon^{ijk}\Sigma^k \quad , \quad \Sigma^k = \begin{pmatrix} \sigma^k & 0 \\ 0 & \sigma^k \end{pmatrix} , \tag{7.2.5}$$

(see Eq. (6.2.19)) one obtains for (7.2.3')

$$
\begin{aligned}
\psi'(x) &= \left(1 + \Delta\omega^{ij}\left(-\frac{i}{4}\epsilon^{ijk}\Sigma^k + x_i\partial_j\right)\right)\psi(x) \\
&= \left(1 - \epsilon^{ij\bar{k}}\Delta\varphi^{\bar{k}}\left(-\frac{i}{4}\epsilon^{ijk}\Sigma^k - x^i\partial_j\right)\right)\psi(x) \\
&= \left(1 - \Delta\varphi^{\bar{k}}\left(-\frac{i}{4}2\delta_{k\bar{k}}\Sigma^k - \epsilon^{ij\bar{k}}x^i\partial_j\right)\right)\psi(x) \\
&= \left(1 + i\Delta\varphi^k\left(\frac{1}{2}\Sigma^k + \epsilon^{kij}x^i\frac{1}{i}\partial_j\right)\right)\psi(x) \\
&\equiv \left(1 + i\Delta\varphi^k J^k\right)\psi(x) \,.
\end{aligned}
\tag{7.2.6}
$$

Here, we have defined the total angular momentum

$$J^k = \epsilon^{kij}x^i\frac{1}{i}\partial_j + \frac{1}{2}\Sigma^k \,. \tag{7.2.7}$$

With the inclusion of $\hbar$, this operator reads:

$$\mathbf{J} = \mathbf{x} \times \frac{\hbar}{i}\boldsymbol{\nabla}1 + \frac{\hbar}{2}\boldsymbol{\Sigma}, \tag{7.2.7'}$$

and is thus the sum of the orbital angular momentum $\mathbf{L} = \mathbf{x} \times \mathbf{p}$ and the spin $\frac{\hbar}{2}\boldsymbol{\Sigma}$.

The total angular momentum (= orbital angular momentum + spin) is the generator of rotations: For a finite angle φ^k one obtains, by combining a succession of infinitesimal rotations,

$$\psi'(x) = e^{i\varphi^k J^k}\,\psi(x)\,. \tag{7.2.8}$$

The operator J^k commutes with the Hamiltonian of the Dirac equation containing a spherically symmetric potential $\Phi(\mathbf{x}) = \Phi(|\mathbf{x}|)$

$$[H, J^i] = 0\,. \tag{7.2.9}$$

A straightforward way to verify (7.2.9) is by an explicit calculation of the commutator (see Problem 7.1). Here, we consider general consequences resulting from the behavior, under rotation, of the structure of commutators of the angular momentum with other operators; Eq. (7.2.9) results as a special case. We consider an operator A, and let the result of its action on ψ_1 be the spinor ψ_2:

$$A\psi_1(x) = \psi_2(x)\,.$$

It follows that

$$e^{i\varphi^k J^k}\,A\,e^{-i\varphi^k J^k}\left(e^{i\varphi^k J^k}\,\psi_1(x)\right) = \left(e^{i\varphi^k J^k}\,\psi_2(x)\right)$$

or, alternatively,

$$e^{i\varphi^k J^k}\,A\,e^{-i\varphi^k J^k}\psi_1'(x) = \psi_2'(x)\,.$$

Thus, in the rotated frame of reference the operator is

$$A' = e^{i\varphi^k J^k}\,A\,e^{-i\varphi^k J^k}\,. \tag{7.2.10}$$

Expanding this for infinitesimal rotations ($\varphi^k \to \Delta\varphi^k$) yields:

$$A' = A - i\Delta\varphi^k[A, J^k] \qquad . \tag{7.2.11}$$

The following special cases are of particular interest:

(i) A is a scalar (rotationally invariant) operator. Then, $A' = A$ and from (7.2.11) it follows that

$$[A, J^k] = 0\,. \tag{7.2.12}$$

   The Hamiltonian of a rotationally invariant system (including a spherically symmetric $\Phi(\mathbf{x}) = \Phi(|\mathbf{x}|)$) is a scalar; this leads to (7.2.9). Hence, in spherically symmetric problems the angular momentum is conserved.

(ii) For the operator A we take the components of a three-vector $\mathbf{v}$. As a vector, $\mathbf{v}$ transforms according to $v'^i = v^i + \epsilon^{ijk}\,\Delta\varphi^j\,v^k$. Equating this, component by component, with (7.2.11), $v^i + \epsilon^{ijk}\Delta\varphi^j v^k = v^i + \frac{i}{\hbar}\Delta\varphi^j\left[J^j, v^i\right]$ which shows that

$$[J^i, v^j] = i\hbar\,\epsilon^{ijk}\,v^k\,. \tag{7.2.13}$$

The commutation relation (7.2.13) implies, among other things,

$$[J^i, J^j] = i\hbar\epsilon^{ijk}J^k \qquad (7.2.14a)$$

$$[J^i, L^j] = i\hbar\epsilon^{ijk}L^k \quad . \qquad (7.2.14b)$$

It is clear from the explicit representation $\Sigma^k = \begin{pmatrix} \sigma^k & 0 \\ 0 & \sigma^k \end{pmatrix}$ that the eigen-values of the 4×4 matrices Σ^k are doubly degenerate and take the values ± 1. The angular momentum $\mathbf{J}$ is the sum of the orbital angular momentum $\mathbf{L}$ and the intrinsic angular momentum or spin $\mathbf{S}$, the components of which have the eigenvalues $\pm\frac{1}{2}$. Thus, particles that obey the Dirac equation have spin $S = 1/2$. The operator $\left(\frac{\hbar}{2}\Sigma\right)^2 = \frac{3}{4}\hbar^2 \mathbb{1}$ has the eigenvalue $\frac{3\hbar^2}{4}$. The eigenvalues of $\mathbf{L}^2$ and L^3 are $\hbar^2 l(l+1)$ and $\hbar m_l$, where $l = 0, 1, 2, \dots$ and m_l takes the values $-l, -l+1, \dots, l-1, l$. The eigenvalues of $\mathbf{J}^2$ are thus $\hbar^2 j(j+1)$, where $j = l \pm \frac{1}{2}$ for $l \neq 0$ and $j = \frac{1}{2}$ for $l = 0$. The eigenvalues of J^3 are $\hbar m_j$, where m_j ranges in integer steps between $-j$ and j. The operators $\mathbf{J}^2$, $\mathbf{L}^2$, Σ^2, and J^3 can be simultaneously diagonalized. The orbital angular momentum operators L^i and the spin operators Σ^i themselves fulfill the angular momentum commutation relations.

Note: One is tempted to ask how it is that the Dirac Hamiltonian, a 4×4 matrix, can be a scalar. In order to see this, one has to return to the transformation (6.2.6′). The transformed Hamiltonian including a central potential $\Phi(|\mathbf{x}|)$

$$(-i\gamma^\nu \partial'_\nu + m + e\Phi(|\mathbf{x}'|)) = S(-i\gamma^\nu \partial_\nu + m + e\Phi(|\mathbf{x}|))S^{-1}$$

has, under rotations, the same form in both systems. The property "scalar" means invariance under rotations, but is not necessarily limited to one-component spherically symmetric functions.

Problems

7.1 Show, by explicit calculation of the commutator, that the total angular momentum

$$\mathbf{J} = \mathbf{x} \times \mathbf{p}\, \mathbb{1} + \frac{\hbar}{2}\Sigma$$

commutes with the Dirac Hamiltonian for a central potential

$$H = c\left(\sum_{k=1}^{3} \alpha^k p^k + \beta mc\right) + e\Phi(|\mathbf{x}|) \ .$$

8. The Coulomb Potential

In this chapter, we shall determine the energy levels in a Coulomb potential. To begin with, we will study the relatively simple case of the Klein–Gordon equation. In the second section, the even more important Dirac equation will be solved exactly for the hydrogen atom.

8.1 Klein–Gordon Equation with Electromagnetic Field

8.1.1 Coupling to the Electromagnetic Field

The coupling to the electromagnetic field in the Klein–Gordon equation

$$-\hbar^2 \frac{\partial^2 \psi}{\partial t^2} = -\hbar^2 c^2 \nabla^2 \psi + m^2 c^4 \psi \,,$$

i.e., the substitution

$$i\hbar \frac{\partial}{\partial t} \longrightarrow i\hbar \frac{\partial}{\partial t} - e\Phi \,, \qquad \frac{\hbar}{i} \nabla \longrightarrow \frac{\hbar}{i} \nabla - \frac{e}{c} \mathbf{A} \,,$$

leads to the Klein–Gordon equation in an electromagnetic field

$$\left(i\hbar \frac{\partial}{\partial t} - e\Phi \right)^2 \psi = c^2 \left(\frac{\hbar}{i} \nabla - \frac{e}{c} \mathbf{A} \right)^2 \psi + m^2 c^4 \psi. \tag{8.1.1}$$

We note that the four-current-density now reads:

$$j_\nu = \frac{i\hbar e}{2m} \left(\psi^* \partial_\nu \psi - \psi \partial_\nu \psi^* \right) - \frac{e^2}{mc} A_\nu \psi^* \psi \tag{8.1.2}$$

with the continuity equation

$$\partial_\nu j^\nu = 0 \,. \tag{8.1.3}$$

One thus finds, in j^0 for example, that the scalar potential $A^0 = c\Phi$ appears.

8.1.2 Klein–Gordon Equation in a Coulomb Field

We assume that $\mathbf{A}$ and Φ are time independent and now seek *stationary solutions* with positive energy

$$\psi(\mathbf{x}, t) = e^{-iEt/\hbar}\psi(\mathbf{x}) \qquad \text{with} \qquad E > 0. \tag{8.1.4}$$

From (8.1.1), one then obtains the time-independent Klein–Gordon equation

$$(E - e\Phi)^2\psi = c^2\left(\frac{\hbar}{i}\boldsymbol{\nabla} - \frac{e}{c}\mathbf{A}\right)^2\psi + m^2c^4\psi. \tag{8.1.5}$$

For a *spherically symmetric potential* $\Phi(\mathbf{x}) \longrightarrow \Phi(r)$ $(r = |\mathbf{x}|)$ and $\mathbf{A} = 0$, it follows that

$$\left(-\hbar^2c^2\boldsymbol{\nabla}^2 + m^2c^4\right)\psi(\mathbf{x}) = (E - e\Phi(r))^2\psi(\mathbf{x}). \tag{8.1.6}$$

The separation of variables in spherical polar coordinates

$$\psi(r, \vartheta, \varphi) = R(r)Y_{\ell m}(\vartheta, \varphi), \tag{8.1.7}$$

where $Y_{\ell m}(\vartheta, \varphi)$ are the spherical harmonic functions, already known to us from nonrelativistic quantum mechanics,[1] leads, analogously to the nonrelativistic theory, to the differential equation

$$\left(-\frac{1}{r}\frac{d}{dr}\frac{d}{dr}r + \frac{\ell(\ell+1)}{r^2}\right)R = \frac{(E - e\Phi(r))^2 - m^2c^4}{\hbar^2c^2}R. \tag{8.1.8}$$

Let us first consider the nonrelativistic limit. If we set $E = mc^2 + E'$ and assume that $E' - e\Phi$ can be neglected in comparison to mc^2, then (8.1.8) yields the nonrelativistic radial Schrödinger equation, since the right-hand side of (8.1.8) becomes

$$\frac{1}{\hbar^2c^2}\left((mc^2)^2 + 2mc^2(E' - e\Phi(r)) + (E' - e\Phi(r))^2 - m^2c^4\right)R(r)$$
$$\approx \frac{2m}{\hbar^2}(E' - e\Phi(r))R(r). \tag{8.1.9}$$

For a π^- *meson* in the *Coulomb field* of a nucleus with charge Z,

$$e\Phi(r) = -\frac{Ze_0^2}{r}. \tag{8.1.10a}$$

Inserting the fine-structure constant $\alpha = \frac{e_0^2}{\hbar c}$, it follows from (8.1.8) that

$$\left[-\frac{1}{r}\frac{d}{dr}\frac{d}{dr}r + \frac{\ell(\ell+1) - Z^2\alpha^2}{r^2} - \frac{2Z\alpha E}{\hbar cr} - \frac{E^2 - m^2c^4}{\hbar^2c^2}\right]R = 0. \tag{8.1.10b}$$

---

[1] QM I, Chap. 5

Remark: The mass of the π meson is $m_{\pi^-} = 273 m_e$ and its half-life $\tau_{\pi^-} = 2.55 \times 10^{-8}$ s. Since the classical orbital period[1] estimated by means of the uncertainty principle is approximately $T \approx \frac{a_{\pi^-}}{\Delta v} \approx \frac{m_{\pi^-} a_{\pi^-}^2}{\hbar} = \frac{m_e^2}{m_{\pi^-}} \frac{a^2}{\hbar} \approx 10^{-21}$s, one can think of well-defined stationary states, despite the finite half-life of the π^-. Even the lifetime of an excited state (see QM I, Sect. 16.4.3) $\Delta T \approx T\alpha^{-3} \approx 10^{-15}$ is still much shorter than τ_{π^-}.

By substituting

$$\sigma^2 = \frac{4(m^2 c^4 - E^2)}{\hbar^2 c^2} \ , \ \gamma = Z\alpha \ , \ \lambda = \frac{2E\gamma}{\hbar c \sigma} \ , \ \rho = \sigma r \qquad \text{(8.1.11a-d)}$$

into (8.1.10b), we obtain

$$\left[\frac{d^2}{d(\rho/2)^2} + \frac{2\lambda}{\rho/2} - 1 - \frac{\ell(\ell+1) - \gamma^2}{(\rho/2)^2} \right] \rho R(\rho) = 0 . \qquad \text{(8.1.12)}$$

This equation has exactly the form of the nonrelativistic Schrödinger equation for the function $u = \rho R$, provided we substitute in the latter

$$\rho_0 \longrightarrow 2\lambda \qquad \text{(8.1.13a)}$$
$$\ell(\ell+1) \longrightarrow \ell(\ell+1) - \gamma^2 \equiv \ell'(\ell'+1) . \qquad \text{(8.1.13b)}$$

Here it should be noted that ℓ' is generally not an integer.

Remark: A similar modification of the centrifugal term is also found in classical relativistic mechanics, where it has as a consequence that the Kepler orbits are no longer closed. Instead of ellipses, one has rosette-like orbits.

The radial Schrödinger equation (8.1.12) can now be solved in the same way as is familiar from the nonrelativistic case: From (8.1.12) one finds for $R(\rho)$ in the limits $\rho \to 0$ and $\rho \to \infty$ the behavior $\rho^{\ell'}$ and $e^{-\rho/2}$ respectively. This suggests the following ansatz for the solution:

$$\rho R(\rho) = \left(\frac{\rho}{2} \right)^{\ell'+1} e^{-\rho/2} w(\rho/2). \qquad \text{(8.1.14)}$$

The resulting differential equation for $w(\rho)$ (Eq. (6.19) in QM I) is solved in terms of a power series. The recursion relation resulting from the differential equation is such that it leads to a function $\sim e^{\rho}$. Taken together with (8.1.14), this means that the function $R(\rho)$ would not be normalizable unless the power series terminated. The condition that the power series for $w(\rho)$ terminates yields[2]:

$$\rho_0 = 2(N + \ell' + 1) ,$$

i.e.,

---

[2] Cf. QM I, Eq. (6.23).

$$\lambda = N + \ell' + 1, \tag{8.1.15}$$

where N is the radial quantum number, $N = 0, 1, 2, \dots$ In order to determine the energy eigenvalues from this, one first needs to use equations (8.1.11a and d) to eliminate the auxiliary quantity σ

$$\frac{4E^2\gamma^2}{\hbar^2 c^2 \lambda^2} = \frac{4(m^2 c^4 - E^2)}{\hbar^2 c^2},$$

which then yields the energy levels as

$$E = mc^2 \left(1 + \frac{\gamma^2}{\lambda^2}\right)^{-\frac{1}{2}}. \tag{8.1.16}$$

Here, one must take the positive root since the rescaling factor is $\sigma > 0$ and, as $\lambda > 0$, it follows from (8.1.11c) that $E > 0$. Thus, for a vanishing attractive potential ($\gamma \to 0$), the energy of these solutions approaches the rest energy $E = mc^2$. For the discussion that follows, we need to calculate ℓ', defined by the quadratic equation (8.1.13b)

$$\ell' = -\frac{1}{2} \pm \sqrt{\left(\ell + \frac{1}{2}\right)^2 - \gamma^2}. \tag{8.1.17}$$

We may convince ourselves that only the positive sign is allowed, i.e.,

$$\lambda = N + \frac{1}{2} + \sqrt{\left(\ell + \frac{1}{2}\right)^2 - \gamma^2}$$

and thus

$$E = \frac{mc^2}{\sqrt{1 + \dfrac{\gamma^2}{\left[N + \frac{1}{2} + \sqrt{\left(\ell + \frac{1}{2}\right)^2 - \gamma^2}\right]^2}}}. \tag{8.1.18}$$

To pursue the parallel with the nonrelativistic case, we introduce the *principal quantum number*

$$n = N + \ell + 1,$$

whereby (8.1.18) becomes

$$E = \frac{mc^2}{\sqrt{1 + \dfrac{\gamma^2}{\left[n - \left(\ell + \frac{1}{2}\right) + \sqrt{\left(\ell + \frac{1}{2}\right)^2 - \gamma^2}\right]^2}}}. \tag{8.1.18'}$$

The principal quantum number has the possible values $n = 1, 2, \dots$; for a given value of n, the possible values of the orbital angular momentum quantum numbers are $\ell = 0, 1, \dots n - 1$. The *degeneracy* that is present in the nonrelativistic theory with respect to the angular momentum is lifted here. The expansion of (8.1.18') in a power series in γ^2 yields:

$$
E = mc^2 \left[1 - \frac{\gamma^2}{2n^2} - \frac{\gamma^4}{2n^4} \left(\frac{n}{\ell + \frac{1}{2}} - \frac{3}{4} \right) \right] + \mathcal{O}(\gamma^6)
$$

$$
= mc^2 - \frac{\mathrm{Ry}}{n^2} - \frac{\mathrm{Ry}\gamma^2}{n^3} \left(\frac{1}{\ell + \frac{1}{2}} - \frac{3}{4n} \right) + \mathcal{O}(\mathrm{Ry}\gamma^4) , \qquad (8.1.19)
$$

with

$$
\mathrm{Ry} = \frac{mc^2 (Z\alpha)^2}{2} = \frac{mZ^2 e^4}{2\hbar^2} .
$$

The first term is the rest energy, the second the nonrelativistic Rydberg formula, and the third term is the relativistic correction. It is identical to the perturbation-theoretical correction due to the relativistic kinetic energy, giving rise to the perturbation Hamiltonian $H_1 = -\frac{(\mathbf{p}^2)^2}{8m^3 c^2}$ (see QM I, Eq. (12.5))[3]. It is this correction that lifts the degeneracy in ℓ:

$$
E_{\ell=0} - E_{\ell=n-1} = -\frac{4\mathrm{Ry}\gamma^2}{n^3} \frac{n-1}{2n-1} . \qquad (8.1.20)
$$

The binding energy E_b is obtained from (8.1.18') or (8.1.19) by subtracting the rest energy

$$
E_b = E - mc^2 .
$$

Further aspects:

(i) We now wish to justify the exclusion of solutions ℓ', for which the negative root was taken in (8.1.17). Firstly, it is to be expected that the solutions should go over continuously into the nonrelativistic solutions and thus that to each ℓ should correspond only one eigenvalue. For the time being we will denote the two roots in (8.1.17) by $\ell'_\pm$. There are a number of arguments for excluding the negative root. The solution ℓ'_- can be excluded on account of the requirement that the kinetic energy be finite. (Here only the lower limit is relevant since the factor $e^{-\rho/2}$ guarantees the convergence at the upper limit):

$$
T \sim - \int dr\, r^2 \frac{\partial^2 R}{\partial^2 r} \cdot R \sim \int dr\, r^2 \left(\frac{\partial R}{\partial r} \right)^2
$$

$$
\sim \int dr\, r^2 \left(r^{\ell'-1} \right)^2 \sim \int dr\, r^{2\ell'} .
$$

This implies that $\ell' > -\frac{1}{2}$ and, hence, only ℓ'_+ is allowed. Instead of the kinetic energy, one can also consider the current density. If solutions with

---

[3] See also Remark (ii) in Sect. 10.1.2

both ℓ'_+ and ℓ'_- were possible, then one would also have linear superpositions of the type $\psi = \psi_{\ell'_+} + i\psi_{\ell'_-}$. The radial current density for this wave function is

$$j_r = \frac{\hbar}{2mi}\left(\psi^*\frac{\partial}{\partial r}\psi - \left(\frac{\partial}{\partial r}\psi^*\right)\psi\right)$$

$$= \frac{\hbar}{2mi}2i\left(\psi_{\ell'_+}\frac{\partial}{\partial r}\psi_{\ell'_-} - \psi_{\ell'_-}\frac{\partial}{\partial r}\psi_{\ell'_+}\right) \sim r^{\ell'_+ + \ell'_- - 1} = \frac{1}{r^2}.$$

The current density would diverge as $\frac{1}{r^2}$ for $r \to 0$. The current through the surface of an arbitrarily small sphere around the origin would then be $\int d\Omega r^2 j_r = $ constant, independent of r. There would have to be a source or a sink for particle current at the origin. The solution ℓ'_+ must certainly be retained as it is the one that transforms into the nonrelativistic solution and, hence, it is the solution with ℓ'_- which must be rejected.

One can confirm this conclusion by solving the problem for a nucleus of finite size, for which the electrostatic potential at $r = 0$ is finite. The solution that is finite at $r = 0$ goes over into the solution of the $\frac{1}{r}$ problem corresponding to the positive sign.

(ii) In order that ℓ' and the energy eigenvalues be real, according to (8.1.17) we must have

$$\ell + \frac{1}{2} > Z\alpha \tag{8.1.21a}$$

(see Fig. 8.1). This condition is most restrictive for s states, i.e., $\ell = 0$:

$$Z < \frac{1}{2\alpha} = \frac{137}{2} = 68.5 . \tag{8.1.21b}$$

For $\gamma > \frac{1}{2}$, we would have a complex value $\ell' = -\frac{1}{2} + is'$ with $s' = \sqrt{\gamma^2 - \frac{1}{4}}$. This would result in complex energy eigenvalues and furthermore, we would have $R(r) \sim r^{-\frac{1}{2}}e^{\pm is' \log r}$, i.e., the solution would oscillate infinitely many times as $r \to 0$ and the matrix element of the kinetic energy would be divergent.

The modification of the centrifugal term into $(\ell(\ell+1) - (Z\alpha)^2)\frac{1}{r^2}$ arises from the relativistic mass increase. Qualitatively speaking, the velocity does not increase so rapidly on approaching the center as in the nonrelativistic case, and thus the centrifugal repulsion is reduced. For the attractive $(-\frac{1}{r^2})$ potential, classical mechanics predicts that the particles spiral into the center. When $Z\alpha > \ell + \frac{1}{2} > \sqrt{\ell(\ell+1)}$, the quantum-mechanical system becomes unstable. The condition $Z\alpha < \frac{1}{2}$ can also be written in the form $Z\frac{e_0^2}{\hbar/m_{\pi^-}c} < \frac{1}{2}m_{\pi^-}c^2$, i.e., the Coulomb energy at a distance of a Compton wavelength $\frac{\hbar}{m_{\pi^-}c} = 1.4 \times 10^{-13}$ cm from the origin should be smaller than $\frac{1}{2}m_{\pi^-}c^2$.

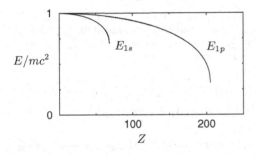

Fig. 8.1. Plot of E_{1s} and E_{1p} for a point-like nucleus according to Eq. (8.1.18) as a function of Z. The curves end at the Z value given by (8.1.21a). For larger Z, the energies become complex

The solutions for the $\left(-\frac{Ze_0^2}{r}\right)$ potential become meaningless for $Z > 68$. Yet, since there exist nuclei with higher atomic number, it must be possible to describe the motion of a π^- meson by means of the Klein–Gordon equation. However, one must be aware of the fact that real nuclei have a finite radius which means that also for large Z, bound states exist.

The Bohr radius for π^- is $a_{\pi^-} = \frac{\hbar^2}{Z m_{\pi^-} e_0^2} = \frac{m_e}{m_{\pi^-}} \frac{a}{Z} \approx \frac{2 \times 10^{-11}}{Z}$ cm, where $a = 0.5 \times 10^{-8}$ cm, the Bohr radius of the electron, and $m_{\pi^-} = 270 m_e$ have been used. Comparison with the nuclear radius $R_N = 1.5 \times 10^{-13} A^{1/3}$ cm reveals that the size of the nucleus is not negligible[4].

For a quantitative comparison of the theory with experiments on π-mesonic atoms, one also has to take the following corrections into account:

(i) The mass m_π must be replaced by the reduced mass $\mu = \frac{m_\pi M}{m_\pi + M}$.

(ii) As already emphasized, one must allow for the finite size of the nucleus.

(iii) The vacuum polarization must be included. This refers to the fact that the photon exchanged between the nucleus and the π-meson transforms virtually into an electron–positron pair, which subsequently recombines into a photon (see Fig. 8.2).

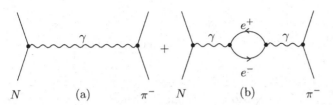

Fig. 8.2. The electromagnetic interaction arises from the exchange of a photon (γ) between the nucleus (N) and the π-meson (π^-). (a) Direct exchange; (b) with vacuum polarization in which a virtual electron–positron pair ($e^- - e^+$) occurs

---

[4] The experimental transition energies for π-mesonic atoms, which lie in the x-ray range, are presented in D.A. Jenkins and R. Kunselman, Phys. Rev. Lett. **17**, 1148 (1966), where they are compared with the result obtained from the Klein–Gordon equation.

(iv) Since the Bohr radius for the π^-, as estimated above, is smaller by approximately a factor $1/300$ than that of the electron, and thus the probability of finding the π^- in the vicinity of the nucleus is appreciable, one must also include a correction for the strong interaction between the nucleus and the π^-.

8.2 Dirac Equation for the Coulomb Potential

In this section, we shall determine the *exact solution* of the Dirac equation for an electron in a Coulomb potential

$$V(r) = -\frac{Ze_0^2}{r} .$$
(8.2.1)

From

$$i\hbar\frac{\partial\psi(\mathbf{x}, t)}{\partial t} = \left(c\boldsymbol{\alpha}\cdot\left(\mathbf{p} - \frac{e}{c}\mathbf{A}\right) + \beta mc^2 + e\Phi\right)\psi(\mathbf{x}, t)$$
(8.2.2)

one finds, for $\mathbf{A} = 0$ and $e\Phi \equiv -\frac{Ze_0^2}{r} \equiv V(r)$, the Dirac Hamiltonian

$$H = c\boldsymbol{\alpha}\cdot\mathbf{p} + \beta mc^2 + V(r)$$
(8.2.3)

and, with $\psi(\mathbf{x}, t) = e^{-iEt/\hbar}\psi(\mathbf{x})$, the time-independent Dirac equation

$$(c\boldsymbol{\alpha}\cdot\mathbf{p} + \beta mc^2 + V(r))\psi(\mathbf{x}) = E\psi(\mathbf{x}).$$
(8.2.4)

Here too, it will turn out to be useful to represent H in spherical polar coordinates. To achieve this end, we first exploit all symmetry properties of H.

The total angular momentum $\mathbf{J}$ from (7.2.7')

$$\mathbf{J} = \mathbf{L}\mathbb{1} + \frac{\hbar}{2}\boldsymbol{\Sigma}$$
(8.2.5)

commutes with H. This implies that H, $\mathbf{J}^2$, and J_z have common eigenstates.

Remarks:

(i) The operators L_z, Σ_z, and $\mathbf{L}^2$ do not commute with H.

(ii) For $\boldsymbol{\Sigma} = \begin{pmatrix} \boldsymbol{\sigma} & 0 \\ 0 & \boldsymbol{\sigma} \end{pmatrix}$, it follows that $\left(\frac{\hbar}{2}\boldsymbol{\Sigma}\right)^2 = \frac{3\hbar^2}{4}\mathbb{1} = \frac{1}{2}\left(1 + \frac{1}{2}\right)\hbar^2\mathbb{1}$ is diagonal.

(iii) $\mathbf{L}^2$, $\boldsymbol{\Sigma}^2$, and $\mathbf{L}\cdot\boldsymbol{\Sigma}$, like H, are scalars and thus commute with $\mathbf{J}$.

As a necessary prerequisite for an exact solution of the Dirac equation, we first discuss the Pauli spinors. As we know from nonrelativistic quantum mechanics[5], the Pauli spinors are common eigenstates of $\mathbf{J}^2, J_z$, and $\mathbf{L}^2$ with the corresponding quantum numbers j, m, and ℓ, where $\mathbf{J} = \mathbf{L} + \frac{\hbar}{2}\boldsymbol{\sigma}$ is now the operator of the total angular momentum in the space of two-component spinors. From the product states

$$
\begin{array}{ll}
|\ell, m_j + 1/2\rangle \, |\downarrow\rangle & \quad Y_{\ell, m_j + \frac{1}{2}} \begin{pmatrix} 0 \\ 1 \end{pmatrix} \\
& \text{or} \\
|\ell, m_j - 1/2\rangle \, |\uparrow\rangle & \quad Y_{\ell, m_j - \frac{1}{2}} \begin{pmatrix} 1 \\ 0 \end{pmatrix}
\end{array}
\tag{8.2.6}
$$

(in Dirac ket space or in the coordinate representation), one forms linear combinations that are eigenstates of $\mathbf{J}^2, J_z$, and $\mathbf{L}^2$. For a particular value of ℓ, one obtains

$$
\varphi_{jm_j}^{(+)} = \begin{pmatrix} \sqrt{\dfrac{\ell + m_j + 1/2}{2\ell + 1}} \, Y_{\ell, m_j - 1/2} \\[2ex] \sqrt{\dfrac{\ell - m_j + 1/2}{2\ell + 1}} \, Y_{\ell, m_j + 1/2} \end{pmatrix} \quad \text{for} \quad j = \ell + \frac{1}{2}
$$

and

$$
\varphi_{jm_j}^{(-)} = \begin{pmatrix} \sqrt{\dfrac{\ell - m_j + 1/2}{2\ell + 1}} \, Y_{\ell, m_j - 1/2} \\[2ex] -\sqrt{\dfrac{\ell + m_j + 1/2}{2\ell + 1}} \, Y_{\ell, m_j + 1/2} \end{pmatrix} \quad \text{for} \quad j = \ell - \frac{1}{2} \,. \tag{8.2.7}
$$

The coefficients that appear here are the Clebsch–Gordan coefficients. Compared to the convention used in QM I, the spinors $\varphi_{jm_j}^{(-)}$ now contain an additional factor -1. The quantum number ℓ takes the values $\ell = 0, 1, 2, \ldots$, whilst j and m_j have half-integer values. For $\ell = 0$, the only states are $\varphi_{jm_j}^{(+)} \equiv \varphi_{\frac{1}{2}m_j}^{(+)}$. The states $\varphi_{jm_j}^{(-)}$ only exist for $\ell > 0$, since $l = 0$ would imply a negative j. The spherical harmonics satisfy

$$
Y_{\ell, m}^* = (-1)^m \, Y_{\ell, -m}. \tag{8.2.8}
$$

The eigenvalue equations for $\varphi_{jm_j}^{(\pm)}$ are (henceforth we set $\hbar = 1$):

$$
\mathbf{J}^2 \varphi_{jm_j}^{(\pm)} = j(j+1)\varphi_{jm_j}^{(\pm)} \quad , \quad j = \frac{1}{2}, \frac{3}{2}, \ldots
$$

$$
\mathbf{L}^2 \varphi_{jm_j}^{(\pm)} = \ell(\ell+1)\varphi_{jm_j}^{(\pm)} \quad , \quad \ell = j \mp \frac{1}{2} \tag{8.2.9}
$$

$$
J_z \varphi_{jm_j}^{(\pm)} = m_j \varphi_{jm_j}^{(\pm)} \quad , \quad m_j = -j, \ldots, j \,.
$$

---

[5] QM I, Chap. 10, Addition of Angular Momenta

Furthermore, we have

$$
\begin{aligned}
\mathbf{L} \cdot \boldsymbol{\sigma} \varphi_{jm_j}^{(\pm)} &= \left(\mathbf{J}^2 - \mathbf{L}^2 - \frac{3}{4} \right) \varphi_{jm_j}^{(\pm)} \\
&= \left(j(j+1) - \ell(\ell+1) - \frac{3}{4} \right) \varphi_{jm_j}^{(\pm)} \qquad (8.2.10) \\
&= \left\{ \begin{array}{c} \ell \\ -\ell - 1 \end{array} \right\} \varphi_{jm_j}^{(\pm)} \\
&= \left\{ \begin{array}{c} -1 + (j+1/2) \\ -1 - (j+1/2) \end{array} \right\} \varphi_{jm_j}^{(\pm)} \quad \text{for } j = \ell \pm \frac{1}{2} \, .
\end{aligned}
$$

The following definition will prove useful

$$
K = (1 + \mathbf{L} \cdot \boldsymbol{\sigma}) \qquad (8.2.11)
$$

whereby, according to (8.2.10), the following eigenvalue equation holds:

$$
K \varphi_{jm_j}^{(\pm)} = \pm \left(j + \frac{1}{2} \right) \varphi_{jm_j}^{(\pm)} \equiv k \varphi_{jm_j}^{(\pm)} \, . \qquad (8.2.12)
$$

The parity of $Y_{\ell m}$ can be seen from

$$
Y_{\ell m}(-\mathbf{x}) = (-1)^{\ell} Y_{\ell m}(\mathbf{x}) \, . \qquad (8.2.13)
$$

For each value of j ($\frac{1}{2}$, $\frac{3}{2}$, ...) there are two Pauli spinors, $\varphi_{jm_j}^{(+)}$ and $\varphi_{jm_j}^{(-)}$, whose orbital angular momenta ℓ differ by 1, and which therefore have opposite parities. We introduce the notation

$$
\varphi_{jm_j}^{\ell} = \left\{ \begin{array}{ll} \varphi_{jm_j}^{(+)} & \ell = j - \frac{1}{2} \\[2mm] \varphi_{jm_j}^{(-)} & \ell = j + \frac{1}{2} \, . \end{array} \right. \qquad (8.2.14)
$$

In place of the index $(\pm)$, one gives the value of ℓ, which yields the quantum number j by the addition (subtraction) of $\frac{1}{2}$. According to (8.2.13), $\varphi_{jm_j}^{\ell}$ has parity $(-1)^{\ell}$, i.e.,

$$
\varphi_{jm_j}^{\ell}(-\mathbf{x}) = (-1)^{\ell} \varphi_{jm_j}^{\ell}(\mathbf{x}) \, . \qquad (8.2.15)
$$

Remark: One may also write

$$
\varphi_{jm_j}^{(+)} = \frac{\boldsymbol{\sigma} \cdot \mathbf{x}}{r} \varphi_{jm_j}^{(-)} \, . \qquad (8.2.16)
$$

This relation can be justified as follows: The operator that generates $\varphi_{jm_j}^{(+)}$ from $\varphi_{jm_j}^{(-)}$ must be a scalar operator of odd parity. Furthermore, due to the difference $\Delta \ell = 1$, the position dependence is of the form $Y_{1,m}(\vartheta, \varphi)$, and thus proportional to $\mathbf{x}$. Therefore, $\mathbf{x}$ must be multiplied by a pseudovector. The only position-independent pseudovector is $\boldsymbol{\sigma}$. A formal proof of (8.2.16) is left as an exercise in Problem 8.2.

The Dirac Hamiltonian for the Coulomb potential is also invariant under spatial reflections, i.e., with respect to the operation (Eq. (6.2.33′))

$$\mathcal{P} = \beta \, \mathcal{P}^{(0)}$$

where $\mathcal{P}^{(0)}$ effects the spatial reflection[6] $\mathbf{x} \to -\mathbf{x}$. One may see this directly by calculating $\beta \mathcal{P}^{(0)} H$ and making use of $\beta \boldsymbol{\alpha} = -\boldsymbol{\alpha}\beta$:

$$\beta \mathcal{P}^{(0)} \left[\frac{1}{i} \boldsymbol{\alpha} \cdot \boldsymbol{\nabla} + \beta m - \frac{Z\alpha}{r} \right] \psi(\mathbf{x})$$

$$= \beta \left[\frac{1}{i} \boldsymbol{\alpha} (-\boldsymbol{\nabla}) + \beta m - \frac{Z\alpha}{r} \right] \psi(-\mathbf{x}) \qquad (8.2.17)$$

$$= \left[\frac{1}{i} \boldsymbol{\alpha} \cdot \boldsymbol{\nabla} + \beta m - \frac{Z\alpha}{r} \right] \beta \mathcal{P}^{(0)} \psi(\mathbf{x}) \,.$$

Therefore, $\beta \mathcal{P}^{(0)}$ commutes with H such that

$$[\beta \mathcal{P}^{(0)}, H] = 0 \,. \qquad (8.2.17')$$

Since $(\beta \mathcal{P}^{(0)})^2 = 1$, it is clear that $\beta \mathcal{P}^{(0)}$ possesses the eigenvalues ± 1. Hence, one can construct even and odd eigenstates of $\beta \mathcal{P}^{(0)}$ and H

$$\beta \mathcal{P}^{(0)} \psi_{jm_j}^{(\pm)}(\mathbf{x}) = \beta \psi_{jm_j}^{(\pm)}(-\mathbf{x}) = \pm \psi_{jm_j}^{(\pm)}(\mathbf{x}) \,. \qquad (8.2.18)$$

Let us remark in passing that the pseudovector $\mathbf{J}$ commutes with $\beta \mathcal{P}^{(0)}$.

In order to solve (8.2.4), we attempt to construct the four-spinors from Pauli spinors. When $\varphi_{jm_j}^{\ell}$ appears in the two upper components, then, on account of β, one must also in the lower components take the other ℓ belonging to j, and hence, according to (8.2.16), $\boldsymbol{\sigma} \cdot \hat{\mathbf{x}} \varphi_{jm_j}^{\ell}$. This gives as solution ansatz the four-spinors[7]

$$\psi_{jm_j}^{\ell} = \begin{pmatrix} \frac{iG_{\ell j}(r)}{r} \varphi_{jm_j}^{\ell} \\ \frac{F_{\ell j}(r)}{r} (\boldsymbol{\sigma} \cdot \hat{\mathbf{x}}) \varphi_{jm_j}^{\ell} \end{pmatrix} \,. \qquad (8.2.19)$$

These spinors have the parity $(-1)^{\ell}$, since

$$\beta P \psi_{jm_j}^{\ell}(\mathbf{x}) = \beta \psi_{jm_j}^{\ell}(-\mathbf{x}) = \beta \begin{pmatrix} \dots (-1)^{\ell} \varphi_{jm_j}^{\ell} \\ \dots (-1)^{\ell+1} \boldsymbol{\sigma} \cdot \hat{\mathbf{x}} \varphi_{jm_j}^{\ell} \end{pmatrix} \qquad (8.2.20)$$

$$= (-1)^{\ell} \psi_{jm_j}^{\ell}(\mathbf{x}) \,.$$

---

[6] This can also be concluded from the covariance of the Dirac equation and the invariance of $\frac{1}{r}$ under spatial reflections (Sect. 6.2.2.4).

[7] Since $[\mathbf{J}, \boldsymbol{\sigma} \cdot \mathbf{x}] = 0$, it is clear that $\frac{J_z}{\mathbf{J}^2} \psi_{jm_j}^{\ell} = \frac{m}{j(j+1)} \psi_{jm_j}^{\ell}$.

The factors $\frac{1}{r}$ and i included in (8.2.19) will turn out to be useful later.

In matrix notation the Dirac Hamiltonian reads

$$H = \begin{pmatrix} m - \frac{Z\alpha}{r} & \boldsymbol{\sigma}\cdot\mathbf{p} \\ \boldsymbol{\sigma}\cdot\mathbf{p} & -m - \frac{Z\alpha}{r} \end{pmatrix}. \tag{8.2.21}$$

In order to calculate $H\psi^\ell_{jm}$, we require the following quantities[8]:

$$\boldsymbol{\sigma}\cdot\mathbf{p}\, f(r)\, \varphi^\ell_{jm_j} = \boldsymbol{\sigma}\cdot\hat{\mathbf{x}}\,\boldsymbol{\sigma}\cdot\hat{\mathbf{x}}\,\boldsymbol{\sigma}\cdot\mathbf{p}\, f(r)\, \varphi^\ell_{jm_j}$$

$$= \frac{\boldsymbol{\sigma}\cdot\hat{\mathbf{x}}}{r}\left(\mathbf{x}\cdot\mathbf{p} + i\boldsymbol{\sigma}\cdot\mathbf{L}\right) f(r)\, \varphi^\ell_{jm_j}$$

$$= -i\frac{\boldsymbol{\sigma}\cdot\hat{\mathbf{x}}}{r}\left\{ r\frac{\partial f(r)}{\partial r} + \left(1 \mp \left(j+\frac{1}{2}\right)\right) f(r) \right\} \varphi^\ell_{jm_j}$$

$$\text{for } j = \ell \pm 1/2 \tag{8.2.22a}$$

and

$$(\boldsymbol{\sigma}\cdot\mathbf{p})(\boldsymbol{\sigma}\cdot\hat{\mathbf{x}})\, f(r)\, \varphi^\ell_{jm_j} = -\frac{i}{r}\left[r\frac{\partial}{\partial r} + 1 \pm \left(j+\frac{1}{2}\right)\right] f(r)\, \varphi^\ell_{jm_j} \tag{8.2.22b}$$

$$\text{for } j = \ell \pm 1/2.$$

By means of (8.2.22a,b), the angle-dependent part of the momentum operator is eliminated, in analogy to the kinetic energy in nonrelativistic quantum mechanics. If one now substitutes (8.2.19), (8.2.21), and (8.2.22) into the time-independent Dirac equation (8.2.4), the radial components reduce to

$$\left(E - m + \frac{Z\alpha}{r}\right) G_{\ell j}(r) = -\frac{dF_{\ell j}(r)}{dr} \mp \left(j+\frac{1}{2}\right)\frac{F_{\ell j}(r)}{r}$$

$$\text{for } j = \ell \pm 1/2$$

$$\left(E + m + \frac{Z\alpha}{r}\right) F_{\ell j}(r) = \frac{dG_{\ell j}(r)}{dr} \mp \left(j+\frac{1}{2}\right)\frac{G_{\ell j}(r)}{r} \tag{8.2.23}$$

$$\text{for } j = \ell \pm 1/2.$$

This system of equations can be solved by making the substitutions

$$\begin{aligned}
\alpha_1 &= m + E, & \alpha_2 &= m - E, & \sigma &= \sqrt{m^2 - E^2} = \sqrt{\alpha_1\alpha_2} \\
\rho &= r\sigma, & k &= \pm\left(j+\tfrac{1}{2}\right), & \gamma &= Z\alpha
\end{aligned} \tag{8.2.24}$$

with the condition $E < m$ for bound states. One obtains

---

[8] $\boldsymbol{\sigma}\cdot\mathbf{a}\,\boldsymbol{\sigma}\cdot\mathbf{b} = \mathbf{a}\cdot\mathbf{b} + i\boldsymbol{\sigma}\cdot\mathbf{a}\times\mathbf{b}, \Rightarrow \boldsymbol{\sigma}\cdot\hat{\mathbf{x}}\,\boldsymbol{\sigma}\cdot\hat{\mathbf{x}} = 1$

$\mathbf{p}\cdot\frac{\mathbf{x}}{r} = \frac{1}{i}\boldsymbol{\nabla}\cdot\frac{\mathbf{x}}{r} = \frac{1}{i}\left(\frac{3}{r} - \mathbf{x}\cdot\frac{\mathbf{x}}{r^3}\right) = -\frac{2i}{r}$

$$\left(\frac{d}{d\rho} + \frac{k}{\rho}\right) F - \left(\frac{\alpha_2}{\sigma} - \frac{\gamma}{\rho}\right) G = 0$$
$$\left(\frac{d}{d\rho} - \frac{k}{\rho}\right) G - \left(\frac{\alpha_1}{\sigma} + \frac{\gamma}{\rho}\right) F = 0 \ . \tag{8.2.25}$$

Differentiating the first equation and inserting it into the second, one sees that, for large ρ, F and G are normalizable solutions that behave like $e^{-\rho}$. Thus, in (8.2.25) we make the ansatz

$$F(\rho) = f(\rho)e^{-\rho}, G(\rho) = g(\rho)e^{-\rho} \ , \tag{8.2.26}$$

which leads to

$$f' - f + \frac{kf}{\rho} - \left(\frac{\alpha_2}{\sigma} - \frac{\gamma}{\rho}\right) g = 0$$
$$g' - g - \frac{kg}{\rho} - \left(\frac{\alpha_1}{\sigma} + \frac{\gamma}{\rho}\right) f = 0 \ . \tag{8.2.27}$$

In order to solve the system (8.2.27), one introduces the power series:

$$g = \rho^s (a_0 + a_1 \rho + \dots) \ , \ a_0 \neq 0$$
$$f = \rho^s (b_0 + b_1 \rho + \dots) \ , \ b_0 \neq 0 \ . \tag{8.2.28}$$

Here, the same power s is assumed for g and f since different values would imply vanishing a_0 and b_0, as can be seen by substitution into (8.2.27) in the limit $\rho \to 0$. For the solution to be finite at $\rho = 0$, s would have to be greater than, or equal to, 1. Our experience with the Klein–Gordon equation, however, prepares us to admit s values that are somewhat smaller than 1. Substituting the power series into (8.2.27) and comparing the coefficients of $\rho^{s+\nu-1}$ yields for $\nu > 0$:

$$(s + \nu + k)b_\nu - b_{\nu-1} + \gamma a_\nu - \frac{\alpha_2}{\sigma} a_{\nu-1} = 0 \tag{8.2.29a}$$

$$(s + \nu - k)a_\nu - a_{\nu-1} - \gamma b_\nu - \frac{\alpha_1}{\sigma} b_{\nu-1} = 0 \ . \tag{8.2.29b}$$

For $\nu = 0$ one finds

$$(s + k)b_0 + \gamma a_0 = 0$$
$$(s - k)a_0 - \gamma b_0 = 0 \ . \tag{8.2.30}$$

This is a system of recursion relations. The coefficients a_0 and b_0 differ from zero, provided the determinant of their coefficients in (8.2.30) disappears, i.e.

$$s = \pm \left(k^2 - \gamma^2\right)^{1/2} \ . \tag{8.2.31}$$

The behavior of the wave function at the origin leads us to take the positive sign. Now, s depends only on k^2, i.e., only on j. Thus, the two states of opposite parity that belong to j turn out to have the same energy. A relationship between a_ν and b_ν is obtained by multiplying the first recursion relation by σ, the second by α_2, and then subtracting

$$b_\nu[\sigma(s + \nu + k) + \alpha_2\gamma] = a_\nu[\alpha_2(s + \nu - k) - \sigma\gamma],\qquad(8.2.32)$$

where we have used $\alpha_1\alpha_2 = \sigma^2$.

In the following we may convince ourselves that the power series obtained, which do not terminate, lead to divergent solutions. To this end, we investigate the asymptotic behavior of the solution. For large ν (and this is also decisive for the behavior at large r) it follows from (8.2.32) that $\sigma\nu b_\nu = \alpha_2\nu a_\nu$, thus

$$b_\nu = \frac{\alpha_2}{\sigma}\,a_\nu\,,$$

and from the first recursion relation (8.2.29a)

$$\nu b_\nu - b_{\nu-1} + \gamma a_\nu - \frac{\alpha_2}{\sigma}\,a_{\nu-1} = 0\,,$$

whence we finally find

$$b_\nu = \frac{2}{\nu}\,b_{\nu-1}\,,\quad a_\nu = \frac{2}{\nu}\,a_{\nu-1}$$

and thus, for the series,

$$\sum_\nu a_\nu\rho^\nu \sim \sum_\nu b_\nu\rho^\nu \sim \sum_\nu \frac{(2\rho)^\nu}{\nu!} \sim e^{2\rho}\,.$$

The two series would approach the asymptotic form $e^{2\rho}$. In order for the solution (8.2.26) to remain well-behaved for large ρ, the series must terminate. Due to the relation (8.2.32), when $a_\nu = 0$ we also have $b_\nu = 0$ and, according to the recursion relations (8.2.29), all subsequent coefficients are also zero, since the determinant of this system of equations does not vanish for $\nu > 0$. Let us assume that the first two vanishing coefficients are $a_{N+1} = b_{N+1} = 0$. The two recursion relations (8.2.29a,b) then yield the termination condition

$$\alpha_2 a_N = -\sigma b_N\,,\quad N = 0, 1, 2, \ldots\,.\qquad(8.2.33)$$

N is termed the "radial quantum number". We now set $\nu = N$ in (8.2.32) and apply the termination condition (8.2.33)

$$b_N\left[\sigma(s + N + k) + \alpha_2\gamma + \sigma(s + N - k) - \frac{\sigma^2}{\alpha_2}\gamma\right] = 0\,,$$

i.e., with (8.2.24)

$$2\sigma(s + N) = \gamma(\alpha_1 - \alpha_2) = 2E\gamma\,.\qquad(8.2.34)$$

We obtain E from this and also see that $E > 0$. According to (8.2.24), the quantity σ also contains the energy E. In the following we reintroduce c and, from (8.2.34), obtain

$$2(m^2c^4 - E^2)^{1/2}(s + N) = 2E\gamma.$$

Solving this equation for E yields the *energy levels:*

$$E = mc^2 \left[1 + \frac{\gamma^2}{(s+N)^2}\right]^{-\frac{1}{2}}. \tag{8.2.35}$$

It still remains to determine which values of k (according to (8.2.12) they are integers) are allowed for a particular value of N. For $N = 0$, the recursion relation (8.2.30) implies:

$$\frac{b_0}{a_0} = -\frac{\gamma}{s+k}$$

and from the termination condition (8.2.33), we have

$$\frac{b_0}{a_0} = -\frac{\alpha_2}{\sigma} < 0.$$

Since, as implied by (8.2.31), $s < |k|$, it follows from the first of these relations that

$$\frac{b_0}{a_0} \begin{cases} < 0 \text{ for } k > 0 \\ > 0 \text{ for } k < 0 \end{cases},$$

whilst from the second relation it always follows that $\frac{b_0}{a_0} < 0$, i.e., for $k < 0$ we arrive at a contradiction. Thus, for $N = 0$, the quantum number k can only take positive integer values. For $N > 0$, all positive and negative integer values are allowed for k. With the definition of the *principal quantum number*

$$n = N + |k| = N + j + \frac{1}{2} \tag{8.2.36}$$

and the value $s = \sqrt{k^2 - \gamma^2}$ from (8.2.31), equation (8.2.35) yields the energy levels

$$E_{n,j} = mc^2 \left[1 + \left(\frac{Z\alpha}{n - |k| + \sqrt{k^2 - (Z\alpha)^2}}\right)^2\right]^{-\frac{1}{2}}$$

$$= mc^2 \left[1 + \left(\frac{Z\alpha}{n - (j + \frac{1}{2}) + \sqrt{(j + \frac{1}{2})^2 - (Z\alpha)^2}}\right)^2\right]^{-\frac{1}{2}}. \tag{8.2.37}$$

Before we discuss the general result, let us look briefly at the nonrelativistic limit together with the leading corrections. This follows from (8.2.37) by expanding as a power series in $Z\alpha$:

$$E_{n,j} = mc^2 \left\{ 1 - \frac{Z^2\alpha^2}{2n^2} - \frac{(Z\alpha)^4}{2n^3}\left(\frac{1}{j+\frac{1}{2}} - \frac{3}{4n}\right) + \mathcal{O}((Z\alpha)^6) \right\}.$$

$$(8.2.38)$$

This expression agrees with the result obtained from the perturbation-theoretic calculation of the relativistic corrections (QM I, Eq. (12.5)).

We now discuss the energy levels given by (8.2.37) and their degeneracies. For the classification of the levels, we note that the quantum number $k = \pm\left(j+\frac{1}{2}\right)$ introduced in (8.2.12) belongs to the Pauli spinors $\varphi_{jm_j}^{(\pm)} = \varphi_{jm_j}^{\ell=j\mp\frac{1}{2}}$. Instead of k, one traditionally uses the quantum number ℓ. Positive k is thus associated with the smaller of the two values of ℓ belonging to the particular j considered. The quantum number k takes the values $k = \pm 1, \pm 2, \ldots$, and the principal quantum number n the values $n = 1, 2, \ldots$. We recall that for $N = 0$ the quantum number k must be positive and thus from (8.2.36), we have $k = n$ and, consequently, $\ell = n-1$ and $j = n-\frac{1}{2}$. Table 8.1 summarizes the values of the quantum numbers $k, j, j+\frac{1}{2}$ and ℓ for a given value of the principal quantum number n.

Table 8.2 gives the quantum numbers for $n = 1, 2$, and 3 and the spectroscopic notation for the energy levels $n L_j$. It should be emphasized that the orbital angular momentum $\mathbf{L}$ is not conserved and that the quantum number ℓ is really only a substitute for k, introduced to characterize parity.

k	± 1	± 2	$\ldots$	$\pm(n-1)$	n
j	1/2	3/2			$n-1/2$
$j+1/2$	1	2			n
ℓ	0	1		$n-2$	$n-1$
	1	2		$n-1$	

Table 8.1. Values of the quantum numbers $k, j, j+\frac{1}{2}$, and ℓ for a given principal quantum number n

n	N	$|k|$	k	j	ℓ	
1	0	1	1	1/2	0	$1S_{1/2}$
2	1	1	+1	1/2	0	$2S_{1/2}$
	1	1	-1	1/2	1	$2P_{1/2}$
	0	2	2	3/2	1	$2P_{3/2}$
3	2	1	1	1/2	0	$3S_{1/2}$
	2	1	-1	1/2	1	$3P_{1/2}$
	1	2	2	3/2	1	$3P_{3/2}$
	1	2	-2	3/2	2	$3D_{3/2}$
	0	3	3	5/2	2	$3D_{5/2}$

Table 8.2. The values of the quantum numbers; principal quantum number n, radial quantum number N, k, angular momentum j and ℓ

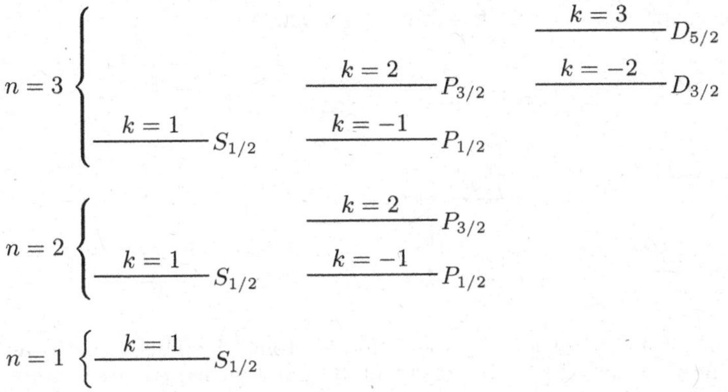

Fig. 8.3. The energy levels of the hydrogen atom according to the Dirac equation for values of the principal quantum number $n = 1, 2$, and 3

Figure 8.3 shows the relativistic energy levels of the hydrogen atom according to (8.2.37) for the values $n = 1, 2$, and 3 of the principal quantum number. The levels $2S_{1/2}$ and $2P_{1/2}$, the levels $3S_{1/2}$ and $3P_{1/2}$, the levels $3P_{3/2}$ and $3D_{3/2}$, etc. are degenerate. These pairs of degenerate levels correspond to opposite eigenvalues of the operator $K = 1 + \mathbf{L}\cdot\boldsymbol{\sigma}$, e.g., $2P_{3/2}$ has the value $k = 2$, whereas $2D_{3/2}$ possesses $k = -2$. The only nondegenerate levels are $1S_{1/2}, 2P_{3/2}, 3D_{5/2}$, etc. These are just the lowest levels for a fixed j, or the levels with radial quantum number $N = 0$, for which it was shown in the paragraph following (8.2.35) that the associated k can only be positive. The lowest energy levels are given in Table 8.3. The energy eigenvalues for $N = 0$ are, according to (8.2.37) and (8.2.36),

$$E = mc^2 \left[1 + \frac{\gamma^2}{k^2 - \gamma^2}\right]^{-\frac{1}{2}} = mc^2 \left[1 + \frac{\gamma^2}{n^2 - \gamma^2}\right]^{-\frac{1}{2}} = mc^2 \sqrt{1 - \gamma^2/n^2}\;.$$

$$(8.2.39)$$

Table 8.3. The lowest energy levels

	n	ℓ	j	$E_{n,j}/mc^2$
$1\,S_{1/2}$	1	0	$\frac{1}{2}$	$\sqrt{1 - (Z\alpha)^2}$
$2\,S_{1/2}$	2	0	$\frac{1}{2}$	$\sqrt{\frac{1+\sqrt{1-(Z\alpha)^2}}{2}}$
$2\,P_{1/2}$	2	1	$\frac{1}{2}$	$\sqrt{\frac{1+\sqrt{1-(Z\alpha)^2}}{2}}$
$2\,P_{3/2}$	2	1	$\frac{3}{2}$	$\frac{1}{2}\sqrt{4 - (Z\alpha)^2}$

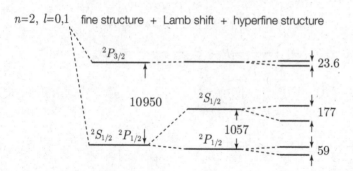

Fig. 8.4. Splitting of the energy levels of the hydrogen atom (MHz) due to the relativistic terms (fine structure, (Fig. 8.3)), the Lamb shift and the hyperfine structure

Figure 8.4 shows how the level $n = 2$, $l = 1$ (a single level according to the Schrödinger equation) splits according to Dirac theory (8.2.37) to yield the *fine structure*. Further weaker splitting, due to the Lamb shift and the hyperfine structure[9], is also shown. It should be noted that all levels are still $(2j + 1)$-fold degenerate since they do not depend on the quantum number m_j. This degeneracy is a general consequence of the spherical symmetry of the Hamiltonian (see the analogous discussion in QM I, Sect. 6.3). The fine-structure splitting between the $2P_{3/2}$ and the $2P_{1/2}$ and $2S_{1/2}$ levels is $10950 \, \text{MHz} \hat{=} 0.45 \times 10^{-4} \text{eV}$.

As has already been mentioned, it is usual to make use of the nonrelativistic notation to classify the levels. One specifies n, j, and ℓ, where ℓ is the index of the Pauli spinor, which really only serves to characterize the parity.

The $2\,S_{1/2}$ and $2\,P_{1/2}$ states are degenerate, as in first-order perturbation theory. This is not surprising since they are the two eigenstates of opposite parity for the same N and j. The $2\,P_{3/2}$ state has a higher energy than the $2\,P_{1/2}$ state. The energy difference arises from the *fine-structure splitting* caused by the spin–orbit interaction. In general, for a given n, the states with larger j have a higher energy. The ground-state energy

$$E_1 = mc^2 \sqrt{1 - (Z\alpha)^2} = mc^2 \left(1 - \frac{(Z\alpha)^2}{2} - \frac{(Z\alpha)^4}{8} \cdots \right) \qquad (8.2.40)$$

---

[9] Section 9.2.2 and QM I, Chap. 12

is doubly degenerate, with the two normalized spinors

$$\psi_{n=1,j=\frac{1}{2},m_j=\frac{1}{2}}(r,\vartheta,\varphi)$$

$$= \frac{(2mZ\alpha)^{3/2}}{\sqrt{4\pi}} \sqrt{\frac{1+\bar{\gamma}}{2\Gamma(1+2\bar{\gamma})}} (2mZ\alpha r)^{\bar{\gamma}-1}$$

$$\times e^{-mZ\alpha r} \begin{pmatrix} 1 \\ 0 \\ \frac{i(1-\bar{\gamma})}{Z\alpha}\cos\vartheta \\ \frac{i(1-\bar{\gamma})}{Z\alpha}\sin\vartheta\, e^{i\varphi} \end{pmatrix}$$

(8.2.41a)

and

$$\psi_{n=1,j=\frac{1}{2},m_j=-\frac{1}{2}}(r,\vartheta,\varphi)$$

$$= \frac{(2mZ\alpha)^{3/2}}{\sqrt{4\pi}} \sqrt{\frac{1+\bar{\gamma}}{2\Gamma(1+2\bar{\gamma})}} (2mZ\alpha r)^{\bar{\gamma}-1}$$

$$\times e^{-mZ\alpha r} \begin{pmatrix} 0 \\ 1 \\ \frac{i(1-\bar{\gamma})}{Z\alpha}\sin\vartheta\, e^{-i\varphi} \\ \frac{-i(1-\bar{\gamma})}{Z\alpha}\cos\vartheta \end{pmatrix}$$

(8.2.41b)

with $\bar{\gamma} = \sqrt{1 - Z^2\alpha^2}$ and the gamma function $\Gamma(x)$. The normalization is given by $\int d^3x \psi^\dagger_{n=1,j=\frac{1}{2},m_j=\pm\frac{1}{2}}(\vartheta,\varphi)\psi_{n=1,j=\frac{1}{2},m_j=\pm\frac{1}{2}}(\vartheta,\varphi) = 1$. The two spinors possess the quantum numbers $m_j = +1/2$ and $m_j = -1/2$. They are constructed from eigenfunctions of the orbital angular momentum: Y_{00} in the components 1 and 2 and $Y_{1,m=0,\pm1}$ in the components 3 and 4. In the nonrelativistic limit $\alpha \to 0, \bar{\gamma} \to 1, \frac{1-\bar{\gamma}}{Z\alpha} \longrightarrow 0$ these solutions reduce to the Schrödinger wave functions multiplied by Pauli spinors in the upper two components.

The solution (8.2.41) displays a weak singularity $r^{\bar{\gamma}-1} = r^{\sqrt{1-Z^2\alpha^2}-1} \approx r^{-Z^2\alpha^2/2}$. However, this only has a noticeable effect in a very tiny region:

$$r < \frac{1}{2mZ\alpha} e^{-\frac{2}{Z^2\alpha^2}} = \frac{10^{-16300/Z^2}}{2mZ\alpha} \quad .$$

Furthermore, for real nuclei with a finite radius, this singularity no longer occurs. For $Z\alpha > 1$, $\bar{\gamma}$ becomes imaginary and the solutions are therefore oscillatory. However, all real nuclei have $Z\alpha < 1$ and, furthermore, this limit is shifted for finite-sized nuclei.

Problems

8.1 Demonstrate the validity of the relation

$$(\boldsymbol{\sigma} \cdot \mathbf{p})(\boldsymbol{\sigma} \cdot \hat{\mathbf{x}}) f(r) \varphi_{jm_j}^{\ell} = -\frac{\mathrm{i}}{r} \left[r \frac{\partial}{\partial r} + 1 \pm \left(j + \frac{1}{2} \right) \right] f(r) \varphi_{jm_j}^{\ell} \ .$$

8.2 Prove the relation (8.2.16)

$$\varphi_{jm_j}^{(+)} = \frac{\boldsymbol{\sigma} \cdot \mathbf{x}}{r} \varphi_{jm_j}^{(-)}$$

that was given in connection with the solution of the Dirac equation for the hydrogen atom.

Hint: Make use of the fact that $\varphi_{jm_j}^{(-)}$ is an eigenfunction of $\boldsymbol{\sigma} \cdot \mathbf{L}$ and calculate the commutator $\left[\boldsymbol{\sigma} \cdot \mathbf{L}, \frac{\boldsymbol{\sigma} \cdot \mathbf{x}}{r} \right]$ (result: $\frac{2}{r} \left(r^2 \boldsymbol{\sigma} \cdot \boldsymbol{\nabla} - (\boldsymbol{\sigma} \cdot \mathbf{x})(\mathbf{x} \cdot \boldsymbol{\nabla}) - \boldsymbol{\sigma} \cdot \mathbf{x} \right)$) or the anticommutator.

8.3 Derive the recursion relations (8.2.29a,b) for the coefficients a_ν and b_ν.

8.4 Calculate the ground-state spinors of the hydrogen atom from the Dirac equation.

8.5 A charged particle is moving in a homogeneous electromagnetic field $\mathbf{B} = (0, 0, B)$ and $\mathbf{E} = (E_0, 0, 0)$. Choose the gauge $\mathbf{A} = (0, Bx, 0)$ and, taking as your starting point the Klein–Gordon equation, determine the energy levels.

9. The Foldy–Wouthuysen Transformation and Relativistic Corrections

In this chapter we present the Foldy–Wouthuysen transformation, by means of which the relativistic corrections may be computed for potentials which are more complex than $\frac{1}{r}$. After their evaluation higher order corrections will be discussed and a simple estimate of the Lamb-shift will be given.

9.1 The Foldy–Wouthuysen Transformation

9.1.1 Description of the Problem

Beyond the Coulomb potential, there are other potentials for which it is also important to be able to calculate the relativistic corrections. Relativistic corrections become increasingly important for nuclei of high atomic number, and it is exactly these, for which the nuclear diameter is no longer negligible, in which the potential deviates from the $1/r$ form. The canonical transformation of Foldy und Wouthuysen[1] transforms the Dirac equation into two decoupled two-component equations. The equation for the components 1 and 2 becomes identical to the Pauli equation in the nonrelativistic limit; it also contains additional terms that give rise to relativistic corrections. The energies for these components are positive. The equation for the components 3 and 4 describes negative energy states.

From the explicit solutions given in previous sections, it is evident that for positive energies the spinor components 1 and 2 are large, and the components 3 and 4 small. We seek a transformation that decouples the small and large components of the spinor from one another. In our treatment of the nonrelativistic limit (Sect. 5.3.5), we achieved this decoupling by eliminating the small components. We now wish to investigate this limit systematically and thereby derive the relativistic corrections. According to a classification that is now established in the literature, the Dirac Hamiltonian contains terms of two types: "odd" operators which couple large and small components (α^i, γ^i, γ_5) and "even" operators which do not couple the large and small components ($\mathbb{1}$, β, $\boldsymbol{\Sigma}$).

The canonical (unitary) transformation that achieves the required decoupling may be written in the form

---

[1] L.L. Foldy and S.A. Wouthuysen, Phys. Rev. **78**, 29 (1950)

$$\psi = e^{-iS}\psi' \; , \tag{9.1.1}$$

where, in general, S can be time dependent. From the Dirac equation, it then follows that

$$i\partial_t \psi = i\partial_t e^{-iS}\psi' = ie^{-iS}\partial_t\psi' + i\left(\partial_t e^{-iS}\right)\psi' = H\psi = He^{-iS}\psi' \tag{9.1.2a}$$

and, thus, we have the equation of motion for ψ':

$$i\partial_t\psi' = \left(e^{iS}(H - i\partial_t)e^{-iS}\right)\psi' \equiv H'\psi' \tag{9.1.2b}$$

with the Foldy–Wouthuysen-transformed Hamiltonian

$$H' = e^{iS}(H - i\partial_t)e^{-iS} \; . \tag{9.1.2c}$$

The time derivative on the right-hand side of this equation only acts on e^{-iS}. One endeavors to construct S such that H' contains no odd operators. For free particles, one can find an exact transformation, but otherwise one has to rely on a series expansion in powers of $\frac{1}{m}$ and, by successive transformations, satisfy this condition to each order of $\frac{1}{m}$. In fact, each power of $\frac{1}{m}$ corresponds to a factor $\frac{p}{mc} \sim \frac{v}{c}$; in the atomic domain this is approximately equal to Sommerfeld's fine-structure constant α, since, from Heisenberg's uncertainty relation, we have $\frac{v}{c} \approx \frac{\hbar}{cm\Delta x} \approx \frac{\hbar}{cma} = \alpha$.

9.1.2 Transformation for Free Particles

For free particles, the Dirac Hamiltonian simplifies to

$$H = \boldsymbol{\alpha} \cdot \mathbf{p} + \beta m \tag{9.1.3}$$

with the momentum operator $\mathbf{p} = -i\boldsymbol{\nabla}$. Since $\{\boldsymbol{\alpha}, \beta\} = 0$, the problem is analogous to that of finding a unitary operator that diagonalizes the Pauli Hamiltonian

$$H = \sigma_x B_x + \sigma_z B_z \; , \tag{9.1.4a}$$

so that, after transformation, H contains only $\mathbb{1}$ and σ_z. This is achieved by a rotation about the y axis through an angle ϑ_0 determined by $(B_x, B_y, 0)$:

$$e^{\frac{1}{2}\sigma_y \vartheta_0} = e^{\frac{1}{2}\sigma_z \sigma_x \vartheta_0} \; . \tag{9.1.4b}$$

This equation suggests the ansatz

$$e^{\pm iS} = e^{\pm \beta \frac{\boldsymbol{\alpha}\cdot\mathbf{p}}{|\mathbf{p}|}\vartheta(\mathbf{p})} = \cos\vartheta \pm \frac{\beta\,\boldsymbol{\alpha}\cdot\mathbf{p}}{|\mathbf{p}|}\sin\vartheta \; . \tag{9.1.5}$$

Here, S is time independent. The last relation results from the Taylor expansion of the exponential function and from

$$(\boldsymbol{\alpha} \cdot \mathbf{p})^2 = \alpha^i \alpha^j \, p^i p^j = \frac{1}{2}\{\alpha^i, \alpha^j\} \, p^i p^j = \delta^{ij} p^i p^j = \mathbf{p}^2 \qquad (9.1.6a)$$

$$(\beta \boldsymbol{\alpha} \cdot \mathbf{p})^2 = \beta \boldsymbol{\alpha} \cdot \mathbf{p} \, \beta \boldsymbol{\alpha} \cdot \mathbf{p} = -\beta^2 (\boldsymbol{\alpha} \cdot \mathbf{p})^2 = -\mathbf{p}^2 \ . \qquad (9.1.6b)$$

Inserting (9.1.5) into (9.1.2c), one obtains H' as

$$
\begin{aligned}
H' &= e^{\beta \frac{\boldsymbol{\alpha} \cdot \mathbf{p}}{|\mathbf{p}|} \vartheta}(\boldsymbol{\alpha} \cdot \mathbf{p} + \beta m)\left(\cos\vartheta - \frac{\beta \boldsymbol{\alpha} \cdot \mathbf{p}}{|\mathbf{p}|}\sin\vartheta\right) \\
&= e^{\beta \frac{\boldsymbol{\alpha} \cdot \mathbf{p}}{|\mathbf{p}|}\vartheta}\left(\cos\vartheta + \frac{\beta \boldsymbol{\alpha} \cdot \mathbf{p}}{|\mathbf{p}|}\sin\vartheta\right)(\boldsymbol{\alpha} \cdot \mathbf{p} + \beta m) \\
&= e^{2\beta \frac{\boldsymbol{\alpha} \cdot \mathbf{p}}{|\mathbf{p}|}\vartheta}(\boldsymbol{\alpha} \cdot \mathbf{p} + \beta m) = \left(\cos 2\vartheta + \frac{\beta \boldsymbol{\alpha} \cdot \mathbf{p}}{|\mathbf{p}|}\sin 2\vartheta\right)(\boldsymbol{\alpha} \cdot \mathbf{p} + \beta m) \\
&= \boldsymbol{\alpha} \cdot \mathbf{p}\left(\cos 2\vartheta - \frac{m}{|\mathbf{p}|}\sin 2\vartheta\right) + \beta m\left(\cos 2\vartheta + \frac{|\mathbf{p}|}{m}\sin 2\vartheta\right) \ . \quad (9.1.7)
\end{aligned}
$$

The requirement that the odd terms disappear yields the condition $\tan 2\vartheta = \frac{|\mathbf{p}|}{m}$, from whence it follows that

$$\sin 2\vartheta = \frac{\tan 2\vartheta}{(1 + \tan^2 2\vartheta)^{1/2}} = \frac{p}{(m^2 + p^2)^{1/2}} \ , \qquad \cos 2\vartheta = \frac{m}{(m^2 + p^2)^{1/2}} \ .$$

Substituting this into (9.1.7) finally yields:

$$H' = \beta m \left(\frac{m}{E} + \frac{\mathbf{p} \cdot \mathbf{p}}{mE}\right) = \beta\sqrt{\mathbf{p}^2 + m^2} \ . \qquad (9.1.8)$$

Thus, H' has now been diagonalized. The diagonal components are nonlocal[2] Hamiltonians $\pm\sqrt{\mathbf{p}^2 + m^2}$. In our first attempt (Sect. 5.2.1) to construct a nonrelativistic theory with a first order time derivative, we encountered the operator $\sqrt{\mathbf{p}^2 + m^2}$. The replacement of $\sqrt{\mathbf{p}^2 + m^2}$ by linear operators necessarily leads to a four-component theory with negative as well as positive energies. Even now, H' still contains the character of the four-component theory due to its dependence on the matrix β, which is different for the upper and lower components. Such an exact transformation is only feasible for free particles.

9.1.3 Interaction with the Electromagnetic Field

Of primary interest, of course, is the case of non-vanishing electromagnetic fields. We assume that the potentials $\mathbf{A}$ and Φ are given, such that the Dirac Hamiltonian reads:

$$
\begin{aligned}
H &= \boldsymbol{\alpha} \cdot (\mathbf{p} - e\mathbf{A}) + \beta m + e\Phi & (9.1.9a) \\
&= \beta m + \mathcal{E} + \mathcal{O} \ . & (9.1.9b)
\end{aligned}
$$

---

[2] They are nonlocal because they contain derivatives of all orders. In a discrete theory, the nth derivative signifies an interation between lattice sites that are n units apart.

Here, we have introduced a decomposition into a term proportional to β, an even term $\mathcal{E}$, and an odd term $\mathcal{O}$:

$$\mathcal{E} = e\Phi \quad \text{and} \quad \mathcal{O} = \boldsymbol{\alpha}(\mathbf{p} - e\mathbf{A}) . \tag{9.1.10}$$

These have different commutation properties with respect to β:

$$\beta\mathcal{E} = \mathcal{E}\beta , \quad \beta\mathcal{O} = -\mathcal{O}\beta . \tag{9.1.11}$$

The solution in the field-free case (9.1.5) implies that, for small ϑ, i.e., in the nonrelativistic limit,

$$iS = \beta \frac{\boldsymbol{\alpha} \cdot \mathbf{p}}{|\mathbf{p}|} \vartheta \sim \beta\boldsymbol{\alpha} \frac{\mathbf{p}}{2m} .$$

We can thus expect that successive transformations of this type will lead to an expansion in $\frac{1}{m}$. In the evaluation of H', we make use of the Baker–Hausdorff identity[3]

$$H' = H + i[S, H] + \frac{i^2}{2}[S, [S, H]] + \frac{i^3}{6}[S, [S, [S, H]]] +$$
$$+ \frac{i^4}{24}[S, [S, [S, [S, H]]]] - \dot{S} - \frac{i}{2}[S, \dot{S}] - \frac{i^2}{6}[S, [S, \dot{S}]] , \tag{9.1.12}$$

given here only to the order required. The odd terms are eliminated up to order m^{-2}, whereas the even ones are calculated up to order m^{-3}.

In analogy to the procedure for free particles, and according to the remark following Eq. (9.1.11), we write for S:

$$S = -i\beta\mathcal{O}/2m . \tag{9.1.13}$$

For the second term in (9.1.12), we find

$$i[S, H] = -\mathcal{O} + \frac{\beta}{2m}[\mathcal{O}, \mathcal{E}] + \frac{1}{m}\beta\mathcal{O}^2 , \tag{9.1.14}$$

obtained using the straightforward intermediate steps

$$[\beta\mathcal{O}, \beta] = \beta\mathcal{O}\beta - \beta\beta\mathcal{O} = -2\mathcal{O}$$
$$[\beta\mathcal{O}, \mathcal{E}] = \beta[\mathcal{O}, \mathcal{E}] \tag{9.1.15}$$
$$[\beta\mathcal{O}, \mathcal{O}] = \beta\mathcal{O}^2 - \mathcal{O}\beta\mathcal{O} = 2\beta\mathcal{O}^2 .$$

Before calculating the higher commutators, let us immediately draw attention to the fact that the first term in (9.1.14) cancels out the term $\mathcal{O}$ in H. Hence, the aim of eliminating the odd operator $\mathcal{O}$ by transformation has been attained; although new odd terms have been generated, e.g., the second term in (9.1.14), these have an additional factor m^{-1}. We now address the other terms in (9.1.12).

The additional commutator with iS can be written immediately by using (9.1.14), (9.1.15), and (9.1.11):

---

[3] $e^A B e^{-A} = B + [A, B] + \ldots + \frac{1}{n!}[A, [A, \ldots, [A, B]\ldots]] + \ldots$

$$\frac{i^2}{2}[S,[S,H]] = -\frac{\beta\mathcal{O}^2}{2m} - \frac{1}{8m^2}[\mathcal{O},[\mathcal{O},\mathcal{E}]] - \frac{1}{2m^2}\mathcal{O}^3 \ ,$$

and likewise,

$$\frac{i^3}{3!}[S,[S,[S,H]]] = \frac{\mathcal{O}^3}{6m^2} - \frac{1}{6m^3}\beta\mathcal{O}^4 - \frac{\beta}{48m^3}[\mathcal{O},[\mathcal{O},[\mathcal{O},\mathcal{E}]]] \ .$$

For the odd operators, it is sufficient to include terms up to order m^{-2} and hence the third term on the right-hand side may be neglected. The next contributions to (9.1.12), written only up to the necessary order in $1/m$, are:

$$\frac{i^4}{4!}[S,[S,[S,[S,H]]]] = \frac{\beta\mathcal{O}^4}{24m^3}$$

$$-\dot{S} = \frac{i\beta\dot{\mathcal{O}}}{2m}$$

$$-\frac{i}{2}[S,\dot{S}] = -\frac{i}{8m^2}[\mathcal{O},\dot{\mathcal{O}}] \ .$$

All in all, one obtains for H':

$$H' = \beta m + \beta\left(\frac{\mathcal{O}^2}{2m} - \frac{\mathcal{O}^4}{8m^3}\right) + \mathcal{E} - \frac{1}{8m^2}[\mathcal{O},[\mathcal{O},\mathcal{E}]] - \frac{i}{8m^2}[\mathcal{O},\dot{\mathcal{O}}]$$

$$+\frac{\beta}{2m}[\mathcal{O},\mathcal{E}] - \frac{\mathcal{O}^3}{3m^2} + \frac{i\beta\dot{\mathcal{O}}}{2m} \equiv \beta m + \mathcal{E}' + \mathcal{O}' \ . \tag{9.1.16}$$

Here, $\mathcal{E}$ and all even powers of $\mathcal{O}$ have been combined into a new even term $\mathcal{E}'$, and the odd powers into a new odd term $\mathcal{O}'$. The odd terms now occur only to orders of at least $\frac{1}{m}$. To reduce them further, we apply another Foldy–Wouthuysen transformation

$$S' = \frac{-i\beta}{2m}\mathcal{O}' = \frac{-i\beta}{2m}\left(\frac{\beta}{2m}[\mathcal{O},\mathcal{E}] - \frac{\mathcal{O}^3}{3m^2} + \frac{i\beta\dot{\mathcal{O}}}{2m}\right) \ . \tag{9.1.17}$$

This transformation yields:

$$H'' = e^{iS'}(H' - i\partial_t)e^{-iS'} = \beta m + \mathcal{E}' + \frac{\beta}{2m}[\mathcal{O}',\mathcal{E}'] + \frac{i\beta\dot{\mathcal{O}}'}{2m} \tag{9.1.18}$$

$$\equiv \beta m + \mathcal{E}' + \mathcal{O}'' \ .$$

Since $\mathcal{O}'$ is of order $1/m$, in $\mathcal{O}''$ there are now only terms of order $1/m^2$. This transformation also generates further even terms, which, however, are all of higher order. For example, $\beta\mathcal{O}'^2/2m = \beta e^2\mathbf{E}^2/8m^3 \sim \beta e^4/m^3r^4 \sim \text{Ry}\,\alpha^4$. By means of the transformation

$$S'' = \frac{-i\beta\mathcal{O}''}{2m} \ , \tag{9.1.19}$$

the odd term $\mathcal{O}'' \sim \mathcal{O}\left(\frac{1}{m^2}\right)$ is also eliminated. The result is the operator

$$H''' = e^{iS''}(H'' - i\partial_t)e^{-iS''} = \beta m + \mathcal{E}'$$

$$= \beta\left(m + \frac{\mathcal{O}^2}{2m} - \frac{\mathcal{O}^4}{8m^3}\right) + \mathcal{E} - \frac{1}{8m^2}\left[\mathcal{O}, [\mathcal{O}, \mathcal{E}] + i\dot{\mathcal{O}}\right], \qquad (9.1.20)$$

which now only consists of even terms.

In order to bring the Hamiltonian H''' into its final form, we have to substitute (9.1.10) and rewrite the individual terms as follows:

2nd term of H''':

$$\frac{\mathcal{O}^2}{2m} = \frac{1}{2m}(\boldsymbol{\alpha} \cdot (\mathbf{p} - e\mathbf{A}))^2 = \frac{1}{2m}(\mathbf{p} - e\mathbf{A})^2 - \frac{e}{2m}\boldsymbol{\Sigma} \cdot \mathbf{B}, \qquad (9.1.21a)$$

since

$$\alpha^i\alpha^j = \alpha^i\beta^2\alpha^j = -\gamma^i\gamma^j = -\frac{1}{2}\left(\{\gamma^i, \gamma^j\} + [\gamma^i, \gamma^j]\right) = -g^{ij} + i\varepsilon^{ijk}\Sigma^k$$

$$= \delta^{ij} + i\varepsilon^{ijk}\Sigma^k$$

and the mixed term with ε^{ijk} yields:

$$-e\left(p^iA^j + A^ip^j\right)i\varepsilon^{ijk}\Sigma^k = -ie\left((p^iA^j) + A^jp^i + A^ip^j\right)\varepsilon^{ijk}\Sigma^k$$

$$= -e\left(\partial_iA^j\right)\varepsilon^{ijk}\Sigma^k = -e\,\mathbf{B}\cdot\boldsymbol{\Sigma}.$$

5th term of H''':

Evaluation of the second argument of the commutator gives

$$\left([\mathcal{O}, \mathcal{E}] + i\dot{\mathcal{O}}\right) = [\alpha^i(p^i - eA^i), e\Phi] - ie\alpha^i\dot{A}^i$$

$$= -ie\alpha^i\left(\partial_i\Phi + \dot{A}^i\right) = ie\alpha^iE^i.$$

It then remains to compute

$$[\mathcal{O}, \boldsymbol{\alpha}\cdot\mathbf{E}] = \alpha^i\alpha^j(p^i - eA^i)E^j - \alpha^jE^j\alpha^i(p^i - eA^i)$$

$$= (p^i - eA^i)E^i - E^i(p^i - eA^i)$$

$$+i\varepsilon^{ijk}\Sigma^k(p^i - eA^i)E^j - i\varepsilon^{jik}\Sigma^kE^j(p^i - eA^i)$$

$$= (p^iE^i) + \boldsymbol{\Sigma}\cdot\boldsymbol{\nabla}\times\mathbf{E} - 2i\boldsymbol{\Sigma}\cdot\mathbf{E}\times(\mathbf{p} - e\mathbf{A}).$$

Hence, the 5th term in H''' reads:

$$-\frac{ie}{8m^2}[\mathcal{O}, \boldsymbol{\alpha}\cdot\mathbf{E}] = -\frac{e}{8m^2}\operatorname{div}\mathbf{E} - \frac{ie}{8m^2}\boldsymbol{\Sigma}\cdot\boldsymbol{\nabla}\times\mathbf{E}$$

$$-\frac{e}{4m^2}\boldsymbol{\Sigma}\cdot\mathbf{E}\times(\mathbf{p} - e\mathbf{A}). \qquad (9.1.21b)$$

Inserting (9.1.10) and (9.1.21a,b) into (9.1.20), one obtains the final expression for H''':

$$H''' = \beta\left(m + \frac{(\mathbf{p} - e\mathbf{A})^2}{2m} - \frac{1}{8m^3}\left[(\mathbf{p} - e\mathbf{A})^2 - e\boldsymbol{\Sigma}\cdot\mathbf{B}]^2\right) + e\Phi\right.$$
$$- \frac{e}{2m}\beta\boldsymbol{\Sigma}\cdot\mathbf{B} - \frac{ie}{8m^2}\boldsymbol{\Sigma}\cdot\operatorname{curl}\mathbf{E} \qquad (9.1.22)$$
$$- \frac{e}{4m^2}\boldsymbol{\Sigma}\cdot\mathbf{E}\times(\mathbf{p} - e\mathbf{A}) - \frac{e}{8m^2}\operatorname{div}\mathbf{E}\,.$$

The Hamiltonian H''' no longer contains any odd operators. Hence, the components 1 and 2 are no longer coupled to the components 3 and 4. The eigenfunctions of H''' can be represented by two-component spinors in the upper and lower components of ψ', which correspond to positive and negative energies. For $\psi' = \binom{\varphi}{0}$, the Dirac equation in the Foldy–Wouthuysen representation acquires the following form:

$$i\frac{\partial\varphi}{\partial t} = \left\{m + e\Phi + \frac{1}{2m}(\mathbf{p} - e\mathbf{A})^2 - \frac{e}{2m}\boldsymbol{\sigma}\cdot\mathbf{B} - \frac{\mathbf{p}^4}{8m^3}\right.$$
$$\left. - \frac{e}{4m^2}\boldsymbol{\sigma}\cdot\mathbf{E}\times(\mathbf{p} - e\mathbf{A}) - \frac{e}{8m^2}\operatorname{div}\mathbf{E}\right\}\varphi\,. \qquad (9.1.23)$$

Here, φ is a two-component spinor and the equation is identical to the Pauli equation plus relativistic corrections. The first four terms on the right-hand side of (9.1.23) are: rest energy, potential, kinetic energy, and coupling of the magnetic moment $\boldsymbol{\mu} = \frac{e}{2m}\boldsymbol{\sigma} = 2\frac{e}{2m}\mathbf{S}$ to the magnetic field $\mathbf{B}$. As was discussed in detail in Sect. 5.3.5.2, the gyromagnetic ratio (Landé factor) is obtained from the Dirac equation as $g = 2$. The three subsequent terms are the relativistic corrections, which will be discussed in the next section.

Remark. Equation (9.1.23) gives only the leading term that follows from $\mathcal{O}^4$, which is still contained in full in (9.1.22), namely $\mathbf{p}^4$. The full expression is

$$-\frac{\beta}{8m^3}\mathcal{O}^4 = -\frac{\beta}{8m^3}((\mathbf{p} - e\mathbf{A})^2 - e\boldsymbol{\Sigma}\mathbf{B})^2 = -\frac{\beta}{8m^3}[(\mathbf{p} - e\mathbf{A})^4 + e^2\mathbf{B}^2 +$$
$$+ e\boldsymbol{\Sigma}\cdot\Delta\mathbf{B} - 2e\boldsymbol{\Sigma}\cdot\mathbf{B}(\mathbf{p} - e\mathbf{A})^2 - 2ie\sigma_j\boldsymbol{\nabla}B_j(\mathbf{p} - e\mathbf{A})]\,.$$

It should also be noted that, in going from (9.1.22) to (9.1.23), it has been assumed that $\operatorname{curl}\mathbf{E} = 0$.

9.2 Relativistic Corrections and the Lamb Shift

9.2.1 Relativistic Corrections

We now discuss the relativistic corrections that emerge from (9.1.22) and (9.1.23). We take $\mathbf{E} = -\boldsymbol{\nabla}\Phi(r) = -\frac{1}{r}\frac{\partial\Phi}{\partial r}\mathbf{x}$ and $\mathbf{A} = 0$. Hence, $\operatorname{curl}\mathbf{E} = 0$ and

$$\boldsymbol{\Sigma}\cdot\mathbf{E}\times\mathbf{p} = -\frac{1}{r}\frac{\partial\Phi}{\partial r}\boldsymbol{\Sigma}\cdot\mathbf{x}\times\mathbf{p} = -\frac{1}{r}\frac{\partial\Phi}{\partial r}\boldsymbol{\Sigma}\cdot\mathbf{L}\,. \qquad (9.2.1)$$

Equation (9.1.23) contains three correction terms:

$$H_1 = -\frac{(\mathbf{p}^2)^2}{8m^3} \qquad \text{\textit{relativistic mass correction}} \qquad (9.2.2\text{a})$$

$$H_2 = \frac{e}{4m^2} \frac{1}{r} \frac{\partial \Phi}{\partial r} \, \boldsymbol{\sigma} \cdot \mathbf{L} \qquad \text{\textit{spin-orbit coupling}} \qquad (9.2.2\text{b})$$

$$H_3 = -\frac{e}{8m^2} \operatorname{div} \mathbf{E} = \frac{e}{8m^2} \boldsymbol{\nabla}^2 \Phi(\mathbf{x}) \qquad \text{\textit{Darwin term.}} \qquad (9.2.2\text{c})$$

Taken together, these terms lead to the perturbation Hamiltonian

$$H_1 + H_2 + H_3 = -\frac{(\mathbf{p}^2)^2}{8m^3 c^2} + \frac{1}{4m^2 c^2} \frac{1}{r} \frac{\partial V}{\partial r} \boldsymbol{\sigma} \cdot \mathbf{L} + \frac{\hbar^2}{8m^2 c^2} \boldsymbol{\nabla}^2 V(\mathbf{x}) \; ;$$
$$(9.2.2\text{d})$$

$(V = e\Phi)$. The order of magnitude of each of these corrections can be obtained from the Heisenberg uncertainty relation

$$\text{Ry} \times \left(\frac{p}{mc}\right)^2 = \text{Ry} \left(\frac{v}{c}\right)^2 = \text{Ry} \, \alpha^2 = mc^2 \alpha^4 \; ,$$

where $\alpha = e_0^2 \; (= e_0^2/\hbar c)$ is the fine-structure constant. The Hamiltonian (9.2.2d) gives rise to the fine structure in the atomic energy levels. The perturbation calculation of the energy shift for hydrogen-like atoms of nuclear charge Z was presented in Chap. 12 of QM I; the result in first-order perturbation theory is

$$\Delta E_{n, j=\ell\pm\frac{1}{2}, \ell} = \frac{\text{Ry} \, Z^2}{n^2} \frac{(Z\alpha)^2}{n^2} \left\{ \frac{3}{4} - \frac{n}{j+\frac{1}{2}} \right\} . \qquad (9.2.3)$$

The energy eigenvalues depend, apart from on n, only on j. Accordingly, the $(n = 2)$ levels ${}^2S_{1/2}$ and ${}^2P_{1/2}$ are degenerate. This degeneracy is also present in the exact solution of the Dirac equation (see (8.2.37) and Fig. 8.3). The determination of the relativistic perturbation terms H_1, H_2, and H_3 from the Dirac theory thus also provides a unified basis for the calculation of the fine-structure corrections $\mathcal{O}(\alpha^2)$.

Remarks:

(i) An heuristic interpretation of the relativistic corrections was given in QM I, Chap. 12. The term H_1 follows from the Taylor expansion of the relativistic kinetic energy $\sqrt{\mathbf{p}^2 + m^2}$. The term H_2 can be explained by transforming into the rest frame of the electron. Its spin experiences the magnetic field that is generated by the nucleus, which, in this frame, orbits around the electron. The term H_3 can be interpreted in terms of the "Zitterbewegung", literally "trembling motion", a fluctuation in the position of the electron with an amplitude $\delta x = \hbar c/m$.

(ii) The occurrence of additional interaction terms in the Foldy–Wouthuysen representation can be understood as follows: An analysis of the transformation from the Dirac representation ψ to ψ' shows that the relationship is nonlocal[4]

---

[4] L.L. Foldy and S.A. Wouthuysen, Phys. Rev. **78**, 29 (1950)

$$\psi'(\mathbf{x}) = \int d^3x' \, K(\mathbf{x}, \mathbf{x}')\psi(\mathbf{x}') \,,$$

where the kernel of the integral $K(\mathbf{x}, \mathbf{x}')$ is of a form such that, at the position $\mathbf{x}$, $\psi'(\mathbf{x})$ consists of ψ contributions stemming from a region of size $\sim \lambda_c$ around the point $\mathbf{x}$; here, λ_c is the Compton wavelength of the particle. Thus, the original sharply localized Dirac spinor transforms in the Foldy–Wouthuysen representation into a spinor which seems to correspond to a particle that extends over a finite region. The reverse is also true: The effective potential that acts on a spinor in the Foldy–Wouthuysen representation at point $\mathbf{x}$ consists of contributions from the original potential $\mathbf{A}(\mathbf{x})$, $\Phi(\mathbf{x})$ averaged over a region around $\mathbf{x}$. The full potential thus has the form of a multipole expansion of the original potential. This viewpoint enables one to understand the interaction of the magnetic moment, the spin–orbit coupling, and the Darwin term.

(iii) Since the Foldy–Wouthuysen transformation is, in general, time dependent, the expectation value of H''' is generally different to the expectation value of H. In the event that $\mathbf{A}(\mathbf{x})$ and $\Phi(\mathbf{x})$ are time independent, i.e., time-independent electromagnetic fields, then S is likewise time independent. This means that the matrix elements of the Dirac Hamiltonian, and in particular its expectation value, are the same in both representations.

(iv) An alternative method[5] of deriving the relativistic corrections takes as its starting point the resolvent $R = \frac{1}{H - mc^2 - z}$ of the Dirac Hamiltonian H. This is analytic in $\frac{1}{c}$ at $c = \infty$ and can be expanded in $\frac{1}{c}$. In zeroth order one obtains the Pauli Hamiltonian, and in $\mathcal{O}(\frac{1}{c^2})$ the relativistic corrections.

9.2.2 Estimate of the Lamb Shift

There are two further effects that also lead to shifts and splitting of the electronic energy levels in atoms. The first is the hyperfine interaction that stems from the magnetic field of the nucleus (see QM I, Chap. 12), and the second is the Lamb shift, for which we shall now present a simplified theory.[6]

The zero-point fluctuations of the quantized radiation field couple to the electron in the atom, causing its position to fluctuate such that it experiences a smeared-out Coulomb potential from the nucleus. This effect is qualitatively similar to the Darwin term, except that the mean square fluctuation in the electron position is now smaller: We consider the change in the potential due to a small displacement $\delta\mathbf{x}$:

$$V(\mathbf{x} + \delta\mathbf{x}) = V(\mathbf{x}) + \delta\mathbf{x}\,\boldsymbol{\nabla}V(\mathbf{x}) + \frac{1}{2}\delta x_i \delta x_j \nabla_i \nabla_j V(\mathbf{x}) + \dots \quad . \qquad (9.2.4)$$

Assuming that the mean value of the fluctuation is $\langle\delta\mathbf{x}\rangle = 0$, we obtain an additional potential

---

[5] F. Gesztesy, B. Thaller, and H. Grosse, Phys. Rev. Lett. **50**, 625 (1983)

[6] Our simple estimate of the Lamb shift follows that of T.A. Welton, Phys. Rev. **74**, 1157 (1948).

$$\Delta H_{\text{Lamb}} = \langle V(\mathbf{x} + \delta\mathbf{x}) - V(\mathbf{x}) \rangle = \frac{1}{6} \langle (\delta\mathbf{x})^2 \rangle \mathbf{\nabla}^2 V(\mathbf{x})$$

$$= \frac{1}{6} \langle (\delta\mathbf{x})^2 \rangle \, 4\pi \, Z\alpha\hbar c \, \delta^{(3)}(\mathbf{x}) \,. \tag{9.2.5}$$

The brackets $\langle \, \rangle$ denote the quantum-mechanical expectation value in the vacuum state of the radiation field. In first order perturbation theory, the additional potential (9.2.5) only influences s waves. These experience an energy shift of

$$\Delta E_{\text{Lamb}} = \frac{2\pi Z\alpha\hbar c}{3} \langle (\delta\mathbf{x})^2 \rangle |\psi_{n,\ell=0}(0)|^2$$

$$= \frac{(2mcZ\alpha)^3}{12\hbar^2} \frac{Z\alpha c}{n^3} \langle (\delta\mathbf{x})^2 \rangle \delta_{\ell,0} \,, \tag{9.2.6}$$

where we have used $\psi_{n,\ell=0}(0) = \frac{1}{\sqrt{\pi}} \left(\frac{m\alpha cZ}{n\hbar} \right)^{3/2}$. The energy shift for the $p, d, \ldots$ electrons is much smaller than that of the s waves due to the fact that they have $\psi(0) = 0$, even when one allows for the finite extent of the nucleus. A more exact theory of the Lamb shift would include, not only the finite size of the nucleus, but also the fact that not all contributing effects can be expressed in the form ΔV, as is assumed in this simplified theory.

We now need to estimate $\langle (\delta\mathbf{x})^2 \rangle$, i.e., find a connection between $\delta\mathbf{x}$ and the fluctuations of the radiation field. To this end, we begin with the nonrelativistic Heisenberg equation for the electron:

$$m \, \delta\ddot{\mathbf{x}} = e\mathbf{E} \,. \tag{9.2.7}$$

The Fourier transformation

$$\delta\mathbf{x}(t) = \int\limits_{-\infty}^{\infty} \frac{d\omega}{2\pi} \, e^{-i\omega t} \delta\mathbf{x}_\omega \tag{9.2.8}$$

yields

$$\langle (\delta\mathbf{x}(t))^2 \rangle = \int\limits_{-\infty}^{\infty} \frac{d\omega}{2\pi} \int\limits_{-\infty}^{\infty} \frac{d\omega'}{2\pi} \langle \delta\mathbf{x}_\omega \delta\mathbf{x}_{\omega'} \rangle \,. \tag{9.2.9}$$

Due to the invariance with respect to translation in time, this mean square fluctuation is time independent, and can thus be calculated at $t = 0$. From (9.2.7) it follows that

$$\delta\mathbf{x}_\omega = -\frac{e}{m} \frac{\mathbf{E}_\omega}{\omega^2} \,. \tag{9.2.10}$$

For the radiation field we use the Coulomb gauge, also transverse gauge, div $\mathbf{A} = 0$. Then, due to the absence of sources, we have

$$\mathbf{E}(t) = -\frac{1}{c}\dot{\mathbf{A}}(0,t) \,. \tag{9.2.11}$$

The vector potential of the radiation field can be represented in terms of the creation and annihilation operators $a^\dagger_{\mathbf{k},\lambda}(a_{\mathbf{k},\lambda})$ for photons with wave vector $\mathbf{k}$, polarization λ, and polarization vector $\varepsilon_{\mathbf{k},\lambda}(\lambda = 1,2)$[7]:

$$\mathbf{A}(\mathbf{x},t) = \sum_{\mathbf{k},\lambda} \sqrt{\frac{\hbar\, 2\pi c}{Vk}} \left(a_{\mathbf{k},\lambda}\varepsilon_{\mathbf{k},\lambda}e^{i(\mathbf{k}\mathbf{x}-ckt)} + a^\dagger_{\mathbf{k},\lambda}\varepsilon^*_{\mathbf{k},\lambda}e^{-i(\mathbf{k}\mathbf{x}-ckt)}\right) \,. \tag{9.2.12}$$

The polarization vectors are orthogonal to one another and to $\mathbf{k}$. From (9.2.12), one obtains the time derivative

$$-\frac{1}{c}\dot{\mathbf{A}}(0,t) = \frac{1}{c}\sum_{\mathbf{k},\lambda} \sqrt{\frac{\hbar\, 2\pi c}{Vk}}\, ick \left(a_{\mathbf{k},\lambda}\varepsilon_{\mathbf{k},\lambda}e^{-ickt} - a^\dagger_{\mathbf{k},\lambda}\varepsilon^*_{\mathbf{k},\lambda}e^{ickt}\right)$$

and the Fourier-transformed electric field

$$\mathbf{E}_\omega = \int_{-\infty}^{\infty} dt\, e^{i\omega t}\mathbf{E}(t)$$

$$= i\sum_{\mathbf{k},\lambda} \sqrt{\frac{\hbar(2\pi)^3 kc}{V}} \left(a_{\mathbf{k},\lambda}\varepsilon_{\mathbf{k},\lambda}\delta(\omega - ck) - a^\dagger_{\mathbf{k},\lambda}\varepsilon^*_{\mathbf{k},\lambda}\delta(\omega + ck)\right) \tag{9.2.13}$$

Now, by making use of (9.2.9), (9.2.10), and (9.2.13), we can calculate the mean square fluctuation of the position of the electron

$$\langle(\delta\mathbf{x}(t))^2\rangle = \int \frac{d\omega\, d\omega'}{(2\pi)^2} \frac{e^2}{m^2} \frac{1}{\omega^2\omega'^2}\langle\mathbf{E}_\omega\mathbf{E}_{\omega'}\rangle$$

$$= -\frac{e^2}{m^2}\left\langle \sum_{\mathbf{k},\lambda}\sum_{\mathbf{k}',\lambda'} \frac{\hbar\, 2\pi\, ck}{V(ck)^2(ck')^2}\left(a_{\mathbf{k},\lambda}\varepsilon_{\mathbf{k},\lambda} - a^\dagger_{\mathbf{k},\lambda}\varepsilon^*_{\mathbf{k},\lambda}\right)\right.$$

$$\left. \times \left(a_{\mathbf{k}',\lambda'}\varepsilon_{\mathbf{k}',\lambda'} - a^\dagger_{\mathbf{k}',\lambda'}\varepsilon^*_{\mathbf{k}',\lambda'}\right)\right\rangle \,.$$

The expectation value is finite only when the photon that is annihilated is the same as that created. We also assume that the radiation field is in its ground state, i.e., the vacuum state $|0\rangle$. Then, with $a_{\mathbf{k},\lambda}a^\dagger_{\mathbf{k},\lambda} = 1 + a^\dagger_{\mathbf{k},\lambda}a_{\mathbf{k},\lambda}$ and $a_{\mathbf{k},\lambda}|0\rangle = 0$, it follows that

---

[7] QM I, Sect. 16.4.2

$$\langle (\delta \mathbf{x}(t))^2 \rangle = \frac{e^2}{m^2} \int \frac{d^3k}{(2\pi)^2} \frac{\hbar}{(ck)^3} \sum_{\lambda=1,2} \left\langle a_{\mathbf{k},\lambda} a_{\mathbf{k},\lambda}^\dagger + a_{\mathbf{k},\lambda}^\dagger a_{\mathbf{k},\lambda} \right\rangle$$

$$= \frac{2}{\pi} \frac{e^2}{\hbar c} \left(\frac{\hbar}{mc} \right)^2 \int \frac{dk}{k} \,,$$
(9.2.14)

where we have also made the replacement $\frac{1}{V} \sum_{\mathbf{k}} \to \int \frac{d^3k}{(2\pi)^3}$. The integral $\int_0^\infty dk \frac{1}{k}$ is ultraviolet ($k \to \infty$) and infrared ($k \to 0$) divergent.

In fact, there are good physical reasons for imposing both an upper and a lower cutoff on this integral. The upper limit genuinely remains finite when one takes relativistic effects into account. The divergence at the lower limit is automatically avoided when the electron is treated, not with the free equation of motion (9.2.7), but quantum mechanically, allowing for the discrete atomic structure. In the following, we give a qualitative estimate of both limits, beginning with the upper one. As a result of the "Zitterbewegung" (the fluctuation in the position of the electron), the electron is spread out over a region the size of the Compton wavelength. Light, because its wavelength is smaller than the Compton wavelength, causes, on average, no displacement of the electron, since the light wave has as many peaks as troughs within one Compton wavelength. Thus, the upper cutoff is given by the Compton wavelength $\frac{1}{m}$, or by the corresponding energy m. For the lower limit, an obvious choice is the Bohr radius $(Z\alpha m)^{-1}$, or the corresponding wave number $Z\alpha m$. The bound electron is not influenced by wavelengths greater than $a = (Z\alpha m)^{-1}$. The lowest frequency of induced oscillations is then $Z\alpha m$. Another plausible choice is the Rydberg energy $Z^2\alpha^2 m$ with the associated length $(Z^2\alpha^2 m)^{-1}$, corresponding to the typical wavelength of the light emitted in an optical transition. Light oscillations at longer wavelengths will not influence the bound electron. In a complete quantum-electrodynamical theory, of course, there are no such heuristic arguments. If we take the first of the above estimates for the lower limit, it follows that

$$\int_{\omega_{\min}}^{\omega_{\max}} d\omega \frac{1}{\omega} = \int_{Z\alpha m}^{m} d\omega \frac{1}{\omega} = \log \frac{1}{Z\alpha} \,,$$

and thus, from (9.2.6) and (9.2.14),

$$\Delta E_{\text{Lamb}} = \frac{(2mc\,Z\alpha)^3}{12\hbar^2} \frac{Z\alpha c}{n^3} \frac{2}{\pi} \frac{e^2}{\hbar c} \left(\frac{\hbar}{mc} \right)^2 \log \frac{1}{Z\alpha} \delta_{\ell,0}$$

$$= \frac{8Z^4\alpha^3}{3\pi n^3} \log \frac{1}{Z\alpha} \frac{1}{2} \alpha^2 mc^2 \delta_{\ell,0} \,.$$
(9.2.15)

This corresponds to a frequency shift[8]

---

[8] T.A. Welton, Phys. Rev. **74**, 1157 (1948)

$$\Delta\nu_{\text{Lamb}} = 667\,\text{MHz} \quad \text{for} \quad n = 2,\ Z = 1,\ \ell = 0\,.$$

The experimentally observed shift[9] is 1057.862 ± 0.020 MHz. The complete quantum-electrodynamical theory of radiative corrections yields 1057.864 ± 0.014 MHz.[10] In comparison with the Darwin term, the radiative corrections are smaller by a factor $\alpha \log \frac{1}{\alpha}$. The full radiative corrections also contain $\alpha(Z\alpha)^4$ terms, which are numerically somewhat smaller. Levels with $\ell \neq 0$ also display shifts, albeit weaker ones than the s levels.

Quantum electrodynamics allows radiative corrections to be calculated with remarkable precision[10,11]. This theory, too, initially encounters divergences: The coupling to the quantized radiation field causes a shift in the energy of the electron that is proportional (in the nonrelativistic case) to $\mathbf{p}^2$, i.e., the radiation field increases the mass of the electron. What one measures, however, is not the bare mass, but the physical (renormalized) mass which contains this coupling effect. Such mass shifts are relevant to both free and bound electrons and are, in both cases, divergent. One now has to reformulate the theory in such a way that it contains only the renormalized mass. For the bound electron, one then finds only a finite energy shift, namely, the Lamb shift[11]. In the calculation by Bethe, which is nonrelativistic and only contains the self-energy effect described above, one finds a lower cutoff of 16.6 Ry and a Lamb shift of 1040 MHz. Simply out of curiosity, we recall the two estimates preceding Eq. (9.2.15) for the lower cutoff wave vector: If one takes the geometrical mean of these two values, for $Z = 1$ one obtains a logarithmic factor in (9.2.15) of $\log \frac{2}{16.55\,\alpha^2}$, which in turn yields $\Delta E = 1040$ MHz.

In conclusion, it is fair to say that the precise theoretical explanation of the Lamb shift represents one of the triumphs of quantum field theory.

Problems

9.1 Verify the expressions given in the text for

$$\frac{i}{2}[S, [S, H]]\,, \quad \frac{i^3}{6}[S, [S, [S, H]]]\,, \quad \frac{1}{24}[S, [S, [S, [S, H]]]] \tag{9.2.16}$$

with $H = \boldsymbol{\alpha}(\mathbf{p} - e\mathbf{A}) + \beta m + e\Phi$ and $S = -\frac{i}{2m}\beta\mathcal{O}$, where $\mathcal{O} \equiv \boldsymbol{\alpha}(\mathbf{p} - e\mathbf{A})$.

---

[9] The first experimental observation was made by W.E. Lamb, Jr. and R.C. Retherford, Phys. Rev. **72**, 241 (1947), and was refined by S. Triebwasser, E.S. Dayhoff, and W.E. Lamb, Phys. Rev. **89**, 98 (1953)

[10] N.M. Kroll and W.E. Lamb, Phys. Rev. **75**, 388 (1949); J.B. French and V.F. Weisskopf, Phys. Rev. **75**, 1240 (1949); G.W. Erickson, Phys. Rev. Lett. **27**, 780 (1972); P.J. Mohr, Phys. Rev. Lett. **34**, 1050 (1975); see also Itzykson and Zuber, op. cit p. 358

[11] The first theoretical (nonrelativistic) calculation of the Lamb shift is due to H.A. Bethe, Phys. Rev. **72**, 339 (1947). See also S.S. Schweber, *An Introduction to Relativistic Quantum Field Theory*, Harper & Row, New York 1961, p. 524.; V.F. Weisskopf, Rev. Mod. Phys. **21**, 305 (1949)

9.2 Here we introduce, for the Klein–Gordon equation, a transformation analogous to Foldy–Wouthuysen's, which leads to the relativistic corrections.

(a) Show that the substitutions

$$\theta = \frac{1}{2}\left(\varphi + \frac{i}{m}\frac{\partial\varphi}{\partial t}\right) \qquad \text{and} \qquad \chi = \frac{1}{2}\left(\varphi - \frac{i}{m}\frac{\partial\varphi}{\partial t}\right)$$

allow the Klein–Gordon equation

$$\frac{\partial^2\varphi}{\partial t^2} = (\nabla^2 - m^2)\varphi$$

to be written as a matrix equation

$$i\frac{\partial\Phi}{\partial t} = H_0\Phi$$

where $\Phi = \begin{pmatrix}\theta\\\chi\end{pmatrix}$ and $H_0 = -\begin{pmatrix}1 & 1\\-1 & -1\end{pmatrix}\frac{\nabla^2}{2m} + \begin{pmatrix}1 & 0\\0 & -1\end{pmatrix}m$.

(b) Show that in the two-component formulation, the Klein–Gordon equation for particles in an electromagnetic field, using the minimal coupling ($p \to \pi = p - eA$), reads:

$$i\frac{\partial\Phi}{\partial t} = \left\{-\begin{pmatrix}1 & 1\\-1 & -1\end{pmatrix}\frac{\pi^2}{2m} + \begin{pmatrix}1 & 0\\0 & -1\end{pmatrix}m + eV(x)\right\}\Phi(x) .$$

(c) Discuss the nonrelativistic limit of this equation and compare it with the corresponding result for the Dirac equation.

Hint: The Hamiltonian of the Klein–Gordon equation in (b) can be brought into the form $H = \mathcal{O} + \mathcal{E} + \eta m$ with $\eta = \begin{pmatrix}1 & 0\\0 & -1\end{pmatrix}$, $\mathcal{O} = \rho\frac{\pi^2}{2m} = \begin{pmatrix}0 & 1\\-1 & 0\end{pmatrix}\frac{\pi^2}{2m}$, and $\mathcal{E} = eV + \eta\frac{\pi^2}{2m}$. Show, in analogy to the procedure for the Dirac equation, that, in the case of static external fields, the Foldy–Wouthuysen transformation $\Phi' = e^{iS}\Phi$ yields the approximate Schrödinger equation $i\frac{\partial\Phi'}{\partial t} = H'\Phi'$, with

$$H' = \eta\left(m + \frac{\pi^2}{2m} - \frac{\pi^4}{8m^3} + \dots\right) + eV + \frac{1}{32m^4}[\pi^2, [\pi^2, eV]] + \dots .$$

The third and the fifth term represent the leading relativistic corrections. In respect to their magnitudes see Eq. (8.1.19) and the remark (ii) in Sect. 10.1.2.

10. Physical Interpretation of the Solutions to the Dirac Equation

In interpreting the Dirac equation as a wave equation, as has been our practice up to now, we have ignored a number of fundamental difficulties. The equation possesses negative energy solutions and, for particles at rest, solutions with negative rest mass. The kinetic energy in these states is negative; the particle moves in the opposite direction to one occupying the usual state of positive energy. Thus, a particle carrying the charge of an electron is repelled by the field of a proton. (The matrix β with the negative matrix elements β_{33} and β_{44} multiplies m and the kinetic energy, but not the potential term $e\Phi$ in Eq. (9.1.9).) States such as these are not realized in nature. The main problem, of course, is their negative energy, which lies below the smallest energy for states with positive rest energy. Thus, one would expect radiative transitions, accompanied by the emission of light quanta, from positive energy into negative energy states. Positive energy states would be unstable due to the infinite number of negative-energy states into which they could fall by emitting light – unless, that is, all of these latter states were occupied.

It is not possible to exclude these states simply by arguing that they are not realized in nature. The positive energy states alone do not represent a complete set of solutions. The physical consequence of this is the following: When an external perturbation, e.g., due to a measurement, causes an electron to enter a certain state, this will in general be a combination of positive and negative energy states. In particular, when the electron is confined to a region that is smaller than its Compton wavelength, the negative energy states will contribute significantly.

10.1 Wave Packets and "Zitterbewegung"

In the previous sections, we for the most part investigated eigenstates of the Dirac Hamiltonian, i.e., stationary states. We now wish to study general solutions of the time-dependent Dirac equation. We proceed analogously to the nonrelativistic theory and consider superpositions of stationary states for free particles. It will emerge that these wave packets have some unusual properties as compared to the nonrelativistic theory (see Sect. 10.1.2).

10.1.1 Superposition of Positive Energy States

We shall first superpose only positive energy states

$$\psi^{(+)}(x) = \int \frac{d^3 p}{(2\pi)^3} \frac{m}{E} \sum_{r=1,2} b(p,r) u_r(p) e^{-ipx} \tag{10.1.1}$$

and investigate the properties of the resulting wave packets. Here, $u_r(p)$ are the free spinors of positive energy and $b(p,r)$ are complex amplitudes. The factor $\frac{m}{(2\pi)^3 E}$ is included so as to satisfy a simple normalization condition.

We note in passing that $\frac{d^3 p}{E}$ is a Lorentz-invariant measure where, as always, $E = \sqrt{\mathbf{p}^2 + m^2}$. We show this by the following rearrangement:

$$\int d^3 p \frac{1}{E} = \int d^3 p \int_0^\infty dp_0 \frac{\delta(p_0 - E)}{E} = \int d^3 p \int_0^\infty dp_0\, 2\delta(p_0^2 - E^2)$$

$$= \int d^3 p \int_{-\infty}^\infty dp_0\, \delta(p_0^2 - E^2) = \int d^4 p\, \delta(p^2 - m^2) . \tag{10.1.2}$$

Both $d^4 p$ and the δ-function are Lorentz covariant. The $d^4 p = \det \Lambda\, d^4 p' = \pm d^4 p'$ transforms as a pseudoscalar, where the Jacobi determinant $\det \Lambda$ would equal 1 for proper Lorentz transformations.

The density corresponding to (10.1.1) is given by

$$j^{(+)0}(t,\mathbf{x}) = \psi^{(+)\dagger}(t,\mathbf{x})\psi^{(+)}(t,\mathbf{x}) . \tag{10.1.3a}$$

Integrated over all of space

$$\int d^3 x\, j^{(+)0}(t,\mathbf{x}) = \int d^3 x \int \frac{d^3 p\, d^3 p'}{(2\pi)^6} \frac{m^2}{EE'} \sum_{r,r'} b^*(p,r)\, b(p',r')$$

$$\times u_r^\dagger(p) u_{r'}(p') e^{i(E-E')t - i(\mathbf{p}-\mathbf{p}')\mathbf{x}} \tag{10.1.3b}$$

$$= \sum_r \int \frac{d^3 p}{(2\pi)^3} \frac{m}{E} |b(p,r)|^2 = 1 ,$$

the density in the sense of a probablity density is normalized to unity. Here, we have used $\int d^3 x\, e^{i(\mathbf{p}-\mathbf{p}')\mathbf{x}} = (2\pi)^3\, \delta^{(3)}(\mathbf{p} - \mathbf{p}')$ and the orthogonality relation (6.3.19a)[1]. The time dependence disappears and the total density is time independent. This equation determines the normalization of the amplitudes $b(p,r)$.

We next calculate the total current, which is defined by

$$\mathbf{J}^{(+)} = \int d^3 x\, \mathbf{j}^{(+)}(t,\mathbf{x}) = \int d^3 x\, \psi^{(+)\dagger}(t,\mathbf{x})\boldsymbol{\alpha}\, \psi^{(+)}(t,\mathbf{x}) . \tag{10.1.4}$$

---

[1] $u_r^\dagger(p)\, u_{r'}(p) = \bar{u}_r(p)\gamma^0\, u_{r'}(p) = \frac{E}{m}\, \delta_{rr'}$

In analogy to the zero component, one obtains

$$\mathbf{J}^{(+)} = \int \frac{d^3x}{(2\pi)^6} \iint d^3p\, d^3p' \frac{m^2}{EE'} \sum_{r,r'} b^*(p,r)b(p',r')$$

$$\times u_r^\dagger(p)\boldsymbol{\alpha}\, u_{r'}(p') e^{\mathrm{i}(E-E')t - \mathrm{i}(\mathbf{p}-\mathbf{p}')\mathbf{x}} \qquad (10.1.4')$$

$$= \int \frac{d^3p}{(2\pi)^3} \sum_{r,r'} \frac{m^2}{E^2} b^*(p,r)b(p,r') u_r^\dagger(p)\boldsymbol{\alpha}\, u_{r'}(p) \ .$$

For further evaluation, we need the Gordon identity (see Problem 10.1)

$$\bar{u}_r(p)\gamma^\mu\, u_{r'}(q) = \frac{1}{2m}\bar{u}_r(p)\left[(p+q)^\mu + \mathrm{i}\sigma^{\mu\nu}(p-q)_\nu\right] u_{r'}(q) \ . \qquad (10.1.5)$$

Taken in conjunction with the orthonormality relations for the u_r given in (6.3.15), $\bar{u}_r(k)u_s(k) = \delta_{rs}$, equation (10.1.4') yields:

$$\mathbf{J}^{(+)} = \sum_r \int \frac{d^3p}{(2\pi)^3} \frac{m}{E}\, |b(p,r)|^2 \frac{\mathbf{P}}{E} = \left\langle \frac{\mathbf{P}}{E} \right\rangle \ . \qquad (10.1.6)$$

This implies that the total current equals the mean value of the group velocity

$$\mathbf{v}_G = \frac{\partial E}{\partial \mathbf{p}} = \frac{\partial\sqrt{\mathbf{p}^2 + m^2}}{\partial \mathbf{p}} = \frac{\mathbf{p}}{E} \ . \qquad (10.1.7)$$

So far, seen from the perspective of nonrelativistic quantum mechanics, nothing appears unusual.

10.1.2 The General Wave Packet

However, on starting with a general wave packet and expanding this using the complete set of solutions of the free Dirac equation, the result contains negative energy states. Let us take as the initial spinor the Gaussian

$$\psi(0,\mathbf{x}) = \frac{1}{(2\pi d^2)^{3/4}} e^{\mathrm{i}\mathbf{x}\mathbf{p}_0 - \mathbf{x}^2/4d^2}\, w \ , \qquad (10.1.8)$$

where, for example, $w = \binom{\varphi}{0}$, i.e., at time zero there are only components with positive energy, and where d characterizes the linear dimension of the wave packet. The most general spinor can be represented by the following superposition:

$$\psi(t,\mathbf{x}) = \int \frac{d^3p}{(2\pi)^3} \frac{m}{E} \sum_r \left(b(p,r)u_r(p)e^{-\mathrm{i}px} + d^*(p,r)v_r(p)e^{\mathrm{i}px} \right) \ . \qquad (10.1.9)$$

We also need the Fourier transform of the Gaussian appearing in the initial spinor (10.1.8)

$$\int d^3x\, e^{i\mathbf{x}\mathbf{p}_0 - \mathbf{x}^2/4d^2 - i\mathbf{p}\cdot\mathbf{x}} = (4\pi d^2)^{3/2} e^{-(\mathbf{p}-\mathbf{p}_0)^2 d^2} \ . \tag{10.1.10}$$

In order to determine the expansion coefficients $b(p,r)$ and $d(p,r)$, we take the Fourier transform at time $t = 0$ of $\psi(0,\mathbf{x})$ and insert (10.1.8) and (10.1.10) on the left-hand side of (10.1.9)

$$(8\pi d^2)^{3/4} e^{-(\mathbf{p}-\mathbf{p}_0)^2 d^2} w = \frac{m}{E} \sum_r (b(p,r)u_r(p) + d^*(\tilde{p},r)v_r(\tilde{p})) \ , \tag{10.1.11}$$

where $\tilde{p} = (p^0, -\mathbf{p})$. After multiplying (10.1.11) by $u_r^\dagger(p)$ and $v_r^\dagger(\tilde{p})$, the orthogonality relations (6.3.19a–c)

$$\bar{u}_r(k)\gamma^0 u_s(k) = \frac{E}{m}\, \delta_{rs} = u_r^\dagger(k)\, u_s(k)$$

$$\bar{v}_r(k)\gamma^0 v_s(k) = \frac{E}{m}\, \delta_{rs} = v_r^\dagger(k)\, v_s(k)$$

$$\bar{v}_r(\tilde{k})\gamma^0 u_s(k) = \quad 0 \quad = v_r^\dagger(\tilde{k})\, u_s(k)$$

yield the Fourier amplitudes

$$b(p,r) = (8\pi d^2)^{3/4}\, e^{-(\mathbf{p}-\mathbf{p}_0)^2 d^2}\, u_r^\dagger(p)w$$
$$d^*(\tilde{p},r) = (8\pi d^2)^{3/4}\, e^{-(\mathbf{p}-\mathbf{p}_0)^2 d^2}\, v_r^\dagger(\tilde{p})w \ , \tag{10.1.12}$$

both of which are finite.

We have thus demonstrated the claim made at the outset that a general wave packet contains both positive and negative energy components. We now wish to study the consequences of this type of wave packet. For the sake of simplicity, we begin with a nonpropagating wave packet, i.e., $\mathbf{p}_0 = 0$. Some of the modifications arising when $\mathbf{p}_0 \neq 0$ will be discussed after Eq. (10.1.14b).

Since we have assumed $w = \binom{\varphi}{0}$, the representation (6.3.11a,b) implies, for the spinors u_r and v_r of free particles, the relation $d^*(p,r)/b(p,r) \sim \frac{|\mathbf{p}|}{m+E}$.

If the wave packet is large, $d \gg \frac{1}{m}$, then $|\mathbf{p}| \lesssim d^{-1} \ll m$ and thus $d^*(p) \ll b(p)$. In this case, the negative energy components are negligible. However, when we wish to confine the particle to a region of dimensions less than a Compton wavelength, $d \ll \frac{1}{m}$, then the negative energy solutions play an important role:

$$|\mathbf{p}| \sim d^{-1} \gg m \ ,$$

i.e., $d^*/b \sim 1$.

The *normalization*

$$\int d^3x\, \psi^\dagger(t,\mathbf{x})\psi(t,\mathbf{x}) = \int \frac{d^3p}{(2\pi)^3}\frac{m}{E}\sum_r \left(|b(p,r)|^2 + |d(p,r)|^2\right) = 1$$

is time independent as a result of the continuity equation. The total current for the spinor (10.1.9) reads:

$$J^i(t) = \int \frac{d^3p}{(2\pi)^3}\frac{m}{E}\left\{\frac{p^i}{E}\sum_r \left[|b(p,r)|^2 + |d(p,r)|^2\right]\right.$$

$$+ i\sum_{r,r'} \left[b^*(\tilde{p},r)d^*(p,r')e^{2iEt}\bar{u}_r(\tilde{p})\sigma^{i0}v_{r'}(p)\right. \tag{10.1.13}$$

$$\left.\left. - b(\tilde{p},r)d(p,r')e^{-2iEt}\bar{v}_{r'}(p)\sigma^{i0}u_r(\tilde{p})\right]\right\} .$$

The first term is a time-independent contribution to the current. The second term contains oscillations at frequencies greater than $\frac{2mc^2}{\hbar} = 2\times 10^{21}\mathrm{s}^{-1}$. This oscillatory motion is known as *"Zitterbewegung"*.

In this derivation, in addition to (10.1.5), we have used

$$\bar{u}_r(\tilde{p})\gamma^\mu v_{r'}(q) = \frac{1}{2m}\bar{u}_r(\tilde{p})\left[(\tilde{p}-q)^\mu + i\sigma^{\mu\nu}(\tilde{p}+q)_\nu\right]v_{r'}(q) , \tag{10.1.14a}$$

from which it follows that

$$u_r^\dagger(\tilde{p})\alpha^i v_{r'}(p) = \bar{u}_r(\tilde{p})\gamma^i v_{r'}(p)$$

$$= \frac{1}{2m}\left[(\tilde{p}^i - p^i)\bar{u}_r(\tilde{p})v_r(p) + \bar{u}_r(\tilde{p})\,\sigma^{i\nu}\,(\tilde{p}+p)_\nu v_{r'}(p)\right] . \tag{10.1.14b}$$

For the initial spinor (10.1.8) with $w = \binom{\varphi}{0}$ and $\mathbf{p}_0 = 0$, the first term of (10.1.14b) contributes nothing to $J^i(t)$ in (10.1.13). If the spinor w also contains components 3 and 4, or if $\mathbf{p}_0 \neq 0$, there are also contributions from Zitterbewegung to the first term of (10.1.14b). One obtains an additional term (see Problem 10.2) to (10.1.13)

$$\Delta J^i(t) = \int \frac{d^3p}{(2\pi)^3}\frac{m}{E}(8\pi d^2)^{3/2}e^{-2(\mathbf{p}-\mathbf{p}_0)^2 d^2}e^{2iEt}p^i w^\dagger \frac{1}{2m^2}(\mathbf{p}^2 - m\mathbf{p}\boldsymbol{\gamma})\gamma_0 w . \tag{10.1.13'}$$

The *amplitude* of the *Zitterbewegung* is obtained as the mean value of $\mathbf{x}$:

$$\langle\mathbf{x}\rangle = \int d^3x\, \psi^\dagger(t,\mathbf{x})\,\mathbf{x}\,\psi(t,\mathbf{x})$$

$$= \int d^3x\, \psi^\dagger(0,\mathbf{x})e^{iHt}\,\mathbf{x}\,e^{-iHt}\psi(0,\mathbf{x}) . \tag{10.1.15a}$$

In order to calculate $\langle\mathbf{x}\rangle$, we first determine the temporal variation of $\langle\mathbf{x}\rangle$, since this can be related to the current, which we have already calculated:

$$\frac{d}{dt}\langle \mathbf{x}\rangle = \frac{d}{dt}\int d^3x\,\psi^\dagger(0,\mathbf{x})e^{iHt}\,\mathbf{x}\,e^{-iHt}\psi(0,\mathbf{x})$$

$$= \int d^3x\,\psi^\dagger(t,\mathbf{x})\,\mathrm{i}\,[H,\mathbf{x}]\,\psi(t,\mathbf{x}) \tag{10.1.15b}$$

$$= \int d^3x\,\psi^\dagger(t,\mathbf{x})\,\boldsymbol{\alpha}\,\psi(t,\mathbf{x}) \equiv \mathbf{J}(t)\,.$$

In evaluating the commutator we have used $H = \boldsymbol{\alpha}\cdot\frac{1}{\mathrm{i}}\boldsymbol{\nabla}+\beta m$. The integration of this relation over the time from 0 to t yields, without (10.1.13'),

$$\langle x^i\rangle = \langle x^i\rangle_{t=0} + \int \frac{d^3p}{(2\pi)^3}\frac{mp^i}{E^2}\sum_r \left[|b(p,r)|^2 + |d(p,r)|^2\right]t$$

$$+ \sum_{r,r'}\int \frac{d^3p}{(2\pi)^3}\frac{m}{2E^2}\left[b^*(\tilde{p},r)d^*(p,r')e^{2iEt}\bar{u}_r(\tilde{p})\sigma^{i0}v_{r'}(p) \tag{10.1.16}\right.$$

$$\left.+ b(\tilde{p},r)d(p,r')e^{-2iEt}\bar{v}_{r'}(p)\sigma^{i0}u_r(\tilde{p})\right]\,.$$

The mean value of x^i contains oscillations with amplitude $\sim \frac{1}{E} \sim \frac{1}{m} \sim \frac{\hbar}{mc} = 3.9 \times 10^{-11}$ cm. The Zitterbewegung stems from the interference between components with positive and negative energy.

Remarks:

(i) If a spinor consists not only of positive-energy, but also of negative energy states, Zitterbewegung follows as a consequence. If one expands bound states in terms of free solutions, these also contain components with negative energy. An example is the ground state of the hydrogen atom (8.2.41).

(ii) A Zitterbewegung also arises from the Klein–Gordon equation. Here too, wave packets with linear dimension less than the Compton wavelength $\lambda_{c\,\pi^-} = \frac{\hbar c}{m_{\pi^-}}$, contain contributions from negative energy solutions, which fluctuate over a region of size $\lambda_{c\,\pi^-}$. The energy shift in a Coulomb potential (Darwin term), however, is a factor α smaller than for spin-$\frac{1}{2}$ particles. (See Problem 9.2)[2].

*10.1.3 General Solution of the Free Dirac Equation in the Heisenberg Representation

The existence of Zitterbewegung can also be seen by solving the Dirac equation in the Heisenberg representation. The Heisenberg operators are defined by

---

[2] An instructive discussion of these phenomena can be found in H. Feshbach and F. Villars, Rev. Mod. Phys. **30**, 24 (1958)

$$O(t) = e^{iHt/\hbar} O e^{-iHt/\hbar} \qquad (10.1.17)$$

which yields the equation of motion

$$\frac{dO(t)}{dt} = \frac{1}{i\hbar}[O(t), H] . \qquad (10.1.18)$$

We assume that the particle is free, i.e., that $\mathbf{A} = 0$ and $\Phi = 0$. In this case, the momentum commutes with

$$H = c\,\boldsymbol{\alpha} \cdot \mathbf{p} + \beta mc^2 , \qquad (10.1.19)$$

that is,

$$\frac{d\mathbf{p}(t)}{dt} = 0 , \qquad (10.1.20)$$

which implies that $\mathbf{p}(t) = \mathbf{p} = \text{const}$. In addition, we see that

$$\mathbf{v}(t) = \frac{d\mathbf{x}(t)}{dt} = \frac{1}{i\hbar}[\mathbf{x}(t), H] = c\,\boldsymbol{\alpha}(t) \qquad (10.1.21a)$$

and

$$\frac{d\boldsymbol{\alpha}}{dt} = \frac{1}{i\hbar}[\boldsymbol{\alpha}(t), H] = \frac{2}{i\hbar}(c\mathbf{p} - H\boldsymbol{\alpha}(t)) . \qquad (10.1.21b)$$

Since $H = \text{const}$ (time independent), the above equation has the solution

$$\mathbf{v}(t) = c\,\boldsymbol{\alpha}(t) = cH^{-1}\mathbf{p} + e^{\frac{2iHt}{\hbar}}\left(\boldsymbol{\alpha}(0) - cH^{-1}\mathbf{p}\right) . \qquad (10.1.22)$$

Integration of (10.1.22) yields:

$$\mathbf{x}(t) = \mathbf{x}(0) + \frac{c^2\,\mathbf{p}}{H}t + \frac{\hbar c}{2iH}\left(e^{\frac{2iHt}{\hbar}} - 1\right)\left(\boldsymbol{\alpha}(0) - \frac{c\mathbf{p}}{H}\right) . \qquad (10.1.23)$$

For free particles, we have

$$\boldsymbol{\alpha} H + H\boldsymbol{\alpha} = 2c\mathbf{p} ,$$

and hence

$$\left(\boldsymbol{\alpha} - \frac{c\mathbf{p}}{H}\right) H + H\left(\boldsymbol{\alpha} - \frac{c\mathbf{p}}{H}\right) = 0 . \qquad (10.1.24)$$

In addition to the initial value $\mathbf{x}(0)$, the solution (10.1.23) also contains a term linear in t which corresponds to the group velocity motion, and an oscillating term that represents the Zitterbewegung. To calculate the mean value $\int \psi^\dagger(0, \mathbf{x})\mathbf{x}(t)\psi(0, \mathbf{x})d^3x$, one needs the matrix elements of the operator $\boldsymbol{\alpha}(0) - \frac{c\mathbf{p}}{H}$. This operator has nonvanishing matrix elements only between states of identical momentum. The vanishing of the anticommutator (10.1.24) implies, furthermore, that the energies must be of opposite sign. Hence, we find that the Zitterbewegung is the result of interference between positive and negative energy states.

*10.1.4 Potential Steps and the Klein Paradox

One of the simplest exactly solvable problems in nonrelativistic quantum mechanics is that of motion in the region of a potential step (Fig. 10.1). If the energy E of the plane wave incident from the left is smaller than the height V_0 of the potential step, i.e., $E < V_0$, then the wave is reflected and penetrates into the classically forbidden region only as a decaying exponential $e^{-\kappa x^3}$ with $\kappa = \sqrt{2m(V_0 - E)}$. Hence, the larger the energy difference $V_0 - E$, the smaller the penetration. The solution of the Dirac equation is also relatively easy to find, but is not without some surprises.

We assume that a plane wave with positive energy is incident from the left. After separating out the common time dependence e^{-iEt}, the solution in region I (cf. Fig.10.1) comprises the incident wave

$$
\psi_{\text{in}}(x^3) = e^{ikx^3}
\begin{pmatrix}
1 \\
0 \\
\frac{k}{E+m} \\
0
\end{pmatrix}
\tag{10.1.25}
$$

and the reflected wave

$$
\psi_{\text{refl}}(x^3) = a\,e^{-ikx^3}
\begin{pmatrix}
1 \\
0 \\
\frac{-k}{E+m} \\
0
\end{pmatrix}
+ b\,e^{-ikx^3}
\begin{pmatrix}
0 \\
1 \\
0 \\
\frac{k}{E+m}
\end{pmatrix} ,
\tag{10.1.26}
$$

i.e., $\psi_{\text{I}}(x^3) = \psi_{\text{in}}(x^3) + \psi_{\text{refl}}(x^3)$. The second term in (10.1.26) represents a reflected wave with opposite spin and will turn out to be zero. In region II, we make a similar ansatz for the transmitted wave

$$
\psi_{\text{II}}(x^3) \equiv \psi_{\text{trans}}(x^3) = c\,e^{iqx^3}
\begin{pmatrix}
1 \\
0 \\
\frac{q}{E-V_0+m} \\
0
\end{pmatrix}
+ d\,e^{iqx^3}
\begin{pmatrix}
0 \\
1 \\
0 \\
\frac{-q}{E-V_0+m}
\end{pmatrix} .
\tag{10.1.27}
$$

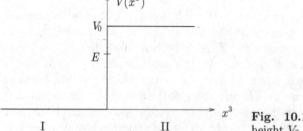

Fig. 10.1. A potential step of height V_0

The wave vector (momentum) in this region is given by

$$q = \sqrt{(E - V_0)^2 - m^2} \,, \tag{10.1.28}$$

and the coefficients a, b, c, d are determined from the requirement that ψ be continuous at the step. If the solution were not continuous, then, upon inserting it into the Dirac equation, one would obtain a contribution proportional to $\delta(x^3)$. From this continuity condition, $\psi_{\mathrm{I}}(0) = \psi_{\mathrm{II}}(0)$, it follows that

$$1 + a = c \,, \tag{10.1.29a}$$

$$1 - a = rc \,, \quad \text{with} \quad r \equiv \frac{q}{k} \frac{E + m}{E - V_0 + m} \,, \tag{10.1.29b}$$

and

$$b = d = 0 \,. \tag{10.1.29c}$$

The latter relation, which stems from components 2 and 4, implies that the spin is not reversed.

As long as $|E - V_0| < m$, i.e. $-m + V_0 < E < m + V_0$, the wave vector q to the right of the step is imaginary and the solution in that case decays exponentially. In particular, when $E, V_0 \ll m$, then the solution $\psi_{\mathrm{trans}} \sim e^{-|q|x^3} \sim e^{-mx^3}$ is localized to within a few Compton wavelengths.

However, when the height of the step V_0 becomes larger, so that finally $V_0 \geq E + m$, then, according to (10.1.28), q becomes real and one obtains an oscillating transmitted plane wave. This is an example of the Klein paradox.

The source of this initially surprising result can be explained as follows: In region I, the positive energy solutions lie in the range $E > m$, and those with negative energy in the range $E < -m$. In region II, the positive energy solutions lie in the range $E > m + V_0$, and those with negative energy in the range $E < -m + V_0$. This means that for $V_0 > m$ the solutions hitherto referred to as "negative energy solutions" actually also possess positive energy. When V_0 becomes so large that $V_0 > 2m$ (see Fig. 10.2), the energy of these "negative energy solutions" in region II eventually becomes larger than m, and thus lies in the same energy range as the solutions of positive energy in region I. The condition for the occurrence of oscillatory solutions given after Eq. (10.1.29c) was $V_0 \geq E + m$, where the energy in region I satisfies $E > m$. This coincides with the considerations above. Instead of complete reflection with exponential penetration into the classically forbidden region, one has a transition into negative energy states for $E > 2m$.

For the transmitted and reflected current density one finds,

$$\frac{j_{\mathrm{trans}}}{j_{\mathrm{in}}} = \frac{4r}{(1 + r)^2} \,, \quad \frac{j_{\mathrm{refl}}}{j_{\mathrm{in}}} = \left(\frac{1 - r}{1 + r}\right)^2 = 1 - \frac{j_{\mathrm{trans}}}{j_{\mathrm{in}}} \,. \tag{10.1.30}$$

However, according to Eq. (10.1.29b), $r < 0$ for positive q and thus the reflected current is greater than the incident current.

If one takes the positive square root for q in (10.1.28), according to (10.1.29b), $r < 0$, and consequently the flux going out to the left exceeds the (from the left) incoming flux. This comes about because, for $V_0 > E + m$, the group velocity

$$v_0 = \frac{1}{E - V_0} q$$

has the opposite sense to the direction of q. That is, wave-packet solutions of this type also contain incident wave packets coming form the right of the step.

If one chooses for q in (10.1.28) the negative square root, $r > 0$, one obtains the regular reflection behavior[3].

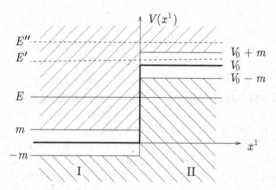

Fig. 10.2. Potential step and energy ranges for $V_0 > 2m$. Potential step (thick line) and energy ranges with positive and negative energy (right- and left-inclined hatching). To the left of the step, the energies E and E' lie in the range of positive energies. To the right of the step, E' lies in the forbidden region, and hence the solution is exponentially decaying. E lies in the region of solutions with negative energy. The energy E'' lies in the positive energy region, both on the right and on the left.

10.2 The Hole Theory

In this section, we will give a preliminary interpretation of the states with negative energy. The properties of positive energy states show remarkable

---

[3] H.G. Dosch, J.H.D. Jensen and V.L. Müller, Physica Norvegica 5, 151 (1971); B. Thaller, *The Dirac Equation*, Springer, Berlin, Heidelberg, 1992, pp. 120, 307; W. Greiner, *Theoretical Physics*, Vol. 3, Relativistic Quantum Mechanics, Wave Equations, 2nd edn., Springer, Berlin, Heidelberg, 1997

agreement with experiment. Can we simply ignore the negative energy states? The answer is: No. This is because an arbitrary wave packet will also contain components of negative energy v_r. Even if we have spinors of positive energy, u_r, to start with, the interaction with the radiation field can cause transitions into negative energy states (see Fig. 10.3). Atoms, and thus all matter surrounding us, would be unstable.

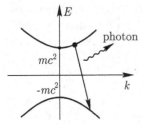

Fig. 10.3. Energy eigenvalues of the Dirac equation and conceivable transitions

A way out of this dilemma was suggested by Dirac in 1930. He postulated that all negative energy states be considered as occupied. Thus, particles with positive energy cannot make transitions into these states because the Pauli principle forbids multiple occupation. In this picture, the vacuum state consists of an infinite sea of particles, all of which are in negative energy states (Fig. 10.4).

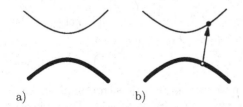

a)   b)

Fig. 10.4. Filled negative energy states (thick line): (a) vacuum state, (b) excited state

An excited state of this vacuum arises as follows: An electron of negative energy is promoted to a state of positive energy, leaving behind a *hole* with charge $-(-e_0) = e_0$! (Fig. 10.4 b). This immediately has an interesting consequence. Suppose that we remove a particle of negative energy from the vacuum state. This leaves behind a hole. In comparison to the vacuum state, this state has positive charge and positive energy. The absence of a negative energy state represents an antiparticle. For the electron, this is the positron. Let us consider, for example, the spinor with negative energy

$$v_{r=1}(p')e^{ip'x} = v_1(p')e^{i(E_{p'}t - p'x)} \ .$$

This is an eigenstate with energy eigenvalue $-E_{\mathbf{p}'}$, momentum $-\mathbf{p}'$ and spin in the rest frame $\frac{1}{2}\Sigma^3$ $1/2$. When this state is *unoccupied*, a positron is present with energy $E_{\mathbf{p}'}$, momentum $\mathbf{p}'$, and spin $\frac{1}{2}\Sigma^3$ $-1/2$. (An analogous situation occurs for the excitation of a degenerate ideal electron gas, as discussed at the end of Sect. 2.1.1.)

The situation described here can be further elucidated by considering the excitation of an electron state by a photon: The γ quantum of the photon with its energy $\hbar\omega$ and momentum $\hbar\mathbf{k}$ excites an electron of negative energy into a positive energy state (Fig. 10.5). In reality, due to the requirements of energy and momentum conservation, this process of pair creation can only take place in the presence of a potential. Let us look at the energy and momentum balance of the process.

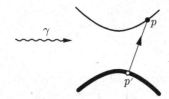

Fig. 10.5. The photon γ excites an electron from a negative energy state into a positive energy state, i.e., $\gamma \rightarrow e^+ + e^-$

The *energy balance* for pair creation reads:

$$\hbar\omega = E_{\text{el. pos. energy}} - E_{\text{el. neg. energy}}$$
$$= E_{\mathbf{p}} - (-E_{\mathbf{p}'}) = E_{\text{el.}} + E_{\text{pos.}} \tag{10.2.1}$$

The energy of the electron is $E_{\text{el.}} = \sqrt{\mathbf{p}^2 c^2 + m^2 c^4}$, and the energy of the positron $E_{\text{pos.}} = \sqrt{\mathbf{p}'^2 c^2 + m^2 c^4}$. The *momentum balance* reads:

$$\hbar\mathbf{k} - \mathbf{p}' = \mathbf{p} \qquad \text{or} \qquad \hbar\mathbf{k} = \mathbf{p} + \mathbf{p}' , \tag{10.2.2}$$

i.e., (photon momentum) = (electron momentum) + (positron momentum). It turns out, however, that this preliminary interpretation of the Dirac theory still conceals a number of problems: The ground state (vacuum state) has an infinitely large (negative) energy. One must also inquire as to the role played by the interaction of the particles in the occupied negative energy states. Furthermore, in the above treatment, there is an asymmetry between electron and positron. If one were to begin with the Dirac equation of the positron, one would have to occupy its negative energy states and the electrons would be holes in the positron sea. In any case one is led to a many-body system.[4] A genuinely adequate description only becomes possible through the quantization of the Dirac field.

---

[4] The simple picture of the hole theory may be used only with care. See, e.g., the article by Gary Taubes, Science **275**, 148 (1997) about spontaneous positron emission.

The original intention was to view the Dirac equation as a generalization of the Schrödinger equation and to interpret the spinor ψ as a sort of wave function. However, this leads to insurmountable difficulties. For example, even the concept of a probability distribution for the localization of a particle at a particular point in space becomes problematic in the relativistic theory. Also connected to this is the fact that the problematic features of the Dirac single-particle theory manifest themselves, in particular, when a particle is highly localized in space (in a region comparable to the Compton wavelength). The appearance of these problems can be made plausible with the help of the uncertainty relation. When a particle is confined to a region of size Δx, it has, according to Heisenberg's uncertainty relation, a momentum spread $\Delta p > \hbar \Delta x^{-1}$. If $\Delta x < \frac{\hbar}{mc}$, then the particle's momentum, and thus energy, uncertainty becomes

$$\Delta E \approx c \Delta p > mc^2 \ .$$

Thus, in this situation, the energy of a single particle is sufficient to create several other particles. This, too, is an indication that the single-particle theory must be replaced by a many-particle theory, i.e. a quantum field theory.

Before finally turning to a representation by means of a quantized field, in the next chapter, we shall first investigate further symmetry properties of the Dirac equation in connection with the relationship of solutions of positive and negative energy to particles and antiparticles.

Problems

10.1 Prove the Gordon identity (10.1.5), which states that, for two positive energy solutions of the free Dirac equation, $u_r(p)$ and $u_{r'}(p)$,

$$\bar{u}_r(p)\, \gamma^\mu u_{r'}(q) = \frac{1}{2m} \bar{u}_r(p)[(p+q)^\mu + \mathrm{i}\sigma^{\mu\nu}(p-q)_\nu] u_{r'}(q) \ .$$

10.2 Derive Eq. (10.1.13) and the additional term (10.1.13').

10.3 Verify the solution for the potential step considered in conjunction with the Klein paradox. Discuss the type of solutions obtained for the energy values E' and E'' indicated in Fig. 10.2. Draw a diagram similar to Fig. 10.2 for a potential step of height $0 < V_0 < m$.

11. Symmetries and Further Properties of the Dirac Equation

Starting from invariance properties and conservation laws, further symmetries of the Dirac equation are presented: the behavior under charge conjugation and time reversal. Finally, the massless Dirac equation is investigated.

*11.1 Active and Passive Transformations, Transformations of Vectors

In this and the following sections we shall investigate the symmetry properties of the Dirac equation in the presence of an electromagnetic potential. We begin by recalling the transformation behavior of spinors under passive and active transformations, as was described in Sect. 7.1. We will then address the transformation of the four-potential, and also investigate the transformation of the Dirac Hamiltonian.

Consider the Lorentz transformation

$$x' = \Lambda x + a \tag{11.1.1}$$

from the coordinate system I into the coordinate system I'. According to Eq. (7.1.2a), a spinor $\psi(x)$ transforms under a passive transformation as

$$\psi'(x') = S\psi(\Lambda^{-1}x') , \tag{11.1.2a}$$

where we have written down only the homogeneous transformation.
An *active* transformation with Λ^{-1} gives rise to the spinor (Eq. (7.1.2b))

$$\psi'(x) = S\psi(\Lambda^{-1}x) . \tag{11.1.2b}$$

The state Z'', which is obtained from Z through the active transformation Λ, appears by definition in I' as the state Z in I, i.e., $\psi(x')$. Since I is obtained from I' by the Lorentz transformation Λ^{-1}, we have (Eq. (7.1.2c))

$$\psi''(x) = S^{-1}\psi(\Lambda x) . \tag{11.1.2c}$$

For a passive transformation Λ, the spinor transforms according to (11.1.2a). For an active transformation Λ, the state is transformed according to (11.1.2c)[1].

---

[1] For inhomogeneous transformations (Λ, a), one has $(\Lambda, a)^{-1} = (\Lambda^{-1}, -\Lambda^{-1}a)$ and in the arguments of Eq. (11.1.2a–c) one must make the replacements $\Lambda x \to \Lambda x + a$ and $\Lambda^{-1}x \to \Lambda^{-1}(x - a)$.

We now consider the transformation of *vector fields* such as the four-potential of the electromagnetic field:

The *passive transformation* of the components of a vector $A^\mu(x)$ under a Lorentz transformation $x'^\mu = \Lambda^\mu{}_\nu x^\nu$ takes the form

$$A'^\mu(x') = \Lambda^\mu{}_\nu A^\nu(x) \equiv \Lambda^\mu{}_\nu A^\nu(\Lambda^{-1}x') \,. \tag{11.1.3a}$$

The inverse of the Lorentz transformation may be established as follows:

$$\Lambda^\lambda{}_\mu g^{\mu\nu} \Lambda^\rho{}_\nu = g^{\lambda\rho} \implies \Lambda^{\lambda\nu} \Lambda_{\rho\nu} = \delta^\lambda{}_\rho \implies \Lambda_\chi{}^\nu \Lambda^\rho{}_\nu = \delta_\chi{}^\rho \,.$$

Since the right inverse of a matrix is equal to its left inverse, together with Eq. (11.1.1), this implies

$$\Lambda^\mu{}_\nu \Lambda_\mu{}^\sigma = \delta^\sigma{}_\nu \implies \Lambda_\mu{}^\sigma x'^\mu = \Lambda_\mu{}^\sigma \Lambda^\mu{}_\nu x^\nu = x^\sigma \,,$$

and so, finally, the inverse of the Lorentz transformation

$$x^\sigma = \Lambda_\mu{}^\sigma x'^\mu \,. \tag{11.1.4}$$

For an *active transformation*, the entire space, along with its vector fields, is transformed and then viewed from the original coordinate system I. For a transformation with Λ, the resulting vector field, when viewed from I', is of the form $A^\mu(x')$ (see Fig. 11.1). The field transformed actively with Λ, which we denote by $A''^\mu(x)$, therefore takes the form

$$A''^\mu(x) = \Lambda^{-1}{}^\mu{}_\nu A^\nu(\Lambda x) = \Lambda_\nu{}^\mu A^\nu(\Lambda x) \quad \text{in } I \,. \tag{11.1.3c}$$

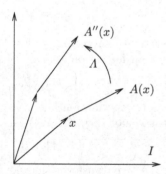

Fig. 11.1. Active transformation of a vector with the Lorentz transformation Λ

For the sake of completeness, we also give the active transformation with respect to the Lorentz transformation Λ^{-1}, which leads to the form

$$A'^\mu(x) = \Lambda^\mu{}_\nu A^\nu(x) \,. \tag{11.1.3b}$$

We now investigate the *transformation of the Dirac equation* in the presence of an electromagnetic field A_μ with respect to a *passive Lorentz transformation:* Starting from the Dirac equation in the system I

$$\left(\gamma^{\mu}(\mathrm{i}\partial_{\mu} - eA_{\mu}(x)) - m\right)\psi(x) = 0 \,, \tag{11.1.5a}$$

one obtains the transformed equation in the system I':

$$\left(\gamma^{\mu}(\mathrm{i}\partial'_{\mu} - eA'_{\mu}(x')) - m\right)\psi'(x') = 0 \,. \tag{11.1.5b}$$

Equation (11.1.5b) is derived by inserting into (11.1.5a) the transformations

$$\partial_{\mu} \equiv \frac{\partial}{\partial x^{\mu}} = \Lambda^{\nu}{}_{\mu}\partial'_{\nu} \,; \quad A_{\mu}(x) = \Lambda^{\nu}{}_{\mu}A'_{\nu}(x') \quad \text{and} \quad \psi(x) = S^{-1}\psi'(x') \,,$$

which yields

$$\left(\gamma^{\mu}\Lambda^{\nu}{}_{\mu}(\mathrm{i}\partial'_{\nu} - eA'_{\nu}(x')) - m\right)S^{-1}\psi'(x') = 0 \,.$$

Multiplying by S,

$$\left(S\gamma^{\mu}\Lambda^{\nu}{}_{\mu}S^{-1}(\mathrm{i}\partial'_{\nu} - eA'_{\nu}(x')) - m\right)\psi'(x') = 0 \,,$$

and making use of $\gamma^{\mu}\Lambda^{\nu}{}_{\mu} = S^{-1}\gamma^{\nu}S$ finally yields the desired result

$$\left(\gamma^{\nu}(\mathrm{i}\partial'_{\nu} - eA'_{\nu}(x')) - m\right)\psi'(x') = 0 \,.$$

Transformation of the Dirac equation with respect to an *active Lorentz transformation*, viz:

$$\psi''(x) = S^{-1}\psi(\Lambda x) \tag{11.1.2c}$$

with

$$A''^{\mu}(x) = \Lambda_{\nu}{}^{\mu}A^{\nu}(\Lambda x) \,. \tag{11.1.3c}$$

Starting from

$$\left(\gamma^{\mu}(\mathrm{i}\partial_{\mu} - eA_{\mu}(x)) - m\right)\psi(x) = 0 \,, \tag{11.1.5a}$$

we take this equation at the point $x' = \Lambda x$, and taking note of the fact that $\frac{\partial}{\partial x'^{\mu}} = \frac{\partial x^{\nu}}{\partial x'^{\mu}}\frac{\partial}{\partial x^{\nu}} = \Lambda_{\mu}{}^{\nu}\partial_{\nu}$,

$$\left(\gamma^{\mu}(\mathrm{i}\Lambda_{\mu}{}^{\nu}\partial_{\nu} - eA_{\mu}(\Lambda x)) - m\right)\psi(\Lambda x) = 0 \,.$$

Multiplying by $S^{-1}(\Lambda)$,

$$\left(S^{-1}\gamma^{\mu}S(\mathrm{i}\Lambda_{\mu}{}^{\nu}\partial_{\nu} - eA_{\mu}(\Lambda x)) - m\right)S^{-1}\psi(\Lambda x) = 0 \,,$$

and using $S^{-1}\gamma^{\mu}S\Lambda_{\mu}{}^{\nu} = \Lambda^{\mu}{}_{\sigma}\gamma^{\sigma}\Lambda_{\mu}{}^{\nu} = \gamma^{\sigma}\delta_{\sigma}{}^{\nu}$ together with Eq. (11.1.4) yields:

$$\left(\gamma^{\nu}(\mathrm{i}\partial_{\nu} - eA''_{\nu}(x)) - m\right)\psi''(x) = 0 \,. \tag{11.1.6}$$

If $\psi(x)$ satisfies the Dirac equation for the potential $A_{\mu}(x)$, then the transformed spinor $\psi''(x)$ satisfies the Dirac equation with the transformed potential $A''_{\mu}(x)$.

In general, the transformed equation is different to the original one. The two equations are the same only when $A''_{\mu}(x) = A_{\mu}(x)$. Then, $\psi(x)$ and $\psi''(x)$ obey the same equation of motion. The equation of motion remains invariant under any Lorentz transformation L that leaves the external potential unchanged. For example, a radially symmetric potential is invariant under rotations.

11.2 Invariance and Conservation Laws

11.2.1 The General Transformation

We write the transformation $\psi''(x) = S^{-1}\psi(\Lambda x)$ in the form

$$\psi'' = T\psi \,, \tag{11.2.1}$$

where the operator T contains both the effect of the matrix S and the transformation of the coordinates. The statement that the Dirac equation transforms under an active Lorentz transformation as above (Eq. (11.1.6)) implies for the operator

$$\mathcal{D}(A) \equiv \gamma^\mu(i\partial_\mu - eA_\mu) \tag{11.2.2}$$

that

$$T\mathcal{D}(A)T^{-1} = \mathcal{D}(A'') \,, \tag{11.2.3}$$

since

$$(\mathcal{D}(A) - m)\psi = 0 \Longrightarrow T(\mathcal{D}(A) - m)\psi = T(\mathcal{D}(A) - m)T^{-1}T\psi$$
$$= (\mathcal{D}(A'') - m)T\psi = 0 \,.$$

As the transformed spinor $T\psi$ obeys the Dirac equation $(\mathcal{D}(A'') - m)T\psi = 0$, and this holds for every spinor, equation (11.2.3) follows.

If A remains unchanged under the Lorentz transformation in question ($A'' = A$), it follows from (11.2.3) that T commutes with $\mathcal{D}(A)$:

$$[T, \mathcal{D}(A)] = 0 \,. \tag{11.2.4}$$

One can construct the operator T for each of the individual transformations, to which we shall now turn our attention.

11.2.2 Rotations

We have already found in Chap. 7 that[2] for *rotations*

$$T = e^{-i\varphi^k J^k} \tag{11.2.5}$$

with

$$\mathbf{J} = \frac{\hbar}{2}\Sigma + \mathbf{x} \times \frac{\hbar}{i}\nabla \,.$$

---

[2] The difference in sign compared to Chap. 7 arises because there the active transformation Λ^{-1} was considered.

The total angular momentum $\mathbf{J}$ is the generator of rotations.

If one takes an infinitesimal φ^k, then, from (11.2.2) and (11.2.4) and after expansion of the exponential function, it follows that for a rotationally invariant potential A,

$$[\mathcal{D}(A), \mathbf{J}] = 0 . \tag{11.2.6}$$

Since $[i\gamma^0\partial_t, \gamma^i\gamma^k] = 0$ and $[i\gamma^0\partial_t, \mathbf{x} \times \mathbf{\nabla}] = 0$, equation (11.2.6) also implies that

$$[\mathbf{J}, H] = 0, \tag{11.2.7}$$

where H is the Dirac Hamiltonian.

11.2.3 Translations

For *translations* we have $S = \mathbb{1}$ and

$$\psi''(x) = \psi(x + a) = e^{a^\mu \partial_\mu} \psi(x) , \tag{11.2.8}$$

and thus the translation operator is

$$T \equiv e^{-ia^\mu i\partial_\mu} = e^{-ia^\mu p_\mu} , \tag{11.2.9}$$

where $p_\mu = i\partial_\mu$ is the momentum operator. The momentum is the generator of translations. The *translational invariance* of a problem means that

$$[\mathcal{D}(A), p_\mu] = 0 \tag{11.2.10}$$

and since $[i\gamma^0\partial_t, p_\mu] = 0$, this also implies that

$$[p_\mu, H] = 0 . \tag{11.2.11}$$

11.2.4 Spatial Reflection (Parity Transformation)

We now turn to the *parity transformation*. The parity operation P, represented by the parity operator $\mathcal{P}$, is associated with a spatial reflection. We use $\mathcal{P}^{(0)}$ to denote the orbital parity operator, which causes a spatial reflection

$$\mathcal{P}^{(0)}\psi(t, \mathbf{x}) = \psi(t, -\mathbf{x}) . \tag{11.2.12}$$

For the total parity operator in Sect. 6.2.2.4, we found, to within an arbitrary phase factor,

$$\mathcal{P} = \gamma^0 \mathcal{P}^{(0)} . \tag{11.2.13}$$

We also have $\mathcal{P}^\dagger = \mathcal{P}$ and $\mathcal{P}^2 = 1$.

If $A^\mu(x)$ is invariant under inversion, then the Dirac Hamiltonian H satisfies

$$[\mathcal{P}, H] = 0 . \tag{11.2.14}$$

There remain two more discrete symmetries of the Dirac equation, charge conjugation and time-reversal invariance.

11.3 Charge Conjugation

The Hole theory suggests that the electron possesses an antiparticle, the positron. This particle was actually discovered experimentally in 1933 by C.D. Anderson. The positron is also a fermion with spin 1/2 and should itself satisfy the Dirac equation with $e \to -e$. There must thus be a connection between negative energy solutions for negative charge and positive energy solutions carrying positive charge. This additional symmetry transformation of the Dirac equation is referred to as "charge conjugation", C.

The Dirac equation of the electron reads:

$$(i\slashed{\partial} - e\slashed{A} - m)\psi = 0\,, \qquad e = -e_0\,, \quad e_0 = 4.8 \times 10^{-10}\text{esu} \qquad (11.3.1)$$

and the Dirac equation for an oppositely charged particle is

$$(i\slashed{\partial} + e\slashed{A} - m)\psi_c = 0\,. \qquad (11.3.2)$$

We seek a transformation that converts ψ into ψ_c. We begin by establishing the effects of complex conjugation on the first two terms of (11.3.1):

$$(i\partial_\mu)^* = -i\partial_\mu \qquad (11.3.3\text{a})$$
$$(A_\mu)^* = A_\mu\,, \qquad (11.3.3\text{b})$$

as the electromagnetic field is real. In the next section, in particular, it will turn out to be useful to define an operator K_0 that has the effect of complex conjugating the operators and spinors upon which it acts. Using this notation, (11.3.3a,b) reads:

$$K_0 i\partial_\mu = -i\partial_\mu K_0 \quad \text{and} \quad K_0 A_\mu = A_\mu K_0\,. \qquad (11.3.3')$$

Thus, when one takes the complex conjugate of the Dirac equation, one obtains

$$\left(-(i\partial_\mu + eA_\mu)\gamma^{\mu*} - m\right)\psi^*(x) = 0\,. \qquad (11.3.4)$$

In comparison with Eq. (11.3.1), not only is the sign of the charge opposite, but also that of the mass term. We seek a nonsingular matrix $C\gamma^0$ with the property

$$C\gamma^0 \gamma^{\mu*}(C\gamma^0)^{-1} = -\gamma^\mu\,. \qquad (11.3.5)$$

With the help of this matrix, we obtain from (11.3.4)

$$\begin{aligned}
C\gamma^0 &\left(-(i\partial_\mu + eA_\mu)\gamma^{\mu*} - m\right)(C\gamma^0)^{-1}C\gamma^0\psi^* \\
&= (i\slashed{\partial} + e\slashed{A} - m)(C\gamma^0\psi^*) = 0\,.
\end{aligned} \qquad (11.3.6)$$

Comparison with (11.3.2) shows that

$$\psi_c = C\gamma^0\psi^* = C\bar\psi^T \tag{11.3.7}$$

since

$$\bar\psi^T = (\psi^\dagger\gamma^0)^T = \gamma^{0T}\psi^{\dagger T} = \gamma^0\psi^* . \tag{11.3.8}$$

Equation (11.3.5) can also be written in the form

$$C^{-1}\gamma^\mu C = -\gamma^{\mu^T} . \tag{11.3.5'}$$

In the standard representation, we have $\gamma^{0T} = \gamma^0$, $\gamma^{2T} = \gamma^2$, $\gamma^{1T} = -\gamma^1$, $\gamma^{3T} = -\gamma^3$, and, hence, C commutes with γ^1 and γ^3 and anticommutes with γ^0 and γ^2. From this, it follows that

$$C = i\gamma^2\gamma^0 = -C^{-1} = -C^\dagger = -C^T , \tag{11.3.9}$$

so that

$$\psi_c = i\gamma^2\psi^* . \tag{11.3.7'}$$

The full charge conjugation operation

$$\mathcal{C} = C\gamma_0 K_0 = i\gamma^2 K_0 \tag{11.3.7''}$$

consists in complex conjugation K_0 and multiplication by $C\gamma_0$.

If $\psi(x)$ describes the motion of a Dirac particle with charge e in the potential $A_\mu(x)$, then ψ_c describes the motion of a particle with charge $-e$ in the same potential $A_\mu(x)$.

Example: For a free particle, for which $A_\mu = 0$,

$$\psi_1^{(-)} = \frac{1}{(2\pi)^{3/2}}\begin{pmatrix}0\\0\\1\\0\end{pmatrix}e^{imt} \tag{11.3.10}$$

and therefore,

$$\left(\psi_1^{(-)}\right)_c = C\gamma^0\left(\psi_1^{(-)}\right)^* = i\gamma^2\left(\psi_1^{(-)}\right)^* = \frac{1}{(2\pi)^{3/2}}\begin{pmatrix}0\\1\\0\\0\end{pmatrix}e^{-imt} = \psi_2^{(+)} \tag{11.3.10'}$$

The charge conjugated state has opposite spin.

We now consider a more general state with momentum k and polarization along n. With respect to the projection operators, this has the property[3]

---

[3] $\not{n} = \gamma^\mu n_\mu$, n_μ space-like unit vector $n^2 = n^\mu n_\mu = -1$ and $n_\mu k^\mu = 0$.
$P(n) = \frac{1}{2}(1 + \gamma_5\not{n})$ projects onto the positive energy spinor $u(k,n)$, which is polarized along $\check{n}$ in the rest frame, and onto the negative energy spinor $v(k,n)$, which is polarized along $-\check{n}$.
$k = \Lambda\check{k}$, $n = \Lambda\check{n}$, $\check{k} = (m,0,0,0)$, $\check{n} = (0,\mathbf{n})$ (see Appendix C). The projection operators $\Lambda_\pm(k) \equiv (\pm\not{k} + m)/2m$ were introduced in Eq. (6.3.21).

$$\psi = \frac{\varepsilon \slashed{k} + m}{2m} \frac{1 + \gamma_5 \slashed{n}}{2} \psi , \quad k^0 > 0 \tag{11.3.11}$$

with $\varepsilon = \pm 1$ indicating the sign of the energy. Applying charge conjugation to this relation, one obtains

$$\psi_c = C \bar{\psi}^T = C \gamma^0 \left(\frac{\varepsilon \slashed{k} + m}{2m} \right)^* \left(\frac{1 + \gamma_5 \slashed{n}}{2} \right)^* \psi^* \tag{11.3.11'}$$

$$= C \gamma_0 \left(\frac{\varepsilon \slashed{k}^* + m}{2m} \right) (C \gamma_0)^{-1} C \gamma_0 \left(\frac{1 + \gamma_5 \slashed{n}^*}{2} \right) (C \gamma_0)^{-1} C \gamma_0 \psi^*$$

$$= \left(\frac{-\varepsilon \slashed{k} + m}{2m} \right) \left(\frac{1 + \gamma_5 \slashed{n}}{2} \right) \psi_c ,$$

where we have used $\gamma_5^* = \gamma_5$ and $\{C\gamma_0, \gamma_5\} = 0$. The state ψ_c is characterized by the same four-vectors, k and n, as ψ, but the energy has reversed its sign. Since the projection operator $\frac{1}{2}(1 + \gamma_5 \slashed{n})$ projects onto spin $\pm \frac{1}{2}$ along $\check{n}$, depending on the sign of the energy, the spin is reversed under charge conjugation. With regard to the momentum, we should like to point out that, for free spinors, complex conjugation yields $e^{-ikx} \to e^{ikx}$, i.e., the momentum $\mathbf{k}$ is transformed into $-\mathbf{k}$. Thus far, we have discussed the transformation of the spinors. In the qualitative description provided by the Hole theory, which finds its ultimate mathematical representation in quantum field theory, the non-occupation of a spinor of negative energy corresponds to an antiparticle with positive energy and exactly the opposite quantum numbers to those of the spinor (Sect. 10.2). Therefore, under charge conjugation, the particles and antiparticles are transformed into one another, having the same energy and spin, but opposite charge.

Remarks:

(i) The Dirac equation is obviously invariant under simultaneous transformation of ψ and A,

$$\psi \longrightarrow \psi_c = \eta_c C \bar{\psi}^T$$
$$A_\mu \longrightarrow A_\mu^c = -A_\mu .$$

With respect to charge conjugation, the four-current density j_μ transforms according to

$$j_\mu = \bar{\psi} \gamma_\mu \psi \longrightarrow j_\mu^c = \bar{\psi}_c \gamma_\mu \psi_c = \bar{\psi}^* C^\dagger \gamma_0 \gamma_\mu C \bar{\psi}^T$$

$$= \psi^T \gamma^0 (-C) \gamma^0 \gamma_\mu C \bar{\psi}^T = \psi^T C \gamma_\mu C \bar{\psi}^T = \psi^T \gamma_\mu^T \bar{\psi}^T$$

$$= \psi_\alpha (\gamma_\mu)_{\beta\alpha} \gamma_{\beta\rho}^0 \psi_\rho^* = \psi_\rho^* \gamma_{\rho\beta}^0 (\gamma_\mu)_{\beta\alpha} \psi_\alpha = \bar{\psi} \gamma_\mu \psi .$$

For the c-number Dirac field one thus obtains $j_\mu^c = j_\mu$. In the quantized form, ψ and $\bar{\psi}$ become anticommuting fields, which leads to an extra minus sign:

$$j_\mu^c = -j_\mu . \tag{11.3.12}$$

Then, under charge conjugation, the combination $ej \cdot A$ remains invariant. As we shall see explicitly in the case of the Majorana representation, the form of the charge conjugation transformation depends on the representation.

(ii) A *Majorana representation* is a representation of the γ matrices with the property that γ^0 is imaginary and antisymmetric, whilst the γ^k are imaginary and symmetric. In a Majorana representation, the Dirac equation

$$(i\gamma^\mu \partial_\mu - m)\psi = 0$$

is a real equation. If ψ is a solution of this equation, then so is

$$\psi_c = \psi^* \ . \tag{11.3.13}$$

In the Majorana representation, the solution related to ψ by charge conjugation is, to within an arbitrary phase factor, given by (11.3.13), since the Dirac equation for the field ψ,

$$(\gamma^\mu (i\partial_\mu - eA_\mu) - m)\psi = , \tag{11.3.14}$$

also leads to

$$(\gamma^\mu (i\partial_\mu + eA_\mu) - m)\psi_c = 0 \ . \tag{11.3.14'}$$

The spinor ψ is the solution of the Dirac equation with a field corresponding to charge e and the spinor ψ_c is the solution for charge $-e$. A spinor that is real, i.e.,

$$\psi^* = \psi \ ,$$

is known as a Majorana spinor. A Dirac spinor consists of two Majorana spinors. An example of a Majorana representation is the set of matrices

$$\gamma_0 = \begin{pmatrix} 0 & \sigma^2 \\ \sigma^2 & 0 \end{pmatrix} \ , \quad \gamma_1 = i \begin{pmatrix} 0 & \sigma^1 \\ \sigma^1 & 0 \end{pmatrix} \ ,$$

$$\gamma_2 = i \begin{pmatrix} \mathbb{1} & 0 \\ 0 & -\mathbb{1} \end{pmatrix} \ , \quad \gamma_3 = i \begin{pmatrix} 0 & \sigma^3 \\ \sigma^3 & 0 \end{pmatrix} \ . \tag{11.3.15}$$

Another example is given in Problem 11.2.

11.4 Time Reversal (Motion Reversal)

Although the more appropriate name for this discrete symmetry transformation would be "motion reversal", the term "time reversal transformation" is so well established that we shall adopt this practice. It should be emphasized from the outset that the time-reversal transformation does not cause a system to evolve backwards in time, despite the fact that it includes a change in the time argument of a state $t \to -t$. One does not need clocks that run backwards in order to study time reversal and the invariance of a theory under this transformation. What one is really dealing with is a reversal of the motion. In quantum mechanics the situation is further complicated by a formal difficulty: In order to describe time reversal, one needs antiunitary operators. In this section we first study the time-reversal transformation in classical mechanics and nonrelativistic quantum mechanics, and then turn our attention to the Dirac equation.

11.4.1 Reversal of Motion in Classical Physics

Let us consider a classical system invariant under time translation, which is described by the generalized coordinates q and momenta p. The time-independent Hamiltonian function is $H(q, p)$. Hamilton's equations of motion are then

$$
\begin{aligned}
\dot{q} &= \frac{\partial H(q, p)}{\partial p} \\
\dot{p} &= -\frac{\partial H(q, p)}{\partial q} \ .
\end{aligned}
\tag{11.4.1}
$$

At $t = 0$, we assume the initial values (q_0, p_0) for the generalized coordinates and momenta. Hence, the solution $q(t), p(t)$ of Hamilton's equations of motion must satisfy the initial conditions

$$
\begin{aligned}
q(0) &= q_0 \\
p(0) &= p_0 \ .
\end{aligned}
\tag{11.4.2}
$$

Let the solution at a later time $t = t_1 > 0$ assume the values

$$
q(t_1) = q_1 \ , \quad p(t_1) = p_1 \ .
\tag{11.4.3a}
$$

The *motion-reversed state* at time t_1 is defined by

$$
q'(t_1) = q_1 \ , \quad p'(t_1) = -p_1 \ .
\tag{11.4.3b}
$$

If, after this motion reversal, the system retraces its path, and after a further time t, returns to its time reversed initial state, the system is said to be time-reversal or motion-reversal invariant (see Fig. 11.2). To test time-reversal invariance there is no need for running backwards in time. In the definition which one encounters above, only motion for the positive time direction arises. As a result, it is possible to test experimentally whether a system is time-reversal invariant.

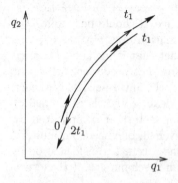

Fig. 11.2. Motion reversal: Shown (displaced for clarity) are the trajectories in real space: $(0, t_1)$ prior to reversal of the motion, and $(t_1, 2t_1)$ after reversal

Let us now investigate the conditions for time-reversal invariance and find the solution for the motion-reversed initial state. We define the functions

$$q'(t) = q(2t_1 - t)$$
$$p'(t) = -p(2t_1 - t) \,.$$

(11.4.4)

These functions obviously satisfy the initial conditions

$$q'(t_1) = q(t_1) = q_1$$ (11.4.5)
$$p'(t_1) = -p(t_1) = -p_1 \,.$$ (11.4.6)

At time $2t_1$ they have the values

$$q'(2t_1) = q(0) = q_0$$
$$p'(2t_1) = -p(0) = -p_0 \,,$$

(11.4.7)

i.e., the motion-reversed initial values. Finally, they satisfy the equation of motion[4]

$$\dot{q}'(t) = -\dot{q}(2t_1 - t) = -\frac{\partial H(q(2t_1 - t), p(2t_1 - t))}{\partial p(2t_1 - t)}$$

$$= \frac{\partial H(q'(t), -p'(t))}{\partial p'(t)}$$

(11.4.8a)

$$\dot{p}'(t) = \dot{p}(2t_1 - t) = -\frac{\partial H(q(2t_1 - t), p(2t_1 - t))}{\partial q(2t_1 - t)}$$

$$= -\frac{\partial H(q'(t), -p'(t))}{\partial q'(t)} \,.$$

(11.4.8b)

The equations of motion of the functions $q'(t), p'(t)$ are described, according to (11.4.8a,b), by a Hamiltonian function $\bar{H}$, which is related to the original Hamiltonian by making the replacement $p \to -p$:

$$\bar{H} = H(q, -p) \,.$$

(11.4.9)

Most Hamiltonians are quadratic in p (e.g., that of particles in an external potential interacting via potentials), and are thus invariant under motion reversal. For these, $\bar{H} = H(q, p)$, and $q'(t), p'(t)$ satisfy the original equation of motion evolving from the motion-reversed starting value $(q_1, -p_1)$ to the motion-reversed initial value $(q_0, -p_0)$ of the original solution $(q(t), p(t))$. This implies that such classical systems are time-reversal invariant.

Motion-reversal invariance in this straightforward fashion does not apply to the motion of particles in a magnetic field, or to any other force that varies linearly with velocity. This is readily seen if one considers Fig. 11.3: In a homogeneous magnetic field, charged particles move along circles, the

---

[4] The dot implies differentiation with respect to the whole argument, e.g.,
$\dot{q}(2t_1 - t) \equiv \frac{\partial q(2t_1 - t)}{\partial (2t_1 - t)}$.

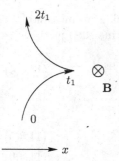

Fig. 11.3. Motion reversal in the presence of a magnetic field $\mathbf{B}$ perpendicular to the plane of the page. The motion is reversed at an instant when the particle is moving in exactly the x-direction

sense of motion depending on the sign of the charge. Thus, when the motion is reversed, the particle does not return along the same circle, but instead moves along the upper arc shown in Fig. 11.3. In the presence of a magnetic field, one can only achieve motion-reversal invariance if the direction of the magnetic field is also reversed:

$$\mathbf{B} \to -\mathbf{B} \,, \tag{11.4.10}$$

as can be seen from the sketch, or from the following calculation. Let the Hamiltonian in cartesian coordinates with no field be written $H = H(\mathbf{x}, \mathbf{p})$, which will be assumed to be invariant with respect to time reversal. The Hamiltonian in the presence of an electromagnetic field is then

$$H = H\left(\mathbf{x}, \mathbf{p} - \frac{e}{c}\mathbf{A}(\mathbf{x})\right) + e\Phi(\mathbf{x}) \,, \tag{11.4.11}$$

where $\mathbf{A}$ is the vector potential and Φ the scalar potential. This Hamiltonian is no longer invariant under the transformation (11.4.4). However, it is invariant under the general transformation

$$\mathbf{x}'(t) = \mathbf{x}(2t_1 - t) \tag{11.4.12a}$$
$$\mathbf{p}'(t) = -\mathbf{p}(2t_1 - t) \tag{11.4.12b}$$
$$\mathbf{A}'(\mathbf{x}, t) = -\mathbf{A}(\mathbf{x}, 2t_1 - t) \tag{11.4.12c}$$
$$\Phi'(\mathbf{x}, t) = \Phi(\mathbf{x}, 2t_1 - t) \,. \tag{11.4.12d}$$

Equations (11.4.12c) and (11.4.12d) imply a change in the sign of the magnetic field, but not of the electric field, as can be seen from

$$\mathbf{B} = \operatorname{curl}\mathbf{A} \to \operatorname{curl}\mathbf{A}' = -\mathbf{B}$$

$$\mathbf{E} = -\boldsymbol{\nabla}\Phi + \frac{1}{c}\frac{\partial}{\partial t}\mathbf{A}(\mathbf{x}, t) \to -\boldsymbol{\nabla}\Phi' + \frac{1}{c}\frac{\partial}{\partial t}\mathbf{A}'(\mathbf{x}, t)$$

$$= -\boldsymbol{\nabla}\Phi + \frac{1}{c}\frac{\partial}{\partial(2t_1 - t)}\mathbf{A}(\mathbf{x}, 2t_1 - t) = \mathbf{E} \,. \tag{11.4.13a}$$

We note in passing that when the Lorentz condition

$$\frac{1}{c}\frac{\partial}{\partial t}\Phi + \boldsymbol{\nabla}\mathbf{A} = 0 \tag{11.4.13b}$$

holds, it also holds for the motion-reversed potentials.

Remark: In the above description we considered motion in the time interval $[0, t_1]$, and then allowed the motion-reversed process to occur in the adjoining time interval $[t_1, 2t_1]$. We could equally well have considered the original motion in the time interval $[-t_1, t_1]$ and, as counterpart, the motion-reversed process also lying in the time interval between $-t_1$ and t_1:

$$\begin{aligned} q''(t) &= q(-t) \\ p''(t) &= -p(-t) \end{aligned} \tag{11.4.14}$$

with the initial conditions

$$\begin{aligned} q''(-t_1) &= q(t_1)\,, \\ p''(-t_1) &= -p(t_1) \end{aligned} \tag{11.4.15}$$

and final values,

$$\begin{aligned} q''(t_1) &= q(-t_1)\,, \\ p''(t_1) &= -p(-t_1)\,. \end{aligned} \tag{11.4.16}$$

$(q''(t), p''(t))$ differs from $(q'(t), p'(t))$ in Eq. (11.4.4) only by a time translation of $2t_1$; in both cases, time runs in the positive sense $-t_1$ to t_1.

11.4.2 Time Reversal in Quantum Mechanics

11.4.2.1 Time Reversal in the Coordinate Representation

Following these classical mechanical preparatory remarks, we now turn to nonrelativistic quantum mechanics (in the coordinate representation). The system is described by the wave function $\psi(\mathbf{x}, t)$, which obeys the Schrödinger equation

$$i\frac{\partial \psi(\mathbf{x}, t)}{\partial t} = H\psi(\mathbf{x}, t)\,. \tag{11.4.17}$$

Let us take the wave function at time $t = 0$ to be given by $\psi_0(\mathbf{x})$, i.e.,

$$\psi(\mathbf{x}, 0) = \psi_0(\mathbf{x})\,. \tag{11.4.18}$$

This initial condition determines $\psi(\mathbf{x}, t)$ at all later times t. Although the Schrödinger equation enables one to calculate $\psi(\mathbf{x}, t)$ at earlier times, this is usually not of interest. The statement that the wave function at $t = 0$ is

$\psi_0(\mathbf{x})$ implies that a measurement that has been made will, in general, have changed the state of the system discontinuously. At the time $t_1 > 0$ we let the wave function be

$$\psi(\mathbf{x}, t_1) \equiv \psi_1(\mathbf{x}) . \tag{11.4.19}$$

What is the motion reversed system for which an initial state $\psi_1(\mathbf{x})$ evolves into the state $\psi_0(\mathbf{x})$ after a time t_1? Due to the presence of the first order time derivative, the function $\psi(\mathbf{x}, 2t_1 - t)$ does not satisfy the Schrödinger equation. However, if, in addition, we take the complex conjugate of the wave function

$$\psi'(\mathbf{x}, t) = \psi^*(\mathbf{x}, 2t_1 - t) \equiv K_0 \psi(\mathbf{x}, 2t_1 - t) , \tag{11.4.20}$$

this satisfies the differential equation

$$\mathrm{i} \frac{\partial \psi'(\mathbf{x}, t)}{\partial t} = H^* \psi'(\mathbf{x}, t) \tag{11.4.21}$$

and the boundary conditions

$$\psi'(\mathbf{x}, t_1) = \psi_1^*(\mathbf{x}) \tag{11.4.22a}$$

$$\psi'(\mathbf{x}, 2t_1) = \psi_0^*(\mathbf{x}) . \tag{11.4.22b}$$

Proof. Omitting the argument $\mathbf{x}$, we have[5]

$$\mathrm{i} \frac{\partial \psi'(t)}{\partial t} = \mathrm{i} \frac{\partial \psi^*(2t_1 - t)}{\partial t} = -K_0 \mathrm{i} \frac{\partial \psi(2t_1 - t)}{\partial t} = K_0 \mathrm{i} \frac{\partial \psi(2t_1 - t)}{\partial(-t)}$$
$$= K_0 H \psi(2t_1 - t) = H^* \psi^*(2t_1 - t) = H^* \psi'(t) .$$

Here, H^* is the complex conjugate of the Hamiltonian, which is not necessarily identical to $H^\dagger$. For the momentum operator, for example, we have

$$\left(\frac{\hbar}{\mathrm{i}} \boldsymbol{\nabla}\right)^\dagger = \frac{\hbar}{\mathrm{i}} \boldsymbol{\nabla} , \text{ but } \left(\frac{\hbar}{\mathrm{i}} \boldsymbol{\nabla}\right)^* = -\frac{\hbar}{\mathrm{i}} \boldsymbol{\nabla} . \tag{11.4.23}$$

When the Hamiltonian is quadratic in $\mathbf{p}$, then $H^* = H$ and thus the system is *time-reversal invariant*.

We now calculate the *expectation values* of momentum, position, and angular momentum (the upper index gives the time and the lower index the wave function):

$$\langle \mathbf{p} \rangle_\psi^t = (\psi, \mathbf{p}\psi) = \int d^3 x \, \psi^* \frac{\hbar}{\mathrm{i}} \boldsymbol{\nabla} \psi \tag{11.4.24a}$$

$$\langle \mathbf{x} \rangle_\psi^t = (\psi, \mathbf{x}\psi) = \int d^3 x \, \psi^*(\mathbf{x}, t) \mathbf{x} \psi(\mathbf{x}, t) \tag{11.4.24b}$$

---

[5] The operator K_0 has the effect of complex conjugation.

$$\langle \mathbf{p} \rangle_{\psi'}^t = (\psi^*, \mathbf{p}\psi^*) = \int d^3x \, \psi \frac{\hbar}{i} \nabla \psi^*$$

$$= -\left(\int d^3x \, \psi^* \frac{\hbar}{i} \nabla \psi \right)^* = -\langle \mathbf{p} \rangle_\psi^{2t_1 - t} \tag{11.4.24c}$$

$$\langle \mathbf{x} \rangle_{\psi'}^t = \langle \mathbf{x} \rangle_\psi^{2t_1 - t} \tag{11.4.24d}$$

$$\langle \mathbf{L} \rangle_{\psi'}^t = \int d^3x \, \psi \, \mathbf{x} \times \frac{\hbar}{i} \nabla \psi^*$$

$$= -\left(\int d^3x \, \psi^* \, \mathbf{x} \times \frac{\hbar}{i} \nabla \psi \right)^* = -\langle \mathbf{L} \rangle_\psi^{2t_1 - t} . \tag{11.4.24e}$$

These results are in exact correspondence to the classical results. The mean value of the position of the motion-reversed state follows the same trajectory backwards, the mean value of the momentum having the opposite sign.

Here, too, we can take $\psi(\mathbf{x}, t)$ in the interval $[-t_1, t_1]$ and likewise,

$$\psi'(\mathbf{x}, t) = K_0 \psi(\mathbf{x}, -t) \tag{11.4.25}$$

in the interval $[-t_1, t_1]$, corresponding to the classical case (11.4.14). In the following, we will represent the time-reversal transformation in this more compact form. The direction of time is always positive.

Since $K_0^2 = 1$, we have $K_0^{-1} = K_0$. Due to the property (11.4.23), and since the spatial coordinates are real, we find the following transformation behavior for $\mathbf{x}$, $\mathbf{p}$, and $\mathbf{L}$:

$$K_0 \mathbf{x} K_0^{-1} = \mathbf{x} \tag{11.4.25'c}$$

$$K_0 \mathbf{p} K_0^{-1} = -\mathbf{p} \tag{11.4.25'd}$$

$$K_0 \mathbf{L} K_0^{-1} = -\mathbf{L} . \tag{11.4.25'e}$$

11.4.2.2 Antilinear and Antiunitary Operators

The transformation $\psi \to \psi'(t) = K_0 \psi(-t)$ is not unitary.

Definition: An operator A is *antilinear* if

$$A(\alpha_1 \psi_1 + \alpha_2 \psi_2) = \alpha_1^* A \psi_1 + \alpha_2^* A \psi_2 . \tag{11.4.26}$$

Definition: An operator A is *antiunitary* if it is antilinear and obeys

$$(A\psi, A\varphi) = (\varphi, \psi) . \tag{11.4.27}$$

K_0 is evidently antilinear,

$$K_0(\alpha_1 \psi_1 + \alpha_2 \psi_2) = \alpha_1^* K_0 \psi_1 + \alpha_2^* K_0 \psi_2 ,$$

and, furthermore,

$$(K_0 \psi, K_0 \varphi) = (\psi^*, \varphi^*) = \int d^3x \, \psi \varphi^* = (\varphi, \psi) . \tag{11.4.28}$$

Hence, K_0 is antiunitary.

If U is unitary, $UU^\dagger = U^\dagger U = 1$, then UK_0 is antiunitary, which can be seen as follows:

$$UK_0(\alpha_1\psi_1 + \alpha_2\psi_2) = U(\alpha_1^* K_0\psi_1 + \alpha_2^* K_0\psi_2) = \alpha_1^* UK_0\psi_1 + \alpha_2^* UK_0\psi_2$$
$$(UK_0\psi, UK_0\varphi) = (K_0\psi, U^\dagger UK_0\varphi) = (K_0\psi, K_0\varphi) = (\varphi, \psi) \ .$$

The converse is also true: Every antiunitary operator can be represented in the form $A = UK_0$.

Proof: We have $K_0^2 = 1$. Let A be a given antiunitary operator; we define $U = AK_0$. The operator U satisfies

$$U(\alpha_1\psi_1 + \alpha_2\psi_2) = AK_0(\alpha_1\psi_1 + \alpha_2\psi_2) = A(\alpha_1^* K_0\psi_1 + \alpha_2^* K_0\psi_2)$$
$$= (\alpha_1 AK_0\psi_1 + \alpha_2 AK_0\psi_2) = (\alpha_1 U\psi_1 + \alpha_2 U\psi_2) \ ,$$

and, hence, U is linear. Furthermore,

$$(U\varphi, U\psi) = (AK_0\varphi, AK_0\psi) = (A\varphi^*, A\psi^*) = (\psi^*, \varphi^*) = \int d^3x\, \psi\varphi^*$$
$$= (\varphi, \psi) \ ,$$

and thus U is also unitary. From $U = AK_0$ it follows that $A = UK_0$, thus proving the assertion.

Notes:

(i) For antilinear operators such as K_0, it is advantageous to work in the coordinate representation. If the Dirac bra and ket notation is used, one must bear in mind that its effect is dependent on the basis employed. If $|a\rangle = \int d^3\xi\, |\xi\rangle\langle\xi|a\rangle$, then in the coordinate representation, and insisting that $K_0|\xi\rangle = |\xi\rangle$,

$$K_0|a\rangle = \int d^3\xi\, (K_0|\xi\rangle)\langle\xi|a\rangle^* = \int d^3\xi\, |\xi\rangle\langle\xi|a\rangle^* \ . \tag{11.4.29}$$

For the momentum eigenstates this implies that

$$K_0|\mathbf{p}\rangle = \int d^3\xi\, |\xi\rangle\langle\xi|\mathbf{p}\rangle^* = |-\mathbf{p}\rangle \ ,$$

since $\langle\xi|\mathbf{p}\rangle = e^{i\mathbf{p}\xi}$ and $\langle\xi|\mathbf{p}\rangle^* = e^{-i\mathbf{p}\xi}$. If one chooses a different basis, e.g., $|n\rangle$ and postulates $K_0|n\rangle = |n\rangle$, then $K_0|a\rangle$ will not be the same as in the basis of position eigenfunctions. When we have cause to use the Dirac notation in the context of time reversal, a basis of position eigenfunctions will be employed.

(ii) In addition, the effect of antiunitary operators is only defined for ket vectors. The relation

$$\langle a|(L|b\rangle) = (\langle a|L)|b\rangle = \langle a|L|b\rangle \ ,$$

valid for linear operators, does not hold in the antiunitary case. This stems from the fact that a bra vector is defined as a linear functional on the ket vectors.[6]

---

[6] See, e.g., QM I, Sect. 8.2, footnote 2.

11.4.2.3 The Time-Reversal Operator $\mathcal{T}$ in Linear State Space

A. General properties, spin 0

Here, and in the next section, we describe the time-reversal transformation in the linear space of the ket and bra vectors, since this is frequently employed in quantum statistics. We give a general analysis of the condition of time reversal and also consider particles with spin. It will emerge anew that time reversal (motion reversal) cannot be represented by a unitary transformation. We denote the time-reversal operator by $\mathcal{T}$. The requirement of time-reversal invariance implies

$$e^{-iHt}\mathcal{T}\,|\psi(t)\rangle = \mathcal{T}\,|\psi(0)\rangle \;, \tag{11.4.30}$$

i.e.,

$$e^{-iHt}\mathcal{T}e^{-iHt}\,|\psi(0)\rangle = \mathcal{T}\,|\psi(0)\rangle \;.$$

Hence, if one carries out a motion reversal after time t and allows the system to evolve for a further period t, the resulting state is identical to the motion-reversed state at time $t = 0$. Since Eq. (11.4.30) is valid for arbitrary $|\psi(0)\rangle$, it follows that

$$e^{-iHt}\mathcal{T}e^{-iHt} = \mathcal{T}$$

whence

$$e^{-iHt}\mathcal{T} = \mathcal{T}e^{iHt} \;. \tag{11.4.31}$$

Differentiating (11.4.31) with respect to time and setting $t = 0$, one obtains

$$\mathcal{T}iH = -iH\mathcal{T} \;. \tag{11.4.32}$$

One might ask whether there could also be a unitary operator $\mathcal{T}$ that satisfies (11.4.32). If $\mathcal{T}$ were unitary and thus also linear, one could then move the i occurring on the left-hand side in front of the $\mathcal{T}$ and cancel it to obtain

$$\mathcal{T}H + H\mathcal{T} = 0 \;.$$

Then, for every eigenfunction ψ_E with

$$H\psi_E = E\psi_E$$

we would also have

$$H\mathcal{T}\psi_E = -E\mathcal{T}\psi_E \;.$$

For every positive energy E there would be a corresponding solution $\mathcal{T}\psi_E$ with eigenvalue $(-E)$. There would be no lower limit to the energy since there are certainly states with arbitrarily large positive energy. Therefore, we can rule out the possibility that there exists a unitary operator $\mathcal{T}$ satisfying (11.4.31). According to a theorem due to Wigner[7], symmetry transformations are either unitary or antiunitary, and hence $\mathcal{T}$ can only be antiunitary. Therefore, $\mathcal{T}iH = -i\mathcal{T}H$ and

$$\mathcal{T}H - H\mathcal{T} = 0 . \tag{11.4.33}$$

Let us now consider a matrix element of a linear operator B:

$$\langle \alpha | B | \beta \rangle = \langle B^\dagger \alpha | \beta \rangle = \langle \mathcal{T}\beta | \mathcal{T}B^\dagger \alpha \rangle$$
$$= \langle \mathcal{T}\beta | \mathcal{T}B^\dagger \mathcal{T}^{-1}\mathcal{T}\alpha \rangle = \langle \mathcal{T}\beta | \mathcal{T}B^\dagger \mathcal{T}^{-1} | \mathcal{T}\alpha \rangle$$

or

$$= \langle \alpha | B\beta \rangle = \langle \mathcal{T}B\beta | \mathcal{T}\alpha \rangle = \langle \mathcal{T}B\mathcal{T}^{-1}\mathcal{T}\beta | \mathcal{T}\alpha \rangle$$
$$= \langle \mathcal{T}\beta | \mathcal{T}B\mathcal{T}^{-1} | \mathcal{T}\alpha \rangle \tag{11.4.34}$$

If we assume that B is hermitian and

$$\mathcal{T}B\mathcal{T}^{-1} = \varepsilon_B B , \quad \text{where } \varepsilon_B \pm 1 , \tag{11.4.35}$$

which is suggested by the results of wave mechanics (Eq. (11.4.24a–e)), it then follows that

$$\langle \alpha | B | \beta \rangle = \varepsilon_B \langle \mathcal{T}\beta | B | \mathcal{T}\alpha \rangle .$$

The quantity ε_B is known as the "signature" of the operator B. Let us take the diagonal element

$$\langle \alpha | B | \alpha \rangle = \varepsilon_B \langle \mathcal{T}\alpha | B | \mathcal{T}\alpha \rangle .$$

Comparing this with (11.4.24c–e) and (11.4.25'c–e) yields the transformation of the operators

$$\mathcal{T} \mathbf{x} \mathcal{T}^{-1} = \mathbf{x} \tag{11.4.36a}$$
$$\mathcal{T} \mathbf{p} \mathcal{T}^{-1} = -\mathbf{p} \tag{11.4.36b}$$
$$\mathcal{T} \mathbf{L} \mathcal{T}^{-1} = -\mathbf{L} , \tag{11.4.36c}$$

i.e., $\varepsilon_{\mathbf{x}} = 1$, $\varepsilon_{\mathbf{p}} = -1$, and $\varepsilon_{\mathbf{L}} = -1$. The last relation is also a consequence of the first two.

---

[7] E.P. Wigner, *Group Theory and its Applications to Quantum Mechanics*, Academic Press, p. 233; V. Bargmann, J. Math. Phys. **5**, 862 (1964)

Remark: If the relations (11.4.36) are considered as the primary defining conditions on the operator $\mathcal{T}$, then, by transforming the commutator $[x, p] = \mathrm{i}$, one obtains

$$\mathcal{T}\mathrm{i}\mathcal{T}^{-1} = \mathcal{T}\,[x, p]\,\mathcal{T}^{-1} = [x, -p] = -\mathrm{i}\ .$$

This yields

$$\mathcal{T}\mathrm{i}\mathcal{T}^{-1} = -\mathrm{i}\ ,$$

which means that $\mathcal{T}$ is antilinear.

We now investigate the effect of $\mathcal{T}$ on coordinate eigenstates $|\boldsymbol{\xi}\rangle$, defined by

$$\mathbf{x}\,|\boldsymbol{\xi}\rangle = \boldsymbol{\xi}\,|\boldsymbol{\xi}\rangle\ ,$$

where $\boldsymbol{\xi}$ is real. Applying $\mathcal{T}$ to this equation and using (11.4.36a), one obtains

$$\mathbf{x}\mathcal{T}\,|\boldsymbol{\xi}\rangle = \boldsymbol{\xi}\mathcal{T}\,|\boldsymbol{\xi}\rangle\ .$$

Hence, with unchanged normalization, $\mathcal{T}\,|\boldsymbol{\xi}\rangle$ equals $|\boldsymbol{\xi}\rangle$ to within a phase factor. The latter is set to 1:

$$\mathcal{T}\,|\boldsymbol{\xi}\rangle = |\boldsymbol{\xi}\rangle\ . \tag{11.4.37}$$

Then, for an arbitrary state $|\psi\rangle$, the antiunitarity implies

$$\mathcal{T}\,|\psi\rangle = \mathcal{T}\int d^3\xi\,\psi(\boldsymbol{\xi})\,|\boldsymbol{\xi}\rangle = \int d^3\xi\,\psi^*(\boldsymbol{\xi})\mathcal{T}\,|\boldsymbol{\xi}\rangle$$
$$= \int d^3\xi\,\psi^*(\boldsymbol{\xi})\,|\boldsymbol{\xi}\rangle\ . \tag{11.4.38}$$

Hence, the operator $\mathcal{T}$ is equivalent to K_0 (cf. Eq. (11.4.29)):

$$\mathcal{T} = K_0\ . \tag{11.4.39}$$

For the momentum eigenstates, it follows from (11.4.38) that

$$|\mathbf{p}\rangle = \int d^3\xi\,\mathrm{e}^{\mathrm{i}\mathbf{p}\boldsymbol{\xi}}\,|\boldsymbol{\xi}\rangle$$
$$\mathcal{T}\,|\mathbf{p}\rangle = \int d^3\xi\,\mathrm{e}^{-\mathrm{i}\mathbf{p}\boldsymbol{\xi}}\,|\boldsymbol{\xi}\rangle = |-\mathbf{p}\rangle\ . \tag{11.4.40}$$

B. Nonrelativistic spin-$\frac{1}{2}$ particles

Up to now we have considered only particles without spin. Here, we will, in analogy to the orbital angular momentum, extend the theory to spin-$\frac{1}{2}$ particles. We demand for the spin operator that

$$\mathcal{T}\mathbf{S}\mathcal{T}^{-1} = -\mathbf{S}\ . \tag{11.4.41}$$

The total angular momentum

$$\mathbf{J} = \mathbf{L} + \mathbf{S} \tag{11.4.42}$$

then also transforms as

$$\mathcal{T}\mathbf{J}\mathcal{T}^{-1} = -\mathbf{J} . \tag{11.4.43}$$

For spin-$\frac{1}{2}$ we assert that the operator $\mathcal{T}$ is given by

$$
\begin{aligned}
\mathcal{T} &= \mathrm{e}^{-\mathrm{i}\pi S_y/\hbar} K_0 \\
&= \mathrm{e}^{-\mathrm{i}\pi\sigma_y/2} K_0 = \left(\cos\frac{\pi}{2} - \mathrm{i}\sin\frac{\pi}{2}\sigma_y\right) K_0 \\
&= -\mathrm{i}\frac{2S_y}{\hbar} K_0 .
\end{aligned}
\tag{11.4.44}
$$

The validity of this assertion is demonstrated by the fact that the proposed form satisfies Eq. (11.4.41) in the form $\mathcal{T}\mathbf{S} = -\mathbf{S}\mathcal{T}$: for the x and z components

$$-\mathrm{i}\sigma_y K_0 \sigma_{x,z} = -\mathrm{i}\sigma_y \sigma_{x,z} K_0 = +\mathrm{i}\sigma_{x,z}\sigma_y K_0 = -\sigma_{x,z}(-\mathrm{i}\sigma_y K_0)$$

and for the y component

$$-\mathrm{i}\sigma_y K_0 \sigma_y = +\mathrm{i}\sigma_y \sigma_y K_0 = -\sigma_y(-\mathrm{i}\sigma_y K_0) .$$

For the square of $\mathcal{T}$, from (11.4.44) one gets

$$
\begin{aligned}
\mathcal{T}^2 &= -\mathrm{i}\sigma_y K_0(-\mathrm{i}\sigma_y K_0) = -\mathrm{i}\sigma_y \mathrm{i}(-\sigma_y)K_0^2 = +\mathrm{i}^2\sigma_y^2 \\
&= -1 .
\end{aligned}
\tag{11.4.45}
$$

For particles with no spin, $\mathcal{T}^2 = K_0^2 = 1$.

For N particles, the time-reversal transformation is given by the direct product

$$\mathcal{T} = \mathrm{e}^{-\mathrm{i}\pi S_y^{(1)}/\hbar} \ldots \mathrm{e}^{-\mathrm{i}\pi S_y^{(N)}/\hbar} K_0 , \tag{11.4.46}$$

where $S_y^{(n)}$ is the y component of the spin operator of the nth particle. The square of $\mathcal{T}$ is now given by

$$\mathcal{T}^2 = (-1)^N . \tag{11.4.45'}$$

In this context, it is worth mentioning *Kramers theorem*.[8] This states that the energy levels of a system with an odd number of electrons must be at least doubly degenerate whenever time-reversal invariance holds, i.e., when no magnetic field is present.
Proof: From $(\mathcal{T}\psi, \mathcal{T}\varphi) = (\varphi, \psi)$ it follows that

---

[8] H.A. Kramers, Koninkl. Ned. Wetenschap. Proc. **33**, 959 (1930)

$$(\mathcal{T}\psi, \psi) = (\mathcal{T}\psi, \mathcal{T}^2\psi) = -(\mathcal{T}\psi, \psi) \, .$$

Thus, $(\mathcal{T}\psi, \psi) = 0$, i.e., $\mathcal{T}\psi$ and ψ are orthogonal to one another. In addition, from

$$H\psi = E\psi$$

and (11.4.33), it follows that

$$H(\mathcal{T}\psi) = E(\mathcal{T}\psi) \, .$$

The states ψ and $\mathcal{T}\psi$ have the same energy. However, the two states are also distinct: If it were the case that $\mathcal{T}\psi = \alpha\psi$, this would imply $\mathcal{T}^2\psi = \alpha^*\mathcal{T}\psi = |\alpha|^2 \psi$, which would contradict the fact that $\mathcal{T}^2 = -1$. However complicated the electric fields acting on the electrons may be, for an odd number of electrons this degeneracy, at least, always remains. It is referred to as "Kramers degeneracy". For an even number of electrons, $\mathcal{T}^2 = 1$, and in this case no degeneracy need exist unless there is some spatial symmetry.

11.4.3 Time-Reversal Invariance of the Dirac Equation

We now turn our attention to the main topic of interest, the time-reversal invariance of the Dirac equation. The time-reversal transformation $\mathcal{T} = \hat{T}\mathcal{T}^{(0)}$, where $\mathcal{T}^{(0)}$ stands for the operation $t \rightarrow -t$ and $\hat{T}$ is a transformation still to be determined, associates to the spinor $\psi(\mathbf{x}, t)$ another spinor

$$\psi'(\mathbf{x}, t) = \hat{T}\mathcal{T}^{(0)}\psi(\mathbf{x}, t) = \hat{T}\psi(\mathbf{x}, -t) \, , \tag{11.4.47}$$

which also satisfies the Dirac equation. If, at a time $-t_1$, the spinor is of the form $\psi(\mathbf{x}, -t_1)$ and evolves, according to the Dirac equation, into the spinor $\psi(\mathbf{x}, t_1)$ at time t_1, then the spinor $\psi'(\mathbf{x}, -t_1) = \hat{T}\psi(\mathbf{x}, t_1)$ at time $-t_1$ evolves into $\psi'(\mathbf{x}, t_1) = \hat{T}\psi(\mathbf{x}, -t_1)$ at time t_1 (see Fig. 11.4).

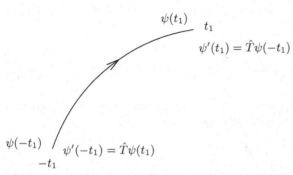

$\psi(t_1)$

t_1

$\psi'(t_1) = \hat{T}\psi(-t_1)$

$\psi(-t_1)$

$\psi'(-t_1) = \hat{T}\psi(t_1)$

$-t_1$

Fig. 11.4. Illustration of time reversal for the spinors ψ and ψ' (space coordinates are suppressed)

Applying $\mathcal{T}^{(0)}$ to the Dirac equation

$$i\frac{\partial \psi(\mathbf{x}, t)}{\partial t} = \left(\boldsymbol{\alpha} \cdot (-i\boldsymbol{\nabla} - e\mathbf{A}(\mathbf{x}, t)) + \beta m + eA_0(\mathbf{x}, t)\right)\psi(\mathbf{x}, t) \qquad (11.4.48)$$

i.e., making the replacement $t \to -t$, yields:

$$i\frac{\partial \psi(\mathbf{x}, -t)}{\partial(-t)} = \left(\boldsymbol{\alpha} \cdot (-i\boldsymbol{\nabla} - e\mathbf{A}(\mathbf{x}, -t)) + \beta m + eA_0(\mathbf{x}, -t)\right)\psi(\mathbf{x}, -t) .$$
$$(11.4.49)$$

Since, in wave mechanics, the time-reversal transformation is achieved by complex conjugation, we set

$$\hat{T} = \hat{T}_0 K_0$$

where $\hat{T}_0$ is to be determined. We now apply $\hat{T}$ to Eq. (11.4.3). The effect of K_0 is to replace i by $-$i, and one obtains

$$i\frac{\partial \psi'(\mathbf{x}, t)}{\partial t} = \hat{T}\left(\boldsymbol{\alpha} \cdot (-i\boldsymbol{\nabla} - e\mathbf{A}(\mathbf{x}, -t)) + \beta m + eA_0(\mathbf{x}, -t)\right)\hat{T}^{-1}\psi'(\mathbf{x}, t) .$$
$$(11.4.49')$$

The motion-reversed vector potential appearing in this equation is generated by current densities, the direction of which is now reversed with respect to the original unprimed current densities. This implies that the vector potential changes its sign, whereas the zero component remains unchanged with respect to motion reversal

$$\mathbf{A}'(\mathbf{x}, t) = -\mathbf{A}(\mathbf{x}, -t) , \quad A'^0(\mathbf{x}, t) = A^0(\mathbf{x}, -t) . \qquad (11.4.50)$$

Hence, the Dirac equation for $\psi'(\mathbf{x}, t)$

$$i\frac{\partial \psi'(\mathbf{x}, t)}{\partial t} = \left(\boldsymbol{\alpha} \cdot (-i\boldsymbol{\nabla} - e\mathbf{A}'(\mathbf{x}, t)) + \beta m + eA_0'(\mathbf{x}, t)\right)\psi'(\mathbf{x}, t) ,$$
$$(11.4.51)$$

is obtained when $\hat{T}$ satisfies the condition

$$\hat{T}\boldsymbol{\alpha}\hat{T}^{-1} = -\boldsymbol{\alpha} \quad \text{and} \quad \hat{T}\beta\hat{T}^{-1} = \beta , \qquad (11.4.52)$$

where the effect of K_0 on i in the momentum operator has been taken into account. With $\hat{T} = \hat{T}_0 K_0$, the last equation implies

$$\hat{T}_0 \boldsymbol{\alpha}^* \hat{T}_0^{-1} = -\boldsymbol{\alpha} \quad \text{and} \quad \hat{T}_0 \beta \hat{T}_0^{-1} = \beta , \qquad (11.4.52')$$

where we have chosen the standard representation for the Dirac matrices in which β is real. Since α_1 and α_3 are real, and α_2 imaginary, we have

$$\hat{T}_0\alpha_1\hat{T}_0^{-1} = -\alpha_1$$
$$\hat{T}_0\alpha_2\hat{T}_0^{-1} = \alpha_2$$
$$\hat{T}_0\alpha_3\hat{T}_0^{-1} = -\alpha_3 \qquad (11.4.52'')$$
$$\hat{T}_0\beta\hat{T}_0^{-1} = \beta\,,$$

which can also be written in the form

$$\{\hat{T}_0, \alpha_1\} = \{\hat{T}_0, \alpha_3\} = 0$$
$$\left[\hat{T}_0, \alpha_2\right] = \left[\hat{T}_0, \beta\right] = 0\,. \qquad (11.4.52''')$$

From (11.4.52'') one finds the representation

$$\hat{T}_0 = -i\alpha_1\alpha_3 \qquad (11.4.53)$$

and hence,

$$\hat{T} = -i\alpha_1\alpha_3 K_0 = i\gamma^1\gamma^3 K_0\,. \qquad (11.4.53')$$

The factor i in (11.4.53) and (11.4.53') is arbitrary.

Proof: $\hat{T}_0$ satisfies (11.4.52'''), since, e.g., $\{\hat{T}_0, \alpha_1\} = \alpha_1\alpha_3\alpha_1 + \alpha_1\alpha_1\alpha_3 = 0$.

The total time-reversal transformation,

$$T = \hat{T}^0 K_0 T^{(0)}\,,$$

can be written in the form

$$\psi'(\mathbf{x}, t) = i\gamma^1\gamma^3 K_0\psi(\mathbf{x}, -t) = i\gamma^1\gamma^3\psi^*(\mathbf{x}, -t) = i\gamma^1\gamma^3\gamma^0\bar{\psi}^T(\mathbf{x}, -t)$$
$$= i\gamma^2\gamma^5\bar{\psi}^T(\mathbf{x}, -t) \qquad (11.4.47')$$

and, as required, $\psi'(\mathbf{x}, t)$ satisfies the Dirac equation

$$i\frac{\partial\psi'(\mathbf{x}, t)}{\partial t} = \left(\boldsymbol{\alpha}\cdot(-i\boldsymbol{\nabla} - e\mathbf{A}'(\mathbf{x}, t)) + \beta m + eA_0'(\mathbf{x}, t)\right)\psi'(\mathbf{x}, t)\,.$$
$$(11.4.51')$$

The transformation of the *current density* under time reversal follows from (11.4.47') as

$$j'^\mu = \bar{\psi}'(x, t)\gamma^\mu\psi'(x, t) = \bar{\psi}(x, -t)\gamma_\mu\psi(x, -t)\,. \qquad (11.4.54)$$

The spatial components of the current density change their signs. Equations (11.4.54) and (11.4.50) show that the d'Alembert equation for the electromagnetic potential $\partial^\nu\partial_\nu A_\mu = j_\mu$ is invariant under time reversal.

In order to investigate the physical properties of a time-reversed spinor, we consider a free spinor

$$\psi = \left(\frac{\epsilon \not{p} + m}{2m}\right)\left(\frac{\mathbb{1} + \gamma_5 \not{n}}{2}\right)\psi \qquad (11.4.55)$$

with momentum p and spin orientation n (in the rest frame). The time-reversal operation yields:

$$\begin{aligned}
\mathcal{T}\psi &= \mathcal{T}\left(\frac{\epsilon\not{p}+m}{2m}\right)\left(\frac{\mathbb{1}+\gamma_5\not{n}}{2}\right)\psi \\
&= \hat{T}_0\left(\frac{\epsilon\not{p}^{*}+m}{2m}\right)\left(\frac{\mathbb{1}+\gamma_5\not{n}^{*}}{2}\right)\psi^{*}(\mathbf{x},-t) \qquad (11.4.56) \\
&= \left(\frac{\epsilon\not{\tilde{p}}+m}{2m}\right)\left(\frac{\mathbb{1}+\gamma_5\not{\tilde{n}}}{2}\right)\mathcal{T}\psi \,,
\end{aligned}$$

where $\tilde{p} = (p^0, -\mathbf{p})$ and $\tilde{n} = (n_0, -\mathbf{n})$. Here, we have used (11.4.52'). The spinor $\mathcal{T}\psi$ has opposite momentum $-\mathbf{p}$ and opposite spin $-\mathbf{n}$.

We have thus far discussed all discrete symmetry transformations of the Dirac equation. We will next investigate the combined action of the parity transformation $\mathcal{P}$, charge conjugation $\mathcal{C}$, and time reversal $\mathcal{T}$. The successive application of these operations to a spinor $\psi(x)$ yields:

$$\begin{aligned}
\psi_{\mathrm{PCT}}(x') &= \mathcal{P}\mathcal{C}\gamma_0 K_0 \hat{T}_0 K_0 \psi(x', -t') \\
&= \gamma^0 i\gamma^2\gamma^0\gamma^0 K_0 i\gamma^1\gamma^3 K_0 \psi(-x') \qquad (11.4.57) \\
&= i\gamma^5\psi(-x') \,.
\end{aligned}$$

If one recalls the structure of γ^5 (Eq. (6.2.48')), it is apparent that the consequence of the $\mathcal{C}$ part of the transformation is to transform a negative-energy electron spinor into a positive-energy positron spinor. This becomes obvious when one begins with a spinor of negative energy and a particular spin orientation $(-n)$, which hence satisfies the projection relation

$$\psi(x) = \left(\frac{-\not{p}+m}{2m}\right)\left(\frac{\mathbb{1}+\gamma_5\not{n}}{2}\right)\psi(x)\,. \qquad (11.4.58)$$

Since $\{\gamma^5, \gamma^\mu\} = 0$, it follows from (11.4.57) and (11.4.58) that

$$\begin{aligned}
\psi_{\mathrm{PCT}}(x') = i\gamma^5\psi(-x') &= i\left(\frac{\not{p}+m}{2m}\right)\left(\frac{\mathbb{1}-\gamma_5\not{n}}{2}\right)\gamma_5\psi_{\mathrm{PCT}}(-x') \\
&= \left(\frac{\not{p}+m}{2m}\right)\left(\frac{\mathbb{1}-\gamma_5\not{n}}{2}\right)\psi_{\mathrm{PCT}}(x')\,. \qquad (11.4.59)
\end{aligned}$$

If $\psi(x)$ is an electron spinor with negative energy, then $\psi_{\mathrm{PCT}}(x)$ is a positron spinor of positive energy. The spin orientation remains unchanged.[9] With

---

[9] To determine the transformation behavior of the quanta from this, one must think of positrons in the context of Hole theory as unoccupied electron states of negative energy. Therefore, under PCT, electrons are transformed into positrons with unchanged momentum and opposite spin.

regard to the first line of (11.4.59), one can interpret a positron spinor with positive energy as an electron spinor with negative energy that is multiplied by $i\gamma_5$ and moves backwards in space and time. This has an equivalent in the Feynman diagrams of perturbation theory (see Fig. 11.5).

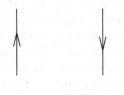

Electron   Positron

Fig. 11.5. Feynman propagators for electrons (arrow pointing upwards, i.e., in positive time direction) and positrons (arrow in negative time direction)

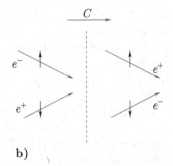

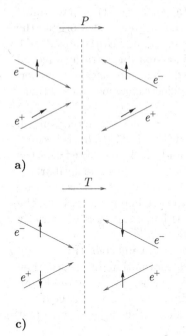

Fig. 11.6. The effect of (**a**) the parity transformation P, (**b**) charge conjugation C, and (**c**) the time-reversal transformation T on an electron and a positron state. The long arrows represent the momentum, and the short arrows the spin orientation. These diagrams represent the transformations not of the spinors, but of the particles and antiparticles, in the sense of Hole theory or in quantum field theory

Figure 11.6a–c illustrates the effect of the transformations P, C, and T on an electron and a positron. According to the Dirac theory, electrons and positrons possess opposite parity. The effect of a parity transformation on a state containing free electrons and positrons is to reverse all momenta while leaving the spins unchanged, and additionally multiplying by a factor (-1) for every positron (Fig. 11.6a). Up until 1956 it was believed that a spatial reflection on the fundamental microscopic level, i.e., a transformation of right-handed coordinate systems into left-handed systems, would lead to an identical physical world with identical physical laws. In 1956 Lee and Yang[10] found convincing arguments indicating the violation of parity conservation in nuclear decay processes involving the weak interaction. The experiments they proposed[11] showed unambiguously that parity is neither conserved in the β decay of nuclei nor in the decay of π mesons. Therefore the Hamiltonian of the weak interaction must, in addition to the usual scalar terms, contain further pseudoscalar terms which change sign under the inversion of all coordinates. This is illustrated in Fig. 11.7 for the experiment of Wu et al. on the β decay of radioactive ^{60}Co nuclei into ^{60}Ni. In this process a neutron within the nucleus decays into a proton, an electron, and a neutrino. Only the electron (β particle) can be readily observed. The nuclei possess a finite spin and a magnetic moment which can be oriented by means of a magnetic field. It is found that the electrons are emitted preferentially in the direction opposite to that of the spin of the nucleus. The essential experimental fact is that the direction of the velocity of the β particle $\mathbf{v}_\beta$ (a polar vector) is determined by the direction of the magnetic field $\mathbf{B}$ (an axial vector), which orients the nuclear spins. Since the inversion P leaves the magnetic field $\mathbf{B}$ unchanged, while reversing $\mathbf{v}_\beta$, the above observation is incompatible with a universal inversion symmetry. Parity is not conserved by the weak interaction. However, in all processes involving only the strong and the electromagnetic interactions, parity is conserved.[12]

Under charge conjugation, electrons and positrons are interchanged, whilst the momenta and spins remain unchanged (Fig. 11.6b). This is because charge conjugation, according to Eqs. (11.3.7′) and (11.3.11′), transforms the spinor into a spinor with opposite momentum and spin. Since the antiparticle (hole) corresponds to the nonoccupation of such a state, it again has opposite values and hence, in total, the same values as the original particle. Even the charge-conjugation invariance present in the free Dirac theory is not strictly valid in nature: it is violated by the weak interaction.[12]

The time-reversal transformation reverses momenta and spins (Fig. 11.6c). The free Dirac theory is invariant under this transformation. In nature, time reversal invariance holds for almost all processes, whereby one should note that time reversal interchanges the initial and final states. It was in the decay processes of neutral K mesons that effects violating T invariance were first observed experimentally.

---

[10] T.D. Lee and C.N. Yang Phys. Rev. **104**, 254 (1956)

[11]   C.S. Wu, E. Ambler, R.W. Hayward, D.D. Hoppes, and R.P. Hudson, Phys. Rev. **105**, 1413 (1957); R.L. Garwin, L.M. Ledermann, and M. Weinrich, Phys. Rev. **105**, 1415 (1957)

[12] A more detailed discussion of experiments testing the invariance of the electromagnetic and strong interactions under C, P, and CP, and their violation by the weak interaction, can be found in D.H. Perkins, *Introduction to High Energy Physics*, 2nd ed., Addison-Wesley, London, 1982.

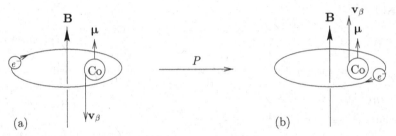

Fig. 11.7. Schematic representation of the parity violation observed in the β-decay experiment of Wu et al. The figure shows the current circulating in a toroidal coil generating the magnetic field **B**, which in turn orients the magnetic moment μ of the cobalt nucleus and the associated angular momentum **I**, together with the velocity $\mathbf{v}_\beta$ of the β particle (electron). The β particles are emitted preferentially in the direction opposite to that of μ. Thus configuration **(a)** corresponds to the experimental result, whereas configuration **(b)** is not observed

The invariances C, P, and T are all violated individually in nature.[12] In relativistic field theory with an arbitrary local interaction, however, the product $\Theta = PCT$ must be an invariance transformation. This theorem, which is known as the PCT theorem[13,14], can be derived from the general axioms of quantum field theory[15]. The PCT theorem implies that particles and antiparticles have the same mass and, if unstable, the same lifetime, although the decay rates for particular channels are not necessarily the same for particles and antiparticles.

*11.4.4 Racah Time Reflection

Here, we determine the spinor transformation corresponding to a pure time reflection. According to Eq. (6.1.9), this is described by the Lorentz transformation

$$
\Lambda^\mu_{\ \nu} = \begin{pmatrix} -1 & 0 & 0 & 0 \\ 0 & 1 & 0 & 0 \\ 0 & 0 & 1 & 0 \\ 0 & 0 & 0 & 1 \end{pmatrix} .
\tag{11.4.60}
$$

One readily sees that the condition for the spinor transformation (6.2.7)

$$
\gamma^\mu S_R = \Lambda^\mu_{\ \nu} S_R \gamma^\nu
$$

[13] G. Lüders, Dan. Mat. Fys. Medd. **28**, 5 (1954); Ann. Phys. (N.Y.) **2**, 1 (1957); W. Pauli, in *Niels Bohr and the development of physics*, ed. by W. Pauli, L. Rosenfeld, and V. Weisskopf, McGraw Hill, New York, 1955

[14] The Lagrangian of a quantum field theory with the properties given in Sect. 12.2 transforms under Θ as $\mathcal{L}(x) \to \mathcal{L}(-x)$, so that the action S is invariant.

[15] R.F. Streater and A.S. Wightman *PCT, Spin Statistics and all that*, W.A. Benjamin, New York, 1964; see also Itzykson, Zuber, op. cit., p. 158.

is satisfied by[16]

$$S_R = \gamma_1 \gamma_2 \gamma_3 \ . \tag{11.4.61}$$

Hence, the transformation for the spinor and its adjoint has the form

$$\begin{aligned}\psi' &= S_R \psi \\ \bar{\psi}' &= \psi^\dagger S_R^\dagger \gamma^0 = -\psi^\dagger \gamma^0 S_R^{-1} = -\bar{\psi} S_R^{-1} \ ,\end{aligned} \tag{11.4.62}$$

in agreement with the general result, Eq. (6.2.34b), $b = -1$ for time reversal, where $S_R^{-1} = -\gamma_3 \gamma_2 \gamma_1$. The current density thus transforms according to

$$(\bar{\psi} \gamma^\mu \psi)' = -\Lambda^\mu_{\ \nu} \bar{\psi} \gamma^\nu \psi \ . \tag{11.4.63}$$

Hence, j^μ transforms as a pseudovector under Racah time reflection. The vector potential $A^\mu(x)$, on the other hand, transforms as

$$A'^\mu(x') = \Lambda^\mu_{\ \nu} A^\nu(\Lambda^{-1} x) \ . \tag{11.4.64}$$

Thus, the field equation for the radiation field

$$\partial_\nu \partial^\nu A^\mu = 4\pi e j^\mu \tag{11.4.65}$$

is *not* invariant under this time reflection
One can combine the Racah transformation with charge conjugation:

$$\psi'(\mathbf{x}, t) = S_R \psi_c(\mathbf{x}, -t) = S(T) \bar{\psi}^T(\mathbf{x}, -t) \ . \tag{11.4.66}$$

Here, the transformation matrix $S(T)$ is related to S_R and $C \equiv \mathrm{i}\gamma^2 \gamma^0$

$$S(T) = S_R C = \gamma_1 \gamma_2 \gamma_3 \mathrm{i}\gamma^2 \gamma^0 = \mathrm{i}\gamma^2 \gamma_5 \ .$$

This is the motion-reversal transformation (= time-reversal transformation), Eq. (11.4.47′). The Dirac equation is invariant under this transformation.

*11.5 Helicity

The *helicity operator* is defined by

$$h(\hat{\mathbf{k}}) = \boldsymbol{\Sigma} \cdot \hat{\mathbf{k}} \ , \tag{11.5.1}$$

where $\hat{\mathbf{k}} = \mathbf{k}/|\mathbf{k}|$ is the unit vector in the direction of the spinor's momentum.

---

[16] S_R is known as the Racah time reflection operator, see J.M. Jauch and F. Rohrlich, *The Theory of Photons and Electrons*, p. 88, Springer, New York, 1980

Since $\boldsymbol{\Sigma} \cdot \hat{\mathbf{k}}$ commutes with the Dirac Hamiltonian, there exist common eigenstates[17] of $\boldsymbol{\Sigma} \cdot \hat{\mathbf{k}}$ and H. The helicity operator $h(\hat{\mathbf{k}})$ has the property $h^2(\hat{\mathbf{k}}) = 1$, and thus possesses the eigenvalues ± 1. The helicity eigenstates with eigenvalue $+1$ (spin parallel to $\mathbf{k}$) are called right-handed, and those with eigenvalue -1 (spin antiparallel to $\mathbf{k}$) are termed left-handed. One can visualize the states of positive and negative helicity as analogous to right- and left-handed screws.

From Eq. (6.3.11a), the effect of the helicity operator on the free spinor $u_r(k)$ is:

$$
\boldsymbol{\Sigma} \cdot \hat{\mathbf{k}}\, u_r(k) = \boldsymbol{\Sigma} \cdot \hat{\mathbf{k}}
\begin{pmatrix}
\left(\dfrac{E+m}{2m}\right)^{\frac{1}{2}} \varphi_r \\
\dfrac{\boldsymbol{\sigma} \cdot \mathbf{k}}{[2m(m+E)]^{\frac{1}{2}}} \varphi_r
\end{pmatrix}
$$

$$
= \begin{pmatrix}
\left(\dfrac{E+m}{2m}\right)^{\frac{1}{2}} \boldsymbol{\sigma} \cdot \hat{\mathbf{k}}\, \varphi_r \\
\dfrac{\boldsymbol{\sigma} \cdot \mathbf{k}}{[2m(m+E)]^{\frac{1}{2}}} \boldsymbol{\sigma} \cdot \hat{\mathbf{k}}\, \varphi_r
\end{pmatrix}
\tag{11.5.2}
$$

with $\varphi_1 = \binom{1}{0}$ and $\varphi_2 = \binom{0}{1}$, and an analogous expression for the spinors $v_r(k)$. The Pauli spinors φ_r are eigenstates of σ_z and thus the $u_r(k)$ and $v_r(k)$ in the rest frame are eigenstates of Σ_z (see Eq. (6.3.4)).

As an example of a simple special case, we now consider free spinors with wave vector along the z axis. Thus $\mathbf{k} = (0,0,k)$, and the helicity operator is

$$
\boldsymbol{\Sigma} \cdot \hat{\mathbf{k}} = \Sigma_z \quad \text{and} \quad \boldsymbol{\sigma} \cdot \hat{\mathbf{k}} = \sigma_z .
\tag{11.5.3}
$$

Furthermore, from Eq. (11.5.2) one sees that the spinors $u_r(k)$ and $v_r(k)$ are eigenstates of the helicity operator. According to Eqs. (6.3.11a) and (6.3.11b), the spinors for $\mathbf{k} = (0,0,k)$, i.e., for $k' = (\sqrt{k^2 + m^2}, 0, 0, k)$ (to distinguish it from the z component, the four-vector is denoted by k'), are

$$
u^{(R)}(k') = u_1(k') = \mathcal{N}
\begin{pmatrix}
1 \\
0 \\
\frac{k}{E+m} \\
0
\end{pmatrix}
, \quad
u^{(L)}(k') = u_2(k') = \mathcal{N}
\begin{pmatrix}
0 \\
1 \\
0 \\
\frac{-k}{E+m}
\end{pmatrix}
,
$$

$$
v^{(R)}(k') = v_1(k') = \mathcal{N}
\begin{pmatrix}
\frac{-k}{E+m} \\
0 \\
1 \\
0
\end{pmatrix}
, \quad
v^{(L)}(k') = v_2(k') = \mathcal{N}
\begin{pmatrix}
0 \\
\frac{k}{E+m} \\
0 \\
1
\end{pmatrix}
,
$$

$$
\tag{11.5.4}
$$

---

[17] In contrast to the nonrelativistic Pauli equation, however, the Dirac equation has no free solutions that are eigenfunctions of $\boldsymbol{\Sigma} \cdot \hat{\mathbf{n}}$ with an arbitrarily oriented unit vector $\hat{\mathbf{n}}$. This is because, except for $\hat{\mathbf{n}} = \pm \hat{\mathbf{k}}$, the product $\boldsymbol{\Sigma} \cdot \hat{\mathbf{n}}$ does not commute with the free Dirac Hamiltonian.

with $\mathcal{N} = \left(\frac{E+m}{2m}\right)^{1/2}$, and satisfy

$$\Sigma_z u_r(k') = \pm u_r(k') \text{ for } r = \begin{cases} 1 & R \\ 2 & L \end{cases}$$

$$\Sigma_z v_r(k') = \pm v_r(k') \text{ for } r = \begin{cases} 1 & R \\ 2 & L \end{cases}. \qquad (11.5.5)$$

The letter R indicates right-handed polarization (positive helicity) and L left-handed polarization (negative helicity).

For $\mathbf{k}$ in an arbitrary direction, the eigenstates $u^{(R)}, u^{(L)}$ with eigenvalues $+1, -1$ are obtained by rotating the spinors (11.5.4). The rotation is through an angle $\vartheta = \arccos\frac{k_z}{|\mathbf{k}|}$ about the axis defined by the vector $(-k_y, k_x, 0)$. It causes the z axis to rotate into the $\mathbf{k}$ direction. According to (6.2.21) and (6.2.29c), the corresponding spinor transformation reads:

$$S = \exp\left(-\mathrm{i}\frac{\vartheta}{2}(-k_y \Sigma_x + k_x \Sigma_y)/\sqrt{k_x^2 + k_y^2}\right)$$

$$= \mathbb{1}\cos\frac{\vartheta}{2} + \mathrm{i}\frac{k_y\Sigma_x - k_x\Sigma_y}{\sqrt{k_x^2 + k_y^2}}\sin\frac{\vartheta}{2}. \qquad (11.5.6)$$

Therefore, the *helicity eigenstates* of positive energy for a wave vector $\mathbf{k}$ are

$$u^{(R)}(k) = \mathcal{N}\begin{pmatrix} \cos\frac{\vartheta}{2} \\ \frac{(k_x + \mathrm{i}k_y)}{\sqrt{k^2 - k_z^2}}\sin\frac{\vartheta}{2} \\ \frac{|\mathbf{k}|}{E+m}\cos\frac{\vartheta}{2} \\ \frac{|\mathbf{k}|}{E+m}\frac{k_x + \mathrm{i}k_y}{\sqrt{k^2 - k_z^2}}\sin\frac{\vartheta}{2} \end{pmatrix} = \frac{\mathcal{N}}{\sqrt{2(\hat{k}_z + 1)}}\begin{pmatrix} \hat{k}_z + 1 \\ \hat{k}_x + \mathrm{i}\hat{k}_y \\ \frac{|\mathbf{k}|}{E+m}(\hat{k}_z + 1) \\ \frac{|\mathbf{k}|}{E+m}(\hat{k}_x + \mathrm{i}\hat{k}_y) \end{pmatrix}$$

$$\qquad (11.5.7)$$

and

$$u^{(L)}(k) = \mathcal{N}\begin{pmatrix} \frac{-k_x + \mathrm{i}k_y}{\sqrt{k^2 - k_z^2}}\sin\frac{\vartheta}{2} \\ \cos\frac{\vartheta}{2} \\ -\frac{|\mathbf{k}|}{E+m}\frac{-k_x + \mathrm{i}k_y}{\sqrt{k^2 - k_z^2}}\sin\frac{\vartheta}{2} \\ -\frac{|\mathbf{k}|}{E+m}\cos\frac{\vartheta}{2} \end{pmatrix}$$

$$= \frac{\mathcal{N}}{\sqrt{2(\hat{k}_z + 1)}}\begin{pmatrix} -\hat{k}_x + \mathrm{i}\hat{k}_y \\ \hat{k}_z + 1 \\ -\frac{|\mathbf{k}|}{E+m}(-\hat{k}_x + \mathrm{i}\hat{k}_y) \\ -\frac{|\mathbf{k}|}{E+m}(\hat{k}_z + 1) \end{pmatrix}.$$

Corresponding expressions are obtained for spinors with negative energy (Problem 11.4).

*11.6 Zero-Mass Fermions (Neutrinos)

Neutrinos are spin-$\frac{1}{2}$ particles and were originally thought to be massless. There is now increasing experimental evidence that they possess a finite albeit very small mass. On neglecting this mass, which is valid for sufficiently high momenta, we may present the standard description by the *Dirac equation having a zero mass term*

$$\not{p}\psi = 0 \,, \qquad (11.6.1)$$

where $p_\mu = i\partial_\mu$ is the momentum operator. In principle, one could obtain the solutions from the plane waves (6.3.11a,b) or the helicity eigenstates by taking the limit $m \to 0$ in the Dirac equation containing a mass term. One merely needs to split off the factor $1/\sqrt{m}$ and introduce a normalization different to (6.3.19a) and (6.3.19b), for example

$$\begin{aligned}
\bar{u}_r(k)\gamma^0 u_s(k) &= 2E\delta_{rs} \\
\bar{v}_r(k)\gamma^0 v_s(k) &= 2E\delta_{rs} \,.
\end{aligned} \qquad (11.6.2)$$

However, it is also interesting to solve the massless Dirac equation directly and study its special properties. We note at the outset that in the representation based on the matrices $\boldsymbol{\alpha}$ and β (5.3.1), for the case of zero mass, β does not appear. However, one could also realize three anticommuting matrices using the two-dimensional representation of the Pauli matrices, a fact that is also reflected in the structure of (11.6.1).

In order to solve (11.6.1), we multiply the Dirac equation by

$$\gamma^5\gamma^0 = -i\gamma^1\gamma^2\gamma^3 \,.$$

With the supplementary calculation

$$\begin{aligned}
\gamma^5\gamma^0\gamma^1 &= -i\gamma^1\gamma^2\gamma^3\gamma^1 = -i\gamma^1\gamma^1\gamma^2\gamma^3 = +i\gamma^2\gamma^3 = \sigma^{23} = \Sigma^1 \,, \\
\gamma^5\gamma^0\gamma^3 &= -i\gamma^1\gamma^2\gamma^3\gamma^3 = i\gamma^1\gamma^2 = \sigma^{12} = \Sigma^3 \,, \gamma^5\gamma^0\gamma^0 = \gamma^5 \,,
\end{aligned}$$

$$(-p^i\Sigma^i + p_0\gamma^5)\psi = 0$$

one obtains

$$\boldsymbol{\Sigma} \cdot \mathbf{p}\,\psi = p^0\gamma^5\psi \,. \qquad (11.6.3)$$

Inserting into (11.6.3) plane waves with positive (negative) energy

$$\psi(x) = e^{\mp ikx}\psi(k) = e^{\mp i(k^0 x^0 - \mathbf{k}\cdot\mathbf{x})}\psi(k) \,, \qquad (11.6.4)$$

this yields

$$\boldsymbol{\Sigma} \cdot \mathbf{k}\psi(k) = k^0\gamma^5\psi(k) \,. \qquad (11.6.5)$$

From (11.6.1) it follows that $\not{p}^2\psi(x) = 0$ and hence, $k^2 = 0$ or $k^0 = E = |\mathbf{k}|$ for solutions of positive (negative) energy. With the unit vector $\hat{\mathbf{k}} = \mathbf{k}/|\mathbf{k}|$, Eq. (11.6.5) takes the form

$$\boldsymbol{\Sigma} \cdot \hat{\mathbf{k}}\psi(k) = \pm\gamma^5\psi(k) \,. \tag{11.6.6}$$

The matrix γ^5, which anticommutes with all γ^μ, commutes with $\boldsymbol{\Sigma}$ and thus has joint eigenfunctions with the helicity operator $\boldsymbol{\Sigma} \cdot \hat{\mathbf{k}}$. The matrix γ^5 is also termed the *chirality operator*. Since $(\gamma^5)^2 = 1$, the eigenvalues of γ^5 are ± 1, and since $\mathrm{Tr}\,\gamma^5 = 0$, they are doubly degenerate. The solutions of Eq. (11.6.6) can thus be written in the form

$$\psi(x) = \begin{cases} e^{-ikx}\ u_\pm(k) \\ e^{ikx}\ \ \ v_\pm(k) \end{cases} \text{ with } k^2 = 0,\, k^0 = |\mathbf{k}| > 0 \,, \tag{11.6.7}$$

where the $u_\pm$ $(v_\pm)$ are eigenstates of the chirality operator

$$\gamma^5 u_\pm(k) = \pm u_\pm(k) \quad \text{und} \quad \gamma^5 v_\pm(k) = \pm v_\pm(k) \,. \tag{11.6.8}$$

The spinors u_+ and v_+ are said to have *positive chirality* (right-handed), and the spinors u_- and v_- to have *negative chirality* (left handed). Using the standard representation $\gamma_5 = \begin{pmatrix} 0 & \mathbb{1} \\ \mathbb{1} & 0 \end{pmatrix}$, Eq. (11.6.8) yields

$$u_\pm(k) = \frac{1}{\sqrt{2}} \begin{pmatrix} a_\pm(k) \\ \pm a_\pm(k) \end{pmatrix} \,, \quad v_\pm(k) = \frac{1}{\sqrt{2}} \begin{pmatrix} b_\pm(k) \\ \pm b_\pm(k) \end{pmatrix} \,. \tag{11.6.9}$$

Inserting(11.6.9) into the Dirac equation (11.6.6), one obtains equations determining $a_\pm(k)$:

$$a_\pm(k) = \pm\boldsymbol{\sigma} \cdot \hat{\mathbf{k}} a_\pm(k) \,. \tag{11.6.10}$$

Their solutions are (cf. Problem 11.7)

$$a_+(k) = \begin{pmatrix} \cos\frac{\vartheta}{2} \\ \sin\frac{\vartheta}{2}e^{i\varphi} \end{pmatrix} \tag{11.6.11a}$$

$$a_-(k) = \begin{pmatrix} -\sin\frac{\vartheta}{2}e^{-i\varphi} \\ \cos\frac{\vartheta}{2} \end{pmatrix} \,, \tag{11.6.11b}$$

where ϑ and φ are the polar angles of $\hat{\mathbf{k}}$. These solutions are consistent with the $m \to 0$ limit of the helicity eigenstates found in (11.5.7). The negative energy solutions $v_\pm(k)$ can be obtained from the $u_\pm(k)$ by charge conjugation (Eqs. (11.3.7) and (11.3.8)):

$$v_+(k) = C\bar{u}_-^T(k) = i\gamma^2 u_-^*(k) = -u_+(k) \tag{11.6.11c}$$

$$v_-(k) = C\bar{u}_+^T(k) = i\gamma^2 u_+^*(k) = -u_-(k) \tag{11.6.11d}$$

i.e., in (11.6.9), $b_\pm(k) = -a_\pm(k)$.

It is interesting in this context to go from the standard representation of the Dirac matrices to the *chiral representation*, which is obtained by the transformation

$$\psi^{ch} = U^{\dagger}\psi \tag{11.6.12a}$$

$$\gamma^{\mu ch} = U^{\dagger}\gamma^{\mu}U \tag{11.6.12b}$$

$$U = \frac{1}{\sqrt{2}}(1+\gamma^5). \tag{11.6.12c}$$

The result is (Problem 11.8):

$$\gamma^{0ch} \equiv \beta^{ch} = -\gamma^5 = \begin{pmatrix} 0 & -\mathbb{1} \\ -\mathbb{1} & 0 \end{pmatrix} \tag{11.6.13a}$$

$$\gamma^{kch} = \gamma^k = \begin{pmatrix} 0 & \sigma^k \\ -\sigma^k & 0 \end{pmatrix} \tag{11.6.13b}$$

$$\gamma^{5ch} = \gamma^0 = \begin{pmatrix} \mathbb{1} & 0 \\ 0 & -\mathbb{1} \end{pmatrix} \tag{11.6.13c}$$

$$\alpha^{kch} = \begin{pmatrix} 0 & \sigma^k \\ \sigma^k & 0 \end{pmatrix} \tag{11.6.13d}$$

$$\sigma_{0i}^{ch} = \frac{i}{2}\left[\gamma_0^{ch}, \gamma_i^{ch}\right] = \frac{1}{i}\begin{pmatrix} \sigma^i & 0 \\ 0 & -\sigma^i \end{pmatrix} \tag{11.6.13e}$$

$$\sigma_{ij}^{ch} = \frac{i}{2}\left[\gamma_i^{ch}, \gamma_j^{ch}\right] = \epsilon^{ijk}\begin{pmatrix} \sigma^k & 0 \\ 0 & \sigma^k \end{pmatrix} \tag{11.6.13f}$$

In the chiral representation, (11.6.13e,f) are diagonal in the space of bispinors, i.e., the upper components (1,2) and the lower components (3,4) of the spinor transform independently of one another under pure Lorentz transformations and under rotations (see (6.2.29b)). This means that the four-dimensional representation of the restricted Lorentz group $\mathcal{L}_+^{\uparrow}$ is reducible to two two-dimensional representations. More precisely, the representation[18] of the group SL(2,C) can be reduced to the two nonequivalent representations $D^{(\frac{1}{2},0)}$ and $D^{(0,\frac{1}{2})}$. When the parity transformation P, which is given by $\mathcal{P} = e^{i\varphi}\gamma^{0ch}\mathcal{P}^0$ (see (6.2.32)), is present as a symmetry element, then the four-dimensional representation is no longer reducible, i.e., it is irreducible.

In the *chiral representation*, the Dirac equation takes the form

$$\begin{aligned} (-i\partial_0 + i\sigma^k\partial_k)\psi_2^{ch} - m\psi_1^{ch} &= 0 \\ (-i\partial_0 - i\sigma^k\partial_k)\psi_1^{ch} - m\psi_2^{ch} &= 0, \end{aligned} \tag{11.6.14}$$

where we have set $\psi^{ch} = \begin{pmatrix} \psi_1^{ch} \\ \psi_2^{ch} \end{pmatrix}$. Equations (11.6.14) are identical to the equations (A.7), but have been obtained in a different way. For $m = 0$ the two equations decouple and one obtains

---

[18] The group SL(2,C) is homomorphic to the group $\mathcal{L}_+^{\uparrow}$ corresponding to the two-valued nature of the spinor representations. For useful group theoretical background we recommend V. Heine, *Group Theory in Quantum Mechanics*, Pergamon Press, Oxford (1960), and R.F. Streater and A.S. Wightman, *PCT, Spin Statistics and all that*, Benjamin, Reading (1964).

$$(i\partial_0 - i\sigma^k\partial_k)\psi_2^{ch} \equiv (p_0 + \boldsymbol{\sigma}\cdot\mathbf{p})\psi_2^{ch} = 0 \tag{11.6.15a}$$

and

$$(i\partial_0 + i\sigma^k\partial_k)\psi_1^{ch} = 0 . \tag{11.6.15b}$$

These are the two *Weyl equations*. A comparison of these with (5.3.1) shows that (11.6.15) contains a two-dimensional representation of the α matrices. As mentioned at the beginning of this section, when β is absent, the algebra of the Dirac α matrices

$$\{\alpha_i, \alpha_j\} = 2\delta_{ij}$$

can be realized by the three Pauli σ^i matrices. The two equations (11.6.15a,b) are not individually parity invariant and in the historical development were initially heeded no further. In fact, it has been known since the experiments of Wu et al.[19] that the weak interaction does not conserve parity. Since the chirality operator in the chiral representation is of the form $\chi_5^{ch} = \begin{pmatrix} \mathbb{1} & 0 \\ 0 & -\mathbb{1} \end{pmatrix}$, spinors of the form $\psi = \begin{pmatrix} \psi_1^{ch} \\ 0 \end{pmatrix}$ have positive chirality, whilst those of the form $\psi = \begin{pmatrix} 0 \\ \psi_2^{ch} \end{pmatrix}$ have negative chirality.

Experimentally, it is found that only neutrinos of negative chirality exist. This means that the first of the two equations (11.6.15) is the one relevant to nature. The solutions of this equation are of the form $\psi_2^{ch(+)}(x) = e^{-ik\cdot x}u(k)$ and $\psi_2^{ch(-)}(x) = e^{ik\cdot x}v(k)$ with $k_0 > 0$, where u and v are now two-component spinors The first state has positive energy and, as directly evident from (11.6.15a), negative helicity since the spin is antiparallel to $\mathbf{k}$. We call this state the *neutrino* state and represent it pictorially by means of a left-handed screw (Fig. 11.8a). Of the solutions shown in (11.6.9), this is $u_-(k)$. The momentum is represented by the straight arrow.

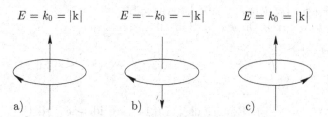

Fig. 11.8. (a) Neutrino state with negative helicity, (b) neutrino state with negative energy and positive helicity, (c) antineutrino with positive helicity

The solution with negative energy $\psi_2^{ch(-)}$ has momentum $-\mathbf{k}$, and hence positive helicity; it is represented by a right-handed screw (Fig. 11.8b). This

[19] See references on p. 234.

solution corresponds to $v_-(k)$ in Eq. (11.6.9). In a hole-theoretical interpretation, the *antineutrino* is represented by an unoccupied state $v_-(k)$. It thus has opposite momentum $(+\mathbf{k})$ and opposite spin, hence the helicity remains positive (Fig. 11.8c). Neutrinos have negative helicity, and their antiparticles, the antineutrinos, have positive helicity. For electrons and other massive particles, it would not be possible for only one particular helicity to occur. Even if only one helicity were initially present, one can reverse the spin in the rest frame of the electron, or, for unchanged spin, accelerate the electron in the opposite direction, in either case generating the opposite helicity. Since massless particles move with the velocity of light, they have no rest frame; for them the momentum $\mathbf{k}$ distinguishes a particular direction.

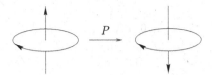

Fig. 11.9. The effect of a parity transformation on a neutrino state

Figure 11.9 illustrates the effect of a parity transformation on a neutrino state. Since this transformation reverses the momentum whilst leaving the spin unchanged, it generates a state of positive energy with positive helicity. As has already been stated, these do not exist in nature.

Although neutrinos have no charge, one can still subject them to charge conjugation. The charge conjugation operation C connects states of positive and negative chirality and changes the sign of the energy. Since only left-handed neutrinos exist in nature, there is no invariance with respect to C. However, since the parity transformation P also connects the two types of solution

$$\psi^{ch}(t, \mathbf{x}) \rightarrow \gamma^0 \psi^{ch}(t, -\mathbf{x}) ,$$

(in the chiral representation γ^0 is nondiagonal), the Weyl equation is invariant under CP. In the chiral representation, C reads

$$C = \begin{pmatrix} -i\sigma_2 & 0 \\ 0 & i\sigma_2 \end{pmatrix} = \begin{pmatrix} -1 & 0 & 0 & 0 \\ 0 & 1 & 0 & 0 \\ 0 & 0 & 1 & 0 \\ 0 & 0 & 0 & -1 \end{pmatrix} .$$

Hence, the effect of CP is

$$\psi^{ch\,CP}(t, \mathbf{x}) = \eta C \psi^{ch\,*}(t, -\mathbf{x}) = \mp i\eta \sigma_2 \psi^{ch\,*}(t, -\mathbf{x})$$

for chirality $\gamma^5 = \pm 1$.

Problems

11.1 Show that equation (11.3.5) implies (11.3.5′).

11.2 In a Majorana representation of the Dirac equation, the γ matrices – indicated here by the subscript M for Majorana – are purely imaginary,

$$\gamma_M^\mu{}^* = -\gamma_M^\mu \,,\ \mu = 0, 1, 2, 3.$$

A special Majorana representation is given by the unitary transformation

$$\gamma_M^\mu = U\gamma^\mu U^\dagger$$

with $U = U^\dagger = U^{-1} = \frac{1}{\sqrt{2}}\gamma^0\left(\mathbb{1} + \gamma^2\right)$.

(a) Show that

$$\gamma_M^0 = \gamma^0\gamma^2 = \begin{pmatrix} 0 & \sigma^2 \\ \sigma^2 & 0 \end{pmatrix} \qquad\qquad \gamma_M^1 = \gamma^2\gamma^1 = \begin{pmatrix} i\sigma^3 & 0 \\ 0 & i\sigma^3 \end{pmatrix}$$

$$\gamma_M^2 = -\gamma^2 = \begin{pmatrix} 0 & -\sigma^2 \\ \sigma^2 & 0 \end{pmatrix} \qquad\qquad \gamma_M^3 = \gamma^2\gamma^3 = \begin{pmatrix} -i\sigma^1 & 0 \\ 0 & -i\sigma^1 \end{pmatrix}.$$

(b) In Eq. (11.3.14′) it was shown that, in a Majorana representation, the charge conjugation transformation (apart from an arbitrary phase factor) has the form $\psi_M^C = \psi_M^*$. Show that application of the transformation U to Eq. (11.3.7′)

$$\psi^C = i\gamma^2\psi$$

leads to

$$\psi_M^C = -i\psi_M \,.$$

11.3 Show that, under a time-reversal operation $\mathcal{T}$, the four-current-density j^μ in the Dirac theory satisfies

$$j'^\mu(\mathbf{x}, t) = j_\mu(\mathbf{x}, -t) \,.$$

11.4 Determine the eigenstates of helicity with negative energy:
(a) as in (11.5.7) by applying a Lorentz transformation to (11.5.4);
(b) by solving the eigenvalue equation for the helicity operator $\boldsymbol{\Sigma} \cdot \hat{\mathbf{k}}$ and taking the appropriate linear combination of the energy eigenstates (6.3.11b).

11.5 Show that $\boldsymbol{\Sigma} \cdot \hat{\mathbf{k}}$ commutes with $(\gamma^\mu k_\mu \pm m)$.

11.6 Prove the validity of Eq. (11.5.7).

11.7 Show that (11.6.11) satisfies the equation (11.6.10).

11.8 Prove the validity of (11.6.13).

Bibliography for Part II

H.A. Bethe and R. Jackiw, *Intermediate Quantum Mechanics*, Benjamin, London, 1968

J.D. Bjorken and S.D. Drell, *Relativistic Quantum Mechanics*, McGraw–Hill, New York, 1964

C. Itzykson and J.-B.Zuber, *Quantum Field Theory*, McGraw–Hill, New York, 1980

A. Messiah, *Quantum Mechanics*, Vol. II, North Holland, Amsterdam, 1964

W. Pauli, *General principles of quantum mechanics*, Springer, Berlin, New York, 1980. Translation of: *Die allgemeinen Prinzipien der Wellenmechanik*, in Encyclopedia of Physics, Vol. V, part I, Springer, Berlin, 1958

J.J. Sakurai, *Advanced Quantum Mechanics*, Addison-Wesley, London, 1967

S.S. Schweber, *An Introduction to Relativistic Quantum Field Theory*, Harper & Row, New York, 1961

Part III

Relativistic Fields

12. Quantization of Relativistic Fields

This chapter is dedicated to relativistic quantum fields. We shall begin by investigating a system of coupled oscillators for which the quantization properties are known. The continuum limit of this oscillator system yields the equation of motion for a vibrating string in a harmonic potential. This is identical in form to the Klein–Gordon equation. The quantized equation of motion of the string and its generalization to three dimensions provides us with an example of a quantized field theory. The quantization rules that emerge here can also be applied to non-material fields. The fields and their conjugate momentum fields are subject to canonical commutation relations. One thus speaks of "canonical quantization". In order to generalize to arbitrary fields, we shall then study the properties of general classical relativistic fields. In particular, we will derive the conservation laws that follow from the symmetry properties (Noether's theorem).

12.1 Coupled Oscillators, the Linear Chain, Lattice Vibrations

12.1.1 Linear Chain of Coupled Oscillators

12.1.1.1 Diagonalization of the Hamiltonian

We consider N particles of mass m with equilibrium positions that lie on a periodic linear chain separated by the lattice constant a. The displacements along the direction of the chain from the equilibrium positions a_n are denoted

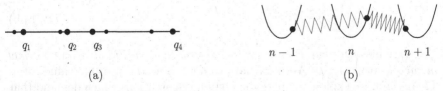

Fig. 12.1. Linear chain: (a) displacement of the point masses (large dots) from their equilibrium positions (small dots); (b) potentials and interactions (represented schematically by springs)

by $q_1, \ldots, q_N$ (Fig.12.1a), and the momenta by $p_1, \ldots, p_N$. It is assumed that each particle is in a harmonic potential and, additionally, is harmonically coupled to its nearest neighbors (Fig.12.1b). The Hamiltonian then reads:

$$H = \sum_{n=1}^{N} \frac{1}{2m} p_n^2 + \frac{m\Omega^2}{2} (q_n - q_{n-1})^2 + \frac{m\Omega_0^2}{2} q_n^2 . \tag{12.1.1}$$

Here, Ω^2 characterizes the strength of the harmonic coupling between nearest neighbors, and Ω_0^2 the harmonic potential of the individual particles (see Fig.12.1b). Since we will eventually be interested in the limiting case of an infinitely large system in which the boundary conditions play no part, we will choose periodic boundary conditions, i.e., $q_0 = q_N$. The x coordinates x_n are represented as $x_n = a_n + q_n = na + q_n$ and, from the commutation relations $[x_n, p_m] = i\delta_{nm}$, etc. ($\hbar = 1$), we have for the canonical commutation relations of the q_n and p_n

$$[q_n, p_m] = i\delta_{nm} , \qquad [q_n, q_m] = 0 , \qquad [p_n, p_m] = 0 . \tag{12.1.2}$$

The *Heisenberg representation*,

$$q_n(t) = e^{iHt} q_n e^{-iHt} , \tag{12.1.3a}$$
$$p_n(t) = e^{iHt} p_n e^{-iHt} , \tag{12.1.3b}$$

yields the two equations of motion

$$\dot{q}_n(t) = \frac{1}{m} p_n(t) \tag{12.1.4a}$$

and

$$\dot{p}_n(t) = m\ddot{q}_n(t) \tag{12.1.4b}$$
$$= m\Omega^2 (q_{n+1}(t) + q_{n-1}(t) - 2q_n(t)) - m\Omega_0^2 q_n(t) .$$

On account of the periodic boundary conditions, we are dealing with a translationally invariant problem (invariant with respect to translations by a). The Hamiltonian can therefore be diagonalized by means of the transformation (Fourier sum)

$$q_n = \frac{1}{(mN)^{1/2}} \sum_k e^{ikan} Q_k \tag{12.1.5a}$$

$$p_n = \left(\frac{m}{N}\right)^{1/2} \sum_k e^{-ikan} P_k . \tag{12.1.5b}$$

The variables Q_k and P_k are termed the *normal coordinates* and *normal momenta*, respectively. We now have to determine the possible values of k. To this end, we exploit the periodic boundary conditions which demand that $q_0 = q_N$, i.e., $1 = e^{ikaN}$; hence, we have $kaN = 2\pi\ell$ and thus

$$k = \frac{2\pi\ell}{Na} , \tag{12.1.6}$$

where ℓ is an integer. The values $k = \frac{2\pi(\ell \pm N)}{Na} = \frac{2\pi\ell}{Na} \pm \frac{2\pi}{a}$ are equivalent to $k = \frac{2\pi\ell}{Na}$ since, for these k values, the phase factors e^{ikan} are equal and, thus, so are the q_n and p_n. The possible k values are therefore reduced to those given by:

$$\text{for even } N \quad : \qquad -\frac{N}{2} < \ell \leq \frac{N}{2} , \qquad \ell = 0, \pm 1, \ldots, \pm \frac{N-2}{2}, \frac{N}{2}$$

$$\text{for odd } N \quad : \qquad -\frac{N-1}{2} \leq \ell \leq \frac{N-1}{2} , \qquad \ell = 0, \pm 1, \ldots, \pm \frac{N-1}{2} .$$

In solid state physics, this reduced interval of k values is also known as the first Brillouin zone. The Fourier coefficients in (12.1.5) satisfy the following orthogonality and completeness relations:

Orthogonality relation :

$$\frac{1}{N} \sum_{n=1}^{N} e^{ikan} e^{-ik'an} = \Delta(k - k') \tag{12.1.7a}$$

$$= \begin{cases} 1 \text{ for } k - k' = \dfrac{2\pi}{a} h, \ h \text{ integer} \\ 0 \text{ otherwise.} \end{cases}$$

In this form, the orthogonality relation is valid for any value of $k = \frac{2\pi\ell}{Na}$. When k is restricted to values in the first Brillouin zone, the generalized Kronecker delta $\Delta(k - k')$ becomes $\delta_{kk'}$.

Completeness relation:

$$\frac{1}{N} \sum_{k} e^{-ikan} e^{ikan'} = \delta_{nn'} . \tag{12.1.7b}$$

Here, the summation variable k is restricted to the first Brillouin zone. (For a proof, see Problem 12.1). The inverse of (12.1.5) reads:

$$Q_k = \sqrt{\frac{m}{N}} \sum_{n} e^{-ikan} q_n \tag{12.1.8a}$$

$$P_k = \frac{1}{\sqrt{mN}} \sum_{n} e^{ikan} p_n . \tag{12.1.8b}$$

Since the operators q_n and p_n are hermitian, it follows that

$$Q_k^\dagger = Q_{-k} , \qquad P_k^\dagger = P_{-k} . \tag{12.1.9}$$

Remark. When N is even, $\ell = \frac{N}{2}$ and $-\frac{N}{2}$ are equivalent and hence only $\frac{N}{2}$ appears. For $k = \frac{2\pi}{Na} \cdot \frac{N}{2} = \frac{\pi}{a}$, we have $Q_k = Q_k^\dagger$ and $P_k = P_k^\dagger$, since $e^{i\frac{\pi}{a}an} = e^{i\pi n} = (-1)^n$.

The commutation relations for the normal coordinates and momenta are obtained from (12.1.2) with the result

$$[Q_k, P_{k'}] = i\delta_{kk'} , \qquad [Q_k, Q_{k'}] = 0 , \qquad [P_k, P_{k'}] = 0 . \tag{12.1.10}$$

Transforming (12.1.1) into normal coordinates according to (12.1.5a,b) yields the Hamiltonian in the form

$$H = \frac{1}{2} \sum_k \left(P_k\, P_k^\dagger + \omega_k^2\, Q_k\, Q_k^\dagger \right) , \tag{12.1.11}$$

where the square of the vibration frequency as a function of k reads:

$$\omega_k^2 = \Omega^2 \left(2\sin\frac{ka}{2} \right)^2 + \Omega_0^2 \tag{12.1.12}$$

(Problem 12.3). The quantity $(\Omega a)^2$ is known as the stiffness constant.
Thus, in Fourier space, one obtains uncoupled oscillators with the frequency $\omega_k = \sqrt{\omega_k^2}$. [It should be noted, however, that the terms in (12.1.11) are of the form $Q_k Q_{-k}$ etc., so that the oscillators with wave numbers k and $-k$ are still interdependent.] The frequency is depicted as a function of k (dispersion relation) in Fig.12.2. In the language of lattice vibrations, $\Omega_0 = 0$ leads to acoustic, and finite Ω_0 to optical, phonons. In order to diagonalize H in Eq.(12.1.11), one introduces *creation* and *annihilation operators*:

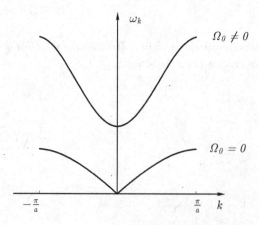

Fig. 12.2. The phonon frequencies for $\Omega_0 \neq 0$ and $\Omega_0 = 0$

$$a_k = \frac{1}{\sqrt{2\omega_k}}\left(\omega_k Q_k + iP_k^\dagger\right) \qquad (12.1.13a)$$

$$a_k^\dagger = \frac{1}{\sqrt{2\omega_k}}\left(\omega_k Q_k^\dagger - iP_k\right) . \qquad (12.1.13b)$$

The inverse of this transformation is given by

$$Q_k = \frac{a_k + a_{-k}^\dagger}{\sqrt{2\omega_k}} \qquad (12.1.14a)$$

and

$$P_k = -i\sqrt{\frac{\omega_k}{2}}\left(a_{-k} - a_k^\dagger\right) . \qquad (12.1.14b)$$

The commutation relations for the normal coordinates (12.1.10) lead to (Problem 12.5)

$$[a_k, a_{k'}^\dagger] = \delta_{k,k'} , \qquad [a_k, a_{k'}] = [a_k^\dagger, a_{k'}^\dagger] = 0 . \qquad (12.1.15)$$

By inserting (12.1.14a,b) into (12.1.11), one obtains

$$H = \sum_k \omega_k \left(a_k^\dagger a_k + \frac{1}{2}\right) , \qquad (12.1.16)$$

a Hamiltonian for N uncoupled oscillators. The summation extends over all N wave numbers in the first Brillouin zone, since

$$\begin{aligned}
H &= \frac{1}{2}\sum_k \frac{\omega_k}{2}(a_{-k} - a_k^\dagger)(a_{-k}^\dagger - a_k) + \frac{\omega_k^2}{2\omega_k}(a_k + a_{-k}^\dagger)(a_k^\dagger + a_{-k}) \\
&= \frac{1}{4}\sum_k \omega_k (a_{-k}a_{-k}^\dagger + a_k^\dagger a_k + a_k a_k^\dagger + a_{-k}^\dagger a_{-k} \\
&\qquad\qquad - a_{-k}a_k - a_k^\dagger a_{-k}^\dagger + a_k a_{-k} + a_{-k}^\dagger a_k^\dagger) \\
&= \frac{1}{2}\sum_k \omega_k (a_k^\dagger a_k + a_k a_k^\dagger) = \sum_k \omega_k \left(a_k^\dagger a_k + \frac{1}{2}\right) . \qquad (12.1.17)
\end{aligned}$$

The energy eigenstates and eigenvalues for the individual oscillators are known. The ground-state energy of the oscillator with the wave vector k is $\frac{1}{2}\omega_k$. The nth excited state of the oscillator with wave vector k is obtained by the n-fold application of the operator $a_k^\dagger$, having energy $(n_k + \frac{1}{2})\omega_k$. The fact that the eigenvalues of the Hamiltonian are, up to the zero-point energy, integer multiples of the eigenfrequencies leads quite naturally to a particle interpretation, although we are dealing here not with material particles but rather with excited states (quasiparticles). In the case of the elastic chain considered here, these quanta are known as phonons. The occupation numbers are $0, 1, 2, \ldots$, hence the quanta are bosons. The operator $a_k^\dagger$ creates a phonon with wave vector k and frequency (energy) ω_k, whilst a_k annihilates a phonon with wave vector k and frequency (energy) ω_k.

Hence, the eigenstates of the Hamiltonian (12.1.17) are of the following form: In the ground state $|0\rangle$, which is determined by the equation

$$a_k |0\rangle = 0, \text{ for all } k ,\tag{12.1.18a}$$

no phonons are present. Its energy represents the zero-point energy

$$E_0 = \sum_k \frac{1}{2}\omega_k .\tag{12.1.18b}$$

A general multiphonon state has the form

$$|n_{k_1}, n_{k_2}, \ldots, n_{k_N}\rangle = \frac{1}{\sqrt{n_{k_1}! n_{k_2}! \ldots n_{k_N}!}}$$
$$\times \left(a_{k_1}^\dagger\right)^{n_{k_1}} \left(a_{k_2}^\dagger\right)^{n_{k_2}} \ldots \left(a_{k_N}^\dagger\right)^{n_{k_N}} |0\rangle \tag{12.1.19a}$$

with energy

$$E = \sum_k n_k \omega_k + E_0 .\tag{12.1.19b}$$

The occupation numbers take the values $n_k = 0, 1, 2, \ldots$ and k runs through the N values of the first Brillouin zone; the n_k are not bounded from above. The operator $\hat{n}_k = a_k^\dagger a_k$ is the occupation number operator for phonons with the wave vector k.
From

$$[\hat{n}_k, a_k] = -a_k \quad \text{und} \quad [\hat{n}_k, a_k^\dagger] = a_k^\dagger \tag{12.1.19c}$$

it follows that

$$a_{k_i} |\ldots, n_{k_i}, \ldots\rangle = \sqrt{n_{k_i}} |\ldots, n_{k_i} - 1, \ldots\rangle ,$$
$$a_{k_i}^\dagger |\ldots, n_{k_i}, \ldots\rangle = \sqrt{n_{k_i} + 1} |\ldots, n_{k_i} + 1, \ldots\rangle . \tag{12.1.19d}$$

Remark. Let us emphasize that the commutation relations (12.1.2) and (12.1.15) are valid even when nonlinear terms are present in the Hamiltonian, since they are a consequence of the general canonical commutation relations of position and momentum operators.

12.1.1.2 Dynamics

Equation (12.1.16) expresses the Hamiltonian of the linear chain in diagonal form. In fact, H is time independent, so that its various representations (12.1.1), (12.1.11), and (12.1.16) are valid at all times. The essential features of the dynamics are most readily described in the Heisenberg picture. Starting from

$$q_n = \frac{1}{\sqrt{mN}} \sum_k \mathrm{e}^{\mathrm{i}kan} Q_k = \frac{1}{\sqrt{mN}} \sum_k \frac{1}{\sqrt{2\omega_k}} \mathrm{e}^{\mathrm{i}kan} \left(a_k + a_{-k}^\dagger\right)$$

$$= \sum_k \frac{1}{\sqrt{2\omega_k mN}} \left(\mathrm{e}^{\mathrm{i}kan} a_k + \mathrm{e}^{-\mathrm{i}kan} a_k^\dagger\right) , \tag{12.1.20}$$

we define the Heisenberg operator

$$q_n(t) = \mathrm{e}^{\mathrm{i}Ht} q_n(0) \mathrm{e}^{-\mathrm{i}Ht} = \mathrm{e}^{\mathrm{i}Ht} q_n \mathrm{e}^{-\mathrm{i}Ht} . \tag{12.1.21}$$

By solving the equation of motion, or by using

$$\mathrm{e}^{\mathrm{i}Ht} a_k \mathrm{e}^{-\mathrm{i}Ht} = a_k + [\mathrm{i}Ht, a_k] + \frac{1}{2!}[\mathrm{i}Ht, [\mathrm{i}Ht, a_k]] + \dots$$

$$= a_k + [\mathrm{i}\omega_k t\, a_k^\dagger a_k, a_k] + \frac{1}{2!}[\mathrm{i}Ht, [\mathrm{i}Ht, a_k]] + \dots$$

$$= a_k - \mathrm{i}\omega_k t a_k + \frac{1}{2!}[\mathrm{i}\omega_k t a_k^\dagger a_k, -\mathrm{i}\omega_k t a_k] + \dots \tag{12.1.22}$$

$$= a_k \left(1 - \mathrm{i}\omega_k t + \frac{1}{2!}(-\mathrm{i}\omega_k t)^2 + \dots\right)$$

$$= a_k \mathrm{e}^{-\mathrm{i}\omega_k t} ,$$

one obtains for the time dependence of the displacements

$$q_n(t) = \sum_k \frac{1}{\sqrt{2\omega_k mN}} \left(\mathrm{e}^{\mathrm{i}(kan-\omega_k t)} a_k + \mathrm{e}^{-\mathrm{i}(kan-\omega_k t)} a_k^\dagger\right) . \tag{12.1.23}$$

Concerning its structure, this solution is identical to the classical solution, although the amplitudes are now the annihilation and creation operators. We will discuss the significance of this solution only in the context of the continuum limit, which we shall now proceed to introduce.

12.1.2 Continuum Limit, Vibrating String

Here, we shall treat the continuum limit for the vibrating string. In this limit the lattice constant becomes $a \to 0$ and the number of oscillators $N \to \infty$, whilst the length of the string $L = aN$ remains finite (Fig. 12.3).

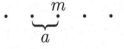

Fig. 12.3. Concerning the continuum limit of the linear chain (see text)

The density $\rho = \frac{m}{a}$ and stiffness constant $v^2 = (\Omega a)^2$ must also remain constant. The positions of the lattice points $x = na$ are then continuously distributed. We also introduce the definitions

$$q(x) = q_n \left(\frac{m}{a}\right)^{1/2} \tag{12.1.24a}$$

$$p(x) = p_n (ma)^{-1/2} \ . \tag{12.1.24b}$$

The equation of motion (12.1.4b)

$$\ddot{q}_n = \Omega^2 (q_{n+1} + q_{n-1} - 2q_n) - \Omega_0^2 \, q_n$$

becomes

$$\ddot{q}(x,t) = \Omega^2 a^2 \frac{(q(x+a,t) - q(x,t)) - (q(x,t) - q(x-a,t))}{a^2} \\ - \Omega_0^2 q(x,t) \tag{12.1.25}$$

and, in the limit $a \to 0$, one has

$$\ddot{q}(x,t) - v^2 \frac{\partial^2}{\partial x^2} q(x,t) + \Omega_0^2 \, q(x,t) = 0 \ . \tag{12.1.26}$$

The form of this equation is identical to that of the one-dimensional Klein–Gordon equation. For $\Omega_0 = 0$, i.e., in the absence of a harmonic potential, Eq. (12.1.26) is the equation of motion for a vibrating string, as is known from classical mechanics.

In the continuum limit, the Hamiltonian (12.1.1) takes the form

$$
\begin{aligned}
H &= \lim_{a \to 0, N \to \infty} \sum_n \left(\frac{1}{2m} p_n^2 + \frac{m\Omega^2}{2} (q_n - q_{n-1})^2 + \frac{m\Omega_0^2}{2} q_n^2 \right) \\
&= \lim_{a \to 0, N \to \infty} \sum_n a \left(\frac{1}{2ma} p_n^2 + \frac{m\Omega^2}{2a} a^2 \left(\frac{q_n - q_{n-1}}{a} \right)^2 + \frac{m\Omega_0^2}{2a} q_n^2 \right) \\
&= \int_0^L dx \, \frac{1}{2} \left[p(x)^2 + v^2 \left(\frac{\partial q}{\partial x} \right)^2 + \Omega_0^2 \, q(x)^2 \right] \ ,
\end{aligned}
\tag{12.1.27}
$$

where $\sum_n a \dots \to \int_0^L dx \dots$. The commutators of the displacements and the momenta are obtained from (12.1.2) and (12.1.24a,b):

$$
\begin{aligned}
[q(x), p(x')] &= \lim_{a \to 0, N \to \infty} \left(\frac{m}{a}\right)^{1/2} (ma)^{-1/2} [q_n, p_{n'}] \\
&= \lim_{a \to 0, N \to \infty} \mathrm{i} \frac{\delta_{nn'}}{a} = \mathrm{i}\delta(x - x')
\end{aligned}
\tag{12.1.28a}
$$

and

$$[q(x), q(x')] = [p(x), p(x')] = 0 \ . \tag{12.1.28b}$$

Next, we will derive the representation in terms of normal coordinates. From (12.1.6), it follows that

$$k = \frac{2\pi\ell}{L} , \quad \text{where } \ell \text{ is an integer with } -\infty \leq \ell \leq \infty . \tag{12.1.29}$$

For a string of finite length, the Fourier space remains discrete in the continuum limit, although the number of wave vectors and thus normal coordinates is now infinite. From (12.1.5a,b) we have

$$q(x) = \frac{1}{L^{1/2}} \sum_k e^{ikx} Q_k \tag{12.1.30a}$$

$$p(x) = \frac{1}{L^{1/2}} \sum_k e^{-ikx} P_k \tag{12.1.30b}$$

and from (12.1.11)

$$H = \sum_k \frac{1}{2} \left(P_k\, P_k^\dagger + \omega_k^2 Q_k\, Q_k^\dagger \right) , \tag{12.1.31}$$

whereby, in the limit $a \to 0$, equation (12.1.12) reduces to

$$\omega_k^2 = v^2 k^2 + \Omega_0^2 . \tag{12.1.32}$$

The commutation relations for the normal coordinates (12.1.10) remain unchanged:

$$[Q_k, P_{k'}] = i\delta_{kk'} , \quad [Q_k, Q_{k'}] = 0 , \quad [P_k, P_{k'}] = 0 . \tag{12.1.33}$$

The transformation to creation and annihilation operators (12.1.14a,b), and also the expression for the Hamiltonian in terms of these quantities (12.1.16) remain correspondingly unchanged. The representation of the displacement field in terms of creation and annihilation operators now takes the form

$$
\begin{aligned}
q(x) &= \frac{1}{L^{1/2}} \sum_k e^{ikx} \frac{a_k + a_{-k}^\dagger}{\sqrt{2\omega_k}} \\
&= \frac{1}{L^{1/2}} \sum_k \left(e^{ikx} a_k + e^{-ikx} a_k^\dagger \right) \frac{1}{\sqrt{2\omega_k}}
\end{aligned}
\tag{12.1.34}
$$

and, from (12.1.23), its time dependence is given by

$$q(x,t) = \frac{1}{L^{1/2}} \sum_k \left(e^{i(kx-\omega_k t)} a_k + e^{-i(kx-\omega_k t)} a_k^\dagger \right) \frac{1}{\sqrt{2\omega_k}} . \tag{12.1.35}$$

We finally obtain for the Hamiltonian

$$H = \sum_k \omega_k \left(a_k^\dagger a_k + \frac{1}{2} \right) , \tag{12.1.36}$$

which is positive definite. The functions $e^{i(kx-\omega_k t)}$ and $e^{-i(kx-\omega_k t)}$ appearing in (12.1.35) are solutions of the free field equation (12.1.26), which, in connection with the Klein–Gordon equation, we had interpreted as solutions with positive and negative energy. In the quantized theory, these solutions appear as amplitude functions, prefactors of the annihilation and creation operators in the expansion of the field operators. The sign of the frequency dependence is of no significance for the value of the energy. This is determined by the Hamiltonian (12.1.36), which is positive definite: there are no states of negative energy. The direct analogy to the vibrating string relates to the real Klein–Gordon field. The complex field will be treated in Eq. (12.1.47a,b) and Sect. 13.2.

12.1.3 Generalization to Three Dimensions, Relationship to the Klein–Gordon Field

12.1.3.1 Generalization to three dimensions

It is now straightforward to generalize the above results to three dimensions. We consider a discrete three-dimensional cubic lattice. Rather than taking an elastic lattice, which would have three-dimensional displacement vectors, we shall assume instead that the displacements are only along one dimension (scalar). In the continuum limit, the one-dimensional coordinate x must be replaced by the three-dimensional vector $\mathbf{x}$

$$x \to \mathbf{x} \,,$$

and the field equation for the displacement $q(\mathbf{x}, t)$ reads:

$$\ddot{q}(\mathbf{x}, t) - v^2 \Delta q(\mathbf{x}, t) + \Omega_0^2 \, q(\mathbf{x}, t) = 0 \,. \tag{12.1.37}$$

Introducing the substitutions

$$v \to c \,, \quad \frac{\Omega_0^2}{v^2} \to m^2 \,, \quad (\mathbf{x}, t) \equiv x \,, \quad \text{and} \quad q(\mathbf{x}, t) \to \phi(x) \,, \tag{12.1.38}$$

we obtain

$$\partial_\mu \partial^\mu \phi(x) + m^2 \phi(x) = 0 \,, \tag{12.1.39}$$

which is precisely the Klein–Gordon equation (5.2.11'). The representation of the solution of the Klein–Gordon equation in terms of annihilation and creation operators (12.1.35), the commutation relations (12.1.15), (12.1.28), and the Hamiltonian (12.1.36) can all be directly translated into three dimensions:

$$\phi(\mathbf{x}, t) = \frac{1}{L^{3/2}} \sum_{\mathbf{k}} \frac{1}{\sqrt{2\omega_{\mathbf{k}}}} \left(e^{i(\mathbf{kx} - \omega_{\mathbf{k}} t)} a_{\mathbf{k}} + e^{-i(\mathbf{kx} - \omega_{\mathbf{k}} t)} a_{\mathbf{k}}^\dagger \right) \tag{12.1.40}$$

$$\equiv \phi^+(x) + \phi^-(x) \,,$$

$$[a_{\mathbf{k}}, a_{\mathbf{k}'}^\dagger] = \delta_{\mathbf{k}, \mathbf{k}'} \,, \quad [a_{\mathbf{k}}, a_{\mathbf{k}'}] = [a_{\mathbf{k}}^\dagger, a_{\mathbf{k}'}^\dagger] = 0, \tag{12.1.41a}$$

$$[\phi(\mathbf{x},t),\dot{\phi}(\mathbf{x}',t)] = i\delta^{(3)}(\mathbf{x}-\mathbf{x}'),$$
$$[\phi(\mathbf{x},t),\phi(\mathbf{x}',t)] = [\dot{\phi}(\mathbf{x},t),\dot{\phi}(\mathbf{x}',t)] = 0,$$

(12.1.41b)

and

$$H = \sum_{\mathbf{k}} \omega_{\mathbf{k}}\left(a_{\mathbf{k}}^{\dagger}a_{\mathbf{k}} + \frac{1}{2}\right).$$

(12.1.42)

Inspired by these mechanical analogies, we arrive at a completely new interpretation of the Klein–Gordon equation. Previously, in Sect. 5.2, an attempt was made to use the Klein–Gordon equation as a relativistic replacement for the Schrödinger equation and to interpret its solutions as probability amplitudes in the same way as for the Schrödinger wave functions in coordinate space. However, $\phi(\mathbf{x},t)$ is not a wave function but an operator in *Fock space*. This field operator is represented as a superposition of single-particle solutions of the Klein–Gordon equation with amplitudes that are themselves operators. The effect of these operators is to create and annihilate the quanta (elementary particles) that are described by the field. The term Fock space describes the state space spanned by the multi-boson states

$$\left(a_{\mathbf{k}_1}^{\dagger}\right)^{n_{\mathbf{k}_1}}\left(a_{\mathbf{k}_2}^{\dagger}\right)^{n_{\mathbf{k}_2}}\ldots|0\rangle,$$

(12.1.43a)

where $|0\rangle$ is the ground state ($\equiv$ vacuum state) of the field. The energy of this state is

$$E = \sum_{\mathbf{k}} \hbar\omega_{\mathbf{k}}\left(n_{\mathbf{k}} + \frac{1}{2}\right).$$

(12.1.43b)

In equation (12.1.40) the field operator was split into positive and negative frequency parts, $\phi^{+}(x)$ and $\phi^{-}(x)$. This notation originates from the positive and negative energy solutions. Due to the hermiticity of the field operator $\phi(x)$, we have $\phi^{+\dagger} = \phi^{-}$, and in the expansion (12.1.40) we encounter the sum of $a_{\mathbf{k}}$ and $a_{\mathbf{k}}^{\dagger}$. This hermitian (real) Klein–Gordon field describes uncharged mesons, as our subsequent investigations will reveal.

12.1.3.2 The infinite-volume limit

Until now, we have based our studies on a finite volume with linear extension L. In order to formulate relativistically invariant theories, it is necessary to include all of space. We thus take the limit $L \to \infty$. In this limit, the previously discrete values of $\mathbf{k}$ move arbitrarily close together, such that $\mathbf{k}$ too becomes a continuous variable. The sums over $\mathbf{k}$ are replaced by integrals according to

$$\sum_{\mathbf{k}}\left(\frac{2\pi}{L}\right)^{3}\ldots \to \int \frac{d^3k}{(2\pi)^3}\ldots.$$

Using the definition

$$a(\mathbf{k}) = \left(\frac{L}{2\pi}\right)^{\frac{3}{2}} a_{\mathbf{k}} ,$$

(12.1.44)

one obtains the field operator from (12.1.39) as

$$\phi(\mathbf{x},t) = \int\limits_{-\infty}^{\infty} \frac{d^3k}{(2\pi)^{3/2}} \frac{1}{\sqrt{2\omega_{\mathbf{k}}}} \left(e^{i(\mathbf{kx}-\omega_{\mathbf{k}}t)}a(\mathbf{k}) + e^{-i(\mathbf{kx}-\omega_{\mathbf{k}}t)}a^{\dagger}(\mathbf{k})\right) ,$$

(12.1.45)

where the $\mathbf{k}$ integration extends in all three spatial dimensions from $-\infty$ to $+\infty$. The commutation relations for the creation and annihilation operators now read:

$$\left[a(\mathbf{k}), a^{\dagger}(\mathbf{k}')\right] = \delta_{\mathbf{kk}'}\left(\frac{L}{2\pi}\right)^3 = \delta(\mathbf{k}-\mathbf{k}'),$$

$$\left[a(\mathbf{k}), a(\mathbf{k}')\right] = 0 \quad , \quad \left[a^{\dagger}(\mathbf{k}), a^{\dagger}(\mathbf{k}')\right] = 0 .$$

(12.1.46)

Proof:

$$1 = \sum_{\mathbf{k}'} \delta_{\mathbf{kk}'} = \sum_{\mathbf{k}'} \left(\frac{2\pi}{L}\right)^3 \left(\left(\frac{L}{2\pi}\right)^3 \delta_{\mathbf{kk}'}\right)$$

$$= \int d^3k' \left(\left(\frac{L}{2\pi}\right)^3 \delta_{\mathbf{kk}'}\right) = \int d^3k' \, \delta(\mathbf{k}-\mathbf{k}') .$$

The *complex Klein–Gordon field* is not hermitian and therefore the expansion coefficients (operators) of the solutions with positive and negative frequency are independent of one another

$$\phi(\mathbf{x},t) = \frac{1}{L^{3/2}} \sum_{\mathbf{k}} \frac{1}{\sqrt{2\omega_{\mathbf{k}}}} \left(e^{-ik\cdot x}a_{\mathbf{k}} + e^{ik\cdot x}b_{\mathbf{k}}^{\dagger}\right) .$$

(12.1.47a)

Here, $k \cdot x = \omega_k t - \mathbf{k} \cdot \mathbf{x}$ is the scalar product of four-vectors. The operators $a_{\mathbf{k}}$ and $b_{\mathbf{k}}$ have the following significance:

$a_{\mathbf{k}}\,(a_{\mathbf{k}}^{\dagger})$ annihilates (creates) a particle with momentum $\mathbf{k}$ and
$b_{\mathbf{k}}\,(b_{\mathbf{k}}^{\dagger})$ annihilates (creates) an antiparticle with momentum $\mathbf{k}$
        and opposite charge,

as will be discussed more fully in subsequent sections. From (12.1.47a), one obtains the hermitian conjugate of the field operator as

$$\phi^{\dagger}(\mathbf{x},t) = \frac{1}{L^{3/2}} \sum_{\mathbf{k}} \frac{1}{\sqrt{2\omega_{\mathbf{k}}}} \left(e^{-ik\cdot x}b_{\mathbf{k}} + e^{ik\cdot x}a_{\mathbf{k}}^{\dagger}\right) .$$

(12.1.47b)

12.2 Classical Field Theory

12.2.1 Lagrangian and Euler–Lagrange Equations of Motion

12.2.1.1 Definitions

In this section we shall study the basic properties of classical (and, in the main, relativistic) field theories. We consider a system described by fields $\phi_r(x)$, where the index r is a number which labels the fields. It can refer to the components of a single field, e.g., the radiation field $A^\mu(x)$ or the four-spinor $\psi(x)$, but it can also serve to enumerate the different fields. To begin with, we define a number of terms and concepts.

We assume the existence of a *Lagrangian density* that depends on the fields ϕ_r and their derivatives $\phi_{r,\mu} \equiv \partial_\mu \phi_r \equiv \frac{\partial}{\partial x^\mu}\phi_r$. The *Lagrangian density* is denoted by

$$\mathcal{L} = \mathcal{L}(\phi_r, \phi_{r,\mu}) \ . \tag{12.2.1}$$

The *Lagrangian* is then defined as

$$L(x^0) = \int d^3x \, \mathcal{L}(\phi_r, \phi_{r,\mu}) \ . \tag{12.2.2}$$

The significance of the Lagrangian in field theory is completely analogous to that in point mechanics. The form of the Lagrangian for various fields will be elucidated in the following sections. We also define the *action*

$$S(\Omega) = \int_\Omega d^4x \, \mathcal{L}(\phi_r, \phi_{r,\mu}) = \int dx^0 \, L(x^0) \ , \tag{12.2.3}$$

where $d^4x = dx^0 \, d^3x \equiv dx^0 \, dx^1 \, dx^2 \, dx^3$. The integration extends over a region Ω in the four-dimensional space–time, which will usually be infinite. We shall use the same notation as in Part II on *relativistic wave equations*, where we set the speed of light $c = 1$ and, thus, $x^0 = t$.

12.2.1.2 Hamilton's principle in point mechanics

As has already been mentioned, the definitions and the procedure needed here are analogous to those of *point mechanics* with n degrees of freedom. We briefly remind the reader of the latter[1,2]. The *Lagrangian* of a system of particles with n degrees of freedom with generalized coordinates $q_i, i = 1, \ldots, n$ has the form:

[1] H. Goldstein, *Classical Mechanics*, 2nd ed., Addison-Wesley, Reading, Mass., 1980

[2] L.D. Landau and E.M. Lifshitz, *Course of Theoretical Physics, Vol. 1*, Pergamon, Oxford, 1960

$$L(t) = \sum_{i=1}^{n} \frac{1}{2} m_i \dot{q}_i^2 - V(q_i) \,. \tag{12.2.4}$$

The first term is the kinetic energy, and the second the negative potential energy due to the interactions between the particles and any external conservative forces. The *action* is defined by

$$S = \int_{t_1}^{t_2} dt \, L(t) \,. \tag{12.2.5}$$

The equations of motion of such a classical system follow from Hamilton's action principle. This states that the actual trajectory $q_i(t)$ of the system is such that the action (12.2.5) is stationary, i.e.,

$$\delta S = 0 \,, \tag{12.2.6}$$

where variations in the trajectory $q_i(t) + \delta q_i(t)$ between the initial and final times t_1 and t_2 are restricted by (see Fig.12.4)

$$\delta q_i(t_1) = \delta q_i(t_2) = 0, \quad i = 1, \dots, n \,. \tag{12.2.7}$$

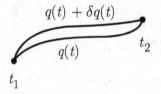

$q(t) + \delta q(t)$

$q(t)$

t_2

t_1

Fig. 12.4. Variation of the solution in the time interval between t_1 and t_2. Here, $q(t)$ stands for $\{q_i(t)\}$

The condition that the action is stationary for the actual trajectory implies

$$
\begin{aligned}
\delta S &= \int_{t_1}^{t_2} dt \left(\frac{\partial L}{\partial q_i(t)} \delta q_i(t) + \frac{\partial L}{\partial \dot{q}_i(t)} \delta \dot{q}_i(t) \right) \\
&= \int_{t_1}^{t_2} dt \left[\left(\frac{\partial L}{\partial q_i(t)} - \frac{d}{dt} \frac{\partial L}{\partial \dot{q}_i(t)} \right) \delta q_i(t) + \frac{d}{dt} \left(\frac{\partial L}{\partial \dot{q}_i(t)} \delta q_i(t) \right) \right] \\
&= \int_{t_1}^{t_2} dt \left[\left(\frac{\partial L}{\partial q_i(t)} - \frac{d}{dt} \frac{\partial L}{\partial \dot{q}_i(t)} \right) \delta q_i(t) \right] + \left(\frac{\partial L}{\partial \dot{q}_i(t)} \delta q_i(t) \right) \Big|_{t_1}^{t_2} = 0 \,.
\end{aligned}
\tag{12.2.8}
$$

The second term on the last line vanishes since, according to (12.2.7), the variation $\delta q(t)$ must be zero at the endpoints. In order for δS to vanish for all $\delta q_i(t)$, we have the condition

$$\frac{\partial L}{\partial q_i(t)} - \frac{d}{dt}\frac{\partial L}{\partial \dot{q}_i(t)} = 0 \,, \qquad i = 1, \ldots, n \,. \tag{12.2.9}$$

These are the *Euler–Lagrange equations of motion*, which are equivalent to Hamilton's equations of motion. We now proceed to extend these concepts to fields.

12.2.1.3 Hamilton's principle in field theory

In field theory the index i is replaced by the continuous variable **x**. The equations of motion (= field equations) are obtained from the variational principle

$$\delta S = 0 \,. \tag{12.2.10}$$

To this end, we consider variations of the fields

$$\phi_r(x) \rightarrow \phi_r(x) + \delta\phi_r(x) \tag{12.2.11}$$

which are required to vanish on the surface $\Gamma(\Omega)$ of the space–time region Ω:

$$\delta\phi_r(x) = 0 \quad \text{on } \Gamma(\Omega) \,. \tag{12.2.12}$$

In analogy to (12.2.9), we now calculate the change in the action (12.2.3)

$$\begin{aligned}
\delta S &= \int_\Omega d^4x \left\{ \frac{\partial \mathcal{L}}{\partial \phi_r}\delta\phi_r + \frac{\partial \mathcal{L}}{\partial \phi_{r,\mu}}\delta\phi_{r,\mu} \right\} \\
&= \int_\Omega d^4x \left\{ \frac{\partial \mathcal{L}}{\partial \phi_r} - \frac{\partial}{\partial x^\mu}\frac{\partial \mathcal{L}}{\partial \phi_{r,\mu}} \right\}\delta\phi_r + \int_\Omega d^4x \frac{\partial}{\partial x^\mu}\left(\frac{\partial \mathcal{L}}{\partial \phi_{r,\mu}}\delta\phi_r \right) \,.
\end{aligned} \tag{12.2.13}$$

Here, we employ the summation convention for the repeated indices r and μ and have also used[3] $\delta\phi_{r,\mu} = \frac{\partial}{\partial x^\mu}\delta\phi_r$. The last term in Eq.(12.2.13) can be re-expressed using Gauss's theorem as the surface integral

$$\int_{\Gamma(\Omega)} d\sigma_\mu \frac{\partial \mathcal{L}}{\partial \phi_{r,\mu}}\delta\phi_r = 0 \,, \tag{12.2.14}$$

where $d\sigma_\mu$ is the μ component of the element of surface area. The condition that δS in Eq.(12.1.13) vanishes for arbitrary Ω and $\delta\phi_r$ yields the *Euler–Lagrange equations* of field theory

$$\frac{\partial \mathcal{L}}{\partial \phi_r} - \frac{\partial}{\partial x^\mu}\frac{\partial \mathcal{L}}{\partial \phi_{r,\mu}} = 0 \,, \qquad r = 1, 2, \ldots \,. \tag{12.2.15}$$

---

[3] $\delta\phi_r(x) = \phi_r'(x) - \phi_r(x)$ and thus $\frac{\partial}{\partial x^\mu}\delta\phi_r(x) = \phi_{r,\mu}'(x) - \phi_{r,\mu}(x) = \delta\phi_{r,\mu}(x)$.

Remark. So far we have considered the case of real fields. Complex fields can be treated as two real fields, for the real and imaginary parts. It is easy to see that this is equivalent to viewing $\phi(x)$ and $\phi^*(x)$ as independent fields. In this sense, the variational principle and the Euler–Lagrange equations also hold for complex fields.

We now introduce two further definitions in analogy to point particles in mechanics. The *momentum field* conjugate to $\phi_r(x)$ is defined by

$$\pi_r(x) = \frac{\delta L}{\delta \dot{\phi}_r(x)} = \frac{\partial \mathcal{L}}{\partial \dot{\phi}_r(x)} \,. \tag{12.2.16}$$

The definition of the *Hamiltonian* reads:

$$H = \int d^3x \left(\pi_r(x)\dot{\phi}_r(x) - \mathcal{L}(\phi_r, \phi_{r,\mu}) \right) = H(\phi_r, \pi_r) \,, \tag{12.2.17}$$

where the $\dot{\phi}_r$ have to be expressed in terms of the π_r.
The Hamiltonian density is defined by

$$\mathcal{H}(x) = \pi_r(x)\dot{\phi}_r(x) - \mathcal{L}(\phi_r, \phi_{r,\mu}) \,. \tag{12.2.18}$$

The Hamiltonian can be expressed in terms of the Hamiltonian density as

$$H = \int d^3x \,\mathcal{H}(x) \,. \tag{12.2.19}$$

The integral extends over all space. H is time independent since $\mathcal{L}$ does not depend explicitly on time.

12.2.1.4 Example: A real scalar field

To illustrate the concepts introduced above, we consider the example of a real scalar field $\phi(x)$. For the Lagrangian density we take the lowest powers of the field and its derivatives that are invariant under Lorentz transformations

$$\mathcal{L} = \frac{1}{2} \left(\phi'_{,\mu}\phi^{,\mu} - m^2\phi^2 \right) \,, \tag{12.2.20}$$

where m is a constant. The derivatives of $\mathcal{L}$ with respect to ϕ and $\phi_{,\mu}$ are

$$\frac{\partial \mathcal{L}}{\partial \phi} = -m^2\phi \,, \quad \frac{\partial \mathcal{L}}{\partial \phi_{,\mu}} = \phi^{,\mu} \,,$$

from which one obtains for the Euler–Lagrange equation (12.2.15)

$$\phi^{,\mu}{}_{\mu} + m^2\phi = 0 \,, \tag{12.2.21}$$

or, in the form previously employed,

$$(\partial^\mu \partial_\mu + m^2)\phi = 0 \, . \tag{12.2.21'}$$

Thus, Eq. (12.2.20) is the Lagrangian density for the Klein–Gordon equation. The conjugate momentum for this field theory is, according to (12.2.16),

$$\pi(x) = \dot{\phi}(x) \, , \tag{12.2.22}$$

and, from (12.2.18), the Hamiltonian density reads:

$$\mathcal{H}(x) = \frac{1}{2} \left[\pi^2(x) + (\nabla \phi)^2 + m^2 \phi^2(x) \right] \, . \tag{12.2.23}$$

If we had included higher powers of ϕ^2 in (12.2.20), for example ϕ^4, the equation of motion (12.2.21') would have contained additional nonlinear terms.

Remarks on the structure of the Lagrangian density

(i) The Lagrangian density may only depend on $\phi_r(x)$ and $\phi_{r,\mu}(x)$; higher derivatives would lead to differential equations of higher than second order. The Lagrangian density can depend on x only via the fields. An additional explicit dependence on x would violate the relativistic invariance.

(ii) The theory must be local, i.e., $\mathcal{L}(x)$ is determined by $\phi_r(x)$ and $\phi_{r,\mu}(x)$ at the position x. Integrals over $\mathcal{L}(x)$ would imply nonlocal terms and could lead to acausal behavior.

(iii) The Lagrangian density $\mathcal{L}$ is not uniquely determined by the action, nor even by the equations of motion. Lagrangian densities that differ from one another by a four-divergence are physically equivalent

$$\mathcal{L}'(x) = \mathcal{L}(x) + \partial_\nu F^\nu(x) \, . \tag{12.2.24}$$

The additional term here leads in the action to a surface integral over the three-dimensional boundary of the four-dimensional integration region. Since the variation of the field vanishes on the surface, this can make no contribution to the equation of motion.

(iv) $\mathcal{L}$ should be real (in quantum mechanics, hermitian) or, in view of remark (iii), equivalent to a real $\mathcal{L}$. This ensures that the equations of motion and the Hamiltonian, when expressed in terms of real fields, are themselves real. $\mathcal{L}$ must be relativistically invariant, i.e., under an inhomogeneous Lorentz transformation

$$\begin{aligned} x \to x' &= \Lambda x + a \\ \phi_r(x) &\to \phi_r'(x') \, , \end{aligned} \tag{12.2.25}$$

$\mathcal{L}$ must behave as a scalar:

$$\mathcal{L}(\phi_r'(x'), \phi_{r,\mu}'(x')) = \mathcal{L}(\phi_r(x), \phi_{r,\mu}(x)) \, . \tag{12.2.26}$$

Since $d^4x = dx^0 dx^1 dx^2 dx^3$ is also invariant, the action is unchanged under the Lorentz transformation (12.2.25) and the equations of motion have the same form in both coordinate systems and are thus covariant.

12.3 Canonical Quantization

Our next task is to quantize the field theory introduced in the previous section. We will allow ourselves to be guided in this by the results of the mechanical elastic continuum model (Sect. 12.1.3) and postulate the following commutation relations for the fields ϕ_r and the momentum fields π_r:

$$[\phi_r(\mathbf{x}, t), \pi_s(\mathbf{x}', t)] = i\delta_{rs}\delta(\mathbf{x} - \mathbf{x}') \,,$$
$$[\phi_r(\mathbf{x}, t), \phi_s(\mathbf{x}', t)] = [\pi_r(\mathbf{x}, t), \pi_s(\mathbf{x}', t)] = 0 \,. \tag{12.3.1}$$

These are known as the *canonical commutation relations* and one speaks of *canonical quantization*. For the real Klein–Gordon field, where according to (12.2.22) $\pi(x) = \dot{\phi}(x)$, this also implies

$$[\phi(\mathbf{x}, t), \dot{\phi}(\mathbf{x}', t)] = i\delta(\mathbf{x} - \mathbf{x}') \,,$$
$$[\phi(\mathbf{x}, t), \phi(\mathbf{x}', t)] = [\dot{\phi}(\mathbf{x}, t), \dot{\phi}(\mathbf{x}', t)] = 0 \,. \tag{12.3.2}$$

In view of the general validity of (12.1.28) and (12.1.41b), one postulates also the canonical commutation relations for interacting fields.

12.4 Symmetries and Conservation Laws, Noether's Theorem

12.4.1 The Energy–Momentum Tensor, Continuity Equations, and Conservation Laws

The invariance of a system under continuous symmetry transformations leads to continuity equations and conservation laws. The derivation of these conservation laws from the invariance of the Lagrangian density is known as Noether's theorem (see below).

Continuity equations can also be derived in an elementary fashion from the equations of motion. This will be illustrated for the case of the *energy–momentum tensor*, which is defined by

$$T^{\mu\nu} = \frac{\partial \mathcal{L}}{\partial \phi_{r,\mu}}\phi_r{}^{,\nu} - \mathcal{L}g^{\mu\nu} \,. \tag{12.4.1}$$

The energy–momentum tensor obeys the *continuity equation*[4]

$$T^{\mu\nu}{}_{,\mu} = 0 \,. \tag{12.4.2}$$

Proof: Differentiation of $T^{\mu\nu}$ yields:

---

[4] In the next section, we shall derive this continuity equation from space–time translational invariance, whence, in analogy to classical mechanics, the term energy–momentum tensor will find its justification.

$$T^{\mu\nu}{}_{,\mu} = \left(\frac{\partial}{\partial x^\mu}\frac{\partial\mathcal{L}}{\partial\phi_{r,\mu}}\right)\phi_r{}^{,\nu} + \frac{\partial\mathcal{L}}{\partial\phi_{r,\mu}}\phi_r{}^{,\nu}{}_\mu - \partial^\nu\mathcal{L} \qquad (12.4.3)$$

$$= \frac{\partial\mathcal{L}}{\partial\phi_r}\phi_r{}^{,\nu} + \frac{\partial\mathcal{L}}{\partial\phi_{r,\mu}}\phi_r{}^{,\nu}{}_\mu - \partial^\nu\mathcal{L} = 0 \,,$$

where we have used the Euler–Lagrange equation (12.2.15) and $\partial^\nu\mathcal{L} = \frac{\partial\mathcal{L}}{\partial\phi_r}\partial^\nu\phi_r + \frac{\partial\mathcal{L}}{\partial\phi_{r,\mu}}\partial^\nu\phi_{r,\mu}$ to obtain the second identity.

If a four-vector g^μ satisfies a continuity equation

$$g^\mu{}_{,\mu} = 0 \,, \qquad (12.4.4)$$

then, assuming that the fields on which g^μ depends vanish rapidly enough at infinity, this leads to the *conservation* of the space integral of its zero component

$$G^0(t) = \int d^3x\, g^0(\mathbf{x},t) \,. \qquad (12.4.5)$$

Proof: The continuity equation, together with the generalized Gauss divergence theorem, leads to

$$\int_\Omega d^4x\, \frac{\partial}{\partial x^\mu} g^\mu = 0 = \int_\sigma d\sigma_\mu\, g^\mu \,. \qquad (12.4.6)$$

This holds for every four-dimensional region Ω with surface σ. One now chooses an integration region whose boundary in the spatial directions extends to infinity. In the time direction, it is bounded by two three-dimensional surfaces $\sigma_1(x^0 = t_1)$ and $\sigma_2(x^0 = t_2)$ (Fig.12.5). In the spatial directions, ϕ_r and $\phi_{r,\mu}$ are zero at infinity:

$$0 = \int_{\sigma_1} d^3x\, g^0 - \int_{\sigma_2} d^3x\, g^0 = \int d^3x\, g^0(\mathbf{x},t_1) - \int d^3x\, g^0(\mathbf{x},t_2)$$

thus,

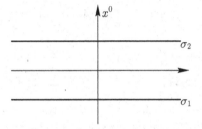

Fig. 12.5. Diagram relating to the derivation of the conservation law (see text)

$$G^0(t_1) = G^0(t_2) \tag{12.4.7a}$$

or, alternatively,

$$\frac{dG^0}{dt} = 0 . \tag{12.4.7b}$$

Applying this result to the continuity equation for the energy–momentum tensor (12.4.1) leads to the conservation of the *energy–momentum four-vector*

$$P^\nu = \int d^3x\, T^{0\nu}(\mathbf{x}, t) . \tag{12.4.8}$$

The components of the energy–momentum vector are

$$P^0 = \int d^3x\, \{\pi_r(x)\dot{\phi}_r(x) - \mathcal{L}(\phi_r, \phi_{r,\mu})\} \tag{12.4.9}$$

$$= \int d^3x\, \mathcal{H} \;=\; H$$

and

$$P^j = \int d^3x\, \pi_r(x)\frac{\partial\phi_r}{\partial x_j} \qquad j = 1, 2, 3 . \tag{12.4.10}$$

The zero component is equal to the Hamiltonian (operator), and the spatial components represent the momentum operator of the field.

12.4.2 Derivation from Noether's Theorem of the Conservation Laws for Four-Momentum, Angular Momentum, and Charge

12.4.2.1 Noether's theorem

Noether's theorem states that every continuous transformation that leaves the action unchanged leads to a conservation law. For instance, the conservation of four-momentum and of angular momentum follows from the invariance of the Lagrangian density $\mathcal{L}$ under translations and rotations, respectively. Since these form continuous symmetry groups, it is sufficient to consider infinitesimal transformations. We therefore consider the infinitesimal Lorentz transformation

$$x_\mu \rightarrow x'_\mu = x_\mu + \delta x_\mu \;=\; x_\mu + \Delta\omega_{\mu\nu}\, x^\nu + \delta_\mu \tag{12.4.11a}$$

$$\phi_r(x) \rightarrow \phi'_r(x') = \phi_r(x) + \frac{1}{2}\Delta\omega_{\mu\nu}\, S^{\mu\nu}_{rs}\, \phi_s(x) . \tag{12.4.11b}$$

Here x and x' represent the same point in space time referred to the two frames of reference, and ϕ_r and ϕ'_r are the field components referred to these coordinate systems. The quantities which appear in these equations should be understood as follows: The constant δ_μ causes an infinitesimal displacement. The homogeneous part of the Lorentz transformation is given by the infinitesimal antisymmetric tensor $\Delta\omega_{\mu\nu} = -\Delta\omega_{\nu\mu}$. The coefficients $S_{rs}^{\mu\nu}$ in the transformation (12.4.11b) of the fields are antisymmetric in μ and ν and are determined by the transformation properties of the fields. For example, in the case of spinors (Eqs.(6.2.13) and (6.2.17)), we have

$$\frac{1}{2}\Delta\omega_{\mu\nu}\, S_{rs}^{\mu\nu}\, \phi_s = -\frac{i}{4}\,\Delta\omega_{\mu\nu}\,\sigma_{rs}^{\mu\nu}\,\phi_s \ , \tag{12.4.12a}$$

i.e.,

$$S_{rs}^{\mu\nu} = -\frac{i}{2}\,\sigma_{rs}^{\mu\nu} \ , \tag{12.4.12b}$$

where r and $s(=1,\dots,4)$ label the four components of the spinor field. Vector fields transform under a Lorentz transformation according to Eq. (11.1.3a) and thus we have

$$S_{rs}^{\mu\nu} = g_r^\mu\, g_s^\nu - g_s^\mu\, g_r^\nu \ , \tag{12.4.12c}$$

where the indices r, s take the values $0, 1, 2, 3$. In Eqs. (12.4.12a,b) summation over the repeated indices μ, ν, and s is implied.

As has already been emphasized, the *invariance* under the transformation (12.4.11a,b) means that the Lagrangian density has the same functional form in the new coordinates and fields as it did in the original ones:

$$\mathcal{L}(\phi_r'(x'), \phi_{r,\mu}'(x')) = \mathcal{L}(\phi_r(x), \phi_{r,\mu}(x)) \ . \tag{12.4.13}$$

From Eq. (12.4.13), the covariance of the equations of motion follows. The variation of $\phi_r(x)$, for unchanged argument, is defined by

$$\delta\phi_r(x) = \phi_r'(x) - \phi_r(x) \ . \tag{12.4.14}$$

Furthermore, we define the *total* variation

$$\Delta\phi_r(x) = \phi_r'(x') - \phi_r(x) \ , \tag{12.4.15}$$

which represents the change due to the form and the argument of the function. These two quantities are related by

$$\begin{aligned}
\Delta\phi_r(x) &= (\phi_r'(x') - \phi_r(x')) + (\phi_r(x') - \phi_r(x)) \\
&= \delta\phi_r(x') + \frac{\partial\phi_r}{\partial x_\nu}\delta x_\nu + \mathcal{O}(\delta^2) \\
&= \delta\phi_r(x) + \frac{\partial\phi_r}{\partial x_\nu}\delta x_\nu + \mathcal{O}(\delta^2) \ ,
\end{aligned} \tag{12.4.16}$$

where $\mathcal{O}(\delta^2)$ stands for terms of second order, which we neglect. In correspondence with Eq.(12.4.16), the difference between the Lagrangian densities in the coordinate systems I and I', i.e., the total variation of the Lagrangian density – which vanishes according to (12.4.13) – can be rewritten as

$$0 = \mathcal{L}(\phi'_r(x'), \phi'_{r,\mu}(x')) - \mathcal{L}(\phi_r(x), \phi_{r,\mu}(x))$$
$$= \mathcal{L}(\phi'_r(x'), \dots) - \mathcal{L}(\phi_r(x'), \dots) + (\mathcal{L}(\phi_r(x'), \dots) - \mathcal{L}(\phi_r(x), \dots))$$
$$= \delta\mathcal{L} + \frac{\partial\mathcal{L}}{\partial x^\mu}\delta x^\mu + O(\delta^2) . \tag{12.4.17}$$

The first term on the right-hand side of (12.4.17) is obtained as

$$\delta\mathcal{L} = \frac{\partial\mathcal{L}}{\partial\phi_r}\delta\phi_r + \frac{\partial\mathcal{L}}{\partial\phi_{r,\mu}}\delta\phi_{r,\mu}$$
$$= \frac{\partial\mathcal{L}}{\partial\phi_r}\delta\phi_r - \left(\frac{\partial}{\partial x^\mu}\frac{\partial\mathcal{L}}{\partial\phi_{r,\mu}}\right)\delta\phi_r + \frac{\partial}{\partial x^\mu}\left(\frac{\partial\mathcal{L}}{\partial\phi_{r,\mu}}\delta\phi_r\right)$$
$$= \frac{\partial}{\partial x^\mu}\left\{\frac{\partial\mathcal{L}}{\partial\phi_{r,\mu}}\left[\Delta\phi_r - \frac{\partial\phi_r}{\partial x_\nu}\delta x_\nu\right]\right\} ,$$

where the Euler–Lagrange equation was used to obtain the second line and Eq.(12.4.16) to perform the last step. Together with $\frac{\partial\mathcal{L}}{\partial x^\mu}\delta x^\mu = \frac{\partial}{\partial x^\mu}(\mathcal{L}\delta x^\mu) = \frac{\partial}{\partial x^\mu}(\mathcal{L}g^{\mu\nu}\delta x_\nu)$, Eq. (12.4.17) leads to the continuity equation

$$g^\mu{}_{,\mu} = 0 \tag{12.4.18a}$$

for the four-vector

$$g^\mu \equiv \frac{\partial\mathcal{L}}{\partial\phi_{r,\mu}}\Delta\phi_r - T^{\mu\nu}\delta x_\nu . \tag{12.4.18b}$$

Here, g^μ depends on the variations $\Delta\phi_r$ and δx_ν, and, according to the choice made, results in different conservation laws.

Equations (12.4.18a) and (12.4.18b), which lead to the conserved quantities (12.4.5), amongst others, represent the general statement of *Noether's theorem*.

12.4.2.2 Application to Translational, Rotational, and Gauge Invariance

We now analyze the result of the previous section for three important special cases.

(i) Pure *translations:*
For translations we have

$$\Delta\omega_{\mu\nu} = 0$$
$$\delta x_\nu = \delta_\nu \tag{12.4.19a}$$

and, hence, (12.4.11b) gives $\phi'_r(x') = \phi_r(x)$; therefore,

$$\Delta\phi_r = 0 . \tag{12.4.19b}$$

Noether's theorem then reduces to the statement $g^\mu = -T^{\mu\nu}\delta_\nu$, and since the four displacements δ_ν are independent of one another, one obtains the four continuity equations

$$T^{\mu\nu}{}_{,\mu} = 0 \tag{12.2.31}$$

for the energy–momentum tensor $T^{\mu\nu}$, $\nu = 0, 1, 2, 3$, defined in (12.4.3). For $\nu \equiv 0$, one obtains the continuity equation for the four-momentum-density $P^\mu = T^{0\mu}$, and for $\nu = i$ that for the quantities $T^{i\mu}$. The conservation laws $T^{i\mu}{}_{,\mu} = 0$ contain as zero components, the spatial momentum densities P^i and as current densities, the components of the so-called stress tensor T^{ij}. (See also the discussion that follows Eq.(12.4.7b).)

(ii) For *rotations* we have, according to (12.4.11a,b),

$$\delta_\mu = 0, \ \delta x_\nu = \Delta\omega_{\nu\sigma} x^\sigma \tag{12.4.20a}$$

and

$$\Delta\phi_r = \frac{1}{2}\Delta\omega_{\nu\sigma} S^{\nu\sigma}_{rs}\phi_s . \tag{12.4.20b}$$

From (12.4.18b), it then follows that

$$g^\mu \equiv \frac{1}{2}\frac{\partial\mathcal{L}}{\partial\phi_{r,\mu}}\Delta\omega_{\nu\sigma} S^{\nu\sigma}_{rs}\phi_s - T^{\mu\nu}\Delta\omega_{\nu\sigma} x^\sigma . \tag{12.4.21}$$

Using the definition

$$M^{\mu\nu\sigma} = \frac{\partial\mathcal{L}}{\partial\phi_{r,\mu}}S^{\nu\sigma}_{rs}\phi_s(x) + (x^\nu T^{\mu\sigma} - x^\sigma T^{\mu\nu}) , \tag{12.4.22}$$

equation (12.4.21) can be re-expressed in the form

$$
\begin{aligned}
g^\mu &= \frac{1}{2}\frac{\partial\mathcal{L}}{\partial\phi_{r,\mu}}S^{\nu\sigma}_{rs}\phi_s\Delta\omega_{\nu\sigma} - \frac{1}{2}T^{\mu\nu}\Delta\omega_{\nu\sigma} x^\sigma - \frac{1}{2}T^{\mu\sigma}\Delta\omega_{\sigma\nu} x^\nu \\
&= \frac{1}{2}\left(\frac{\partial\mathcal{L}}{\partial\phi_{r,\mu}}S^{\nu\sigma}_{rs}\phi_s + x^\nu T^{\mu\sigma} - x^\sigma T^{\mu\nu}\right)\Delta\omega_{\nu\sigma} \qquad (12.4.20') \\
&= \frac{1}{2}M^{\mu\nu\sigma}\Delta\omega_{\nu\sigma} .
\end{aligned}
$$

Since the six nonvanishing elements of the antisymmetric matrix $\Delta\omega_{\nu\sigma}$ are independent of one another, it follows that the quantities $M^{\mu\nu\sigma}$ satisfy the six continuity equations

$$\partial_\mu M^{\mu\nu\sigma} = 0 . \tag{12.4.23}$$

This yields the six quantities

$$
\begin{aligned}
M^{\nu\sigma} &= \int d^3x \, M^{0\nu\sigma} \\
&= \int d^3x \left(\pi_r(x)S^{\nu\sigma}_{rs}\phi_s(x) + x^\nu T^{0\sigma} - x^\sigma T^{0\nu}\right) .
\end{aligned}
\tag{12.4.24}
$$

For the spatial components, one obtains the *angular-momentum* operator

$$M^{ij} = \int d^3x \; \left(\pi_r \, S^{ij}_{rs} \, \phi_s + x^i \, T^{0j} - x^j T^{0i} \right) . \tag{12.4.25}$$

Here, the angular-momentum vector $(I^1, I^2, I^3) \equiv (M^{23}, M^{31}, M^{12})$ is conserved. The sum of the second and third terms in the integral represents the vector product of the coordinate vector with the spatial momentum density and can thus be considered as the angular momentum of the field. The first term can be interpreted as intrinsic angular momentum or spin (see below (13.3.13′) and (E.31c)). The space–time components $(0\,i)$

$$M^{0i} = \int d^3x \, M^{00i}$$

can be combined into the three-component boost vector (boost generator)

$$\mathbf{K} = (M^{01}, M^{02}, M^{03}) . \tag{12.4.26}$$

(iii) *Gauge transformations* (gauge transformation of the first kind).
As a final application of Noether's theorem we consider the consequences of *gauge invariance* .
Assuming that the Lagrangian density contains a subset of fields ϕ_r and $\phi_r^\dagger$ only in combinations of the type $\phi_r^\dagger(x)\phi_r(x)$ and $\phi_{r,\mu}^\dagger(x)\phi_r{}^{,\mu}(x)$, then it is invariant with respect to gauge transformations of the first kind. These are defined by

$$\begin{aligned} \phi_r(x) &\rightarrow \phi_r'(x) = \mathrm{e}^{\mathrm{i}\varepsilon}\phi_r(x) \approx (1 + \mathrm{i}\varepsilon)\phi_r(x) \\ \phi_r^\dagger(x) &\rightarrow \phi_r^{\dagger'}(x) = \mathrm{e}^{-\mathrm{i}\varepsilon}\phi_r^\dagger(x) \approx (1 - \mathrm{i}\varepsilon)\phi_r^\dagger(x) , \end{aligned} \tag{12.4.27}$$

where ϵ is an arbitrary real number. The coordinates are not transformed and hence, according to Eq. (12.4.14),

$$\begin{aligned} \delta\phi_r(x) &= \mathrm{i}\varepsilon \, \phi_r(x) \\ \delta\phi_r^\dagger(x) &= -\mathrm{i}\varepsilon \, \phi_r^\dagger(x) \end{aligned} \tag{12.4.28}$$

and [cf. (12.4.16)]

$$\Delta\phi_r(x) = \delta\phi_r(x) , \quad \Delta\phi_r^\dagger(x) = \delta\phi_r^\dagger(x) . \tag{12.4.29}$$

The four-current-density follows from Noether's theorem (12.4.18b) as

$$g^\mu \propto \frac{\partial \mathcal{L}}{\partial \phi_{r,\mu}} \mathrm{i}\varepsilon \, \phi_r + \frac{\partial \mathcal{L}}{\partial \phi_{r,\mu}^\dagger} (-\mathrm{i}\varepsilon)\phi_r^\dagger ,$$

i.e.,

$$\begin{aligned} g^\mu(x) &= \mathrm{i}\left(\frac{\partial \mathcal{L}}{\partial \phi_{r,\mu}} \phi_r - \frac{\partial \mathcal{L}}{\partial \phi_{r,\mu}^\dagger} \phi_r^\dagger \right) \\ g^0(x) &= \mathrm{i}\left(\pi_r(x)\phi_r(x) - \pi_r^\dagger(x)\phi_r^\dagger(x) \right) \end{aligned} \tag{12.4.30}$$

satisfies a continuity equation. This implies that

$$Q = -\mathrm{i}q \int d^3x \left(\pi_r(x)\phi_r(x) - \pi_r^\dagger(x)\phi_r^\dagger(x) \right) \tag{12.4.31}$$

is conserved. Thus, in quantized form,

$$\frac{dQ}{dt} = 0, \quad [Q, H] = 0 . \tag{12.4.32}$$

The quantity q will turn out to be the charge. We can already see this by calculating the commutator of Q and ϕ_r with the commutation relations (12.3.1):

$$[Q, \phi_r(x)] = -\mathrm{i}q \int d^3x' \underbrace{[\pi_s(x'), \phi_r(x)]}_{-\mathrm{i}\delta_{sr}\delta(\mathbf{x}' - \mathbf{x})} \phi_s(x') = -q\phi_r(x) . \tag{12.4.33}$$

If $|Q'\rangle$ is an eigenstate of Q,

$$Q|Q'\rangle = Q'|Q'\rangle , \tag{12.4.34}$$

then $\phi_r(x)|Q'\rangle$ is an eigenstate with the eigenvalue $Q' - q$ and
$\phi_r^\dagger(x)|Q'\rangle$ is an eigenstate with the eigenvalue $Q' + q$,
as follows from (12.4.33):

$$(Q\phi_r(x) - \phi_r(x)Q)|Q'\rangle = -q\phi_r(x)|Q'\rangle$$
$$Q\phi_r(x)|Q'\rangle - \phi_r(x)Q'|Q'\rangle = -q\phi_r(x)|Q'\rangle \tag{12.4.35}$$
$$Q\phi_r(x)|Q'\rangle = (Q' - q)\phi_r(x)|Q'\rangle .$$

Hence, by using complex, i.e., nonhermitian, fields, one can represent charged particles. The conservation of charge is a consequence of the invariance under gauge transformations of the first kind (i.e., ones in which the phase is independent of x). In theories in which the field is coupled to a gauge field, one can also have gauge transformations of the second kind $\psi \to \psi' = \psi \mathrm{e}^{\mathrm{i}\alpha(x)}$, $A^\mu \to A'^\mu = A^\mu + \frac{1}{e}\partial^\mu \alpha(x)$.

12.4.2.3 Generators of Symmetry Transformations in Quantum Mechanics

We assume that the Hamiltonian H is time independent and consider constants of the motion that do not depend explicitly on time. The Heisenberg equations of motion

$$\frac{dA(t)}{dt} = \mathrm{i}[H, A(t)] \tag{12.4.36}$$

imply that such constants of the motion commute with H

$$[H, A] = 0 . \tag{12.4.37}$$

Symmetry transformations can in general be represented by unitary, or, in the case of time reversal, by antiunitary, transformations[1]. In the case of a continuous symmetry group, every element of which is continuously connected with the identity, e.g., rotations, the transformations are represented by unitary operators. This means that the states and operators transform as

$$|\psi\rangle \rightarrow |\psi'\rangle = U|\psi\rangle \tag{12.4.38a}$$

and

$$A \rightarrow A' = UAU^\dagger \,. \tag{12.4.38b}$$

The unitarity guarantees that transition amplitudes and matrix elements of operators remain invariant, and that operator equations are covariant, i.e., the equations of motion and the commutation relations have the same form, regardless of whether they are expressed in the original or in the transformed operators.

For a continuous transformation, we can represent the unitary operator in the form

$$U = e^{i\alpha T} \tag{12.4.39}$$

where $T^\dagger = T$ and α is a real continuous parameter. The hermitian operator T is called the generator of the transformation. For $\alpha = 0$, we have $U(\alpha = 0) = 1$. For an infinitesimal transformation ($\alpha \rightarrow \delta\alpha$), it is possible to expand U as

$$U = 1 + i\,\delta\alpha\,T + O(\delta\alpha^2) \,, \tag{12.4.39'}$$

and the transformation rule for an operator A has the form

$$A' = A + \delta A = (1 + i\,\delta\alpha\,T)A(1 - i\,\delta\alpha\,T) + O(\delta\alpha^2)$$
$$\text{and thus} \quad \delta A = i\,\delta\alpha\,[T, A] \,. \tag{12.4.37b'}$$

When the physical system remains invariant under the transformation considered, then the Hamiltonian must remain invariant, $\delta H = 0$, and from (12.4.37b') it follows that

$$[T, H] = 0 \,. \tag{12.4.40}$$

Since T commutes with H, the generator of the symmetry transformation is a constant of the motion. Conversely, every conserved quantity G^0 generates a symmetry transformation through the unitary operator

$$U = e^{i\alpha G^0} \,, \tag{12.4.41}$$

---

[1] E.P. Wigner, *Group Theory and its Application to the Quantum Mechanics of Atomic Spectra*, Academic Press, New York, 1959, Appendix to Chap. 20, p. 233; V. Bargmann, J. Math. Phys. **5**, 862 (1964)

since G^0 commutes with H on account of $[H, G^0] = \frac{1}{i}\dot{G}^0 = 0$, and hence $UHU^\dagger = H$, signifying that H is invariant. Not unnaturally, this is exactly the same transformation from which one derives the corresponding conserved four-current-density, which satisfies a continuity equation. This can be confirmed explicitly for P^μ, Q, and $M^{\mu\nu}$. See Problem 13.2(b) for the Klein–Gordon field and 13.10 for the Dirac field. See also Problems 13.5 and 13.12 referring to the charge conjugation operator.

The boost vector, $K^i \equiv M^{0i}$, (12.4.26)

$$K^i = tP^i - \int d^3x \left(x^i T^{00}(\mathbf{x}, t) - \pi_r(x) S_{rs}^{0i} \phi_s(x) \right) \tag{12.4.42}$$

is a constant, but it depends explicitly on time. From the Heisenberg equation of motion $\dot{\mathbf{K}} = 0 = i[H, \mathbf{K}] + \mathbf{P}$, it follows that $\mathbf{K}$ does not commute with H

$$[H, \mathbf{K}] = i\mathbf{P} . \tag{12.4.43}$$

For the Dirac-field one finds

$$K^i = tP^i - \int d^3x \left(x^i \mathcal{H}(x) - \frac{i}{2}\bar{\psi}(x)\gamma^i\psi(x) \right) . \tag{12.4.44}$$

Problems

12.1 Prove the completeness relation (12.1.7b) and the orthogonality relation (12.1.7b).

12.2 Demonstrate the validity of the commutation relation (12.1.10).

12.3 Show that the Hamiltonian (12.1.1) for the coupled oscillators can be transformed into (12.1.11) and gives the dispersion relation (12.1.12).

12.4 Prove the inverse transformations given in (12.1.14a,b).

12.5 Prove the commutation relations for the creation and annihilation operators (12.1.15).

12.6 Prove the conservation law (12.4.7b) by calculating $\frac{dG^0}{dt}$ using the three-dimensional Gauss's law and by using in the definition of G^0 the integral over all space.

12.7 The coherent states for the linear chain are defined as eigenstates of the annihilation operators a_k. Calculate the expectation value of the operator

$$q_n(t) = \sum_k \frac{1}{\sqrt{2Nm\omega_k}} \left[e^{i(kna - \omega_k t)} a_k(0) + e^{-i(kna - \omega_k t)} a_k^\dagger(0) \right]$$

for coherent states.

12.8 Show for vector fields A_s, $s = 0, 1, 2, 3$, the validity of Eq. (12.4.12c).

13. Free Fields

We shall now apply the results of the previous chapter to the free real and complex Klein–Gordon fields, as well as to the Dirac and radiation fields. We shall thereby derive the fundamental properties of these free field theories. The spin-statistics theorem will also be proved.

13.1 The Real Klein–Gordon Field

Since the Klein–Gordon field was found as the continuum limit of coupled oscillators, the most important properties of this quantized field theory have already been encountered in Sects. 12.1 and 12.2.1.4. Nevertheless, here we shall once more present the essential relations in a closed, deductive manner.

13.1.1 The Lagrangian Density, Commutation Relations, and the Hamiltonian

The Lagrangian density of the free real Klein–Gordon field is of the form

$$\mathcal{L} = \frac{1}{2} \left(\phi_{,\mu} \phi^{,\mu} - m^2 \phi^2 \right) . \tag{13.1.1}$$

The equation of motion (12.2.21) reads:

$$\left(\partial_\mu \partial^\mu + m^2 \right) \phi = 0 . \tag{13.1.2}$$

The conjugate momentum field follows from (13.1.1) as

$$\pi(x) = \frac{\partial \mathcal{L}}{\partial \dot{\phi}} = \dot{\phi}(x) . \tag{13.1.3}$$

The quantized real Klein–Gordon field is represented by the hermitian operators

$$\phi^\dagger(x) = \phi(x) \qquad \text{and} \qquad \pi^\dagger(x) = \pi(x) .$$

The canonical quantization prescription (12.3.1) yields for the Klein–Gordon field

$$\left[\phi(\mathbf{x}, t), \dot{\phi}(\mathbf{x}', t)\right] = i\,\delta(\mathbf{x} - \mathbf{x}')$$

$$\left[\phi(\mathbf{x}, t), \phi(\mathbf{x}', t)\right] = \left[\dot{\phi}(\mathbf{x}, t), \dot{\phi}(\mathbf{x}', t)\right] = 0 \,. \tag{13.1.4}$$

Remarks:

(i) Since $\phi(x)$ transforms as a scalar under Lorentz transformations and possesses no intrinsic degrees of freedom, the coefficients $S_{rs}^{\mu\nu}$ in (12.4.11b) and (12.4.25) are zero. The spin of the Klein–Gordon field is therefore zero.

(ii) Since the field operator ϕ is hermitian, the $\mathcal{L}$ of Eq. (13.1.1) is not gauge invariant. Thus the particles described by ϕ carry no charge.

(iii) Not all electrically neutral mesons with spin 0 are described by a real Klein–Gordon field. For example, the K_0 meson has an additional property known as the hypercharge Y. At the end of the next section we will see that the K_0, together with its antiparticle $\bar{K}_0$, can be described by a complex Klein–Gordon field.

(iv) For the case of quantized fields, too, it is still common practice to speak of real and complex fields.

The expansion of $\phi(x)$ in terms of a complete set of solutions of the Klein–Gordon equation is of the form

$$\phi(x) = \phi^+(x) + \phi^-(x) \tag{13.1.5}$$

$$= \sum_{\mathbf{k}} \frac{1}{\sqrt{2V\omega_{\mathbf{k}}}} \left(e^{-ikx} a_{\mathbf{k}} + e^{ikx} a_{\mathbf{k}}^\dagger\right)$$

with

$$k^0 = \omega_{\mathbf{k}} = (m^2 + \mathbf{k}^2)^{1/2} \,, \tag{13.1.6}$$

where ϕ^+ and ϕ^- represent the contributions of positive (e^{-ikx}) and negative frequency (e^{ikx}), respectively. Inverting (13.1.5) yields:

$$a_{\mathbf{k}} = \sqrt{\frac{1}{2V\omega_{\mathbf{k}}}} \int d^3x\, e^{ikx} \left(\omega_{\mathbf{k}}\phi(\mathbf{x}, 0) + i\dot{\phi}(\mathbf{x}, 0)\right)$$

$$a_{\mathbf{k}}^\dagger = \sqrt{\frac{1}{2V\omega_{\mathbf{k}}}} \int d^3x\, e^{-ikx} \left(\omega_{\mathbf{k}}\phi(\mathbf{x}, 0) - i\dot{\phi}(\mathbf{x}, 0)\right) \,. \tag{13.1.5'}$$

From the canonical commutation relations of the fields (13.1.4), one obtains the commutation relations for the $a_{\mathbf{k}}$ and $a_{\mathbf{k}}^\dagger$:

$$\left[a_{\mathbf{k}}, a_{\mathbf{k}'}^\dagger\right] = \delta_{\mathbf{k}\mathbf{k}'} \,, \qquad \left[a_{\mathbf{k}}, a_{\mathbf{k}'}\right] = \left[a_{\mathbf{k}}^\dagger, a_{\mathbf{k}'}^\dagger\right] = 0 \,. \tag{13.1.7}$$

These are the typical commutation relations for uncoupled oscillators, i.e., for bosons. The operators

$$\hat{n}_{\mathbf{k}} = a_{\mathbf{k}}^\dagger a_{\mathbf{k}} \tag{13.1.8}$$

have the eigenvalues

$$n_{\mathbf{k}} = 0, 1, 2, \ldots$$

and can thus be interpreted as occupation-number, or particle-number, operators. The operators $a_{\mathbf{k}}$ and $a_{\mathbf{k}}^{\dagger}$ annihilate and create particles with momentum $\mathbf{k}$.

From the energy–momentum four-vector (12.4.8), one obtains the Hamiltonian of the scalar field as

$$H = \int d^3x \, \frac{1}{2} \left[\dot{\phi}^2(x) + (\boldsymbol{\nabla}\phi(x))^2 + m^2\phi^2(x) \right] \tag{13.1.9}$$

$$= \int d^3x \frac{1}{2} \left[\pi^2(x) + (\boldsymbol{\nabla}\phi(x))^2 + m^2\phi^2(x) \right] \, ,$$

and the momentum operator of the Klein–Gordon field as

$$\mathbf{P} = - \int d^3x \, \dot{\phi}(x) \, \boldsymbol{\nabla}\phi(x) \, . \tag{13.1.10}$$

Remark. The quantum-mechanical field equations also follow from the Heisenberg equations, and the commutation relations (13.1.4) and (12.3.1):

$$\dot{\phi}(x) = \mathrm{i}[H, \phi(x)] = \pi(x) \tag{13.1.11}$$

$$\dot{\pi}(x) = \mathrm{i}[H, \pi(x)] = (\boldsymbol{\nabla}^2 - m^2)\phi(x) \, , \tag{13.1.12}$$

from which we have, in accordance with Eq. (13.1.2),

$$\ddot{\phi}(x) = (\boldsymbol{\nabla}^2 - m^2)\phi(x) \, . \tag{13.1.13}$$

Substitution of the expansion (13.1.5) yields H and $\mathbf{P}$ as

$$H = \sum_{\mathbf{k}} \frac{1}{2}\omega_{\mathbf{k}} \left(a_{\mathbf{k}}^{\dagger}a_{\mathbf{k}} + a_{\mathbf{k}}a_{\mathbf{k}}^{\dagger} \right) = \sum_{\mathbf{k}} \omega_{\mathbf{k}} \left(a_{\mathbf{k}}^{\dagger}a_{\mathbf{k}} + \frac{1}{2} \right) \tag{13.1.14}$$

$$\mathbf{P} = \sum_{\mathbf{k}} \frac{1}{2}\mathbf{k} \left(a_{\mathbf{k}}^{\dagger}a_{\mathbf{k}} + a_{\mathbf{k}}a_{\mathbf{k}}^{\dagger} \right) = \sum_{\mathbf{k}} \mathbf{k} \left(a_{\mathbf{k}}^{\dagger}a_{\mathbf{k}} + \frac{1}{2} \right) \, . \tag{13.1.15}$$

The state of lowest energy, the ground state or vacuum state $|0\rangle$, is characterized by the fact that it contains no particles, i.e., $n_{\mathbf{k}} = 0$, or

$$a_{\mathbf{k}} |0\rangle = 0 \quad \text{for all } \mathbf{k} \, . \tag{13.1.16a}$$

Thus

$$\phi^{+}(x) |0\rangle = 0 \quad \text{for all } x. \tag{13.1.16b}$$

The energy of the vacuum state

$$E_0 = \frac{1}{2} \sum_{\mathbf{k}} \omega_{\mathbf{k}} \, , \tag{13.1.17}$$

also known as the zero-point energy, is divergent. In itself, this is not a problem, since only energy differences are measurable and these are finite. However, it is desirable and possible to eliminate the zero-point energy from the outset by the use of normal ordering of operators. In a *normal ordered product* all *annihilation operators* are placed to the *right* of all *creation operators*. For *bosons*, we illustrate the definition of normal order, symbolized by two colons : ... :, by means of the following examples:

(i) $: a_{\mathbf{k}_1} a_{\mathbf{k}_2} a^\dagger_{\mathbf{k}_3} := a^\dagger_{\mathbf{k}_3} a_{\mathbf{k}_1} a_{\mathbf{k}_2}$ (13.1.18a)

(ii) $: a^\dagger_{\mathbf{k}} a_{\mathbf{k}} + a_{\mathbf{k}} a^\dagger_{\mathbf{k}} := 2 a^\dagger_{\mathbf{k}} a_{\mathbf{k}}$ (13.1.18b)

and

$$
\begin{aligned}
\text{(iii)} \ : \phi(x)\phi(y) : &= : (\phi^+(x) + \phi^-(x))(\phi^+(y) + \phi^-(y)) : \\
&= : \phi^+(x)\phi^+(y) : + : \phi^+(x)\phi^-(y) : \\
&\quad + : \phi^-(x)\phi^+(y) : + : \phi^-(x)\phi^-(y) : \\
&= \phi^+(x)\phi^+(y) + \phi^-(y)\phi^+(x) \\
&\quad + \phi^-(x)\phi^+(y) + \phi^-(x)\phi^-(y).
\end{aligned}
$$
(13.1.18c)

One treats the Bose operators in a normal product as if they had vanishing commutators. The order of the creation (annihilation) operators among themselves is irrelevant since their commutators are all zero. The positive frequency parts are placed to the right of the negative frequency parts. The vacuum expectation value of any normal product vanishes.

We now redefine the Lagrangian density and the observables such as energy–momentum vector, angular momentum, etc. as normal products :   :. This means, for example, that the momentum operator (13.1.10) is replaced by

$$
\mathbf{P} = - \int d^3x \ : \dot{\phi}(x) \boldsymbol{\nabla} \phi(x) : .
$$
(13.1.10′)

It follows from this that the energy–momentum vector, instead of being given by (13.1.14) and (13.1.15), now takes the form

$$
P^\mu = \sum_{\mathbf{k}} k^\mu \, a^\dagger_{\mathbf{k}} a_{\mathbf{k}} .
$$
(13.1.19)

This no longer contains any zero-point terms. We shall illustrate this for the Hamiltonian operator H. In the calculation leading to (13.1.14), the first step involved no permutation of operators. If the original H is now replaced by $: H :$, then, corresponding to example (ii) above, normal ordering gives $H = \sum_{\mathbf{k}} \omega_{\mathbf{k}} a^\dagger_{\mathbf{k}} a_{\mathbf{k}}$, i.e., the zero component of (13.1.19).

The normalized particle states and their energy eigenvalues are:

The vacuum	$|0\rangle$	$E_0 = 0$
single-particle states	$a_{\mathbf{k}}^\dagger |0\rangle$	$E_{\mathbf{k}} = \omega_{\mathbf{k}}$
two-particle states	$a_{\mathbf{k_1}}^\dagger a_{\mathbf{k_2}}^\dagger |0\rangle$ for arbitrary $\mathbf{k_1} \neq \mathbf{k_2}$	$E_{\mathbf{k_1},\mathbf{k_2}}$ $= \omega_{\mathbf{k_1}} + \omega_{\mathbf{k_2}}$
	$\dfrac{1}{\sqrt{2}}\left(a_{\mathbf{k}}^\dagger\right)^2 |0\rangle$ for arbitrary $\mathbf{k}$	$E_{\mathbf{k},\mathbf{k}} = 2\omega_{\mathbf{k}}$

One obtains a general two-particle state by a linear superposition of these states. As a result of (13.1.7), we have $a_{\mathbf{k_1}}^\dagger a_{\mathbf{k_2}}^\dagger |0\rangle = a_{\mathbf{k_2}}^\dagger a_{\mathbf{k_1}}^\dagger |0\rangle$. The particles described by the Klein–Gordon field are bosons: each of the occupation numbers takes the values $n_{\mathbf{k}} = 0, 1, 2, \ldots$. The operator $\hat{n}_{\mathbf{k}} = a_{\mathbf{k}}^\dagger a_{\mathbf{k}}$ is the particle-number operator for particles with the wave vector $\mathbf{k}$ whose eigenvalues are the occupation numbers $n_{\mathbf{k}}$.

We now turn to the angular momentum of the scalar field. This single-component field contains no intrinsic degrees of freedom and the coefficients S_{rs} in Eq. (12.4.25) vanish, $S_{rs} = 0$. The angular momentum operator (12.4.25) therefore contains no spin component; it comprises only orbital angular momentum

$$\mathbf{J} = \int d^3x\, \mathbf{x} \times \mathbf{P}(x) \tag{13.1.20}$$

$$=: \int d^3x\, \mathbf{x} \times \dot{\phi}(x)\frac{1}{i}\boldsymbol{\nabla}\phi(x) : .$$

The spin of the particles is thus zero. Since the Lagrangian density (13.1.1) and the Hamiltonian (13.1.9) are not gauge invariant, there is no charge operator. The real Klein–Gordon field can only describe uncharged particles. An example of a neutral meson with zero spin is the π^0.

13.1.2 Propagators

For perturbation theory, and also for the spin-statistic theorem to be discussed later, one requires the vacuum expectation values of bilinear combinations of the field operators. To calculate these, we first consider the commutators

$$[\phi^+(x), \phi^+(y)] = [\phi^-(x), \phi^-(y)] = 0$$

$$[\phi^+(x), \phi^-(y)] = \frac{1}{2V}\sum_{\mathbf{k}}\sum_{\mathbf{k'}}\frac{1}{(\omega_{\mathbf{k}}\omega_{\mathbf{k'}})^{1/2}}\left[a_{\mathbf{k}}, a_{\mathbf{k'}}^\dagger\right] e^{-ikx+ik'y}$$

$$= \frac{1}{2}\int\frac{d^3k}{(2\pi)^3}\frac{e^{-ik(x-y)}}{\omega_{\mathbf{k}}} \qquad , \quad k_0 = \omega_{\mathbf{k}} . \tag{13.1.21}$$

Using the definitions

$$\Delta^{\pm}(x) = \mp \frac{i}{2} \int \frac{d^3k}{(2\pi)^3} \frac{e^{\mp ikx}}{\omega_{\mathbf{k}}} \quad , \quad k_0 = \omega_{\mathbf{k}} \tag{13.1.22a}$$

$$\Delta(x) = \frac{1}{2i} \int \frac{d^3k}{(2\pi)^3} \frac{1}{\omega_{\mathbf{k}}} \left(e^{-ikx} - e^{ikx} \right) \quad , \quad k_0 = \omega_{\mathbf{k}} \tag{13.1.22b}$$

one can represent the commutators as follows:

$$\left[\phi^+(x), \phi^-(y)\right] = i\,\Delta^+(x-y) \tag{13.1.23a}$$

$$\left[\phi^-(x), \phi^+(y)\right] = i\,\Delta^-(x-y) = -i\,\Delta^+(y-x) \tag{13.1.23b}$$

$$\left[\phi(x), \phi(y)\right] = \left[\phi^+(x), \phi^-(y)\right] + \left[\phi^-(x), \phi^+(y)\right] \tag{13.1.23c}$$
$$= i\,\Delta(x-y)\,.$$

We also have the obvious relations

$$\Delta(x-y) = \Delta^+(x-y) + \Delta^-(x-y)) \tag{13.1.24a}$$

$$\Delta^-(x) = -\Delta^+(-x)\,. \tag{13.1.24b}$$

In order to emphasize the relativistic covariance of the commutators of the field, it is convenient to introduce the following four-dimensional integral representations:

$$\Delta^{\pm}(x) = -\int_{C^{\pm}} \frac{d^4k}{(2\pi)^4} \frac{e^{-ikx}}{k^2 - m^2} \tag{13.1.25a}$$

$$\Delta(x) = -\int_{C} \frac{d^4k}{(2\pi)^4} \frac{e^{-ikx}}{k^2 - m^2}\,, \tag{13.1.25b}$$

for which the contours of integration in the complex k_0 plane are shown in Fig. 13.1.

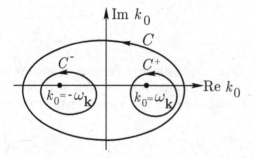

Fig. 13.1. Contours of integration $C^{\pm}$ and C in the complex k_0 plane for the propagators $\Delta^{\pm}(x)$ and $\Delta(x)$

The expressions (13.1.25a,b) can be verified by evaluating the path integrals in the complex k_0 plane using the residue theorem. The integrands are proportional to $[(k_0 - \omega_{\mathbf{k}})(k_0 + \omega_{\mathbf{k}})]^{-1}$ and have poles at the positions $\pm\omega_{\mathbf{k}}$.

Depending on the integration path, these poles may contribute to the integrals. The right-hand sides of (13.1.25a,b) are manifestly Lorentz covariant. This was shown in (10.1.2) for the volume element, and for the integrand is self-evident.

We now turn to the evaluation of the vacuum expectation values and propagators. Taking the vacuum expectation value of (13.1.23a) and using $\phi \left|0\right> = 0$, one obtains

$$i\,\Delta^+(x - x') = \langle 0| \left[\phi^+(x), \phi^-(x')\right] |0\rangle = \langle 0| \phi^+(x)\phi^-(x') |0\rangle$$
$$= \langle 0| \phi(x)\phi(x') |0\rangle \;. \tag{13.1.26}$$

In perturbation theory (Sect. 15.2) we will encounter time-ordered products of the perturbation Hamiltonian. For their evaluation we will need vacuum expectation values of time-ordered products. The *time-ordered product* T is defined for bosons as follows:

$$T\,\phi(x)\phi(x') = \begin{cases} \phi(x)\phi(x') & t > t' \\ \phi(x')\phi(x) & t < t' \end{cases} \tag{13.1.27}$$
$$= \Theta(t - t')\phi(x)\phi(x') + \Theta(t' - t)\phi(x')\phi(x) \;.$$

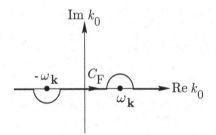

Fig. 13.2. Contour of integration in the k_0 plane for the Feynman propagator $\Delta_F(x)$

The *Feynman propagator* is defined in terms of the expectation value of the time-ordered product:

$$i\,\Delta_F(x - x') \equiv \langle 0| T(\phi(x)\phi(x')) |0\rangle \tag{13.1.28}$$
$$= i\left(\Theta(t - t')\Delta^+(x - x') - \Theta(t' - t)\Delta^-(x - x')\right) \;.$$

This is related to $\Delta^\pm(x)$ through

$$\Delta_F(x) = \pm\Delta^\pm(x) \qquad \text{for } t \gtrless 0 \tag{13.1.29}$$

and has the integral representation

$$\Delta_F(x) = \int_{C_F} \frac{d^4k}{(2\pi)^4} \frac{e^{-ikx}}{k^2 - m^2} \;. \tag{13.1.30}$$

The latter can be seen by adding an infinite half-circle to the integration contour in the upper or lower half-plane of k_0 and comparing it with

Eq.(13.1.25a). The integration along the path C_F defined in Fig. 13.2 is identical to the integration along the real k_0 axis, whereby the infinitesimal displacements η and ε in the integrands serve to shift the poles $k_0 = \pm(\omega_k - i\eta) = \pm(\sqrt{\mathbf{k}^2 + m^2} - i\eta)$ away from the real axis:

$$\Delta_\mathrm{F}(x) = \lim_{\eta \to 0^+} \int \frac{d^4k}{(2\pi)^4} \frac{e^{-ikx}}{k_0^2 - (\omega_\mathbf{k} - i\eta)^2} \tag{13.1.31}$$

$$= \lim_{\varepsilon \to 0^+} \int \frac{d^4k}{(2\pi)^4} \frac{e^{-ikx}}{k^2 - m^2 + i\varepsilon} .$$

As a preparation for the perturbation-theoretical representation in terms of Feynman diagrams, it is useful to give a pictorial description of the processes represented by propagators. Plotting the time axis to the right in Fig. 13.3, diagram (a) means that a meson is created at x' and subsequently annihilated at x, i.e., it is the process described by $\langle 0| \phi(x)\phi(x') |0\rangle = i\Delta^+(x - x')$. Diagram (b) represents the creation of a particle at x and its annihilation at x', i.e., $\langle 0| \phi(x')\phi(x) |0\rangle = -i\Delta^-(x - x')$. Both processes together are described by the Feynman propagator for the mesons of the Klein–Gordon field, which is thus often called, for short, the meson propagator.

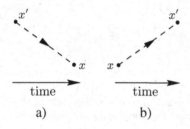

time

a)

time

b)

Fig. 13.3. Propagation of a particle (a) from x' to x and (b) from x to x'

As an example, we consider the scattering of two nucleons, which are represented in Fig. 13.4 by the full lines. The scattering arises due to the exchange of mesons. The two processes are represented jointly, and independently of their temporal sequence, by the Feynman propagator.

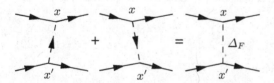

Fig. 13.4. Graphical representation of the meson propagator $\Delta_\mathrm{F}(x - x')$. In the first diagram, a meson is created at x' and annihilated at x. In the second diagram, a meson is created at x and annihialted at x'. Full lines represent nucleons, and dashed lines mesons

13.2 The Complex Klein–Gordon Field

The complex Klein–Gordon field is very similar to the real Klein–Gordon field, except that now the particles created and annihilated by the field carry a charge. Our starting point is the Lagrangian density

$$\mathcal{L} =: \phi^\dagger_{,\mu}(x)\phi^{,\mu}(x) - m^2\phi^\dagger(x)\phi(x) : . \tag{13.2.1}$$

In line with the remark following Eq. (12.2.24), $\phi(x)$ and $\phi^\dagger(x)$ are treated as independent fields. Hence, we have, for example, $\frac{\partial \mathcal{L}}{\partial \phi^\dagger_{,\mu}(x)} = \phi^{,\mu}(x)$, and, from the Euler–Lagrange equations (12.2.15), the equations of motion

$$(\partial^\mu\partial_\mu + m^2)\phi(x) = 0 \quad \text{and} \quad (\partial^\mu\partial_\mu + m^2)\phi^\dagger(x) = 0 . \tag{13.2.2}$$

The conjugate fields of $\phi(x)$ and $\phi^\dagger(x)$ are, according to (12.2.16),

$$\pi(x) = \dot\phi^\dagger(x) \quad \text{and} \quad \pi^\dagger(x) = \dot\phi(x) . \tag{13.2.3}$$

Since the complex Klein–Gordon field also behaves as a scalar under Lorentz transformations, it has spin = 0. Due to the gauge invariance of $\mathcal{L}$, this field possesses an additional conserved quantity, namely the charge Q. The equal time commutators of the fields and their adjoints are, according to canonical quantization (12.3.1),

$$\begin{aligned} \left[\phi(\mathbf{x},t), \dot\phi^\dagger(\mathbf{x}',t)\right] &= i\,\delta(\mathbf{x} - \mathbf{x}') \\ \left[\phi^\dagger(\mathbf{x},t), \dot\phi(\mathbf{x}',t)\right] &= i\,\delta(\mathbf{x} - \mathbf{x}') \end{aligned} \tag{13.2.4}$$

and

$$\begin{aligned} \left[\phi(\mathbf{x},t), \phi(\mathbf{x}',t)\right] &= \left[\phi(\mathbf{x},t), \phi^\dagger(\mathbf{x}',t)\right] \\ = \left[\dot\phi(\mathbf{x},t), \dot\phi(\mathbf{x}',t)\right] &= \left[\dot\phi(\mathbf{x},t), \dot\phi^\dagger(\mathbf{x}',t)\right] = 0 . \end{aligned}$$

The solutions of the field equations (13.2.2) for the complex Klein–Gordon field are also of the form $e^{\pm ikx}$, so that the expansion of the field operator takes the form

$$\phi(x) = \phi^+(x) + \phi^-(x) = \sum_{\mathbf{k}} \frac{1}{(2V\omega_{\mathbf{k}})^{1/2}} \left(a_{\mathbf{k}}e^{-ikx} + b^\dagger_{\mathbf{k}}e^{ikx}\right) , \tag{13.2.5a}$$

where, in contrast to the real Klein–Gordon field, the amplitudes $b^\dagger_{\mathbf{k}}$ and $a_{\mathbf{k}}$ are now independent of one another. From (13.2.5a) we have

$$\phi^\dagger(x) = \phi^{\dagger^+}(x) + \phi^{\dagger^-}(x) = \sum_{\mathbf{k}} \frac{1}{(2V\omega_{\mathbf{k}})^{1/2}} \left(b_{\mathbf{k}}e^{-ikx} + a^\dagger_{\mathbf{k}}e^{ikx}\right) .$$

$$\tag{13.2.5b}$$

In (13.2.5a,b), the operators $\phi(x)$ and $\phi^\dagger(x)$ are split into their positive (e^{-ikx}) and negative (e^{ikx}) frequency components. Taking the inverse of the Fourier series (13.2.5a,b) and using (13.2.4), one finds the commutation relations

$$\left[a_{\mathbf{k}}, a_{\mathbf{k}'}^\dagger\right] = \left[b_{\mathbf{k}}, b_{\mathbf{k}'}^\dagger\right] = \delta_{\mathbf{kk}'}$$
$$\left[a_{\mathbf{k}}, a_{\mathbf{k}'}\right] = \left[b_{\mathbf{k}}, b_{\mathbf{k}'}\right] = \left[a_{\mathbf{k}}, b_{\mathbf{k}'}\right] = \left[a_{\mathbf{k}}, b_{\mathbf{k}'}^\dagger\right] = 0 \ .$$

(13.2.6)

One now has two occupation-number operators, for particles a and for particles b

$$\hat{n}_{a\mathbf{k}} = a_{\mathbf{k}}^\dagger a_{\mathbf{k}} \qquad \text{and} \qquad \hat{n}_{b\mathbf{k}} = b_{\mathbf{k}}^\dagger b_{\mathbf{k}} \ .$$

(13.2.7)

The operators $a_{\mathbf{k}}^\dagger, a_{\mathbf{k}}$ create and annihilate particles of type a, whilst $b_{\mathbf{k}}^\dagger, b_{\mathbf{k}}$ create and annihilate particles of type b, in each case the wave vector being $\mathbf{k}$. The vacuum state $|0\rangle$ is defined by

$$a_{\mathbf{k}} |0\rangle = b_{\mathbf{k}} |0\rangle = 0 \qquad \text{for all } \mathbf{k} \ ,$$

(13.2.8a)

or, equivalently,

$$\phi^+(x) |0\rangle = \phi^{\dagger^+}(x) |0\rangle = 0 \qquad \text{for all } x \ .$$

(13.2.8b)

One obtains for the four-momentum

$$P^\mu = \sum_{\mathbf{k}} k^\mu \left(\hat{n}_{a\mathbf{k}} + \hat{n}_{b\mathbf{k}}\right) \ ,$$

(13.2.9)

whose zero component, having $k^0 = \omega_{\mathbf{k}}$, represents the Hamiltonian. On account of the invariance of the Lagrangian density under gauge transformations of the first kind, the charge

$$Q = -iq \int d^3x \ : \dot{\phi}^\dagger(x)\phi(x) - \dot{\phi}(x)\phi^\dagger(x) \ :$$

(13.2.10)

is conserved. The corresponding four-current-density is of the form

$$j^\mu(x) = -iq \left(: \frac{\partial \phi^\dagger}{\partial x_\mu} \phi - \frac{\partial \phi}{\partial x_\mu} \phi^\dagger :\right)$$

(13.2.11)

and satisfies the continuity equation

$$j^\mu_{,\mu} = 0 \ .$$

(13.2.12)

Substituting the expansions (13.2.5a,b) into Q, one obtains

$$Q = q \sum_{\mathbf{k}} (\hat{n}_{a\mathbf{k}} - \hat{n}_{b\mathbf{k}}) \ . \tag{13.2.13}$$

The charge operator commutes with the Hamiltonian. The a particles have charge q, and the b particles charge $-q$. Except for the sign of their charge, these particles have identical properties. The interchange $a \leftrightarrow b$ changes only the sign of Q. In relativistic quantum field theory, every charged particle is automatically accompanied by an *antiparticle* carrying opposite charge. This is a general result in field theory and also applies to particles with other spin values. It is also confirmed by experiment.

An example of a particle–antiparticle pair are the charged π mesons π^+ and π^- which have electric charges $+e_0$ and $-e_0$. However, the charge need not necessarily be an electrical charge: The electrically neutral K^0 meson has an antiparticle $\bar{K}^0$, which is also electrically neutral. These two particles carry opposite *hypercharge*: $Y = 1$ for the K^0 and $Y = -1$ for the $\bar{K}^0$, and are described by a complex Klein–Gordon field. The hypercharge[1] is a charge-like intrinsic degree of freedom, which is related to other intrinsic quantum numbers, namely the electrical charge Q, the isospin I_z, the strangeness S, and the baryon number N, by

$$Y = 2(Q - I_z)$$

and

$$S = Y - N \ .$$

The hypercharge is conserved for the strong, but not for the weak interactions. However, since the latter is weaker by a factor of about 10^{-12}, the hypercharge is very nearly conserved. The electrical charge is always conserved perfectly! The physical significance of the charge of a free field will become apparent when we consider the interaction with other fields. The sign and magnitude of the charge will then play a role.

13.3 Quantization of the Dirac Field

13.3.1 Field Equations

The quantized Klein–Gordon equation provides a description of mesons and, simultaneously, difficulties in its interpretation as a quantum-mechanical wave equation were overcome. Similarly, we shall consider the Dirac equation (5.3.20) as a classical field equation to be quantized:

---

[1] See, e.g., E. Segrè, *Nuclei and Particles*, 2nd ed., Benjamin/Cummins, London (1977), O. Nachtmann, *Elementary Particle Physics*, Springer, Heidelberg (1990)

$$(i\gamma\partial - m)\psi = 0 \quad \text{and} \quad \bar{\psi}(i\gamma\overleftarrow{\partial} + m) = 0 \ . \tag{13.3.1}$$

The arrow above the ∂ in the second equation signifies that the differentiation acts to the left on the $\bar{\psi}$. The second equation is obtained by taking the adjoint of the first, and using the relations $\bar{\psi} = \psi^{\dagger}\gamma^{0}$ and $\gamma^{0}\gamma_{\mu}^{\dagger}\gamma_{0} = \gamma_{\mu}$. A possible Lagrangian density for these field equations is

$$\mathcal{L} = \bar{\psi}(x)(i\gamma^{\mu}\partial_{\mu} - m)\psi(x) \ , \tag{13.3.2}$$

which may be verified from

$$\frac{\partial\mathcal{L}}{\partial\bar{\psi}} - \partial_{\mu}\frac{\partial\mathcal{L}}{\partial(\partial_{\mu}\bar{\psi})} = (i\gamma^{\mu}\partial_{\mu} - m)\psi = 0$$

and

$$\frac{\partial\mathcal{L}}{\partial\psi} - \partial_{\mu}\frac{\partial\mathcal{L}}{\partial(\partial_{\mu}\psi)} = -m\bar{\psi} - \partial_{\mu}\bar{\psi}\,i\gamma^{\mu} = 0 \ . \tag{13.3.3}$$

The Lagrangian density (13.3.2) is not real, but differs from a real one only by a four-divergence:

$$\mathcal{L} = \frac{i}{2}\left[\bar{\psi}\gamma^{\mu}\partial_{\mu}\psi - (\partial_{\mu}\bar{\psi})\gamma^{\mu}\psi\right] - m\bar{\psi}\psi + \frac{i}{2}\partial_{\mu}(\bar{\psi}\gamma^{\mu}\psi)$$

$$\mathcal{L}^{*} = -\frac{i}{2}\left[(\partial_{\mu}\psi^{\dagger})\gamma_{0}^{2}\gamma^{\mu\dagger}\gamma_{0}\psi - \psi^{\dagger}\gamma_{0}^{2}\gamma^{\mu\dagger}\partial_{\mu}\gamma^{0}\psi\right]$$

$$\qquad - m\bar{\psi}\psi + (\frac{i}{2}\partial_{\mu}(\bar{\psi}\gamma^{\mu}\psi))^{\dagger}$$

$$= -\frac{i}{2}\left[(\partial_{\mu}\bar{\psi})\gamma^{\mu}\psi - \bar{\psi}\gamma^{\mu}\partial_{\mu}\psi\right] - m\bar{\psi}\psi - (\frac{i}{2}\partial_{\mu}(\bar{\psi}\gamma^{\mu}\psi)) \ . \tag{13.3.4}$$

The first three terms in (13.3.4), taken together, are real and could also be used as the Lagrangian density, since the last non-real term is a four-divergence and makes no contribution to the Euler–Lagrange equations of motion.

The conjugate fields following from (13.3.2) are:

$$\pi_{\alpha}(x) = \frac{\partial\mathcal{L}}{\partial\dot{\psi}_{\alpha}} = i\psi_{\alpha}^{\dagger}$$

$$\bar{\pi}_{\alpha}(x) = \frac{\partial\mathcal{L}}{\partial\dot{\bar{\psi}}_{\alpha}} = 0 \ . \tag{13.3.5}$$

Here, there is already an indication that the previously applied canonical quantization is not going to work for the Dirac equation because

$$[\bar{\psi}_{\alpha}(x), \bar{\pi}_{\alpha}(x')] = \bar{\psi}_{\alpha} \cdot 0 - 0 \cdot \bar{\psi}_{\alpha} = 0 \neq \delta(\mathbf{x} - \mathbf{x}') \ .$$

Furthermore, particles with $S = \frac{1}{2}$ are fermions and not bosons and, in the nonrelativistic limit, these were quantized by means of anticommutation relations. The Hamiltonian density resulting from (13.3.2) is

$$\mathcal{H} = \pi_\alpha \dot{\psi}_\alpha - \mathcal{L} = i\psi_\alpha^\dagger \dot{\psi}_\alpha - \bar{\psi}(i\gamma^\mu \partial_\mu - m)\psi \tag{13.3.6}$$
$$= -i\bar{\psi}\gamma^j \partial_j \psi + m\bar{\psi}\psi$$

and the Hamiltonian reads:

$$H = \int d^3x \, \bar{\psi}(x) \left(-i\gamma^j \partial_j + m\right) \psi(x) . \tag{13.3.7}$$

13.3.2 Conserved Quantities

For the energy–momentum tensor (12.4.1) one obtains from (13.3.2)

$$T^{\mu\nu} = (\partial^\nu \bar{\psi}) \frac{\partial \mathcal{L}}{\partial(\partial_\mu \bar{\psi})} + \frac{\partial \mathcal{L}}{\partial(\partial_\mu \psi)} \partial^\nu \psi - g^{\mu\nu} \mathcal{L}$$
$$= 0 + \bar{\psi} i\gamma^\mu \partial^\nu \psi - g^{\mu\nu} \bar{\psi}(i\gamma\partial - m)\psi \tag{13.3.8}$$
$$= i\bar{\psi}\gamma^\mu \partial^\nu \psi .$$

Since the Lagrangian density does not contain the derivative $\partial_\mu \bar{\psi}$, the first term in (13.3.8) vanishes. To obtain the final line, we have made use of the fact that the Lagrangian density vanishes for every solution of the Dirac equation. The order of the factors in (13.3.8) is arbitrary. As long as we are dealing only with a classical field theory, the order is irrelevant. Later, we shall introduce normal ordering.

According to (12.4.8), the momentum density follows from (13.3.8) as

$$\mathcal{P}^\mu = T^{0\mu} \tag{13.3.9}$$

and the momentum as

$$P^\mu = \int d^3x \, T^{0\mu} = i \int d^3x \, \bar{\psi}(x)\gamma^0 \partial^\mu \psi(x) . \tag{13.3.10}$$

The zero component, in particular, is given by

$$P^0 = i \int d^3x \, \bar{\psi}(x)\gamma^0 \partial_0 \psi = i \int d^3x \, \psi^\dagger \partial_0 \psi = H . \tag{13.3.11}$$

This result is identical to the Hamiltonian H of Eq. (13.3.7), as can be seen by using the Dirac equation.

Finally, we consider the angular momentum determined by (12.4.24): If, in the general relation

$$M^{\nu\sigma} = \int d^3x \left(\pi_r(x) S_{rs}^{\nu\sigma} \phi_s + x^\nu T^{0\sigma} - x^\sigma T^{0\nu} \right) ,$$

one substitutes the spinor field for ϕ_s, Eq. (13.3.5) for π_r, and Eq. (12.4.12b) for S, one then obtains

$$M^{\nu\sigma} = \int d^3x \left(i\psi_\alpha^\dagger \left(-\frac{i}{2}\right) \sigma_{\alpha\beta}^{\nu\sigma} \psi_\beta + x^\nu i\psi^\dagger \partial^\sigma \psi - x^\sigma i\psi^\dagger \partial^\nu \psi \right)$$

$$= \int d^3x\, \psi^\dagger \left(ix^\nu \partial^\sigma - ix^\sigma \partial^\nu + \frac{1}{2}\sigma^{\nu\sigma} \right) \psi \,.$$

$$(13.3.12)$$

For the spatial components, this yields:

$$M^{ij} = \int d^3x\, \psi^\dagger \left(\underbrace{x^i \frac{1}{i}\frac{\partial}{\partial x^j} - x^j \frac{1}{i}\frac{\partial}{\partial x^i}}_{\text{orbital angular momentum}} + \underbrace{\frac{1}{2}\sigma^{ij}}_{\text{spin}} \right) \psi \,, \qquad (13.3.13)$$

which can be combined to form the angular momentum vector

$$\mathbf{M} = (M^{23}, M^{31}, M^{12})$$

$$= \int d^3x\, \psi^\dagger(x) \left(\mathbf{x} \times \frac{1}{i}\nabla + \frac{1}{2}\mathbf{\Sigma} \right) \psi(x) \,.$$

$$(13.3.13')$$

The first term represents the orbital angular momentum and the second the spin, with $\mathbf{\Sigma}$ represented by the Pauli spin matrices in (6.2.29d).

13.3.3 Quantization

It will prove useful here to modify the definition of plane wave spinors. Instead of the spinors $v_r(k)$, $r = 1, 2$, we will now adopt the notation

$$w_r(k) = \begin{cases} v_2(k) & \text{for } r = 1 \\ -v_1(k) & \text{for } r = 2, \end{cases} \qquad (13.3.14)$$

where the $v_r(k)$ are given in Eq. (6.3.11b), and hence

$$u_r(k) = \left(\frac{E+m}{2m} \right)^{\frac{1}{2}} \begin{pmatrix} \chi_r \\ \frac{\sigma\cdot\mathbf{k}}{m+E}\chi_r \end{pmatrix} \qquad (13.3.15a)$$

$$w_r(k) = -\left(\frac{E+m}{2m} \right)^{\frac{1}{2}} \begin{pmatrix} \frac{\sigma\cdot\mathbf{k}}{m+E}i\sigma^2\chi_r \\ i\sigma^2\chi_r \end{pmatrix}, \qquad (13.3.15b)$$

with $i\sigma^2 \equiv \begin{pmatrix} 0 & 1 \\ -1 & 0 \end{pmatrix}$ and $\chi_1 = \begin{pmatrix} 1 \\ 0 \end{pmatrix}$, $\chi_2 = \begin{pmatrix} 0 \\ 1 \end{pmatrix}$. This definition is motivated by the ideas of Hole theory (Sect. 10.2) and implies that relations involving the spin have the same form for both electrons and positrons. An electron with spinor $u_{\frac{1}{2}}(m, \mathbf{0})$ and a positron with spinor $w_{\frac{1}{2}}(m, \mathbf{0})$ both have spin $\pm\frac{1}{2}$, i.e., for the operator $\frac{1}{2}\Sigma^3$ they have the eigenvalues $\pm\frac{1}{2}$ and the effect of $\frac{1}{2}\mathbf{\Sigma}$ on electron and positron states is of the same form. Given this definition, the charge conjugation operation $\mathcal{C}$ transforms the spinors $u_r(k)$ into $w_r(k)$ and vice versa:

$$\mathcal{C}u_r(k) = i\gamma^2 u_r(k)^* = w_r(k), \qquad r = 1,2$$
$$\mathcal{C}w_r(k) = i\gamma^2 w_r(k)^* = u_r(k), \qquad r = 1,2 \, . \tag{13.3.15c}$$

In the new notation, the orthogonality relations (6.3.15) and (6.3.19a–c) acquire the form

$$\bar{u}_r(k)u_s(k) = \delta_{rs} \qquad \bar{u}_r(k)w_s(k) = 0$$
$$\bar{w}_r(k)w_s(k) = -\delta_{rs} \qquad \bar{w}_r(k)u_s(k) = 0 \tag{13.3.16}$$

and

$$\bar{u}_r(k)\gamma^0 u_s(k) = \frac{E}{m}\delta_{rs} \qquad \bar{u}_r(\tilde{k})\gamma^0 w_s(k) = 0$$
$$\bar{w}_r(k)\gamma^0 w_s(k) = \frac{E}{m}\delta_{rs} \qquad \bar{w}_r(\tilde{k})\gamma^0 u_s(k) = 0 \, , \tilde{k} = (k^0, -\mathbf{k}) \, . \tag{13.3.17}$$

Relations that are bilinear in $v_r(k)$, e.g., the projections (6.3.23), have the same form in $w_r(k)$. We now turn to the representation of the field as a superposition of free solutions in a finite volume V:

$$\psi(x) = \sum_{\mathbf{k},r} \left(\frac{m}{VE_\mathbf{k}}\right)^{1/2} \left(b_{r\mathbf{k}} u_r(k)\, e^{-ikx} + d^\dagger_{r\mathbf{k}} w_r(k)\, e^{ikx}\right) \tag{13.3.18a}$$
$$\equiv \psi^+(x) + \psi^-(x) \, ,$$

with

$$E_\mathbf{k} = (\mathbf{k}^2 + m^2)^{1/2} \, , \tag{13.3.19}$$

where the last line indicates the decomposition into positive and negative frequency contributions. In classical field theory, the amplitudes $b_{r\mathbf{k}}$ and $d_{r\mathbf{k}}$ are complex numbers, as in (10.1.9), and hermitian conjugation becomes simply complex conjugation, i.e., $d^\dagger_{r\mathbf{k}} = d^*_{r\mathbf{k}}$. Below, we shall quantize $\psi(x)$ and $\bar{\psi}(x)$, and then the amplitudes $b_{r\mathbf{k}}$ and $d_{r\mathbf{k}}$ will be replaced by operators. The relations (13.3.18a,b) are written in such a way that they also remain valid as an operator expansion. For the adjoint field (the adjoint field operator) $\bar{\psi}(x) = \psi^\dagger(x)\gamma^0$, one obtains from (13.3.18a)

$$\bar{\psi}(x) = \sum_{\mathbf{k},r} \left(\frac{m}{VE_\mathbf{k}}\right)^{1/2} \left(d_{r\mathbf{k}} \bar{w}_r(k)\, e^{-ikx} + b^\dagger_{r\mathbf{k}} \bar{u}_r(k)\, e^{ikx}\right) \, . \tag{13.3.18b}$$

Inserting (13.3.18a,b) into (13.3.10) yields for the momentum

$$P^\mu = \sum_{\mathbf{k},r} k^\mu \left(b^\dagger_{r\mathbf{k}} b_{r\mathbf{k}} - d_{r\mathbf{k}} d^\dagger_{r\mathbf{k}}\right) \, , \tag{13.3.20}$$

as the following algebra shows:

$$P^\mu = i \int d^3x\, \bar\psi \gamma^0 \partial^\mu \psi = i \int d^3x \sum_{\mathbf{k},r} \sum_{\mathbf{k}',r'} \left(\frac{m}{VE_\mathbf{k}}\right)^{\frac{1}{2}} \left(\frac{m}{VE_{\mathbf{k}'}}\right)^{\frac{1}{2}}$$

$$\times \left[b^\dagger_{r'\mathbf{k}'} \bar{u}_{r'}(k') e^{ik'x} + d_{r'\mathbf{k}'} \bar{w}_{r'}(k') e^{-ik'x}\right]$$

$$\times \gamma^0 \partial^\mu \left[b_{r\mathbf{k}} u_r(k) e^{-ikx} + d^\dagger_{r\mathbf{k}} w_r(k) e^{ikx}\right]$$

$$= i \sum_{\mathbf{k},r} \sum_{\mathbf{k}',r'} \left(\frac{m}{E_\mathbf{k}} \frac{m}{E_{\mathbf{k}'}}\right)^{\frac{1}{2}} \Big\{ \delta_{\mathbf{k}\mathbf{k}'} \Big(-ik^\mu b^\dagger_{r'\mathbf{k}'} b_{r\mathbf{k}} \bar{u}_{r'}(k') \gamma^0 u_r(k)$$

$$+ ik^\mu d_{r'\mathbf{k}'} d^\dagger_{r\mathbf{k}} \bar{w}_{r'}(k') \gamma^0 w_r(k) \Big)$$

$$+ \delta_{\mathbf{k},-\mathbf{k}'} \Big(ik^\mu e^{i(k_0+k_0')x_0} b^\dagger_{r'\mathbf{k}'} d^\dagger_{r\mathbf{k}} \bar{u}_{r'}(k') \gamma^0 w_r(k)$$

$$- ik^\mu e^{-i(k_0+k_0')x_0} d_{r'\mathbf{k}'} b_{r\mathbf{k}} \bar{w}_{r'}(k') \gamma^0 u_r(k) \Big) \Big\} \,.$$

$$(13.3.21)$$

In the first term after the last identity we have immediately set $e^{\pm i(k_0-k_0')x_0} = 1$, on account of $\delta_{\mathbf{k}\mathbf{k}'}$, (and thus $k_0' = \sqrt{\mathbf{k}' + m^2} = k_0$). The orthogonality relations for the u and w yield the assertion (13.3.20).

In quantized field theory, $b_{r\mathbf{k}}$ and $d_{r\mathbf{k}}$ are operators. What are their algebraic properties? These we can determine using the result (13.3.20) for the Hamiltonian

$$H = P^0 = \sum_{\mathbf{k},r} k_0 \left(b^\dagger_{r\mathbf{k}} b_{r\mathbf{k}} - d_{r\mathbf{k}} d^\dagger_{r\mathbf{k}}\right) \,. \qquad (13.3.22)$$

If, as in the Klein–Gordon theory, commutation rules of the form

$$\left[d_{r\mathbf{k}}, d^\dagger_{r\mathbf{k}'}\right] = \delta_{\mathbf{k}\mathbf{k}'} \,,$$

held, there would be no lower bound to the energy. (It would not help to replace $d_r^\dagger$ in the expansion of the field by an annihilation operator e_r, as H would still not be positive definite.) A system described by this Hamiltonian would not be stable; the excitation of particles by the operator $d^\dagger_{r\mathbf{k}}$ would reduce its energy! The way out of this dilemma is to demand anticommutation rules:

$$\left\{b_{r\mathbf{k}}, b^\dagger_{r'\mathbf{k}'}\right\} = \delta_{rr'} \delta_{\mathbf{k}\mathbf{k}'}$$

$$\left\{d_{r\mathbf{k}}, d^\dagger_{r'\mathbf{k}'}\right\} = \delta_{rr'} \delta_{\mathbf{k}\mathbf{k}'} \qquad (13.3.23)$$

$$\left\{b_{r\mathbf{k}}, b_{r'\mathbf{k}'}\right\} = \left\{d_{r\mathbf{k}}, d_{r'\mathbf{k}'}\right\} = \left\{d_{r\mathbf{k}}, b_{r'\mathbf{k}'}\right\} = \left\{b_{r\mathbf{k}}, d^\dagger_{r'\mathbf{k}'}\right\} = 0 \,.$$

That anticommutation rules should apply for fermions comes as no surprise in view of the nonrelativistic many-particle theory discussed in Part I. The second term in (13.3.22) then becomes $-d_{r\mathbf{k}} d^\dagger_{r\mathbf{k}} = d^\dagger_{r\mathbf{k}} d_{r\mathbf{k}} - 1$, so that the creation of a d particle makes a positive energy contribution. The anticommutation relations (13.3.23) imply that each state can be, at most, singly occupied, i.e., the occupation-number operators $\hat{n}^{(b)}_{r\mathbf{k}} = b^\dagger_{r\mathbf{k}} b_{r\mathbf{k}}$ and $\hat{n}^{(d)}_{r\mathbf{k}} = d^\dagger_{r\mathbf{k}} d_{r\mathbf{k}}$ have the eigenvalues (occupation numbers) $n^{(b,d)}_{r\mathbf{k}} = 0, 1$. To avoid zero-point terms, we have also introduced normal products for the Dirac field. The definition of the normal ordering for fermions reads: All annihilation operators are written to the right of all creation operators, whereby each permutation contributes a factor (-1). Let us illustrate this definition with an example:

$$: \psi_\alpha \psi_\beta : = : \left(\psi^+_\alpha + \psi^-_\alpha \right) \left(\psi^+_\beta + \psi^-_\beta \right) :$$
$$= \psi^+_\alpha \psi^+_\beta - \psi^-_\beta \psi^+_\alpha + \psi^-_\alpha \psi^+_\beta + \psi^-_\alpha \psi^-_\beta \ . \tag{13.3.24}$$

All observables, e.g., (13.3.10) or (13.3.22) are defined as normal products, i.e., the final Hamiltonian is defined by $H = \; : H_{\mathrm{original}} : $ and $\mathbf{P} = \; : \mathbf{P}_{\mathrm{original}} :$, where H_{original} and $\mathbf{P}_{\mathrm{original}}$ refer to the expressions in (13.3.22) and (13.3.10). We thus have

$$H = \sum_{\mathbf{k},r} E_\mathbf{k} \left(b^\dagger_{r\mathbf{k}} b_{r\mathbf{k}} + d^\dagger_{r\mathbf{k}} d_{r\mathbf{k}} \right) \tag{13.3.25}$$

$$\mathbf{P} = \sum_{\mathbf{k},r} \mathbf{k} \left(b^\dagger_{r\mathbf{k}} b_{r\mathbf{k}} + d^\dagger_{r\mathbf{k}} d_{r\mathbf{k}} \right) \ . \tag{13.3.26}$$

The operators $b^\dagger_{r\mathbf{k}}$ ($b_{r\mathbf{k}}$) create (annihilate) an electron with spinor $u_r(k)e^{-ikx}$ and the operators $d^\dagger_{r\mathbf{k}}$ ($d_{r\mathbf{k}}$) create (annihilate) a positron in the state $w_r(k)e^{ikx}$. From (13.3.25) and (13.3.26) and the corresponding representation of the angular momentum operator, it is clear that the d particles – now already called positrons – have the same energy, momentum, and spin degrees of freedom as the electrons. For their complete characterization, we still have to consider their charge.

13.3.4 Charge

We start from the general formula (12.4.31) which yields for the *charge* Q

$$Q = -\mathrm{i}\, q \int d^3 x (\pi \psi - \bar\psi \bar\pi) = -\mathrm{i}\, q \int d^3 x \mathrm{i} \psi^\dagger_\alpha \psi_\alpha$$
$$= q \int d^3 x \, \bar\psi \gamma_0 \psi \ . \tag{13.3.27a}$$

The associated four-current-density is of the form

$$j^\mu(x) = q \; : \bar\psi(x) \gamma^\mu \psi(x) : \tag{13.3.27b}$$

and satisfies the continuity equation

$$j^\mu_{,\mu} = 0 \ . \tag{13.3.27c}$$

Setting $q = -e_0$ for the electron, we obtain from (13.3.27a) the modified definition of Q in terms of normal products

$$
\begin{aligned}
Q &= -e_0 \int d^3x : \bar{\psi}(x)\gamma^0\psi(x) : \equiv \int d^3x j^0(x) \\
&= -e_0 \sum_{\mathbf{k}} \sum_{r=1,2} \left(b^\dagger_{r\mathbf{k}} b_{r\mathbf{k}} - d^\dagger_{r\mathbf{k}} d_{r\mathbf{k}} \right) .
\end{aligned}
\tag{13.3.28}
$$

In the last identity the expansions (13.3.18a) and (13.3.18b) have been inserted and evaluated as in Eq.(13.3.21). The difference in sign between (13.3.28) and the Hamiltonian (13.3.25) stems from the fact that in (13.3.21) the differential operators ∂^μ produce different signs for the positive and negative frequency components, which, however, are compensated by the anti-commutation. From (13.3.28), it is already evident that the d particles, i.e., the positrons, have opposite charge to the electrons. This is further confirmed by the following argument:

$$
\begin{aligned}
\left[Q, b^\dagger_{r\mathbf{k}} \right] &= -e_0 b^\dagger_{r\mathbf{k}} \\
\left[Q, d^\dagger_{r\mathbf{k}} \right] &= e_0 d^\dagger_{r\mathbf{k}} .
\end{aligned}
\tag{13.3.29}
$$

We consider a state $|\Psi\rangle$, which is taken to be an eigenstate of the charge operator with eigenvalue q:

$$
Q|\Psi\rangle = q|\Psi\rangle .
\tag{13.3.30}
$$

From (13.3.29) it then follows that

$$
\begin{aligned}
Q b^\dagger_{r\mathbf{k}} |\Psi\rangle &= (q - e_0) b^\dagger_{r\mathbf{k}} |\Psi\rangle \\
Q d^\dagger_{r\mathbf{k}} |\Psi\rangle &= (q + e_0) d^\dagger_{r\mathbf{k}} |\Psi\rangle .
\end{aligned}
\tag{13.3.31}
$$

The state $b^\dagger_{r\mathbf{k}} |\Psi\rangle$ has the charge $(q - e_0)$ and the state $d^\dagger_{r\mathbf{k}} |\Psi\rangle$ the charge $(q + e_0)$. Hence, using (13.3.18b), we also conclude that

$$
Q \bar{\psi}(x) |\Psi\rangle = (q - e_0) \bar{\psi}(x) |\Psi\rangle .
\tag{13.3.32}
$$

The creation of an electron or the annihilation of a positron reduces the charge by e_0. The vacuum state $|0\rangle$ has zero charge.

The charge operator, as a conserved quantity, commutes with the Hamiltonian and is time independent. As can be seen directly from (13.3.28) and (13.3.26), it also commutes with the momentum vector $\mathbf{P}$, which can be written

$$
[Q, P^\mu] = 0 .
\tag{13.3.33}
$$

Hence, there exist joint eigenfunctions of the charge and the momentum operators. In the course of our attempt, in Part II, to construct a relativistic wave equation, and to interpret ψ in analogy to the Schrödinger wave function as a probability amplitude, we interpreted $j^0 = \psi^\dagger \psi$ as a positive density; H, however, was indefinite. In the quantum field theoretic form, Q is indefinite, which is acceptable for the charge, and the Hamiltonian is positive definite. This leads to a physically meaningful picture: $\psi(x)$ is not the state, but rather a field operator that creates and annihilates particles. The states are given by the states in Fock space, i.e., $|0\rangle$, $b^\dagger_{r\mathbf{k}}|0\rangle$, $b^\dagger_{r_1\mathbf{k}_1}d^\dagger_{r_2\mathbf{k}_2}|0\rangle$, $b^\dagger_{r_1\mathbf{k}_1}b^\dagger_{r_2\mathbf{k}_2}d^\dagger_{r_3\mathbf{k}_3}|0\rangle$, etc. The operator $b^\dagger_{r\mathbf{k}=0}$ with $r = 1(r = 2)$ creates an electron at rest with spin in the z direction $s_z = \frac{1}{2}(s_z = -\frac{1}{2})$. Likewise, $d^\dagger_{r\mathbf{k}=0}$ creates a positron at rest with $s_z = \frac{1}{2}(s_z = -\frac{1}{2})$. Correspondingly, $b^\dagger_{r\mathbf{k}}$ ($d^\dagger_{r\mathbf{k}}$) creates an electron (positron) with momentum $\mathbf{k}$, which, in its rest frame, possesses the spin $\frac{1}{2}$ for $r = 1$ and $-\frac{1}{2}$ for $r = 2$ (see Problem 13.11).

*13.3.5 The Infinite-Volume Limit

When using the Dirac field operators, we will always consider a finite volume, i.e., in their expansion in terms of creation and annihilation operators we have sums rather than integrals over the momentum. The infinite-volume limit will only be introduced in the final results; for example, in the scattering cross-section. However, there are some circumstances in which it is convenient to work with an infinite volume from the outset. Equation (13.3.18a) then becomes[2]

$$\psi(x) = \int \frac{d^3k}{(2\pi)^3} \frac{\sqrt{m}}{k_0} \sum_{r=1,2} \left(b_r(\mathbf{k})u_r(k)e^{-ikx} + d^\dagger_r(\mathbf{k})w_r(k)e^{ikx} \right) \ . \tag{13.3.34}$$

The annihilation and creation operators are related to their finite-volume counterparts by

$$b_r(\mathbf{k}) = \sqrt{k_0 V}\, b_{r\mathbf{k}} \quad , \quad d_r(\mathbf{k}) = \sqrt{k_0 V}\, d_{r\mathbf{k}} \ . \tag{13.3.35}$$

These operators thus satisfy the anticommutation relations

$$\left\{ b_r(\mathbf{k}), b^\dagger_{r'}(\mathbf{k}') \right\} = (2\pi)^3 k_0 \delta_{rr'} \delta^{(3)}(\mathbf{k} - \mathbf{k}')$$
$$\left\{ d_r(\mathbf{k}), d^\dagger_{r'}(\mathbf{k}') \right\} = (2\pi)^3 k_0 \delta_{rr'} \delta^{(3)}(\mathbf{k} - \mathbf{k}') \ , \tag{13.3.36}$$

and all other anticommutators vanish. The momentum operator has the form

$$P^\mu = \int \frac{d^3k}{(2\pi)^3} \frac{k^\mu}{k_0} \sum_{r=1,2} b^\dagger_r(\mathbf{k})b_r(\mathbf{k}) + d^\dagger_r(\mathbf{k})d_r(\mathbf{k}) \ . \tag{13.3.37}$$

---

[2] The factor $\sqrt{m}$ in (13.3.34) is chosen, as in (13.3.18a), in order to cancel the corresponding factor $1/\sqrt{m}$ in the spinors, so that the limit $m \to 0$ exists.

We have

$$\{P^\mu, b_r^\dagger(\mathbf{k})\} = k^\mu b_r^\dagger(\mathbf{k}) , \quad \{P^\mu, b_r(\mathbf{k})\} = -k^\mu b_r(\mathbf{k}) ,$$
$$\{P^\mu, d_r^\dagger(\mathbf{k})\} = k^\mu d_r^\dagger(\mathbf{k}) , \quad \{P^\mu, d_r(\mathbf{k})\} = -k^\mu d_r^\dagger(\mathbf{k}) . \qquad (13.3.38)$$

From (13.3.38), one sees directly that the one-electron (positron) state $b_r^\dagger(\mathbf{k}) |0\rangle$ ($d_r^\dagger(\mathbf{k}) |0\rangle$) possesses momentum k^μ.

13.4 The Spin Statistics Theorem

13.4.1 Propagators and the Spin Statistics Theorem

We are now in a position to prove the spin statistics theorem, which relates the values of the spin to the statistics (i.e., to the commutation properties and, hence, to the possible occupation numbers). By way of preparation we calculate the anticommutator of the Dirac field operators, where α and β stand for the spinor indices $1, \dots, 4$. Making use of the anticommutation relations (13.3.23), the projectors (6.3.23), and Eq. (6.3.21) we find

$$\{\psi_\alpha(x), \bar\psi_{\alpha'}(x')\} = \frac{1}{V} \sum_{\mathbf{k}} \sum_{\mathbf{k}'} \left(\frac{mm}{E_{\mathbf{k}} E_{\mathbf{k}'}}\right)^{1/2} \sum_r \sum_{r'} \delta_{rr'} \delta_{\mathbf{k}\mathbf{k}'}$$

$$\times \left(u_{r\alpha}(k) \bar u_{r'\alpha'}(k') e^{-ikx} e^{ik'x'} + w_{r\alpha}(k) \bar w_{r'\alpha'}(k') e^{ikx} e^{-ik'x'} \right)$$

$$= \frac{1}{V} \sum_{\mathbf{k}} \frac{m}{E_{\mathbf{k}}} \left(e^{-ik(x-x')} \sum_r u_{r\alpha}(k) \bar u_{r\alpha'}(k) \right.$$

$$\left. + e^{ik(x-x')} \sum_r w_{r\alpha}(k) \bar w_{r\alpha'}(k) \right)$$

$$= \int \frac{d^3k}{(2\pi)^3} \frac{m}{E_{\mathbf{k}}} \left(e^{-ik(x-x')} \left(\frac{\slashed{k} + m}{2m}\right)_{\alpha\alpha'} \right. \qquad (13.4.1)$$

$$\left. + e^{ik(x-x')} \left(\frac{\slashed{k} - m}{2m}\right)_{\alpha\alpha'} \right)$$

$$= (i\slashed\partial + m)_{\alpha\alpha'} \frac{1}{2} \int \frac{d^3k}{(2\pi)^3} \frac{1}{E_{\mathbf{k}}} \left(e^{-ik(x-x')} - e^{ik(x-x')} \right)$$

$$= (i\slashed\partial + m)_{\alpha\alpha'} i\Delta(x - x') ,$$

where the function (13.1.22b)

$$\Delta(x - x') = \frac{1}{2i} \int \frac{d^3k}{(2\pi)^3} \frac{1}{E_{\mathbf{k}}} \left(e^{-ik(x-x')} - e^{ik(x-x')} \right) , \quad k_0 = E_{\mathbf{k}} \qquad (13.4.2)$$

has already been encountered in (13.1.23c) as the commutator of free bosons, namely

$$[\phi(x), \phi(x')] = i\Delta(x - x') .$$

The anticommutator of the free field operators thus has the form

$$\{\psi_\alpha(x), \bar{\psi}_{\alpha'}(x')\} = (i\not{\partial} + m)_{\alpha\alpha'} i\Delta(x - x') . \tag{13.4.1'}$$

In order to proceed further with the analysis, we require certain properties of $\Delta(x)$, which we summarize below.

Properties of $\Delta(x)$

(i) It is possible to represent $\Delta(x)$ in the form

$$\Delta(x) = \frac{1}{i} \int \frac{d^4k}{(2\pi)^3} \delta(k^2 - m^2)\epsilon(k^0)e^{-ikx} \tag{13.4.3a}$$

with

$$\epsilon(k^0) = \Theta(k^0) - \Theta(-k^0) .$$

See Problem 13.16.

(ii)

$$\Delta(-x) = -\Delta(x) . \tag{13.4.3b}$$

This can be seen directly from Eq.(13.4.3a).

(iii)

$$(\Box + m^2)\Delta(x) = 0 . \tag{13.4.3c}$$

The functions $\Delta(x), \Delta^+(x), \Delta^-(x)$ are solutions of the free Klein–Gordon equation, since they are linear superpositions of its solutions. The propagator $\Delta_F(x)$ and the retarded and advanced Green's functions[3] $\Delta_R(x), \Delta_A(x)$ satisfy the inhomogeneous Klein–Gordon equation with a source $\delta^{(4)}(x) : (\Box + m^2)\Delta_F(x) = -\delta^{(4)}(x)$.
    See Problem 13.17.

(iv)

$$\partial_0\Delta(x)|_{x_0=0} = -\delta^{(3)}(x) . \tag{13.4.3d}$$

This follows by taking the derivative of (13.4.2).

(v) $\Delta(x)$ is Lorentz invariant.

This can be shown by considering a Lorentz transformation Λ

$$\Delta(\Lambda x) = \frac{1}{i} \int \frac{d^4k}{(2\pi)^3} \delta(k^2 - m^2)\epsilon(k^0)e^{-ik\cdot\Lambda x} .$$

---

[3] The retarded and advanced Green's functions are defined by $\Delta_R(x) \equiv \Theta(x_0)\Delta(x)$ and $\Delta_A(x) \equiv -\Theta(-x_0)\,\Delta(x)$.

Using $k \cdot \Lambda x = \Lambda^{-1} k \cdot x$ and the substitution $k' = \Lambda^{-1} k$, we have $d^4 k = d^4 k'$ and $k'^2 = k^2$. Furthermore, the δ function in (13.4.3a) vanishes for space-like vectors k, i.e., $k^2 < 0$. Since for time-like k and orthochronous Lorentz transformations $\epsilon(k^{0'}) = \epsilon(k^0)$, it follows that

$$\Delta(\Lambda x) = \Delta(x) . \tag{13.4.3e}$$

For $\Lambda \in \mathcal{L}^{\downarrow}$ on the other hand, $\Delta(\Lambda x) = -\Delta(x)$.

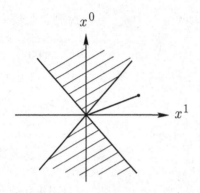

Fig. 13.5. Minkowski diagram: past and future light cone of the origin, and a space-like vector (outside the light cone), are shown

(vi) For space-like vectors one has

$$\Delta(-x) = \Delta(x) . \tag{13.4.3f}$$

Proof. The assertion is valid for purely space-like vectors as seen from the representation (13.4.3a) with the substitutions $\mathbf{x} \rightarrow -\mathbf{x}$ and $\mathbf{k} \rightarrow -\mathbf{k}$. However, all space-like vectors can be transformed into purely space-like vectors by means of an orthochronous Lorentz transformation (Fig. 13.5).

(vii) Thus, by combining (13.4.3b) and (13.4.3f), it follows for space-like vectors that

$$\Delta(x) = 0 \quad \text{for} \quad x^2 < 0 . \tag{13.4.3g}$$

For purely space-like vectors this can be seen directly from the definition (13.4.2) of $\Delta(x)$.

We now turn to the proof of the *spin statistics theorem*. First, we show that two local observables of the type $\bar{\psi}(x)\psi(x)$, etc., commute for space-like separations, e.g.:

$$
\begin{aligned}
[\bar{\psi}(x)\psi(x), \bar{\psi}(x')\psi(x')] &= \bar{\psi}_\alpha(x)\,[\psi_\alpha(x), \bar{\psi}(x')\psi(x')] \\
&\quad + [\bar{\psi}_\alpha(x), \bar{\psi}(x')\psi(x')]\,\psi_\alpha(x) \\
&= \bar{\psi}_\alpha(x)\,(\{\psi_\alpha(x), \bar{\psi}_\beta(x')\}\,\psi_\beta(x') \\
&\quad - \bar{\psi}_\beta(x')\,\{\psi_\alpha(x), \psi_\beta(x')\}) \\
&\quad + (\{\bar{\psi}_\alpha(x), \bar{\psi}_\beta(x')\}\,\psi_\beta(x') \\
&\quad - \bar{\psi}_\beta(x')\,\{\bar{\psi}_\alpha(x), \psi_\beta(x')\})\,\psi_\alpha(x) \\
&= \bar{\psi}_\alpha(x)\left((i\!\!\not{\partial} + m)_{\alpha\beta}\mathrm{i}\Delta(x - x')\right)\psi_\beta(x') \\
&\quad + \bar{\psi}_\beta(x')\left((-i\!\!\not{\partial} + m)_{\beta\alpha}\mathrm{i}\Delta(x - x')\right)\psi_\alpha(x).
\end{aligned} \tag{13.4.4}
$$

Because of (13.4.3g), this commutator vanishes for space-like separations. Thus, causality is satisfied since no signal can be transmitted between x and x' when $(x - x')^2 < 0$.[4]

What would have been the result if we had used commutators instead of anticommutators for the quantization? Apart from the absence of a lower energy bound, we would encounter a violation of causality. We would then have

$$
[\psi_\alpha(x), \bar{\psi}_{\alpha'}(x')] = (i\!\!\not{\partial} + m)_{\alpha\alpha'}\mathrm{i}\Delta_1(x - x')\,, \tag{13.4.5a}
$$

where

$$
\Delta_1(x - x') = \frac{1}{2\mathrm{i}}\int \frac{d^3k}{(2\pi)^3}\frac{1}{k_0}\left(\mathrm{e}^{-\mathrm{i}k(x-x')} + \mathrm{e}^{\mathrm{i}k(x-x')}\right)\,. \tag{13.4.5b}
$$

The function $\Delta_1(x) = \Delta_+(x) - \Delta_-(x)$ is an even solution of the homogeneous Klein–Gordon equation, which does not vanish for space-like separations $(x - x')^2 < 0$. Likewise, for space-like separations, $(i\!\!\not{\partial} + m)\mathrm{i}\Delta_1(x - x') \neq 0$. For this kind of quantization, local operators at the same point in time but different points in space would *not* commute. This would amount to a violation of locality or *microcausality*. Based on these arguments, we can formulate the spin statistics theorem as follows:

Spin statistics theorem: Particles with spin $\frac{1}{2}$ and, more generally, all particles with half-integral spin are fermions, whose field operators are quantized by anticommutators. Particles with integral spin are bosons, and their field operators are quantized by commutators.

Remarks:

(i) *Microcausality:* Two physical observables at positions with a space-like separation must be simultaneously measurable; the measurements cannot

---

[4] If it were possible to transmit signals between space–time points with space-like separations, then this could only occur with speeds greater than the speed of light. In a different coordinate system, this would correspond to a movement into the past, i.e., to acausal behavior.

influence one another. This property is known as microcausality. Without it, space-like separations would have to be linked by a signal which, in violation of special relativity, would need to travel faster than light in order for the observables to influence one another. This situation would also apply at arbitrarily small separations, hence the expression microcausality. Instead of microcausality, the term *locality* is also used synonymously. We recall the general result that two observables do not interfere (are simultaneously diagonalizable) if, and only if, they commute.

(ii) The prediction of the spin statistics theorem for *free particles* with spin $S = 0$ can be demonstrated in analogy to (13.4.5): Commutation rules lead to $[\phi(x), \phi(x')] = i\Delta(x - x')$. The fields thus satisfy microcausality and, by calculating the commutators of products, one can show that the observables $\phi(x)^2$ etc. also do. If, on the other hand, one were to quantize the Klein–Gordon field with Fermi commutation relations, then, as is readily seen, neither $[\phi(x), \phi(x')]_+$ nor $[\phi(x), \phi(x')]_-$ could possess the microcausality property $[\phi(x), \phi(x')]_\pm = 0$ for $(x - x')^2 < 0$. Therefore, also composite operators would violate the requirement of *microcausality*.

(iii) Perturbation theory leads one to expect that the property of microcausality can be extended from the free propagator to the interacting case[5]. For the interacting Klein–Gordon field one can derive the spectral representation

$$\langle 0| \, [\phi(x), \phi(x')] \, |0\rangle = \int_0^\infty d\sigma^2 \varrho(\sigma^2) \Delta(x - x', \sigma) \qquad (13.4.6)$$

for the expectation value of the commutator[6]. Here $\Delta(x - x', \sigma)$ is the free commutator given in Eq. (13.4.2) with explicit reference to the mass, which in Eq. (1.3.4) has been integrated over. Hence, microcausality is also fulfilled for the interacting field. If on the other hand the Klein–Gordon field had been quantized using Fermi anti-commutation rules one would find

$$\langle 0| \, \{\phi(x), \phi(x')\} \, |0\rangle = \int_0^\infty d\sigma^2 \varrho(\sigma^2) \Delta_1(x - x', \sigma) \,, \qquad (13.4.7)$$

where $\Delta_1(x - x', \sigma)$ given in Eq. (13.4.5b) does not vanish for space-like separations. Microcausality would then be violated. Analogously one obtains for fermions, if they are quantized with commutators, a spectral representation containing Δ_1, which is again a contradiction to microcausality.

---

[5] A general proof for interacting fields on the basis of axiomatic field theory can be found in R.F. Streater, A.S. Wightman, *PCT, Spin & Statistics and all that*, W.A. Benjamin, New York, 1964, p. 146 f.

[6] J.D. Bjorken and S.D. Drell, *Relativistic Quantum Fields*, McGraw Hill, New York, 1965, p. 171.

(iv) The reason why the observables of the Dirac field can only be bilinear quantities $\bar{\psi}\psi$ as well as powers or derivatives thereof is the following. The field $\psi(x)$ itself is not measurable, since it is changed by a gauge transformation of the first kind,

$$\psi(x) \rightarrow \psi'(x) = e^{i\alpha}\psi(x) ,$$

and observables may only be gauge-invariant quantities. Measurable quantities must, as the Lagrangian density, remain unchanged under a gauge transformation. Furthermore, there are no other fields that couple to $\psi(x)$ alone. The electromagnetic vector potential A_μ, for example, couples to a bilinear combination of ψ.

Another reason why $\psi(x)$ is not an observable quantity follows from the transformation behavior of a spinor under a rotation through 2π, Eq. (6.2.23a). Since the physically observable world is unchanged by a rotation through 2π, whereas a spinor ψ becomes $-\psi$, one must conclude that spinors themselves are not directly observable. This does not contradict the fact that, under rotation in a spatial subdomain, one can observe the phase change of a spinor with respect to a reference beam by means of an interference experiment, since the latter is determined by a bilinear quantity (see the remarks and references following Eq. (6.2.23a)).

13.4.2 Further Properties of Anticommutators and Propagators of the Dirac Field

We summarize here for later use a few additional properties of anticommutators and propagators of the Dirac field.

According to Eq. (13.4.1) and using the properties (13.4.3d) and (13.4.3g) of $\Delta(x)$, the equal time anticommutator of the Dirac field is given by

$$\begin{aligned}
\{\psi_\alpha(t, \mathbf{x}), \bar{\psi}_{\alpha'}(t, \mathbf{y})\} &= -\gamma^0_{\alpha\alpha'}\partial_0\Delta(x^0 - y^0, \mathbf{x} - \mathbf{y})|_{y_0 = x_0} \\
&= \gamma^0_{\alpha\alpha'}\delta^3(\mathbf{x} - \mathbf{y}) .
\end{aligned}$$

Multiplying this by $\gamma^0_{\alpha'\beta}$ and summing over α' yields:

$$\left\{\psi_\alpha(t, \mathbf{x}), \psi^\dagger_\beta(t, \mathbf{y})\right\} = \delta_{\alpha\beta}\,\delta^3(\mathbf{x} - \mathbf{y}) . \tag{13.4.8}$$

Hence, $i\psi^\dagger$ is thus sometimes called the anticommutating conjugate operator to $\psi(x)$.

Fermion propagators
In analogy to (13.1.23a–c), one defines for the Dirac field

$$\left[\psi^\pm(x), \bar{\psi}^\mp(x')\right]_+ = iS^\pm(x - x') \tag{13.4.9a}$$

$$[\psi(x), \bar{\psi}(x')]_+ = iS(x - x') . \qquad (13.4.9b)$$

The anticommutator (13.4.9b) has already been calculated in (13.4.1). From this calculation one sees that $iS^+(x-x')$ $(iS^-(x-x'))$ is given by the first (second) term in the penultimate line of (13.4.1). Hence, on account of (13.1.10a–c), we have

$$S^{\pm}(x) = (i\partial\!\!\!/ + m)\Delta^{\pm}(x) \qquad (13.4.10a)$$

$$S(x) = S^+(x) + S^-(x) = (i\partial\!\!\!/ + m)\Delta(x) . \qquad (13.4.10b)$$

Starting from the integral representation (13.1.25a,b) of $\Delta^{\pm}$ and Δ, one obtains from (13.4.9a,b)

$$S^{\pm}(x) = \int_{C^{\pm}} \frac{d^4 p}{(2\pi)^4} e^{-ipx} \frac{p\!\!\!/ + m}{p^2 - m^2} = \int_{C^{\pm}} \frac{d^4 p}{(2\pi)^4} \frac{e^{-ipx}}{p\!\!\!/ - m} \qquad (13.4.11a)$$

and

$$S(x) = \int_C \frac{d^4 p}{(2\pi)^4} \frac{e^{-ipx}}{p\!\!\!/ - m} , \qquad (13.4.11b)$$

where we have used $(p\!\!\!/ \pm m)(p\!\!\!/ \mp m) = p^2 - m^2$. The paths $C^{\pm}$ and C are the same as defined in Fig. 13.1. For Fermi operators one also introduces a time-ordered product. The definition of the *time-ordered product* for Fermi fields reads:

$$T\left(\psi(x)\bar{\psi}(x')\right) \equiv \begin{cases} \psi(x)\bar{\psi}(x') & \text{for } t > t' \\ -\bar{\psi}(x')\psi(x) & \text{for } t < t' \end{cases}$$

$$\equiv \Theta(t - t')\psi(x)\bar{\psi}(x') - \Theta(t' - t)\bar{\psi}(x')\psi(x) . \qquad (13.4.12)$$

For use in the perturbation theory to be developed later, we also introduce the following definition of the Feynman fermion propagator

$$\langle 0| T(\psi(x)\bar{\psi}(x')) |0\rangle \equiv iS_F(x - x') . \qquad (13.4.13)$$

For its evaluation, we note that

$$\langle 0| \psi(x)\bar{\psi}(x') |0\rangle = \langle 0| \psi^+(x)\bar{\psi}^-(x') |0\rangle = \langle 0| \left[\psi^+(x), \bar{\psi}^-(x')\right]_+ |0\rangle$$
$$= iS^+(x - x')$$

$$(13.4.14a)$$

and, similarly,

$$\langle 0| \bar{\psi}(x')\psi(x) |0\rangle = iS^-(x - x') , \qquad (13.4.14b)$$

from which it follows that the Feynman fermion propagator (see Problem 13.18) is

$$S_F(x) = \Theta(t)S^+(x) - \Theta(-t)S^-(x) = (i\gamma^\mu \partial_\mu + m)\Delta_F(x) . \qquad (13.4.15)$$

By exploiting (13.1.31), one can also write the Feynman fermion propagator
in the form

$$S_F(x) = \int \frac{d^4 p}{(2\pi)^4} e^{-ipx} \frac{\not{p} + m}{p^2 - m^2 + i\epsilon} \ . \tag{13.4.16}$$

Problems

13.1 Confirm the validity of (13.1.5′).

13.2 (a) For the scalar field, show that the four-momentum operator

$$P^\mu =: \int d^3 x \left\{ \pi \phi^{'\mu} - \delta_0^\mu \mathcal{L} \right\} :$$

can be written in the form (13.1.19)

$$P^\mu = \sum_{\mathbf{k}} k^\mu a_{\mathbf{k}}^\dagger a_{\mathbf{k}} \ .$$

(b) Show that the four-momentum operator is the generator of the translation
operator:

$$e^{ia_\mu P^\mu} F(\phi(x)) e^{-ia_\mu P^\mu} = \phi(x + a) \ .$$

13.3 Confirm formula (13.1.25a) for $\Delta^\pm(x)$.

13.4 Confirm the formula (13.1.31) for $\Delta_F(x)$, taking into account Fig. 13.2.

13.5 Verify the commutation relations (13.2.6).

13.6 For the quantized, complex Klein–Gordon field, the charge conjugation operation is defined by

$$\phi'(x) = \mathcal{C}\phi(x)\mathcal{C}^\dagger = \eta_{\mathcal{C}} \phi^\dagger(x) \ ,$$

where the charge conjugation operator $\mathcal{C}$ is unitary and leaves the vacuum state
invariant: $\mathcal{C} |0\rangle = |0\rangle$.
(a) Show for the annihilation operators that

$$\mathcal{C} a_{\mathbf{k}} \mathcal{C}^\dagger = \eta_{\mathcal{C}} b_{\mathbf{k}}, \quad \mathcal{C} b_{\mathbf{k}} \mathcal{C}^\dagger = \eta_{\mathcal{C}}^* a_{\mathbf{k}}$$

and derive for the single-particle states

$$|a, \mathbf{k}\rangle \equiv a_{\mathbf{k}}^\dagger |0\rangle \ , \quad |b, \mathbf{k}\rangle \equiv b_{\mathbf{k}}^\dagger |0\rangle \ ,$$

the transformation property

$$\mathcal{C} |a, \mathbf{k}\rangle = \eta_{\mathcal{C}}^* |b, \mathbf{k}\rangle \ , \quad \mathcal{C} |b, \mathbf{k}\rangle = \eta_{\mathcal{C}} |a, \mathbf{k}\rangle \ .$$

(b) Also show that the Lagrangian density (13.2.1) is invariant under charge conjugation and that the current density (13.2.11) changes sign:

$$\mathcal{C}j(x)\mathcal{C}^\dagger = -j(x).$$

Hence, particles and antiparticles are interchanged, with the four-momentum remaining unchanged.

(c) Find a representation for the operator $\mathcal{C}$.

13.7 Show for the Klein–Gordon field that:

$$[P^\mu, \phi(x)] = -i\partial^\mu \phi(x)$$

$$[\mathbf{P}, \phi(x)] = i\nabla \phi(x).$$

13.8 Derive the equations of motion for the Dirac field operator $\psi(x)$, starting from the Heisenberg equations of motion with the Hamiltonian (13.3.7).

13.9 Calculate the expectation value of the quantized angular momentum operator (13.3.13) in a state with a positron at rest.

13.10 Show that the spinors $u_r(k)$ and $w_r(k)$ transform into one another under charge conjugation.

13.11 Consider an electron at rest and a stationary positron

$$\left| e^{\mp}, \mathbf{k} = 0, s \right\rangle = \begin{cases} b^\dagger_{s\mathbf{k}=0} \left| 0 \right\rangle \\ d^\dagger_{s\mathbf{k}=0} \left| 0 \right\rangle \end{cases}.$$

Show that

$$\mathbf{J}\left| e^{\mp}, \mathbf{k} = 0, s \right\rangle = \sum_{r=1,2} \left| e^{\mp}, \mathbf{k} = 0, r \right\rangle \frac{1}{2}(\boldsymbol{\sigma})_{rs}$$

and

$$J^3 \left| e^{\mp}, \mathbf{k} = 0, s \right\rangle = \pm \frac{1}{2} \left| e^{\mp}, \mathbf{k} = 0, s \right\rangle,$$

where $(\boldsymbol{\sigma})_{rs}$ are the matrix elements of the Pauli matrices in the Pauli spinors χ_r and χ_s.

13.12 Prove that the momentum operator of the Dirac field (13.3.26)

$$P^\mu = \sum_{\mathbf{k},r} k^\mu \left[b^\dagger_{r\mathbf{k}} b_{r\mathbf{k}} + d^\dagger_{r\mathbf{k}} d_{r\mathbf{k}} \right]$$

is the generator of the translation operator:

$$e^{ia_\mu P^\mu} \psi(x) e^{-ia_\mu P^\mu} = \psi(x + a).$$

13.13 From the gauge invariance of the Lagrangian density, derive the expression (13.3.27b) for the current-density operator of the Dirac field.

13.14 Show that the operator of the charge conjugation transformation

$$\mathcal{C} = \mathcal{C}_1 \mathcal{C}_2$$

$$\mathcal{C}_1 = \exp\left[-i \sum_{k,r} \frac{m}{V E_k} \lambda (b_{kr}^\dagger b_{kr} - d_{kr}^\dagger d_{kr})\right]$$

$$\mathcal{C}_2 = \exp\left[\frac{i\pi}{2} \sum_{k,r} \frac{m}{V E_k} (b_{kr}^\dagger - d_{kr}^\dagger)(b_{kr} - d_{kr})\right]$$

transforms the creation and annihilation operators of the Dirac field and the field operator as follows:

$$\mathcal{C} b_{kr} \mathcal{C}^\dagger = \eta_{\mathcal{C}} d_{kr} \ ,$$
$$\mathcal{C} d_{kr}^\dagger \mathcal{C}^\dagger = \eta_{\mathcal{C}}^* b_{kr}^\dagger \ ,$$
$$\mathcal{C} \psi(x) \mathcal{C}^\dagger = \eta_{\mathcal{C}} C \bar{\psi}^T(x),$$

where the transpose only refers to the spinor indices, and $C = i\gamma^2\gamma^0$. The factor $\mathcal{C}_1$ yields the phase factor $\eta_{\mathcal{C}} = e^{i\lambda}$. The transformation $\mathcal{C}$ exchanges particles and antiparticles with the same momentum, energy, and helicity.
Show also that the vacuum is invariant with respect to this transformation and that the current density $j^\mu = e : \bar{\psi}\gamma^\mu\psi :$ changes sign.

13.15 (a) Show for the spinor field that

$$Q = -e_0 \sum_k \sum_r \left(b_{rk}^\dagger b_{rk} - d_{rk}^\dagger d_{rk}\right)$$

(Eq. (13.3.28)), by starting from

$$Q = -e_0 \int d^3x \ : \ \bar{\psi}(x)\gamma^0\psi(x) \ : \ .$$

(b) Show, furthermore, that

$$\left[Q, b_{rk}^\dagger\right] = -e_0 b_{rk}^\dagger \quad \text{and} \quad \left[Q, d_{rk}^\dagger\right] = e_0 d_{rk}^\dagger$$

(Eq.(13.3.29)).

13.16 Show that (13.4.2) can also be written in the form (13.4.3a).

13.17 Show that $\Delta_F(x), \Delta_R(x)$, and $\Delta_A(x)$ satisfy the inhomogeneous Klein–Gordon equation

$$(\partial_\mu \partial^\mu + m^2)\Delta_F(x) = -\delta^{(4)}(x) \ .$$

13.18 Prove equation (13.4.15).

Hint: Use

$$\delta \left(k^2 - m^2\right) = \left(\delta \left(k^0 - \sqrt{\mathbf{k}^2 + m^2}\right) + \delta \left(k^0 + \sqrt{\mathbf{k}^2 + m^2}\right)\right) \Big/ \sqrt{\mathbf{k}^2 + m^2} \ .$$

13.19 Show for the Dirac-field that:

$$[P^\mu, \psi(x)] = -i\partial^\mu \psi(x)$$

$$[\mathbf{P}, \psi(x)] = i\boldsymbol{\nabla}\psi(x) \ .$$

14. Quantization of the Radiation Field

This chapter describes the quantization of the free radiation field. Since, for certain aspects, it is necessary to include the coupling to external current densities, a separate chapter is devoted to this subject. Starting from the classical Maxwell equations and a discussion of gauge transformations, the quantization will be carried out in the Coulomb gauge. The principal aim in this chapter is to calculate the propagator for the radiation field. In the Coulomb gauge, one initially obtains a propagator that is not Lorentz invariant. However, when one includes the effect of the instantaneous Coulomb interaction in the propagator and notes that the terms in the propagator that are proportional to the wave vector yield no contribution in perturbation theory, one then finds that the propagator is equivalent to using a covariant one. The difficulty in quantizing the radiation field arises from the massless nature of the photons and from gauge invariance. Therefore, the vector potential $A^\mu(x)$ has, in effect, only two dynamical degrees of freedom and the instantaneous Coulomb interaction.

14.1 Classical Electrodynamics

14.1.1 Maxwell Equations

We begin by recalling classical electrodynamics for the electric and magnetic fields $\mathbf{E}$ and $\mathbf{B}$. The *Maxwell* equations in the presence of a charge density $\rho(\mathbf{x}, t)$ and a current density $\mathbf{j}(\mathbf{x}, t)$ read[1]:

---

[1] Here and in the following we shall use rationalized units, also known as Heaviside–Lorentz units. In these units the fine-structure constant is $\alpha = \frac{\hat{e}_0^2}{4\pi\hbar c} = \frac{1}{137}$, whereas in Gaussian units it is given by $\alpha = \frac{e_0^2}{\hbar c}$, i.e., $\hat{e}_0 = e_0\sqrt{4\pi}$. Correspondingly, we have $\mathbf{E} = \mathbf{E}_{\text{Gauss}}/\sqrt{4\pi}$ and $\mathbf{B} = \mathbf{B}_{\text{Gauss}}/\sqrt{4\pi}$, and the Coulomb law $V(\mathbf{x}) = \frac{e^2}{4\pi|\mathbf{x}-\mathbf{x}'|}$. In the following, we shall also set $\hbar = c = 1$.

$$\nabla \cdot \mathbf{E} = \rho \tag{14.1.1a}$$

$$\nabla \times \mathbf{E} = -\frac{\partial \mathbf{B}}{\partial t} \tag{14.1.1b}$$

$$\nabla \cdot \mathbf{B} = 0 \tag{14.1.1c}$$

$$\nabla \times \mathbf{B} = \frac{\partial \mathbf{E}}{\partial t} + \mathbf{j} . \tag{14.1.1d}$$

If we introduce the antisymmetric field tensor

$$F^{\mu\nu} = \begin{pmatrix} 0 & E_x & E_y & E_z \\ -E_x & 0 & B_z & -B_y \\ -E_y & -B_z & 0 & B_x \\ -E_z & B_y & -B_x & 0 \end{pmatrix} , \tag{14.1.2}$$

whose components can also be written in the form

$$E^i = F^{0i}$$
$$B^i = \frac{1}{2}\epsilon^{ijk}F_{jk} , \tag{14.1.3}$$

the Maxwell equations then acquire the form

$$\partial_\nu F^{\mu\nu} = j^\mu \tag{14.1.4a}$$

and

$$\partial^\lambda F^{\mu\nu} + \partial^\mu F^{\nu\lambda} + \partial^\nu F^{\lambda\mu} = 0 , \tag{14.1.4b}$$

where the four-current-density is

$$j^\mu = (\rho, \mathbf{j}) , \tag{14.1.5}$$

satisfying the continuity equation

$$j^\mu_{,\mu} = 0 . \tag{14.1.6}$$

The homogeneous equations (14.1.1b,c) or (14.1.4b) can be satisfied automatically by expressing $F^{\mu\nu}$ in terms of the four-potential A^μ:

$$F^{\mu\nu} = A^{\mu,\nu} - A^{\nu,\mu} . \tag{14.1.7}$$

The inhomogeneous equations (14.1.1a,d) or (14.1.4a) imply that

$$\Box A^\mu - \partial^\mu \partial_\nu A^\nu = j^\mu . \tag{14.1.8}$$

In Part II, where relativistic wave equations were discussed, $j^\mu(x)$ was the particle current density. However, in quantum field theory, and particularly in quantum electrodynamics, it is usual to use $j^\mu(x)$ to denote the electrical current density. In the following, we will have, for the Dirac field for example, $j^\mu(x) = e\bar{\psi}(x)\gamma^\mu\psi(x)$, where e is the charge of the particle, i.e., for the electron $e = -\hat{e}_0$.

14.1.2 Gauge Transformations

Equation (14.1.8) is not sufficient to determine the four-potential uniquely, since, for an arbitrary function $\lambda(x)$, the transformation

$$A^\mu \rightarrow A'^\mu = A^\mu + \partial^\mu \lambda \tag{14.1.9}$$

leaves the electromagnetic field tensor $F^{\mu\nu}$ and hence, also the fields $\mathbf{E}$ and $\mathbf{B}$, as well as Eq. (14.1.8) invariant. One refers to (14.1.9) as a gauge transformation of the second kind. It is easy to see that not all components of A_μ are independent dynamical variables and, by a suitable choice of the function $\lambda(x)$, one can impose certain conditions on the components A_μ, or, in other words, transform to a certain gauge. Two particularly important gauges are the *Lorentz gauge*, for which one requires

$$A^\mu{}_{,\mu} = 0 \, , \tag{14.1.10}$$

and the *Coulomb gauge*, for which

$$\boldsymbol{\nabla} \cdot \mathbf{A} = 0 \tag{14.1.11}$$

is specified. Other gauges include the time gauge $A^0 = 0$ and the axial gauge $A^3 = 0$. The advantage of the coulomb gauge is that it yields only two transverse photons, or, after an appropriate transformation, two photons with helicity ± 1. The advantage of the Lorentz gauge consists in its obvious Lorentz invariance. In this gauge, however, there are, in addition to the two transverse photons, a longitudinal and a scalar photon. In the physical results of the theory these latter photons, apart from mediating the Coulomb interaction, will play no role.[2]

14.2 The Coulomb Gauge

We shall be dealing here mainly with the *Coulomb gauge* (also called transverse or radiation gauge). It is always possible to transform into the Coulomb gauge. If A^μ does not satisfy the Coulomb gauge, then one takes instead the gauge-transformed field $A^\mu + \partial^\mu \lambda$, where λ is determined through $\boldsymbol{\nabla}^2 \lambda = -\boldsymbol{\nabla} \cdot \mathbf{A}$. In view of the Coulomb gauge condition (14.1.11), for the zero components ($\mu = 0$), equation (14.1.8) simplifies to

$$(\partial_0^2 - \boldsymbol{\nabla}^2)A_0 - \partial_0(\partial_0 A_0 - \boldsymbol{\nabla} \cdot \mathbf{A}) = j_0 \, ,$$

and due to (14.1.11) we thus have

$$\boldsymbol{\nabla}^2 A_0 = -j_0 \, . \tag{14.2.1}$$

---

[2] The most important aspects of the covariant quantization by means of the Gupta–Bleuler method are summarized in Appendix E.

This is just the Poisson equation, well known from electrostatics, which has the solution

$$A_0(\mathbf{x}, t) = \int d^3x' \frac{j_0(\mathbf{x}', t)}{4\pi|\mathbf{x} - \mathbf{x}'|} . \qquad (14.2.2)$$

Since the charge density $j^0(x)$ depends only on the matter fields and their conjugate fields, Eq. (14.2.2) represents an explicit solution for the zero components of the vector potential. Therefore, in the Coulomb gauge, the scalar potential is determined by the Coulomb field of the charge density and is thus not an independent dynamical variable. The remaining spatial components A^i are subject to the gauge condition (14.1.11), and there are thus only two independent field components.

We now turn to the spatial components of the wave equation (14.1.8), thereby taking account of (14.1.11),

$$\Box A_j - \partial_j \partial_0 A_0 = j_j . \qquad (14.2.3)$$

From (14.2.2), using the continuity equation (14.1.6) and partial integration, we have

$$\partial_j \partial_0 A_0(x) = \partial_j \int \frac{d^3x' \partial_0 j_0(x')}{4\pi|\mathbf{x} - \mathbf{x}'|} = -\partial_j \int \frac{d^3x' \partial_k' j_k(x')}{4\pi|\mathbf{x} - \mathbf{x}'|}$$
$$= -\partial_j \partial_k \int \frac{d^3x' j_k(x')}{4\pi|\mathbf{x} - \mathbf{x}'|} = \frac{\partial_j \partial_k}{\boldsymbol{\nabla}^2} j_k(x) , \qquad (14.2.4)$$

where $-\frac{1}{\boldsymbol{\nabla}^2}$ is a short-hand notation for the integral over the Coulomb Green's function[3]. If we insert (14.2.4) into (14.2.3), we obtain

$$\Box A_j = j_j^{\text{trans}} \equiv \left(\delta_{jk} - \frac{\partial_j \partial_k}{\boldsymbol{\nabla}^2}\right) j_k . \qquad (14.2.5)$$

The wave equation for A_j (14.2.5) contains the transverse part of the current density j_j^{trans}. The significance of the transversality will become more evident later on when we work in Fourier space.

---

[3] In this way, the solution of the Poisson equation

$$\boldsymbol{\nabla}^2 \Phi = -\rho \quad \text{is represented as} \quad \Phi = -\frac{1}{\boldsymbol{\nabla}^2}\rho \equiv \int \frac{d^3x' \rho(\mathbf{x}', t)}{4\pi|\mathbf{x} - \mathbf{x}'|} \};.$$

For the special case $\rho(\mathbf{x}, t) = -\delta^3(\mathbf{x})$, we have

$$\boldsymbol{\nabla}^2 \Phi = \delta^3(\mathbf{x}) \quad \text{and thus} \quad \Phi = \frac{1}{\boldsymbol{\nabla}^2}\delta^3(\mathbf{x}) = -\int \frac{d^3x' \delta^3(\mathbf{x}')}{4\pi|\mathbf{x} - \mathbf{x}'|} = -\frac{1}{4\pi|\mathbf{x}|} .$$

14.3 The Lagrangian Density for the Electromagnetic Field

The Lagrangian density of the electromagnetic field is not unique. One can derive the Maxwell equations from

$$\mathcal{L} = -\frac{1}{4} F_{\mu\nu} F^{\mu\nu} - j_\mu A^\mu \tag{14.3.1}$$

with $F_{\mu\nu} = A_{\mu,\nu} - A_{\nu,\mu}$. This is because

$$\partial_\nu \frac{\partial \mathcal{L}}{\partial A_{\mu,\nu}} = \partial_\nu \left(-\frac{1}{4} \right) (F^{\mu\nu} - F^{\nu\mu}) \times 2 = -\partial_\nu F^{\mu\nu} \tag{14.3.2}$$

and

$$\frac{\partial \mathcal{L}}{\partial A_\mu} = -j^\mu$$

yield the Euler–Lagrange equations

$$\partial_\nu F^{\mu\nu} = j^\mu , \tag{14.3.3}$$

i.e. (14.1.4a). As was noted before Eq. (14.1.7), equation (14.1.4b) is automatically satisfied. From (14.3.1), one finds for the momentum conjugate to A_μ

$$\Pi^\mu = \frac{\partial \mathcal{L}}{\partial \dot{A}_\mu} = -F^{\mu 0} . \tag{14.3.4}$$

Hence, the momentum conjugate to A_0 vanishes,

$$\Pi^0 = 0$$

and

$$\Pi^j = -F^{j0} = E^j . \tag{14.3.5}$$

The vanishing of the momentum component Π^0 shows that it is not possible to apply the canonical quantization procedure to all four components of the radiation field without modification.

Another Lagrangian density for the four-potential $A^\mu(x)$, which leads to the wave equation in the Lorentz gauge, is

$$\mathcal{L}_L = -\frac{1}{2} A_{\mu,\nu} A^{\mu,\nu} - j_\mu A^\mu . \tag{14.3.6}$$

Here, we have

$$\Pi_L^\mu = \frac{\mathcal{L}_L}{\partial \dot{A}_\mu} = -A^{\mu,0} = -\dot{A}^\mu , \tag{14.3.7}$$

and the equation of motion reads:

$$\Box A^\mu = j^\mu . \tag{14.3.8}$$

This equation of motion is only identical to (14.1.8) when the potential A^μ satisfies the Lorentz condition

$$\partial_\mu A^\mu = 0 \ . \tag{14.3.9}$$

The Lagrangian density $\mathcal{L}_L$ of equation (14.3.6) differs from the $\mathcal{L}$ of (14.3.1) in the occurrence of the term $-\frac{1}{2}(\partial_\lambda A^\lambda)^2$, which fixes the gauge:

$$\mathcal{L}_L = -\frac{1}{4}F_{\mu\nu}F^{\mu\nu} - \frac{1}{2}(\partial_\lambda A^\lambda)^2 - j_\mu A^\mu \ . \tag{14.3.6'}$$

This can easily be seen when one rewrites $\mathcal{L}_L$ as follows:

$$\mathcal{L}_L = -\frac{1}{4}(A_{\mu,\nu} - A_{\nu,\mu})(A^{\mu,\nu} - A^{\nu,\mu}) - \frac{1}{2}\partial_\lambda A^\lambda \partial_\sigma A^\sigma - j_\mu A^\mu$$

$$= -\frac{1}{2}A_{\mu,\nu}A^{\mu,\nu} + \frac{1}{2}A_{\mu,\nu}A^{\nu,\mu} - \frac{1}{2}\partial_\lambda A^\lambda \partial_\sigma A^\sigma - j_\mu A^\mu$$

$$= -\frac{1}{2}A_{\mu,\nu}A^{\mu,\nu} - j_\mu A^\mu \ .$$

In the last line, a total derivative has been omitted since it disappears in the Lagrangian through partial integration. If one adds the term $-\frac{1}{2}(\partial_\lambda A^\lambda)^2$ to the Lagrangian density, one must choose the Lorentz gauge in order that the equations of motion be consistent with electrodynamics

$$\Box A^\mu = j^\mu \ .$$

Remarks:

(i) Unlike the differential operator in (14.1.8), the d'Alembert operator appearing in (14.3.8) can be inverted.

(ii) With or without the term fixing the gauge, the longitudinal part of the vector potential $\partial_\lambda A^\lambda$ satisfies the d'Alembert equation

$$\Box\left(\partial_\lambda A^\lambda\right) = 0 \ .$$

This also holds in the presence of a current density $j^\mu(x)$.

14.4 The Free Electromagnetic Field and its Quantization

When $j^\mu = 0$, i.e., in the absence of external sources, the solution of the Poisson equation which vanishes at infinity is $A^0 = 0$, and the electromagnetic fields read:

$$\mathbf{E} = -\dot{\mathbf{A}} \ , \qquad \mathbf{B} = \nabla \times \mathbf{A} \ . \tag{14.4.1}$$

From the Lagrangian density of the free radiation field

$$\mathcal{L} = -\frac{1}{4}F^{\mu\nu}F_{\mu\nu} = \frac{1}{2}(\mathbf{E}^2 - \mathbf{B}^2) \ , \tag{14.4.2}$$

where (14.1.2) is used to obtain the last expression, one obtains the Hamiltonian density of the radiation field as

$$
\begin{aligned}
\mathcal{H}_\gamma &= \Pi^j \dot{A}_j - \mathcal{L} = \mathbf{E}^2 - \frac{1}{2}(\mathbf{E}^2 - \mathbf{B}^2) \\
&= \frac{1}{2}(\mathbf{E}^2 + \mathbf{B}^2) .
\end{aligned}
\tag{14.4.3}
$$

Since the zero component of A^μ vanishes and the spatial components satisfy the free d'Alembert equation and $\nabla \cdot \mathbf{A} = 0$, the general free solution is given by

$$
A^\mu(x) = \sum_{\mathbf{k}} \sum_{\lambda=1}^{2} \frac{1}{\sqrt{2|\mathbf{k}|V}} \left(e^{-ikx} \epsilon^\mu_{\mathbf{k},\lambda} a_{\mathbf{k}\lambda} + e^{ikx} \epsilon^\mu_{\mathbf{k},\lambda}{}^* a^\dagger_{\mathbf{k}\lambda} \right) ,
\tag{14.4.4}
$$

where $k_0 = |\mathbf{k}|$, and the two polarization vectors have the properties

$$
\mathbf{k} \cdot \boldsymbol{\epsilon}_{\mathbf{k},\lambda} = 0 \quad , \qquad \epsilon^0_{\mathbf{k},\lambda} = 0
$$
$$
\boldsymbol{\epsilon}_{\mathbf{k},\lambda} \cdot \boldsymbol{\epsilon}_{\mathbf{k},\lambda'} = \delta_{\lambda\lambda'} .
\tag{14.4.5}
$$

In the classical theory, the amplitudes $a_{\mathbf{k}\lambda}$ are complex numbers. In (14.4.4), we chose a notation such that this expansion also remains valid for the quantized theory in which the $a_{\mathbf{k}\lambda}$ are replaced by the operators $a_{\mathbf{k}\lambda}$. The form of (14.4.4) guarantees that the vector potential is real.

Remarks:

(i) The different factor in QM I Eq. (16.49) arises due to the use there of Gaussian units, in which the energy density, for example, is given by $\mathcal{H} = \frac{1}{8\pi}(\mathbf{E}^2 + \mathbf{B}^2)$.

(ii) In place of the two photons polarized transversely to $\mathbf{k}$, one can also use helicity eigenstates whose polarization vectors have the form

$$
\epsilon^\mu_{\mathbf{p},\pm1} = R(\hat{p}) \begin{pmatrix} 0 \\ 1/\sqrt{2} \\ \pm i/\sqrt{2} \\ 0 \end{pmatrix} ,
\tag{14.4.6}
$$

where $R(\hat{p})$ is a rotation that rotates the z axis into the direction of $\mathbf{p}$.

(iii) The first attempt at quantization could lead to

$$
\left[A^i(\mathbf{x},t), \dot{A}^j(\mathbf{x}',t) \right] = i\delta_{ij}\delta(\mathbf{x} - \mathbf{x}')
$$

i.e.,
$$
\tag{14.4.7}
$$

$$
\left[A^i(\mathbf{x},t), E^j(\mathbf{x}',t) \right] = -i\delta_{ij}\delta(\mathbf{x} - \mathbf{x}') .
$$

This relation, however, contradicts the condition for the Coulomb gauge $\partial_i A^i = 0$ and the Maxwell equation $\partial_i E^i = 0$.

We will carry out the quantization of the theory in the following way: It is already clear that the quanta of the radiation field – the photons – are bosons. This is a consequence of the statistical properties (the strict validity of Planck's radiation law) and of the fact that the intrinsic angular momentum (spin) has the value $S = 1$. The spin statistics theorem tells us that this spin value corresponds to a Bose field. Therefore, the amplitudes of the field, the $a_{k\lambda}$, are quantized by means of Bose commutation relations.

We begin by expressing the Hamiltonian (function or operator) (14.4.3) in terms of the expansion (14.4.4). Using the fact that the three vectors $\mathbf{k}, \epsilon_{\mathbf{k}_1}$, and $\epsilon_{\mathbf{k}_2}$ form an orthogonal triad, we obtain

$$H = \sum_{\mathbf{k},\lambda} \frac{|\mathbf{k}|}{2} \left(a_{\mathbf{k}\lambda}^\dagger a_{\mathbf{k}\lambda} + a_{\mathbf{k}\lambda} a_{\mathbf{k}\lambda}^\dagger \right) . \tag{14.4.8}$$

We postulate the *Bose commutation relations*

$$\left[a_{\mathbf{k}\lambda}, a_{\mathbf{k}'\lambda'}^\dagger \right] = \delta_{\lambda\lambda'} \delta_{\mathbf{k}\mathbf{k}'} \text{ and}$$
$$\left[a_{\mathbf{k}\lambda}, a_{\mathbf{k}'\lambda'} \right] = \left[a_{\mathbf{k}\lambda}^\dagger, a_{\mathbf{k}'\lambda'}^\dagger \right] = 0 . \tag{14.4.9}$$

The Hamiltonian (14.4.8) then follows as

$$H = \sum_{\mathbf{k},\lambda} |\mathbf{k}| \left(a_{\mathbf{k}\lambda}^\dagger a_{\mathbf{k}\lambda} + \frac{1}{2} \right) . \tag{14.4.8'}$$

The divergent zero-point energy appearing here will be eliminated later by a redefinition of the Hamiltonian using normal ordering. We now calculate the commutators of the field operators. Given the definition

$$\Lambda^{\mu\nu} \equiv \sum_{\lambda=1}^{2} \epsilon_{\mathbf{k},\lambda}^\mu \epsilon_{\mathbf{k},\lambda}^\nu , \tag{14.4.10}$$

it follows from (14.4.5) that

$$\Lambda^{00} = 0 \quad , \quad \Lambda^{0j} = 0 \tag{14.4.11}$$

and

$$\Lambda^{ij} = \sum_{\lambda=1}^{2} \epsilon_{\mathbf{k},\lambda}^i \epsilon_{\mathbf{k},\lambda}^j = \delta^{ij} - \frac{k^i k^j}{k^2}$$

($\mathbf{k}, \epsilon_{\mathbf{k},\lambda}, \lambda = 1, 2$ form an orthogonal triad, i.e., $\hat{k}^i \hat{k}^j + \sum_{\lambda=1}^{2} \epsilon_{\mathbf{k},\lambda}^i \epsilon_{\mathbf{k},\lambda}^j = \delta^{ij}$).

For the commutator, we now have

$$
\begin{aligned}
\left[A^i(\mathbf{x},t), \dot{A}^j(\mathbf{x}',t)\right] &= \sum_{\mathbf{k}\lambda} \sum_{\mathbf{k}'\lambda'} \frac{1}{2V\sqrt{kk'}} \Big\{ e^{-ik x} e^{ik'x'} \epsilon^i_{\mathbf{k},\lambda} \epsilon^{j\,*}_{\mathbf{k}',\lambda'} (ik'_0)\delta_{\lambda\lambda'}\delta_{\mathbf{k}\mathbf{k}'} \\
&\quad - e^{ik x} e^{-ik'x'} \epsilon^{i\,*}_{\mathbf{k},\lambda} \epsilon^j_{\mathbf{k}',\lambda'} (-ik'_0)\delta_{\lambda\lambda'}\delta_{\mathbf{k}\mathbf{k}'} \Big\} \\
&= \frac{i}{2}\sum_{\mathbf{k}\lambda}\left(\epsilon^i_{\mathbf{k},\lambda}\epsilon^{j\,*}_{\mathbf{k},\lambda}\, e^{ik(\mathbf{x}-\mathbf{x}')} + \epsilon^{i\,*}_{\mathbf{k},\lambda}\epsilon^j_{\mathbf{k}',\lambda'} e^{-ik(\mathbf{x}-\mathbf{x}')}\right) \\
&= \frac{i}{2}\sum_{\mathbf{k}}\left(\delta^{ij} - \frac{k^i k^j}{k^2}\right)\left(e^{ik(\mathbf{x}-\mathbf{x}')} + e^{-ik(\mathbf{x}-\mathbf{x}')}\right) \\
&= i\left(\delta^{ij} - \frac{\partial^i\partial^j}{\nabla^2}\right)\sum_{\mathbf{k}} e^{ik(\mathbf{x}-\mathbf{x}')} \\
&= i\left(\delta^{ij} - \frac{\partial^i\partial^j}{\nabla^2}\right)\delta(\mathbf{x}-\mathbf{x}')\,.
\end{aligned}
$$

The commutator of the canonical variables thus reads:

$$
\left[A^i(\mathbf{x},t), \dot{A}^j(\mathbf{x}',t)\right] = i\left(\delta^{ij} - \frac{\partial^i\partial^j}{\nabla^2}\right)\delta(\mathbf{x}-\mathbf{x}') \tag{14.4.12a}
$$

or, on account of (14.4.1),

$$
\left[A^i(\mathbf{x},t), E^j(\mathbf{x}',t)\right] = -i\left(\delta^{ij} - \frac{\partial^i\partial^j}{\nabla^2}\right)\delta(\mathbf{x}-\mathbf{x}') \tag{14.4.12b}
$$

and is consistent with the transversality condition that must be fulfilled by $\mathbf{A}$ and $\mathbf{E}$. For the two remaining commutators we find

$$
\left[A^i(\mathbf{x},t), A^j(\mathbf{x}',t)\right] = 0 \tag{14.4.12c}
$$

$$
\left[\dot{A}^i(\mathbf{x},t), \dot{A}^j(\mathbf{x}',t)\right] = 0\,. \tag{14.4.12d}
$$

These quantization properties are dependent on the gauge. However, the resulting commutators for the fields $\mathbf{E}$ and $\mathbf{B}$ are independent of the gauge chosen. Since $\mathbf{E} = -\dot{\mathbf{A}}$ and $\mathbf{B} = \operatorname{curl}\mathbf{A}$, one finds

$$
\left[E^i(\mathbf{x},t), E^j(\mathbf{x}',t)\right] = \left[B^i(\mathbf{x},t), B^j(\mathbf{x}',t)\right] = 0 \tag{14.4.12e}
$$

$$
\begin{aligned}
\left[E^i(\mathbf{x},t), B^j(\mathbf{x}',t)\right] &= \left[E^i(\mathbf{x},t), \epsilon^{jkm}\frac{\partial}{\partial x'^k}A^m(\mathbf{x}',t)\right] \\
&= \epsilon^{jkm}\frac{\partial}{\partial x'^k}(-i)\left(\delta^{im} - \frac{\partial'^i\partial'^m}{\nabla'^2}\right)\delta(\mathbf{x}-\mathbf{x}') \\
&= -i\epsilon^{jki}\frac{\partial}{\partial x'^k}\delta(\mathbf{x}-\mathbf{x}') \\
&= i\epsilon^{ijk}\frac{\partial}{\partial x^k}\delta(\mathbf{x}-\mathbf{x}')\,. \tag{14.4.12f}
\end{aligned}
$$

Whereas the commutator (14.4.12b) contains the nonlocal term ∇^{-2}, the commutators (14.4.12e,f) of the fields $\mathbf{E}$ and $\mathbf{B}$ are local.

In order to eliminate the divergent zero-point energy in the Hamiltonian, we introduce the following definition

$$H =: \frac{1}{2} \int d^3x \, (\mathbf{E}^2 + \mathbf{B}^2) : = \sum_{\mathbf{k},\lambda} k_0 \, a^\dagger_{\mathbf{k}\lambda} a_{\mathbf{k}\lambda} \qquad (14.4.13)$$

where $k_0 = |\mathbf{k}|$. Similarly, for the momentum operator of the radiation field, we have

$$\mathbf{P} = : \int d^3x \, \mathbf{E} \times \mathbf{B} := \sum_{\mathbf{k},\lambda} \mathbf{k} \, a^\dagger_{\mathbf{k}\lambda} a_{\mathbf{k}\lambda} \ . \qquad (14.4.14)$$

The normal ordered product for the components of the radiation field is defined in exactly the same way as for Klein–Gordon fields.

14.5 Calculation of the Photon Propagator

The photon propagator is defined by

$$iD_F^{\mu\nu}(x - x') = \langle 0| \, T(A^\mu(x)A^\nu(x')) \, |0\rangle \ . \qquad (14.5.1)$$

In its most general form, this second-rank tensor can be written as

$$D_F^{\mu\nu}(x) = g^{\mu\nu} D(x^2) - \partial^\mu \partial^\nu D^{(l)}(x^2) \ , \qquad (14.5.2)$$

where $D(x^2)$ and $D^{(l)}(x^2)$ are functions of the Lorentz invariant x^2. In momentum space, (14.5.2) yields:

$$D_F^{\mu\nu}(k) = g^{\mu\nu} D(k^2) + k^\mu k^\nu D^{(l)}(k^2) \ . \qquad (14.5.3)$$

In perturbation theory, the photon propagator always occurs in the combination $j_\mu D_F^{\mu\nu}(k) j_\nu$, where j_μ and j_ν are electron–positron current densities. As a result of current conservation, $\partial_\mu j^\mu = 0$, in Fourier space we have

$$k_\mu j^\mu = 0 \qquad (14.5.4)$$

and, hence, the physical results are unchanged when one replaces $D_F^{\mu\nu}(k)$ by

$$D_F^{\mu\nu}(k) \longrightarrow D_F^{\mu\nu}(k) + \chi^\mu(k)k^\nu + \chi^\nu(k)k^\mu \ , \qquad (14.5.5)$$

where the $\chi^\mu(k)$ are arbitrary functions of k.

Once a particular gauge is specified, e.g., the Coulomb gauge, the resulting $D_F^{\mu\nu}(k)$ is not of the Lorentz-invariant form (14.5.3); the physical results are, however, the same. The change of gauge (14.5.5) can be carried out simply with a view to convenience. We will now calculate the propagator in the Coulomb gauge and then deduce from this other equivalent representations. It is clear that $D(k^2)$ in (14.5.3) is of the form

$$D(k^2) \propto \frac{1}{k^2} \ ,$$

since $D_F^{\mu\nu}(k)$ must satisfy the inhomogeneous d'Alembert equation with a four-dimensional δ-source. We can adopt the same relations as for the Klein–Gordon propagators, but now it is also necessary to introduce the polarization vectors of the photon field. Introducing, in addition to (14.5.1),

$$i D_+^{\mu\nu}(x - x') = \langle 0 | A^\mu(x) A^\nu(x') | 0 \rangle \; , \tag{14.5.6a}$$

we obtain from (13.1.25a) and (13.1.31)

$$
\begin{aligned}
D_\pm^{\mu\nu}(x) &= \int_{C\pm} \frac{d^4 k}{(2\pi)^4} \frac{e^{-ikx}}{k^2} \sum_{\lambda=1}^{2} \epsilon_{\mathbf{k},\lambda}^\mu \, \epsilon_{\mathbf{k},\lambda}^\nu \\
&= \mp \frac{i}{2} \int \frac{d^3 k}{(2\pi)^3} \frac{1}{|\mathbf{k}|} \sum_\lambda \epsilon_{\mathbf{k},\lambda}^\mu \, \epsilon_{\mathbf{k},\lambda}^\nu e^{\mp ikx}
\end{aligned}
\tag{14.5.6b}
$$

and

$$
\begin{aligned}
D_F^{\mu\nu}(x - x') &= \Theta(t - t') D_+^{\mu\nu}(x - x') - \Theta(t' - t) D_-^{\mu\nu}(x - x') \\
&= -i \int \frac{d^3 k}{(2\pi)^3} \frac{1}{2|\mathbf{k}|} \sum_{\lambda=1}^{2} \epsilon_{\mathbf{k},\lambda}^\mu \epsilon_{\mathbf{k},\lambda}^\nu \left(\Theta(x^0 - x'^0) e^{-ik(x - x')} \right. \\
&\quad \left. + \Theta(x'^0 - x^0) e^{ik(x - x')} \right) ,
\end{aligned}
\tag{14.5.6c}
$$

i.e.,

$$D_F^{\mu\nu}(x - x') = \lim_{\epsilon \to 0} \int \frac{d^4 k}{(2\pi)^4} \Lambda^{\mu\nu}(k) \frac{e^{-ik(x-x')}}{k^2 + i\epsilon} \; . \tag{14.5.6d}$$

Here, we have

$$\Lambda^{\mu\nu}(k) = \sum_{\lambda=1}^{2} \epsilon_{\mathbf{k},\lambda}^\mu \, \epsilon_{\mathbf{k},\lambda}^\nu \tag{14.5.7}$$

with the components

$$\Lambda^{00} = 0 \quad , \quad \Lambda^{0k} = \Lambda^{k0} = 0 \quad , \quad \Lambda^{lk} = \delta^{lk} - \frac{k^l k^k}{k^2} \; .$$

Making use of the tetrad

$$
\begin{aligned}
\epsilon_0^\mu(\mathbf{k}) &= n^\mu \equiv (1, 0, 0, 0) \\
\epsilon_1^\mu(\mathbf{k}) &= (0, \epsilon_{\mathbf{k},1}) \quad , \quad \epsilon_2^\mu(\mathbf{k}) = (0, \epsilon_{\mathbf{k},2}) \\
\epsilon_3^\mu(\mathbf{k}) &= (0, \mathbf{k}/|\mathbf{k}|) = \frac{k^\mu - (nk)n^\mu}{\left((kn)^2 - k^2 \right)^{1/2}} \; ,
\end{aligned}
\tag{14.5.8}
$$

one can also write $\Lambda^{\mu\nu}$ in the form

$$
\begin{aligned}
\Lambda^{\mu\nu}(k) &= -g^{\mu\nu} - \frac{\left(k^\mu - (kn)n^\mu\right)\left(k^\nu - (kn)n^\nu\right)}{(kn)^2 - k^2} + n^\mu n^\nu \\
&= -g^{\mu\nu} - \frac{k^\mu k^\nu - (kn)\left(n^\mu k^\nu + k^\mu n^\nu\right)}{(kn)^2 - k^2} - \frac{k^2 n^\mu n^\nu}{(kn)^2 - k^2} .
\end{aligned}
\tag{14.5.9}
$$

As was noted in connection with Eq. (14.5.5), the middle term on the second line of (14.5.9) makes no contribution in perturbation theory and can thus be omitted. The third term in (14.5.9) makes a contribution to the Feynman propagator $iD_F(x - x')$ (14.5.6d) of the form,

$$
\begin{aligned}
&- \lim_{\epsilon \to 0} \int \frac{d^4 k}{(2\pi)^4} e^{-ik(x-x')} \frac{i}{k^2 + i\epsilon} \frac{k^2}{\mathbf{k}^2} n^\mu n^\nu \\
&= -in^\mu n^\nu \int \frac{d^3 k}{(2\pi)^3} \frac{e^{i\mathbf{k}(\mathbf{x}-\mathbf{x}')}}{\mathbf{k}^2} \delta(x^0 - x'^0) \\
&= -i\delta(x^0 - x'^0) \frac{n^\mu n^\nu}{4\pi|\mathbf{x} - \mathbf{x}'|} .
\end{aligned}
\tag{14.5.10}
$$

In perturbation theory, this term is compensated by the Coulomb interaction, which appears explicitly when one works in the Coulomb gauge. To see this in more detail, we must also consider the Hamiltonian. The Lagrangian density $\mathcal{L}$ (14.3.1)

$$
\mathcal{L} = -\frac{1}{4} F_{\mu\nu} F^{\mu\nu} - j_\mu A^\mu
\tag{14.3.1}
$$

can also be written in the form

$$
\mathcal{L} = \frac{1}{2}(\mathbf{E}^2 - \mathbf{B}^2) - j_\mu A^\mu ,
\tag{14.5.11}
$$

where

$$
\mathbf{E} = \mathbf{E}^{\text{tr}} + \mathbf{E}^{\text{l}}
\tag{14.5.12a}
$$

with the transverse and longitudinal components

$$
\mathbf{E}^{\text{tr}} = -\dot{\mathbf{A}}
\tag{14.5.12b}
$$

and

$$
\mathbf{E}^{\text{l}} = -\boldsymbol{\nabla} A^0 .
\tag{14.5.12c}
$$

In the Lagrangian, the mixed term

$$
\int d^3x\, \mathbf{E}^{\text{tr}} \cdot \mathbf{E}^{\text{l}} = \int d^3x\, \dot{\mathbf{A}} \cdot \boldsymbol{\nabla} A^0 ,
$$

vanishes, as can be seen by partial integration and use of $\boldsymbol{\nabla} \cdot \mathbf{A} = 0$. Thus, the Lagrangian density (14.5.11) is equivalent to

$$\mathcal{L} = \frac{1}{2}\left(\left(\dot{\mathbf{A}}^{\,\mathrm{tr}}\right)^2 + \left(\mathbf{E}^{\mathrm{l}}\right)^2 - \left(\boldsymbol{\nabla}\times\mathbf{A}\right)^2\right) - j_\mu A^\mu\ . \tag{14.5.13}$$

For the momentum conjugate to the electromagnetic potential $\mathbf{A}$, this yields:

$$\boldsymbol{\Pi}^{\mathrm{tr}} \equiv \frac{\partial\mathcal{L}}{\partial\dot{\mathbf{A}}} = -\dot{\mathbf{A}}\ . \tag{14.5.14}$$

This, in turn, yields the Hamiltonian density

$$\begin{aligned}
\mathcal{H} &= \mathcal{H}_\gamma + \mathcal{H}_{\mathrm{int}} \\
&= \frac{1}{2}\left(\boldsymbol{\Pi}^{\mathrm{tr}}\right)^2 + \frac{1}{2}(\boldsymbol{\nabla}\times\mathbf{A})^2 - \frac{1}{2}\left(\mathbf{E}^{\mathrm{l}}\right)^2 + j_\mu A^\mu\ ,
\end{aligned} \tag{14.5.15}$$

where the first two terms are the Hamiltonian density of the radiation field (14.4.3), and

$$\mathcal{H}_{\mathrm{int}} = -\frac{1}{2}\left(\mathbf{E}^{\mathrm{l}}\right)^2 + j_\mu A^\mu$$

is the interaction term. It will be helpful to separate out from the interaction term $\mathcal{H}_{\mathrm{int}}$ the part corresponding to the Coulomb interaction of the charge density

$$\mathcal{H}_{\mathrm{Coul}} = -\frac{1}{2}\left(\mathbf{E}^{\mathrm{l}}\right)^2 + j_0 A^0\ . \tag{14.5.16}$$

When integrated over space, this yields:

$$\begin{aligned}
H_{\mathrm{Coul}} &= \int d^3x\,\mathcal{H}_{\mathrm{Coul}} = \int d^3x\left(-\frac{1}{2}(\boldsymbol{\nabla}A_0)^2 + j_0 A^0\right) \\
&= \int d^3x\left(\frac{1}{2}A_0\boldsymbol{\nabla}^2 A_0 + j_0 A^0\right) = \frac{1}{2}\int d^3x\,j_0 A_0 \\
&= \frac{1}{2}\int d^3x\,d^3x'\,\frac{j_0(\mathbf{x},t)j_0(\mathbf{x}',t)}{4\pi|\mathbf{x}-\mathbf{x}'|}\ ,
\end{aligned} \tag{14.5.17}$$

which is exactly the Coulomb interaction between the charge densities $j_0(\mathbf{x},t)$. Thus, the total interaction now takes the form

$$H_{\mathrm{int}} = H_{\mathrm{Coul}} - \int d^3x\,\mathbf{j}(\mathbf{x},t)\mathbf{A}(\mathbf{x},t)\ . \tag{14.5.18}$$

The propagator of the transverse photons (14.5.6d), together with the Coulomb interaction, is thus equivalent to the following covariant propagator:

$$D_F^{\mu\nu}(x) = -g^{\mu\nu}\lim_{\epsilon\to 0}\int\frac{d^4k}{(2\pi)^4}\frac{e^{-ikx}}{k^2 + i\epsilon}\ . \tag{14.5.19}$$

As already stated at the beginning of this chapter, there are various ways to treat the quantized radiation field: In the Coulomb gauge, for every wave vector, one has as dynamical degrees of freedom the two transverse photons and, in addition, there is the instantaneous Coulomb interaction. Neither of these two descriptions on its own is covariant, but they can be combined to yield a covariant propagator, as in Eq. (14.5.19). In the Lorentz gauge, one has four photons that automatically lead to the covariant propagator (14.5.19) or (E.10b). As a result of the Lorentz condition, the longitudinal and scalar photons can only be excited in such a way that Eq. (E.20a) is satisfied for every state. They therefore make no contribution to any physically observable quantities, except the Coulomb interaction, which they mediate.

Problems

14.1 Derive the commutation relations (E.11) from (E.8).

14.2 Calculate the energy–momentum tensor for the radiation field. Show that the normal ordered momentum operator has the form

$$\mathbf{P} = : \int d^3x \, \mathbf{E} \times \mathbf{B} :$$
$$= \sum_{\mathbf{k},\lambda} \mathbf{k} \, a_{\mathbf{k}\lambda}^\dagger a_{\mathbf{k}\lambda} \ .$$

14.3 Using the results of Noether's theorem, deduce the form of the angular momentum tensor of the electromagnetic field starting from the Lagrangian density

$$\mathcal{L} = -\frac{1}{4} F^{\mu\nu} F_{\mu\nu} \ .$$

(a) Write down the orbital angular momentum density.
(b) Give the spin density.
(c) Explain the fact that, although $S = 1$, only the values ± 1 occur for the projection of the spin onto the direction $\mathbf{k}$.

15. Interacting Fields, Quantum Electrodynamics

In this chapter, the Lagrangians for interacting fields are presented and important concepts are discussed, e.g.: the interaction representation, perturbation theory, the S matrix, Wick's theorem and the Feynman rules. Some important scattering processes and fundamental physical phenomena in quantum electrodynamics are studied.

15.1 Lagrangians, Interacting Fields

15.1.1 Nonlinear Lagrangians

We now turn to the treatment of interacting fields. When there are nonlinear terms in the Lagrangian density, or in the Hamiltonian, transitions and reactions between particles become possible. The simplest example of a model demonstrating this is a neutral scalar field with a self-interaction,

$$\mathcal{L} = \frac{1}{2} \left(\partial_\mu \phi \right) \left(\partial^\mu \phi \right) - \frac{m^2}{2} \phi^2 - \frac{g}{4!} \phi^4 . \tag{15.1.1}$$

This so-called ϕ^4 theory is a theoretical model whose special significance lies in the fact that it enables one to study the essential phenomena of a nonlinear field theory in a particularly clear form. The division of ϕ into creation and annihilation operators shows that the ϕ^4 term leads to a number of transition processes. For example, two incoming particles with the momentum vectors $\mathbf{k}_1$ and $\mathbf{k}_2$ can scatter from one another to yield outgoing particles with the momenta $\mathbf{k}_3$ and $\mathbf{k}_4$, the total momentum being conserved.

As another example we consider the Lagrangian density for the interaction of charged fermions, described by the Dirac field ψ, with the radiation field A_μ

$$\mathcal{L} = \bar{\psi}(i\gamma^\mu \partial_\mu - m)\psi - \frac{1}{2} \left(\partial^\mu A^\nu \right) \left(\partial_\mu A_\nu \right) - e\bar{\psi}\gamma_\mu \psi A^\mu . \tag{15.1.2}$$

The interaction term is the lowest nonlinear term in A^μ and ψ that is bilinear in ψ (see remark (iv) at the end of Sect. 13.4.1) and Lorentz invariant. A physical justification for this form will be given in Sect. 15.1.2 making use of the known interaction with the electromagnetic field (5.3.40).

Quantum electrodynamics (QED), which is based on the Lagrangian density (15.1.2), is a theory describing the electromagnetic interaction between electrons, positrons, and photons. It serves as an excellent example of an interacting field theory for the following reasons:

(i) It contains a small expansion parameter, the Sommerfeld fine-structure constant $\alpha \approx \frac{1}{137}$, so that perturbation theory can be successfully applied.

(ii) Quantum electrodynamics is able to explain, among other things, the Lamb shift and the anomalous magnetic moment of the electron.

(iii) The theory is renormalizable.

(iv) Quantum electrodynamics is a simple (abelian) gauge theory.

(v) It admits a description of all essential concepts of quantum field theory (perturbation theory, S matrix, Wick's theorem, etc.).

15.1.2 Fermions in an External Field

Here, we consider the simplest case of the interaction of an electron field with a known electromagnetic field $A_{e\mu}$, which varies in space and time. The Dirac equation for this case reads:

$$(i\gamma^\mu \partial_\mu - m)\psi = e\gamma^\mu A_{e\mu}\psi \tag{15.1.3}$$

and has the Lagrangian density

$$\mathcal{L} = \bar{\psi}(\gamma^\mu(i\partial_\mu - eA_{e\mu}) - m)\psi \tag{15.1.4}$$
$$\equiv \mathcal{L}_0 + \mathcal{L}_1 \,,$$

where $\mathcal{L}_0$ is the free Lagrangian density and $\mathcal{L}_1$ the interaction with the field $A_{e\mu}$,

$$\begin{aligned}\mathcal{L}_0 &= \bar{\psi}(i\gamma^\mu \partial_\mu - m)\psi \\ \mathcal{L}_1 &= -e\bar{\psi}\gamma^\mu\psi A_{e\mu} \equiv -ej^\mu A_{e\mu} \,.\end{aligned} \tag{15.1.5}$$

The momentum conjugate to ψ_α is $\pi_\alpha = \frac{\partial \mathcal{L}}{\partial \dot{\psi}_\alpha} = i\psi_\alpha^\dagger$, as in (13.3.5), so that the Hamiltonian density is given by

$$\begin{aligned}\mathcal{H} &= \mathcal{H}_0 + \mathcal{H}_1 \\ &= \bar{\psi}(-i\gamma^j \partial_j + m)\psi + e\bar{\psi}\gamma^\mu\psi A_{e\mu} \,.\end{aligned} \tag{15.1.6}$$

In the above, $A_{e\mu}$ was an external field. In the next section we will consider the coupling to the radiation field, which is itself a quantized field.

15.1.3 Interaction of Electrons with the Radiation Field: Quantum Electrodynamics (QED)

15.1.3.1 The Lagrangian and the Hamiltonian Densities

The Hamiltonian and the Lagrangian densities of the interacting Dirac and radiation fields are obtained by replacing $A_{e\mu}$ in (15.1.5) by the quantized radiation field and adding the Lagrangian density of the free radiation field

$$\mathcal{L} = \bar{\psi}(i\partial\!\!\!/ - m)\psi - \frac{1}{2}\left(\partial_\nu A_\mu\right)\left(\partial^\nu A^\mu\right) - e\bar{\psi}A\!\!\!/\psi \ . \tag{15.1.7}$$

This is identical to the form postulated for formal reasons in (15.1.2). It leads to the conjugate momenta to the Dirac and radiation fields:

$$\pi_\alpha = \frac{\partial \mathcal{L}}{\partial \dot{\psi}_\alpha} = i\psi_\alpha^\dagger \quad , \quad \Pi_\mu = \frac{\partial \mathcal{L}}{\partial \dot{A}^\mu} = -\dot{A}_\mu \ , \tag{15.1.8}$$

and the Hamiltonian density operator

$$\mathcal{H} = \mathcal{H}_0^{\text{Dirac}} + \mathcal{H}_0^{\text{photon}} + \mathcal{H}_1 \ , \tag{15.1.9}$$

where $\mathcal{H}_0^{\text{Dirac}}$ and $\mathcal{H}_0^{\text{photon}}$ are the Hamiltonian densities of the free Dirac and radiation fields (Eq. (13.3.7) and (E.14)). Here, $\mathcal{H}_1$ represents the interaction between these fields

$$\mathcal{H}_1 = e\bar{\psi}A\!\!\!/\psi \ . \tag{15.1.10}$$

15.1.3.2 Equations of Motion
of Interacting Dirac and Radiation Fields

For the Lagrangian density (15.1.7), the equations of motion of the field operators in the Heisenberg picture read:

$$(i\partial\!\!\!/ - m)\psi = e A\!\!\!/\psi \tag{15.1.11a}$$

$$\Box A^\mu = e\bar{\psi}\gamma^\mu\psi \ . \tag{15.1.11b}$$

These are nonlinear field equations which, in general, cannot be solved exactly. An exception occurs for the simplified case of one space and one time dimension: a few such $(1+1)$-dimensional field theories can be solved exactly. An interesting example is the Thirring model

$$(i\partial\!\!\!/ - m)\psi = g\bar{\psi}\gamma^\mu\psi\gamma_\mu\psi \ . \tag{15.1.12}$$

This can also be obtained as a limiting case of Eq. (15.1.11a) with a massive radiation field, i.e.,

$$(\Box + M)A^\mu = e\bar{\psi}\gamma^\mu\psi \ , \tag{15.1.13}$$

in the limit of infinite M. In general, however, one is obliged to use the methods of perturbation theory. These will be treated in the next sections.

15.2 The Interaction Representation,
Perturbation Theory

Experimentally, one is primarily interested in *scattering processes*. In this section we derive the S-matrix formalism necessary for the theoretical description of such processes. We begin by reiterating a few essential points from quantum mechanics[1] concerning the interaction representation. These will facilitate our perturbation treatment of scattering processes.

15.2.1 The Interaction Representation (Dirac Representation)

We divide the Lagrangian density and the Hamiltonian into a free and an interaction part, where H_0 is time independent:

$$\mathcal{L} = \mathcal{L}_0 + \mathcal{L}_1 \tag{15.2.1}$$

$$H = H_0 + H_1 . \tag{15.2.2}$$

When the interaction $\mathcal{L}_1$ contains no derivatives, the density corresponding to the interaction Hamiltonian $H_1 = \int d^3x \mathcal{H}_1$ is given by

$$\mathcal{H}_1 = -\mathcal{L}_1 \tag{15.2.3}$$

We shall make use of the *Schrödinger representation* in which the states $|\psi, t\rangle$ are time dependent and satisfy the Schrödinger equation

$$i\frac{\partial}{\partial t} |\psi, t\rangle = H |\psi, t\rangle . \tag{15.2.4}$$

The operators are denoted by A in the subsequent equations. The fundamental operators such as the momentum, and the field operators such as $\psi(\mathbf{x})$, are time independent in the Schrödinger picture. (Note that the field equations (13.1.13), (13.3.1), etc. were equations of motion in the Heisenberg picture.) When external forces are present, one may also encounter Schrödinger operators with explicit time dependence (e.g., in Sect. 4.3 on linear response theory). The definition of the *interaction representation* reads:

$$|\psi, t\rangle_I = e^{iH_0 t} |\psi, t\rangle , \quad A_I(t) = e^{iH_0 t} A e^{-iH_0 t} . \tag{15.2.5}$$

In the interaction representation, due to Eq. (15.2.4), the states and the operators satisfy the equations of motion

$$i\frac{\partial}{\partial t} |\psi, t\rangle_I = H_{1I}(t) |\psi, t\rangle_I \tag{15.2.6a}$$

$$\frac{d}{dt} A_I(t) = i [H_0, A_I(t)] + \frac{\partial}{\partial t} A_I(t) . \tag{15.2.6b}$$

The final term in (15.2.6b) only occurs when the Schrödinger operator A depends explicitly on time. In the following, we will make use of the abbreviated notation

$$|\psi(t)\rangle \equiv |\psi, t\rangle_I \tag{15.2.7a}$$

and

$$H_I(t) \equiv H_{1I}(t) . \tag{15.2.7b}$$

---

[1] See, e.g., QM I, Sects. 8.5.3 and 16.3.1.

The equation of motion for $|\psi(t)\rangle$ has the form of a Schrödinger equation with time-dependent Hamiltonian $H_I(t)$. When the interaction is switched off, i.e., when $H_I(t) = 0$, the state vector in the interaction picture is time independent. The field operators in this representation satisfy the equations of motion

$$\frac{d\phi_{rI}(\mathbf{x}, t)}{dt} = \mathrm{i} \left[H_0, \phi_{rI}(\mathbf{x}, t) \right] \tag{15.2.8}$$

i.e., the free equations of motion. The field operators in the interaction representation are thus identical to the Heisenberg operators of free fields. Since $\mathcal{L}_1$ contains no derivatives, the canonical conjugate fields have the same form as the free fields, e.g.,

$$\frac{\partial \mathcal{L}}{\partial \dot{\psi}_\alpha} = \frac{\partial \mathcal{L}_0}{\partial \dot{\psi}_\alpha}$$

in quantum electrodynamics. Hence, the equal time commutation relations of the interacting fields are the same as those for the free fields.

Since the interaction representation arises from the Schrödinger representation, and hence also from the Heisenberg representation, through a unitary transformation, the interacting fields obey the same commutation relations as the free fields. Since the equations of motion in the interaction picture are identical to the free equations of motion, the operators have the same simple form, the same time dependence, and the same representation in terms of creation and annihilation operators as the free operators. The plane waves (spinor solutions, free photons, and free mesons) are still solutions of the equations of motion and lead to the same expansion of the field operators as in the free case. The Feynman propagators are again $\mathrm{i}\Delta_F(x - x')$ etc., where the vacuum is defined here with reference to the operators $a_{r\mathbf{k}'}, b_{r\mathbf{k}'}, d_{\lambda\mathbf{k}}$. The time evolution of the states is determined by the interaction Hamiltonian.

Let us once more draw attention to the differences between the various representations in quantum mechanics. In the Schrödinger representation the states are time dependent. In the Heisenberg representation the state vector is time independent, whereas the operators are time dependent and satisfy the Heisenberg equation of motion. In the interaction representation the time dependence is shared between the operators and the states. The free part of the Hamiltonian determines the time dependence of the operators. The states change in time as a result of the interaction. Thus, in the interaction representation, the field operators of an interacting nonlinear field theory satisfy the free field equations: for the real Klein–Gordon field these are given by Eq. (13.1.2), for the complex Klein–Gordon field by (13.2.2), for the Dirac field by (13.3.1), and for the radiation field by (14.1.8). For the time dependence of these fields one thus has the corresponding plane-wave expansions (13.1.5), (13.2.5), (13.3.18), and (14.4.4) or (E.5) (see also (15.3.12a–c)). We also recall the relation between Schrödinger and Heisenberg operators in the interacting field theory

$$\psi_{\text{Heisenb.}}(\mathbf{x}, t) = e^{iHt} \psi_{\text{Schröd.}}(\mathbf{x}) e^{-iHt}$$
$$A_{\text{Heisenb.}}(\mathbf{x}, t) = e^{iHt} A_{\text{Schröd.}}(\mathbf{x}) e^{-iHt} .$$

(15.2.9)

In the interaction representation, one obtains

$$\psi_I(x) \equiv e^{iH_0 t} \psi_{\text{Schröd.}}(\mathbf{x}) e^{-iH_0 t} = \psi(x)$$
$$A_I^\mu(x) \equiv e^{iH_0 t} A_{\text{Schröd.}}^\mu(\mathbf{x}) e^{-iH_0 t} = A^\mu(x) ,$$

(15.2.10)

where $\psi(x)$ $(A^\mu(x))$ is the free Dirac field (radiation field) in the Heisenberg representation, $x \equiv (\mathbf{x}, t)$. Since the interaction Hamiltonian is a polynomial in the fields, e.g., in quantum electrodynamics, in the Schrödinger picture,

$$H_1 = e \int d^3 x \bar{\psi} \gamma^\mu \psi A_\mu ,$$

and in the interaction representation one has

$$H_I(t) \equiv H_{1I}(t) \equiv e^{iH_0 t} H_1 e^{-iH_0 t}$$
$$= e \int d^3 x \bar{\psi}(x) \gamma^\mu \psi(x) A_\mu(x) ,$$

(15.2.11)

$x \equiv (\mathbf{x}, t)$. Here, the field operators are identical to the Heisenberg operators of the free field theories, as given in (13.3.18) and (14.4.4) or (E.5).

One finds the *time-evolution operator in the interaction picture* by starting from the formal solution of the Schrödinger equation (15.2.4), $|\psi, t\rangle = e^{-iH(t-t_0)} |\psi, t_0\rangle$. This leads, in the interaction representation, to

$$|\psi(t)\rangle = e^{iH_0 t} e^{-iH(t-t_0)} |\psi, t_0\rangle$$
$$= e^{iH_0 t} e^{-iH(t-t_0)} e^{-iH_0 t_0} |\psi(t_0)\rangle$$
$$\equiv U'(t, t_0) |\psi(t_0)\rangle$$

(15.2.12)

with the time-evolution in the interaction picture

$$U'(t, t_0) = e^{iH_0 t} e^{-iH(t-t_0)} e^{-iH_0 t_0} .$$

(15.2.13)

From this relation, one immediately recognizes the group property

$$U'(t_1, t_2) U'(t_2, t_0) = U'(t_1, t_0)$$

(15.2.14a)

and the unitarity

$$U'^\dagger(t, t_0) = U'(t_0, t) = U'^{-1}(t, t_0)$$

(15.2.14b)

of the time-evolution operator. Unitarity requires the hermiticity of H and H_0. The equation of motion for this time-evolution operator is obtained from

$$i \frac{\partial}{\partial t} U'(t, t_0) = e^{iH_0 t} (-H_0 + H) e^{-iH(t-t_0)} e^{-iH_0 t_0}$$
$$= e^{iH_0 t} H_1 e^{-iH_0 t} e^{iH_0 t} e^{-iH(t-t_0)} e^{-iH_0 t_0}$$

(or from the equation of motion (15.2.6a) for $|\psi(t)\rangle$):

$$i\frac{\partial}{\partial t}U'(t,t_0) = H_I(t)U'(t,t_0) . \qquad (15.2.15)$$

Remark. This equation of motion also holds in the case where H, and hence H_1, have an explicit time dependence: Then, in Eqs. (15.2.12)–(15.2.15), one must replace $e^{-iH(t-t_0)}$ by the general Schrödinger time-evolution operator $U(t,t_0)$, which satisfies the equation of motion $i\frac{\partial}{\partial t}U(t,t_0) = HU(t,t_0)$.

15.2.2 Perturbation Theory

The equation of motion for the time evolution operator (15.2.15) in the interaction picture can be solved formally using the initial condition

$$U'(t_0,t_0) = 1 \qquad (15.2.16)$$

in the form

$$U'(t,t_0) = 1 - i\int_{t_0}^{t} dt_1 H_I(t_1)U'(t_1,t_0) , \qquad (15.2.17)$$

i.e., it is now given by an integral equation. The iteration of (15.2.17) yields:

$$U'(t,t_0) = 1 + (-i)\int_{t_0}^{t} dt_1 H_I(t_1) + (-i)^2 \int_{t_0}^{t} dt_1 \int_{t_0}^{t_1} dt_2 H_I(t_1)H_I(t_2)$$

$$+ (-i)^3 \int_{t_0}^{t} dt_1 \int_{t_0}^{t_1} dt_2 \int_{t_0}^{t_2} dt_3 H_I(t_1)H_I(t_2)H_I(t_3) + \dots ,$$

i.e.,

$$U'(t,t_0) = \sum_{n=0}^{\infty} (-i)^n \int_{t_0}^{t} dt_1 \int_{t_0}^{t_1} dt_2 \cdots \qquad (15.2.18)$$

$$\times \int_{t_0}^{t_{n-1}} dt_n \, H_I(t_1)H_I(t_2)\dots H_I(t_n) .$$

By making use of the time-ordering operator T, this infinite series can be written in the form

$$U'(t,t_0) = \sum_{n=0}^{\infty} \frac{(-i)^n}{n!} \int_{t_0}^{t} dt_1 \int_{t_0}^{t} dt_2 \cdots \qquad (15.2.19)$$

$$\times \int_{t_0}^{t} dt_n \, T\left(H_I(t_1)H_I(t_2)\dots H_I(t_n)\right)$$

or, still more compactly, as

$$U'(t,t_0) = T\exp\left(-i\int_{t_0}^{t} dt' H_I(t')\right) . \qquad (15.2.19')$$

One can readily convince oneself of the equivalence of expressions (15.2.18) and (15.2.19) by considering the nth-order term: In (15.2.19) the times fulfil either the inequality sequence $t_1 \geq t_2 \geq \ldots \geq t_n$, or a permutation of this inequality sequence. In the former case, the contribution to (15.2.19) is

$$\frac{(-\mathrm{i})^n}{n!} \int_{t_0}^{t} dt_1 \int_{t_0}^{t_1} dt_2 \ldots \int_{t_0}^{t_{n-1}} dt_n \, (H_I(t_1) \ldots H_I(t_n)) \; .$$

In the latter case, i.e., when a permutation of the inequality sequence applies, one can rename the integration variables and, thereby, return once more to the case $t_1 \geq t_2 \geq \ldots t_n$. One thus obtains the same contribution $n!$ times. This proves the equivalence of (15.2.18) and (15.2.19). The contribution to (15.2.19) with n factors of H_I is referred to as the nth-order term.

The time-ordering operator in (15.2.19) and (15.2.19'), also known as Dyson's time-ordering operator or the chronological operator, signifies at this stage the time ordering of the composite operators $H_I(t)$. If, as is the case in quantum electrodynamics, the Hamiltonian contains only even powers of Fermi operators, it can be replaced by what is known as the Wick's time-ordering operator, which time-orders the field operators. It is in this sense that we shall use T in the following. The time-ordered product $T(\ldots)$ orders the factors so that later times appear to the left of earlier times. All Bose operators are treated as if they commute, and all Fermi operators as if they anticommute.

We conclude this section with a remark concerning the significance of the time-evolution operator $U'(t, t_0)$, which, in the interaction picture, according to Eq. (15.2.12), gives the state $|\psi(t)\rangle$ from a specified state $|\psi(t_0)\rangle$. If at time t_0 the system is in the state $|i\rangle$, then the probability of finding the system at a later time t in the state $|f\rangle$ is given by

$$|\langle f| U'(t, t_0) |i\rangle|^2 \; . \tag{15.2.20}$$

From this, one obtains the transition rate, i.e., the probability per unit time of a transition from an initial state $|i\rangle$, to a final state $|f\rangle$ differing from the initial state ($\langle i|f\rangle = 0$) as,

$$w_{i \to f} = \frac{1}{t - t_0} |\langle f| U'(t, t_0) |i\rangle|^2 \; . \tag{15.2.21}$$

15.3 The S Matrix

15.3.1 General Formulation

We now turn our attention to the description of scattering processes. The typical situation in a scattering experiment is the following: At the initial time (idealized as $t = -\infty$), we have widely separated and thus noninteracting

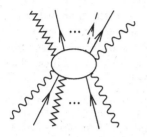

Fig. 15.1. Schematic representation of a general scattering process. A number of particles are incident upon one another, interact, and scattered particles leave the interaction region. The number of scattered particles can be greater or smaller than the number of incoming particles

particles. These particles approach one another and interact for a short time corresponding to the range of the forces. The particles, and possibly newly created ones, that remain after this interaction then travel away from one another and cease to interact. At a much later time (idealized as $t = \infty$), these are observed. The scattering process is represented schematically in Fig. 15.1. The time for which the particles interact is very much shorter than the time taken for the particles to travel from the source to the point of observation (detector); hence, it is reasonable to take the final and initial times as $t = \pm\infty$, respectively.

At the initial time $t_i = -\infty$ of the scattering process, we have a state $|i\rangle$ corresponding to free, noninteracting particles

$$|\psi(-\infty)\rangle = |i\rangle \ .$$

After the scattering, the particles that remain are again well separated from one another and are described by

$$|\psi(\infty)\rangle = U'(\infty, -\infty)\,|i\rangle \ . \tag{15.3.1}$$

The transition amplitude into a particular final state $|f\rangle$ is given by

$$\langle f|\psi(\infty)\rangle = \langle f|\,U'(\infty, -\infty)\,|i\rangle = \langle f|\,S\,|i\rangle = S_{fi} \ . \tag{15.3.2}$$

The states $|i\rangle$ and $|f\rangle$ are eigenstates of H_0. One imagines that the interaction is switched off at the beginning and the end. Here, we have introduced the scattering matrix, or S matrix for short, by way of $S = U(\infty, -\infty)$,

$$S = \sum_{n=0}^{\infty} \frac{(-\mathrm{i})^n}{n!} \int_{-\infty}^{\infty} dt_1 \int_{-\infty}^{\infty} dt_2 \dots \tag{15.3.3}$$

$$\times \int_{-\infty}^{\infty} dt_n \, T\left(H_I(t_1) H_I(t_2) \dots H_I(t_n)\right) \ .$$

If the Hamiltonian is expressed in terms of the Hamiltonian density, one obtains

$$S = \sum_{n=0}^{\infty} \frac{(-\mathrm{i})^n}{n!} \int \dots \int d^4 x_1 \dots d^4 x_n \, T\left(\mathcal{H}_I(x_1) \dots \mathcal{H}_I(x_n)\right)$$

$$= T\left(\exp\left(-\mathrm{i}\int d^4 x (\mathcal{H}_I(x))\right)\right) \ . \tag{15.3.4}$$

Since the interaction operator is Lorentz invariant, and the time ordering does not change under orthochronous Lorentz transformations, the scattering matrix is itself invariant with respect to Lorentz transformations, i.e., it is a relativistically invariant quantity. In quantum electrodynamics, the interaction Hamiltonian density appearing in (15.3.4) is

$$\mathcal{H}_I = e : \bar{\psi}(x)\gamma^\mu \psi(x)A_\mu(x) : . \tag{15.3.5}$$

The unitarity of $U(t, t_0)$ (see (15.2.14b)) implies that the S matrix is also unitary

$$SS^\dagger = 1 \tag{15.3.6a}$$
$$S^\dagger S = 1 \tag{15.3.6b}$$

or, equivalently,

$$\sum_n S_{fn}S_{in}^* = \delta_{fi} \tag{15.3.7a}$$

$$\sum_n S_{nf}^* S_{ni} = \delta_{fi} . \tag{15.3.7b}$$

To appreciate the significance of the unitarity, we expand the asymptotic state that evolves from the initial state $|i\rangle$ through

$$|\psi(\infty)\rangle = S|i\rangle \tag{15.3.8}$$

in terms of a complete set of final states $\{|f\rangle\}$:

$$|\psi(\infty)\rangle = \sum_f |f\rangle \langle f|\psi(\infty)\rangle = \sum_f |f\rangle S_{fi} . \tag{15.3.9}$$

We now form

$$\langle \psi(\infty)|\psi(\infty)\rangle = \sum_f S_{fi}^* S_{fi} = \sum_f |S_{fi}|^2 = 1 , \tag{15.3.10}$$

where we have used (15.3.7b). The unitarity of the S matrix expresses conservation of probability. If the initial state is $|i\rangle$, then the probability of finding the final state $|f\rangle$ in an experiment is given by $|S_{fi}|^2$. The unitarity of the S matrix guarantees that the sum of these probabilites over all possible final states is equal to one. Since particles may be created or annihilated, the possible final states may contain particles different to those in the initial states.

The states $|i\rangle$ and $|f\rangle$ have been assumed to be eigenstates of the unperturbed Hamiltonian H_0, i.e., the interaction was assumed to be switched off. In reality, the physical states of real particles differ from these free states. The interaction turns the "bare" states into "dressed" states. An electron in such a state is surrounded by a cloud of virtual photons that are continually being emitted and reabsorbed, as illustrated in Fig. 15.2. [2]

---

[2] In order that the energy spectrum of the bare (free) particle be identical to that of the physical particle, the Lagrangian density of the Dirac field is reparameterized

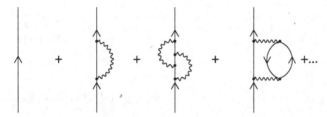

Fig. 15.2. The propagation of a real (physical) electron involves the free propagation and the propagation that includes the additional emission and reabsorption of virtual photons. The significance of the different lines is explained in Fig. 15.3.

The calculation of the transition elements between bare states $|i\rangle$ and $|f\rangle$ can be justified by appealing to the *adiabatic hypothesis*. The interaction Hamiltonian $H_I(t)$ is replaced by $H_I(t)\zeta(t)$, where

$$\lim_{t\to\pm\infty} \zeta(t) = 0 \quad \text{and} \quad \zeta(t) = 1 \quad \text{for} \quad -T < t < T ,$$

i.e., at time $t = -\infty$ one has free particles. During the time interval $-\infty < t < T$, the interaction turns the free particles into physical particles. Thus, in the time interval $[-T, T]$, we have real particles that experience the total interaction $H_I(t)$. Since the particles involved in a scattering process are initially widely separated, they only interact during the time interval $[-\tau, \tau]$, which is determined by the range of the interaction and the speed of the particles. The time T must, of course, be much larger than τ: $T \gg \tau$. The assumption of the adiabatic hypothesis is that the scattering cannot depend on the description of the states long before, or long after, the interaction. At the end of the calculation one takes the limit $T \to \infty$. If one is calculating a process only in the lowest order perturbation theory at which it occurs, one uses the entire interaction for the transition and not to convert the bare state into a physical state. In this case, one can take the limit $T \to \infty$ from the outset, and use the full interaction Hamiltonian in the whole time interval.

The types of transition processes are determined by the form of the interaction Hamiltonian. If the initial state contains a certain number of particles, then the effect of the term of nth order in S (Eq.(15.3.4)) is the following: The application of $\mathcal{H}_I(x_n)$ causes some of the original particles to be annihilated and new ones to be created. The next factor $\mathcal{H}_I(x_{n-1})$ leads to further annihilation and creation processes, etc. It is necessary here to integrate over the space–time position of these processes. We will elucidate this for a few examples taken from quantum electrodynamics. Here, the interaction Hamiltonian

---

as

$$\mathcal{L} = \bar{\psi}(i\partial\!\!\!/ - m_R)\psi - e\bar{\psi}A\!\!\!/\psi + \delta m\bar{\psi}\psi ,$$

with the renormalized (physical) mass $m_R = m + \delta m$. The Hamiltonian density then includes an additional perturbation term $-\delta m\bar{\psi}\psi$. In the lowest order processes treated in the next section, this plays no role. It will be analysed further when we come to the topic of radiative corrections in Sect. 15.6.1.2.

density is

$$\mathcal{H}_I(x) = e : \bar{\psi}(x)\slashed{A}(x)\psi(x) : \tag{15.3.11}$$

with the field operators

$$\psi(x) = \sum_{\mathbf{p}, r=1,2} \left(\frac{m}{VE_\mathbf{p}}\right)^{1/2} \left(b_{r\mathbf{p}} u_r(p)\, e^{-ipx} + d_{r\mathbf{p}}^\dagger w_r(p)\, e^{ipx}\right) \tag{15.3.12a}$$

$$\bar{\psi}(x) = \sum_{\mathbf{p}, r=1,2} \left(\frac{m}{VE_\mathbf{p}}\right)^{1/2} \left(b_{r\mathbf{p}}^\dagger \bar{u}_r(p)\, e^{ipx} + d_{r\mathbf{p}} \bar{w}_r(p)\, e^{-ipx}\right) \tag{15.3.12b}$$

$$A^\mu(x) = \sum_{\mathbf{k}} \sum_{\lambda=0}^{3} \left(\frac{1}{2V|\mathbf{k}|}\right)^{1/2} \epsilon_\lambda^\mu(\mathbf{k}) \left(a_\lambda(\mathbf{k})e^{-ikx} + a_\lambda^\dagger(\mathbf{k})e^{ikx}\right). \tag{15.3.12c}$$

As in previous chapters, at this stage it is useful to introduce a graphical representation, shown in Fig. 15.3. A photon is represented by a wavy line, an electron by a full line, and a positron by a full line with an arrow in the opposite time direction. If the particle interacts with an external electromagnetic field, it is similarly represented to a photon line, with a cross[3]

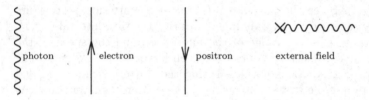

photon   electron   positron   external field

Fig. 15.3. The lines of the Feynman diagrams; the time axis points upwards

15.3.2 Simple Transitions

We first discuss the basic processes that are brought about by a single factor $\mathcal{H}_I(x)$. The three field operators in $\mathcal{H}_I(x)$ can be split into components of positive and negative frequency, yielding a total of eight terms. For example, the term ψ^+ annihilates an electron, while ψ^- creates a positron. The term $e\bar{\psi}^-(x)A^+(x)\psi^+(x)$ annihilates a photon and an electron originally present at x, and once more creates an electron at x. This process is represented by the first diagram in Fig. 15.4. One could also describe this as the absorption of a photon by an electron. If one instead takes the summand A^- from the photon field, one obtains a transition in which a photon is created at the position x, in other words, a process in which an electron at x emits a photon (the first of the lower series of diagrams in Fig. 15.4). The other six elementary processes

---

[3] As already mentioned elsewhere, these graphical representations are more than just simple illustrations: In the form of Feynman diagrams they prove to be unambiguous representations of the analytical expressions of perturbation theory.

The elementary processes of the QED vertex

$$\mathcal{H}_I(x) = e : (\ \underbrace{\bar{\psi}^+}_{\text{ann } e^+} + \underbrace{\bar{\psi}^-}_{\text{cr } e^-}\ )(\ \underbrace{A\!\!\!/^+}_{\text{ann } \gamma} + \underbrace{A\!\!\!/^-}_{\text{cr } \gamma}\ )(\ \underbrace{\psi^+}_{\text{ann } e^-} + \underbrace{\psi^-}_{\text{cr } e^+}\ ) :$$

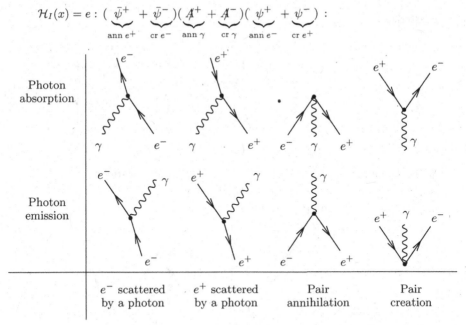

Fig. 15.4. The elementary processes of the QED vertex; the time axis points upwards

are also shown in this figure. It is not necessary to discuss each of these in detail; we shall select just one. The third diagram of the lower series stems from $e\bar{\psi}^+ A^- \psi^+$ and represents the annihilation of an electron–positron pair, i.e., the transition of an electron and a positron into a photon. The range of possible processes is determined by the form of $\mathcal{H}_I(x)$ and its powers. The points on the diagrams at which particles are incident or are emitted (i.e., are created or annihilated) are also known as vertices.

Figure 15.5 shows processes of various order, all of which are possible for an initial state consisting of an incoming electron and an incoming positron. Figure 15.5a shows the noninteracting motion described by the zeroth-order term. Figure 15.5b shows the second-order interaction in which the electron emits a photon which is absorbed by the positron. This process likewise contains the emission of a photon by the positron and its absorption by the electron. This process leads to a final state that once again contains an electron and a positron, hence it describes the scattering of an electron–positron pair. The diagram is of second order with two vertices. At higher orders of perturbation theory, the electron and positron can interact to produce the fourth-order scattering process represented in Fig. 15.5c. In the diagram shown in Fig. 15.5d, the positron propagates without interaction. The electron first emits a photon and then experiences a deflection due to an external

Initial state $e^- + e^+$

Scattering:

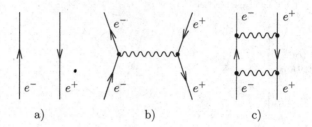

a)   b)   c)

Bremsstrahlung: Pair annihilation:

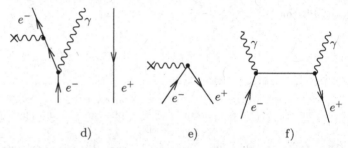

d)   e)   f)

Fig. 15.5. Examples of reactions having as the initial state electron plus positron: a) motion of noninteracting electron and positron; b) scattering of electron and positron; c) fourth-order scattering, exchange of two photons; d) Bremsstrahlung of the electron in the presence of an external field; the positron here propagates without interaction; e) pair annihilation in the presence of an external time-varying field whose frequency is equal to the energy of the e^+e^- pair; f) pair annihilation with a final state containing two photons

potential. The final state consists of an electron, a positron, and a photon. One refers here to the bremsstrahlung of the electron in the presence of the external field. In Fig. 15.5e an external field causes the annihilation of the electron–positron pair. In order that this process may really happen, the frequency of the external field must be high enough that it at least equals the energy of the electron–positron pair. The diagram given by Fig. 15.5f represents pair annihilation, the final state consisting of two photons.

We should now like to establish the transition probabilities for these processes. In order to determine the energy, momentum, and angular dependence of the individual transitions, one has to calculate the matrix elements. The procedure is similar to that for calculating the correlation functions in non-relativistic many-particle physics. One has to re-order the creation and annihilation operators, using the commutation and anticommutation relations, in such a way that all annihilation operators are on the right, and all creation operators on the left. The effect of an annihilation operator on the vacuum state to the right is to yield zero, as is the effect of a creation operator acting to the left. In a transition from a state $|i\rangle$ to a state $|f\rangle$, there are contributions only from those summands of products in $T\left(\mathcal{H}(x_1)\ldots\mathcal{H}(x_n)\right)$ for which the creation and annihilation operators exactly compensate one another[4]. The commutation of annihilation operators to the right then yields finite contributions from commutators or anticommutators. As will emerge, these can be expressed in terms of propagators. Thus, the result for the transition amplitude has the following structure: If the diagram consists of vertices at positions $x_1,\ldots,x_n$ and of incoming and outgoing particles, then the result is a product of propagators and this product is to be integrated over all positions of the vertices $x_1,\ldots,x_n$.

For simple processes it is easy to carry out the procedure that we have sketched step by step. In so doing, we find a set of rules which enable an analytical expression to be assigned to every diagram. These are known as the Feynman rules. To derive the Feynman rules systematically, one needs Wick's theorem, which allows an arbitrary time-ordered product to be represented by a sum of normal ordered products. The lines in the Feynman diagrams that correspond to incoming and outgoing particles are known as external lines, and the others are called internal lines. The particles represented by internal lines are termed virtual particles.

In the diagram of Fig. 15.5f, the internal line can be viewed either as the motion of a virtual electron from the left- to the right-hand vertex, or as the motion of a positron from the right- to the left-hand vertex. To obtain the total transition probability, one must integrate over all space–time positions of the two vertices. Both processes are described by the Feynman propagator that analytically represents this internal line (see also the discussion at the end of Sect. 13.1).

*15.4 Wick's Theorem

In order to calculate the transition amplitude from the state $|i\rangle$ to the state $|f\rangle$, one needs to determine the matrix element $\langle f|\,S\,|i\rangle$. If one considers a particular order of perturbation theory, one has to evaluate the matrix element of a time-ordered product of interaction Hamiltonians. Of the many

---

[4] If, for example, the initial state $|i\rangle$ contains an electron with quantum numbers $(\mathbf{p}, r)$ and a photon with $(\mathbf{k}, \lambda)$, then it is of the form

$$|i\rangle = b_{r\mathbf{p}}^{\dagger} a_{\lambda}^{\dagger}(\mathbf{k})\,|0\rangle \ ,$$

whereas the final state $|f\rangle$ with particles $(\mathbf{p}', r')$ and $(\mathbf{k}', \lambda')$ appears in bra form as

$$\langle f| = \langle 0|\,a_{\lambda'}(\mathbf{k}') b_{r'\mathbf{p}'} \ .$$

terms in the perturbation expansion, contributions come only from those whose application to $|i\rangle$ yields the state $|f\rangle$. Hence, (apart from the possibility of individual particles moving without interaction) the corresponding perturbation-theoretical contribution to the S matrix must contain those annihilation operators that annihilate the particles in $|i\rangle$ and those creation operators that create the particles in $|f\rangle$. In addition, a general term in S will also contain further creation and annihilation operators responsible for the creation and subsequent annihilation of *virtual* particles. These particles are termed virtual because they are not present in the initial or final states; they are emitted and reabsorbed in intermediate processes, e.g., the photon in Fig. 15.5b. The virtual particles do not obey the energy–momentum relation, $p^2 = m^2$, valid for real particles, i.e., they do not lie on the mass shell. As already mentioned in the previous section, one can calculate the value of such matrix elements of S by using the commutation relations to move the annihilation operators to the right. Instead of carrying out this calculation for every single case individually, it is helpful to rewrite the time-ordered products so that they are normal ordered from the start, i.e., so that all annihilation operators are to the left of all creation operators. *Wick's theorem* tells us how an arbitrary time-ordered product can be represented as a sum of normal ordered products. Wick's theorem is the basis for the systematic calculation of pertubation-theoretical contributions and their representation by means of Feynman diagrams.

Since the Hamiltonian density of a normal ordered product is

$$\mathcal{H}(x) = e : \bar{\psi}(x)\slashed{A}(x)\psi(x) : , \tag{15.4.1}$$

the nth-order term of S has the form

$$\begin{aligned} S^{(n)} &= \frac{(-ie)^n}{n!} \int d^4 x_1 \ldots d^4 x_n \\ &\times T\left(: \bar{\psi}(x_1)\slashed{A}(x_1)\psi(x_1) : \ldots : \bar{\psi}(x_n)\slashed{A}(x_n)\psi(x_n) :\right) ; \end{aligned} \tag{15.4.2}$$

this type of time-ordered product of partially normal ordered factors is known as a *mixed* time-ordered product.

In order to facilitate the formulation of Wick's theorem, we summarize a few properties of the time-ordered and normal ordered products that were introduced in Eqs. (13.1.28) and (13.4.12).

For given field operators $A_1, A_2, A_3, \ldots$, one has (see Eq. (13.1.18c)) the distributive law

$$\begin{aligned} : (A_1 + A_2)A_3 A_4 : &=: A_1 A_3 A_4 + A_2 A_3 A_4 : \\ &=: A_1 A_3 A_4 : + : A_2 A_3 A_4 : . \end{aligned} \tag{15.4.3}$$

The *contraction* of two field operators A and B, such as $\psi(x_1), \psi^\dagger(x_2)$, or $A^\mu(x_3), \ldots$, is defined by

$$\underline{AB} \equiv T(AB) - : AB : . \tag{15.4.4}$$

It is easy to convince oneself that such contractions are c numbers: According to the general definition, $T(AB)$ orders the operators A and B chronologically, and, for the case of two Fermi operators, introduces a factor (-1). Since the commutator or the anticommutator of free fields is a c number, $T(AB) - AB$ is also a c number, and the same is true of $: AB : -AB$ and the difference (15.4.4). Since the vacuum expectation value of a normal ordered product vanishes, it follows from (15.4.4) that

$$\contraction{}{A}{}{B} AB = \langle 0| T(AB) |0\rangle . \tag{15.4.5}$$

With this, the most important contractions are already known as a result of the Feynman propagators evaluated in (13.1.31), (13.4.16), and (14.5.19). For the real and the complex Klein–Gordon field, for the Dirac field, and for the radiation field, respectively, we find the following:

$$\begin{aligned}
\contraction{}{\phi}{(x_1)}{\phi} \phi(x_1)\phi(x_2) &= \mathrm{i}\Delta_F(x_1 - x_2) \\
\contraction{}{\phi}{(x_1)}{\phi^\dagger} \phi(x_1)\phi^\dagger(x_2) &= \contraction{}{\phi^\dagger}{(x_2)}{\phi} \phi^\dagger(x_2)\phi(x_1) = \mathrm{i}\Delta_F(x_1 - x_2) \\
\contraction{}{\psi_\alpha}{(x_1)}{\bar\psi_\beta} \psi_\alpha(x_1)\bar\psi_\beta(x_2) &= -\contraction{}{\bar\psi_\beta}{(x_2)}{\psi_\alpha}\bar\psi_\beta(x_2)\psi_\alpha(x_1) = \mathrm{i}S_{F\alpha\beta}(x_1 - x_2) \\
\contraction{}{A^\mu}{(x_1)}{A^\nu} A^\mu(x_1)A^\nu(x_2) &= \mathrm{i}D_F^{\mu\nu}(x_1 - x_2) .
\end{aligned} \tag{15.4.6}$$

Furthermore, we also have

$$\begin{aligned}
\contraction{}{\psi}{(x_1)}{\psi}\psi(x_1)\psi(x_2) &= \contraction{}{\bar\psi}{(x_1)}{\bar\psi}\bar\psi(x_1)\bar\psi(x_2) = 0 \\
\contraction{}{\psi}{(x_1)}{\phi}\psi(x_1)\phi(x_2) &= 0 , & \contraction{}{\psi}{(x_1)}{A^\mu}\psi(x_1)A^\mu(x_2) = 0 & & \text{etc.} \\
\contraction{}{\phi^\pm}{(x_1)}{\phi^\pm}\phi^\pm(x_1)\phi^\pm(x_2) &= 0 , & \contraction{}{\psi^\pm}{(x_1)}{\psi^\pm}\psi^\pm(x_1)\psi^\pm(x_2) = 0 , & & \contraction{}{\bar\psi^\pm}{(x_1)}{\bar\psi^\pm}\bar\psi^\pm(x_1)\bar\psi^\pm(x_2) = 0 \\
\contraction{}{\phi}{(x_1)}{\bar\psi}\phi(x_1)\bar\psi(x_2) &= 0 ,
\end{aligned} \tag{15.4.7}$$

since all these pairs of operators either commute or anticommute with one another. We recall that, in the interaction representation, the fields in $\mathcal{H}(x)$ are free Heisenberg fields. According to Eq. (15.4.4), the time-ordered product of two field operators can be represented in normal ordered form as follows:

$$T(AB) = : AB : + \contraction{}{A}{}{B}AB . \tag{15.4.8}$$

We now define what is known as the *generalized normal product* (normal product with contractions) of the field operators $A = A(x_1), B = B(x_2), \ldots$, which also contains within itself contractions of these operators:

$$: ABC\,D\,EF\ldots KL\ldots : = (-1)^P\, AC\,DL\,EF\ldots : BK\ldots : , \tag{15.4.9}$$

where P is the number of the individual permutations of Fermi operators that is necessary to obtain the order $ACDLEF\ldots BK\ldots$. For example,

$$\begin{aligned}
&: \bar\psi_\alpha(x_1)A^\mu(x_2)\psi_\beta(x_3)\psi_\gamma(x_4)A^\nu(x_5)\bar\psi_\delta(x_6) : \\
&= (-1)\psi_\beta(x_3)\bar\psi_\delta(x_6) : \bar\psi_\alpha(x_1)A^\mu(x_2)\psi_\gamma(x_4)A^\nu(x_5) : .
\end{aligned} \tag{15.4.10}$$

We are now in a position to formulate *Wick's theorem* both for pure time-ordered products and for mixed time-ordered products.

1st Theorem: The time-ordered product of the field operators is equal to the sum of their normal products in which the operators are linked by all different possible contractions:

$$T(A_1 A_2 A_3 \ldots A_n) =: A_1 A_2 A_3 \ldots A_n :$$
$$+ : A_1 A_2 A_3 \ldots A_n : + : A_1 A_2 A_3 \ldots A_n : + \ldots + : A_1 A_2 \ldots A_{n-1} A_n :$$
$$+ : A_1 A_2 A_3 A_4 \ldots A_n : + \ldots + : A_1 A_2 \ldots A_{n-3} A_{n-2} A_{n-1} A_n :$$
$$+ \ldots .$$

$$(15.4.11)$$

In the first line there are no contractions, in the second, one, in the third, two, etc.

2nd Theorem: A mixed T product of field operators is equal to the sum of their normal products in the form (15.4.11), with the difference that the sum does not include contractions between operators that occur within one and the same normal product factor. For example, we have

$$T(\psi_1 : \psi_2 \psi_3 \psi_4 :) =$$
$$: \psi_1 \psi_2 \psi_3 \psi_4 : + : \psi_1 \psi_2 \psi_3 \psi_4 : + : \psi_1 \psi_2 \psi_3 \psi_4 : + : \psi_1 \psi_2 \psi_3 \psi_4 : .$$

$$(15.4.12)$$

The proof of Wick's theorem is not essential for its application. Hence, the remainder of this section, which presents a simple proof, could be omitted.

We first prove the 1st theorem, Eq. (15.4.11), for the case in which the operators of the product $A_1 A_2 \ldots A_n$ are time ordered from the outset. We will show that the general case can be reduced to this special case. We now express the field operators in terms of their positive and negative frequency components. We divide this product, which occurs in time-ordered form from the outset, into a sum of products of positive and negative frequency parts. We select one such term arbitrarily; this is time ordered but, in general, not normal ordered. We then re-order its factors in the following way: The leftmost creation operator that is not in normal order is moved step-by-step to the left by permuting it successively – through commutation or anticommutation – with each of the annihilation operators that appear to the left of it. This procedure is then repeated for the next non-normal ordered creation operator and continued until all operators have become normal ordered. For each of these permutations one obtains, from (15.4.8) and the definition of the normal ordered product,

$$A_i^+ A_k^- = T(A_i^+ A_k^-) =: A_i^+ A_k^- : + A_i^+ A_k^-$$
$$= \pm A_k^- A_i^+ + A_i^+ A_k^- ,$$

$$(15.4.13)$$

where the lower sign applies when both operators are Fermi operators. In the final result, each of the summands is in normal ordered form with a sign that is determined by the number of pairs of Fermi operators that have been permuted. These signs can be eliminated from the expression if one writes each of the summands in the time-ordered (original) sequence and subjects it to the normal-ordering operation : ... : (see, e.g., (15.4.13), where one can write $\pm A_k^- A_i^+ =: A_i^+ A_k^-$:). The result now closely resembles the expression (15.4.11) except for the fact that not all contractions appear; it includes only those between the "wrongly" positioned (non-normal ordered) operators. However, since the contraction of two operators that are both time ordered and normal ordered vanishes, we can add all such contractions to the result and thus, using the distributive law, we obtain the sum of normal ordered products with all contractions. We have thus proved (15.4.11) for the time sequence $t_1 > ... > t_n$. We now consider the operators $A_1 A_2 ... A_n$ and an arbitrary permutation $P(A_1 A_2 ... A_n)$ of these operators. On account of the definition of the time-ordering and normal-ordering operations, we have

$$T\left(P(A_1 A_2 ... A_n)\right) = (-1)^P T(A_1 A_2 ... A_n) \qquad (15.4.14a)$$

and

$$: P(A_1 A_2 ... A_n) := (-1)^P : A_1 A_2 ... A_n : \qquad (15.4.14b)$$

with the same power P. Hence, we have demonstrated theorem 1, Eq. (15.4.11), for arbitrary time ordering of the operators $A_1, ... , A_n$.

Theorem 2 is obtained from the proof of theorem 1 as follows: The partial factors : AB ... : within the mixed time-ordered product are already normal ordered. In the procedure described above for constructing normal ordering there is no permutation, and thus no contraction, of these simultaneous operators. The contractions of these already normal ordered, simultaneous operators – which would not vanish – do not occur. This proves theorem 2.

15.5 Simple Scattering Processes, Feynman Diagrams

We shall now investigate a number of simple scattering processes, for which we will calculate the matrix elements of the S matrix. In so doing, we will encounter the most important features of the Feynman rules, which have already been mentioned on several occasions. In order of increasing complexity, we will study the first-order processes of the emission of a photon by an electron and the scattering of an electron by an external potential (Mott scattering) and, as examples of second-order processes, the scattering of two electrons (Møller scattering) and the scattering of a photon from an electron (Compton scattering).

15.5.1 The First-Order Term

Taking the simplest possible example, we consider the first-order contribution to the S matrix, Eq. (15.4.14)

$$S^{(1)} = -\mathrm{i}e \int d^4x \; : \; \bar\psi(x)\slashed{A}(x)\psi(x) \; : \tag{15.5.1}$$

with the field operators from Eq. (15.3.12). The possible elementary processes that result have already been discussed in Sect. 15.3.2 and are represented in Fig. 15.4. Of the eight possible transitions, we consider here the emission of a photon γ by an electron (Fig. 15.6), or, in other words, the transition of an electron into an electron and a photon:

$$e^- \to e^- + \gamma \; .$$

Fig. 15.6. The emission of a photon γ from an electron e^-. This process is virtual, i.e., only possible within a diagram of higher order

The initial state containing an electron with momentum $\mathbf{p}$

$$|i\rangle = |e^- \mathbf{p}\rangle = b^\dagger_{r\mathbf{p}}|0\rangle \tag{15.5.2a}$$

goes into the final state containing an electron with momentum $\mathbf{p}'$ and a photon with momentum $\mathbf{k}'$

$$|f\rangle = |e^- \mathbf{p}', \gamma\mathbf{k}'\rangle = b^\dagger_{r'\mathbf{p}'} a^\dagger_\lambda(\mathbf{k}')|0\rangle \; . \tag{15.5.2b}$$

The spinor index r and the polarization index λ are given for the creation operators, but for the sake of brevity are not indicated for the states. The first-order contribution to the scattering amplitude is given by the matrix element of (15.5.1). Contributions to $\langle f|S^{(1)}|i\rangle$ come from $\psi(x)$ only through the term with $b_{r\mathbf{p}}$, from $\bar\psi(x)$ only through the term $b^\dagger_{r'\mathbf{p}'}$, and from $A(x)$ only through $a_\lambda(\mathbf{k}')$:

$$\langle f|S^{(1)}|i\rangle = -\mathrm{i}e \int d^4x \left[\left(\frac{m}{VE_{\mathbf{p}'}}\right)^{\frac{1}{2}} \bar u_{r'}(p')\mathrm{e}^{\mathrm{i}p'x}\right]$$

$$\times \gamma^\mu \left[\left(\frac{1}{2V|\mathbf{k}'|}\right)^{\frac{1}{2}} \epsilon_{\lambda\mu}(\mathbf{k}')\mathrm{e}^{\mathrm{i}k'x}\right] \tag{15.5.3}$$

$$\times \left[\left(\frac{m}{VE_{\mathbf{p}}}\right)^{\frac{1}{2}} u_r(p)\mathrm{e}^{-\mathrm{i}px}\right] \; .$$

The integration over x leads to the conservation of four-momentum and thus to the matrix element

$$\langle f|\, S^{(1)}\, |i\rangle = -(2\pi)^4 \delta^{(4)} (p' + k' - p) \left(\frac{m}{V E_{\mathbf{p}}} \right)^{\frac{1}{2}} \left(\frac{m}{V E_{\mathbf{p}'}} \right)^{\frac{1}{2}} \left(\frac{1}{2V|\mathbf{k}'|} \right)^{\frac{1}{2}}$$

$$\times i e \bar{u}_{r'}(p') \gamma^\mu \epsilon_{\lambda\mu} (\mathbf{k}' = \mathbf{p} - \mathbf{p}') u_r(p) \,. \tag{15.5.4}$$

The four-dimensional δ function imposes conservation of momentum $\mathbf{p}' = \mathbf{p} - \mathbf{k}'$ and of energy $E_{\mathbf{p}-\mathbf{k}'} + |\mathbf{k}'| = E_{\mathbf{p}}$. For electrons and photons, the latter condition leads to $\mathbf{k}' \cdot \mathbf{p}/|\mathbf{k}'||\mathbf{p}| = \sqrt{1 + m^2/\mathbf{p}^2}$. In general, this cannot be satisfied since the two end products would always have a lower energy than the incident electron. The condition of *energy–momentum conservation* cannot be satisfied for real electrons and photons. Thus, this process can only occur as a component of higher-order diagrams. A preliminary comparison of Fig. 15.6 with Eq. (15.5.3) shows that the following analytical expressions can be assigned to the elements of the Feynman diagram: To the incident electron $u_r(p)e^{-ipx}$, to the outgoing electron $\bar{u}_{r'}(p')e^{ip'x}$, to the vertex point $-ie\gamma^\mu$, to the outgoing photon $\epsilon_{\lambda\mu}(\mathbf{k}')e^{ik'x}$, and, in addition, one has to integrate over the position of the interaction point x, i.e., the vertex point is associated with the integration $\int d^4x$. Carrying out this integration over x, one obtains from the exponential functions the conservation of four-momentum $(2\pi)^4 \delta^{(4)}(p' + k' - p)$, and hence one obtains finally the following rules in momentum space: Assigned to the incident electron is $u_r(p)$, to the outgoing electron $\bar{u}_{r'}(p')$, to the outgoing photon $\epsilon_\lambda^\alpha(\mathbf{k}')$, and to the vertex point $-ie\gamma^\alpha (2\pi)^4 \delta^{(4)}(p'+k'-p)$.

15.5.2 Mott Scattering

Mott scattering is the term used to describe the scattering of an electron by an external potential. In practice, this is usually the Coulomb potential of a nucleus. The external vector potential then has the form

$$A_e^\mu = (V(\mathbf{x}), 0, 0, 0) \,. \tag{15.5.5}$$

The initial state

$$|i\rangle = b_{r\mathbf{p}}^\dagger |0\rangle \tag{15.5.6a}$$

and the final state

$$|f\rangle = b_{r'\mathbf{p}'}^\dagger |0\rangle \tag{15.5.6b}$$

each contains a single electron. The scattering process is represented diagrammatically in Fig. 15.7.

Fig. 15.7. Mott scattering: An electron e^- is scattered by an external potential

The S-matrix element that follows from $S^{(1)}$ in Eq. (15.5.1) has the form

$$\langle f | S^{(1)} | i \rangle = -ie \int d^4x \left(\frac{m}{V E_{p'}} \right)^{\frac{1}{2}} \bar{u}_{r'}(p') e^{ip'x} \gamma^0$$

$$\times V(\mathbf{x}) \left(\frac{m}{V E_{\mathbf{p}}} \right)^{\frac{1}{2}} u_r(p) e^{-ipx} \tag{15.5.7}$$

$$= -ie\tilde{V}(\mathbf{p} - \mathbf{p}') \left(\frac{m}{V E_{p'}} \right)^{\frac{1}{2}} \left(\frac{m}{V E_{\mathbf{p}}} \right)^{\frac{1}{2}}$$

$$\times \mathcal{M} 2\pi \delta(p^0 - p'^0) \, .$$

Here,

$$\mathcal{M} = \bar{u}_{r'}(p') \gamma^0 u_r(p) \tag{15.5.8}$$

is the spinor matrix element. In the calculation of the transition probability $|\langle f | S^{(1)} | i \rangle|^2$, formally the square of a δ-function appears. In order to give meaning to this quantity, one should recall that the scattering experiment is carried out over a very long, but nonetheless finite, time interval T and that $2\pi\delta(E)$ should be replaced by

$$2\pi\delta(E) \to \int_{-T/2}^{T/2} dt \, e^{iEt} \, . \tag{15.5.9}$$

The square of this function is also encountered in the derivation of Fermi's golden rule

$$\left(\int_{-T/2}^{T/2} dt \, e^{iEt} \right)^2 = \left(\frac{2}{E} \sin \frac{ET}{2} \right)^2 = 2\pi T \left(\frac{\sin^2 ET/2}{\pi E^2 T/2} \right) = 2\pi T \delta(E) \, . \tag{15.5.10}$$

Here we have used the fact, detailed in QM I, Eqs. (16.34)–(16.35), that the final expression in brackets is a representation of the δ-function:

$$\lim_{T \to \infty} \delta_T(E) = \lim_{T \to \infty} \left(\frac{\sin^2 ET/2}{\pi E^2 T/2} \right) = \delta(E)$$

In briefer form, this justification may also be presented as

$$\lim_{T \to \infty} \int_{-T/2}^{T/2} dt \, e^{iEt} \int_{-T/2}^{T/2} dt \, e^{iEt} = 2\pi T \delta(E) \, , \tag{15.5.10'}$$

where the limit of the first factor is expressed by $2\pi\delta(E)$, which for the second integral then yields $\int_{-T/2}^{T/2} dt \, e^0 = T$.

The transition probability per unit time Γ_{if} is obtained by dividing $|\langle f | S^{(1)} | i \rangle|^2$ by T (Eq. (15.5.7)):

$$\Gamma_{if} = 2\pi\delta(E - E')\left(\frac{m}{VE}\right)^2 |\mathcal{M}|^2 e^2 |\tilde{V}(\mathbf{p} - \mathbf{p}')|^2 .$$ (15.5.11)

The *differential scattering cross-section* is defined by

$$\frac{d\sigma}{d\Omega} = \frac{dN(\Omega)}{N_{in}d\Omega} ,$$ (15.5.12)

where $dN(\Omega)$ is the number of particles scattered into the element of solid angle $d\Omega$, and N_{in} is the number of particles incident per unit area (see Fig. 15.8). The differential scattering cross-section (15.5.12) is also equal to the number of particles scattered per unit time into $d\Omega$ divided by the incident current density j_{in} and by $d\Omega$:

$$\frac{d\sigma}{d\Omega} = \frac{dN(\Omega)/dt}{j_{in}d\Omega} .$$ (15.5.12')

In addition to the transition rate Γ_{if} that has already been found, we also need to know the incident flux and the number of final states in the element of solid angle $d\Omega$. We will first show that the flux of incident electrons is given by $\frac{|\mathbf{p}|}{EV}$. In order to do so, we need to calculate the expectation value of the current density

$$j^\mu(x) = : \bar\psi(x)\gamma^\mu\psi(x) :$$ (15.5.13)

in the initial state

$$|i\rangle = |e^-, \mathbf{p}\rangle = b_{r\mathbf{p}}^\dagger |0\rangle .$$

We obtain

$$\langle e^-, \mathbf{p}| j^\mu(x) |e^-, \mathbf{p}\rangle = \frac{m}{VE_\mathbf{p}}\bar{u}_r(p)\gamma^\mu u_r(p) = \frac{p^\mu}{VE_\mathbf{p}} ,$$ (15.5.14)

where we have used the Gordon identity (Eq. (10.1.5))

$$\bar{u}_r(p)\gamma^\mu u_{r'}(q) = \frac{1}{2m}\bar{u}_r(p)\left[(p+q)^\mu + i\sigma^{\mu\nu}(p-q)_\nu\right]u_{r'}(q) .$$

The incident current density j_{in} is thus equal to

$$j_{in} = \frac{|\mathbf{p}|}{VE_\mathbf{p}} ,$$ (15.5.15)

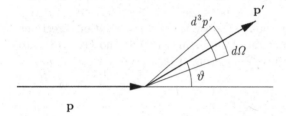

Fig. 15.8. Scattering of an electron with momentum $\mathbf{p}$ by a potential. The momentum of the scattered electron is $\mathbf{p}'$, the angle of deflection ϑ, and the solid angle $d\Omega$

which, as might be expected, is the product of the particle number density $\frac{1}{V}$ and the relative velocity $\frac{|\mathbf{p}|}{E_\mathbf{p}}$. To determine $dN(\Omega)$ per unit time, we need the number of final states in the interval d^3p' around $\mathbf{p}'$. Since the volume of momentum space per momentum value is $(2\pi)^3/V$, the number of momentum states in the interval d^3p' is

$$\frac{d^3p'}{(2\pi)^3/V} = \frac{V|\mathbf{p}'|^2d|\mathbf{p}'|d\Omega}{(2\pi)^3} = \frac{V|\mathbf{p}'|E'dE'd\Omega}{(2\pi)^3} , \tag{15.5.16a}$$

where we have used

$$E' = \sqrt{\mathbf{p}'^2 + m^2} \quad \text{and} \quad dE' = \frac{|\mathbf{p}'|d|\mathbf{p}'|}{\sqrt{\mathbf{p}'^2 + m^2}} = \frac{|\mathbf{p}'|d|\mathbf{p}'|}{E'} \tag{15.5.16b}$$

Inserting Eqs. (15.5.11), (15.5.15) and (15.5.16a) into the differential scattering cross-section (15.5.12')[5], one obtains the cross-section per element of solid angle $d\Omega$, by keeping $d\Omega$ fixed and integrating over the remaining variable E'

$$\begin{aligned}\frac{d\sigma}{d\Omega} &= \int 2\pi\delta(E-E')\left(\frac{m}{VE}\right)^2|\mathcal{M}|^2e^2|\tilde{V}(\mathbf{p}-\mathbf{p}')|^2\frac{1}{\frac{|\mathbf{p}|}{EV}}\frac{V|\mathbf{p}'|E'dE'}{(2\pi)^3}\\ &= \left(\frac{em}{2\pi}\right)^2|\mathcal{M}|^2|\tilde{V}(\mathbf{p}-\mathbf{p}')|^2\big|_{|\mathbf{p}'|=|\mathbf{p}|} ,\end{aligned}$$

$$\tag{15.5.17}$$

where the conservation of energy, expressed by $\delta(E-E')$, yields the condition $|\mathbf{p}'| = |\mathbf{p}|$ for the momentum of the scattered particle. The Fourier transform of the Coulomb potential of a nucleus with charge Z, in Heaviside–Lorentz units,[6]

$$V(\mathbf{x}) = \frac{Ze}{4\pi|\mathbf{x}|}$$

reads:

$$\tilde{V}(\mathbf{p}-\mathbf{p}') = \frac{Ze}{|\mathbf{p}-\mathbf{p}'|^2} . \tag{15.5.18}$$

For the sake of simplicity, we assume that the incident electron beam is unpolarized. This corresponds to a sum over both polarization directions with the weight $\frac{1}{2}$, i.e., $\frac{1}{2}\sum_{r=1,2}$. The polarization of the scattered particles is likewise not resolved, giving a sum $\sum_{r'}$ over the two polarization directions of the final state. Under this condition, inserting (15.5.8) and (15.5.18) into (15.5.17), yields for the differential scattering cross-section

---

[5] $\frac{dN(\Omega)}{dt} = \sum_{f \in d\Omega} \Gamma_{if} = \frac{V}{(2\pi)^3} \int\limits_{\mathbf{p}' \in d\Omega} d^3p'\Gamma_{if}$.

[6] See footnote 1 in Chap. 14.

$$\frac{d\sigma}{d\Omega} = \left(\frac{em}{2\pi}\right)^2 \frac{1}{2} \sum_{r'} \sum_{r} |\bar{u}_{r'}(p')\gamma^0 u_r(p)|^2 \left. \frac{(Ze)^2}{|\mathbf{p}-\mathbf{p}'|^4}\right|_{|\mathbf{p}'|=|\mathbf{p}|} . \tag{15.5.19}$$

Hence, we are led to the calculation of

$$\sum_{r'} \sum_{r} |\bar{u}_{r'}(p')\gamma^0 u_r(p)|^2$$

$$= \sum_{r'} \sum_{r} \bar{u}_{r'\alpha'}(p')\gamma^0_{\alpha'\alpha} u_{r\alpha}(p)\bar{u}_{r\beta}(p)\gamma^0_{\beta\beta'} u_{r'\beta'}(p')$$

$$= \gamma^0_{\alpha'\alpha}\left(\frac{\not{p}+m}{2m}\right)_{\alpha\beta}\gamma^0_{\beta\beta'}\left(\frac{\not{p}'+m}{2m}\right)_{\beta'\alpha'} \tag{15.5.20}$$

$$= \frac{1}{4m^2}\operatorname{Tr}\gamma^0(\not{p}+m)\gamma^0(\not{p}'+m) ,$$

where we have used the representations (6.3.21) and (6.3.23) of the projection operator Λ_+, which leads to the trace of the product of γ matrices.

Making use of the cyclic invariance of the trace, $\operatorname{Tr}\gamma^\nu = 0$, $\{\gamma^\mu, \gamma^\nu\} = 2g^{\mu\nu}\mathbb{1}$, and $\operatorname{Tr}\gamma^0\gamma^\mu\gamma^0\gamma^\nu = 0$ for $\mu \neq \nu$, one obtains

$$\begin{aligned}
\operatorname{Tr}\gamma^0(\not{p}+m)\gamma^0(\not{p}'+m) &= \operatorname{Tr}\gamma^0\not{p}\gamma^0\not{p}' + m\operatorname{Tr}\gamma^0\not{p}\gamma^0 + m\operatorname{Tr}\gamma^0\not{p}'\gamma^0 + m^2\operatorname{Tr}\mathbb{1}\\
&= \operatorname{Tr}\gamma^0\not{p}\gamma^0\not{p}' + 4m^2 = p_\mu p'_\nu \operatorname{Tr}\gamma^0\gamma^\mu\gamma^0\gamma^\nu + 4m^2\\
&= p_0 p'_0 \operatorname{Tr}\mathbb{1} + p_k p'_k \operatorname{Tr}\gamma^0\gamma^k\gamma^0\gamma^k + 4m^2\\
&= 4(p_0^2 + \mathbf{p}\mathbf{p}' + m^2)\\
&= 4(E_\mathbf{p}^2 + \mathbf{p}\mathbf{p}' + m^2) .
\end{aligned} \tag{15.5.21}$$

From the expression for the velocity[7],

$$\mathbf{v} = \frac{\partial E}{\partial \mathbf{p}} = \frac{\mathbf{p}}{\sqrt{\mathbf{p}^2+m^2}} = \frac{\mathbf{p}}{E} \tag{15.5.22a}$$

and $|\mathbf{p}| = Ev$, $\mathbf{p}\mathbf{p}' = |\mathbf{p}|^2\cos\vartheta$ (for $|\mathbf{p}'| = |\mathbf{p}|$)

we gain the following relations:

$$|\mathbf{p}-\mathbf{p}'|^2 = 2\mathbf{p}^2(1-\cos\vartheta) = 4\mathbf{p}^2\sin^2\frac{\vartheta}{2} \tag{15.5.22b}$$

and

$$E^2 + \mathbf{p}\cdot\mathbf{p}' + m^2 = 2E^2 - p^2(1-\cos\vartheta) = 2E^2 - 2p^2\sin^2\frac{\vartheta}{2}$$

$$= 2E^2\left(1 - v^2\sin^2\frac{\vartheta}{2}\right) . \tag{15.5.22c}$$

Inserting (15.5.20), (15.5.21), and (15.5.22a–c) into (15.5.19), one finally obtains the differential scattering cross-section for *Mott scattering*

---

[7] E and $E_\mathbf{p}$ are used interchangeably.

$$\frac{d\sigma}{d\Omega} = \frac{(\alpha Z)^2 (1 - v^2 \sin^2 \frac{\vartheta}{2})}{4E^2 v^4 \sin^4 \frac{\vartheta}{2}} , \tag{15.5.23}$$

where α is Sommerfeld's fine-structure constant[6] $\alpha = \frac{\hat{e}_0^2}{4\pi}$. In the nonrelativistic limit, (15.5.23) yields the *Rutherford scattering law*, see Eq. (18.37) in QM I,

$$\frac{d\sigma}{d\Omega} = \frac{(Z\alpha)^2}{4m^2 v^4 \sin^4 \frac{\vartheta}{2}} . \tag{15.5.24}$$

For the scattering of Klein–Gordon particles one has, instead of Eq. (15.5.23),

$$\frac{d\sigma}{d\Omega} = \frac{(\alpha Z)^2}{4E^2 v^4 \sin^4 \frac{\vartheta}{2}} ,$$

see Problem 15.2.

In addition to the elements of the Feynman diagrams encountered in the preceding section, here we also have a static external field $A_e^\mu(\mathbf{x})$, represented as a wavy line with a cross. According to Eq. (15.5.7), in the transition amplitude this is assigned the factor $A_e^\mu(\mathbf{x})$, or in momentum space

$$A_e^\mu(\mathbf{q}) = \int d^3 x e^{-i\mathbf{q}\cdot\mathbf{x}} A_e^\mu(\mathbf{x}) . \tag{15.5.25}$$

15.5.3 Second-Order Processes

15.5.3.1 Electron–electron scattering

Our next topic is the scattering of two electrons, also known as Møller scattering. The corresponding Feynman diagram is shown in Fig. 15.9. This is a second-order process following from the S-matrix term:

$$\begin{aligned}
S^{(2)} &= \frac{(-i)^2}{2!} \int d^4 x_1 \, d^4 x_2 \, T \left(\mathcal{H}_I(x_1) \mathcal{H}_I(x_2) \right) \\
&= \frac{(ie)^2}{2!} \int d^4 x_1 \, d^4 x_2 \\
&\quad \times T \left(: \bar{\psi}(x_1) \mathcal{A}(x_1) \psi(x_1) : \; : \bar{\psi}(x_2) \mathcal{A}(x_2) \psi(x_2) : \right) .
\end{aligned} \tag{15.5.26}$$

Application of Wick's theorem leads to one term without contraction, three terms each with one contraction, three terms each with two contractions, and, finally, to one term with three contractions. The term that contains the two external incident and outgoing fermions is

$$-\frac{e^2}{2}\int d^4x_1\, d^4x_2\; :\; \bar{\psi}(x_1)\underline{A(x_1)\psi(x_1)\bar{\psi}(x_2)A(x_2)}\psi(x_2)\; :$$

$$=-\frac{e^2}{2}\int d^4x_1\, d^4x_2\; :\; \bar{\psi}(x_1)\gamma^\mu\psi(x_1)\bar{\psi}(x_2)\gamma^\nu\psi(x_2)\mathrm{i}D_{F\mu\nu}(x_1-x_2)\; :\,,$$

$$(15.5.27)$$

where $D_{F\mu\nu}(x_1-x_2)$ is the photon propagator (14.5.19). Depending on the initial state, this term leads to the scattering of two electrons, of two positrons, or of one electron and one positron.

We shall consider the scattering of two electrons

$$e^- + e^- \rightarrow e^- + e^-\;,$$

from the initial state

$$|i\rangle = \left|e^-(\mathbf{p}_1 r_1), e^-(\mathbf{p}_2 r_2)\right\rangle = b^\dagger_{r_1\mathbf{p}_1} b^\dagger_{r_2\mathbf{p}_2}|0\rangle \qquad (15.5.28\mathrm{a})$$

into the final state

$$|f\rangle = \left|e^-(\mathbf{p}'_1 r'_1), e^-(\mathbf{p}'_2 r'_2)\right\rangle = b^\dagger_{r'_1\mathbf{p}'_1} b^\dagger_{r'_2\mathbf{p}'_2}|0\rangle\;. \qquad (15.5.28\mathrm{b})$$

Here, there are clearly two contributions to the matrix element of $S^{(2)}$. The direct scattering contribution, in which the operator $\bar{\psi}(x_1)\gamma^\mu\psi(x_1)$ annihilates the particle labeled 1 with spinor $u_{r_1}(p_1)$ at the position x_1 and creates the particle with spinor $u_{r'_1}(p'_1)$. The operator $\bar{\psi}(x_2)\gamma^\nu\psi(x_2)$ has the same effect on the particle labeled 2. The other contribution is the exchange scattering. This is obtained when the effect of the annihilation operators remains as just described, whilst the operator $\bar{\psi}(x_1)\gamma^\mu\psi(x_1)$ creates the particle in the final state $u_{r'_2}(p'_2)$ and the second operator creates the particle in the state $u_{r'_1}(p'_1)$. These two contributions are shown diagrammatically in Fig. 15.9. Exactly the same contributions arise when one, instead, exchanges the positions x_1 and x_2 of the two interaction operators. Since one has to integrate over x_1 and x_2, one obtains twice the contribution of the two diagrams in Fig. 15.9. The factor 2 arising from the permutation of the two vertex positions cancels with the factor $\frac{1}{2!}$ in $S^{(2)}$. This is a general property of Feynman diagrams. The factor $\frac{1}{n!}$ in $S^{(n)}$ can be omitted when one is summing only over topologically distinct diagrams.

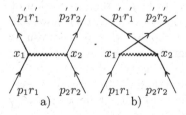

Fig. 15.9. Electron–electron scattering: (a) direct scattering, (b) exchange scattering

The S-matrix element for the direct scattering in Fig. 15.9a is

$$\langle f | S^{(2)} | i \rangle_a = -e^2 \int d^4x_1 \, d^4x_2 \left(\frac{m^4}{V^4 E_{\mathbf{p}_1} E_{\mathbf{p}_2} E_{\mathbf{p}'_1} E_{\mathbf{p}'_2}} \right)^{\frac{1}{2}}$$
$$\times e^{-ip_1x_1 + ip'_1x_1 - ip_2x_2 + ip'_2x_2}$$
$$\times (\bar{u}_{r'_1}(p'_1)\gamma^\mu u_{r_1}(p_1))(\bar{u}_{r'_2}(p'_2)\gamma^\nu u_{r_2}(p_2)) i D_{F\mu\nu}(x_1 - x_2) .$$
$$(15.5.29)$$

The exchange scattering contribution (b) is obtained by exchanging the wave functions of the final states in $-\langle f | S^{(2)} | i \rangle_a$, so that the wave-function part has the form

$$e^{-ip_1x_1 + ip'_1x_2 - ip_2x_2 + ip'_2x_1} (\bar{u}_{r'_2}(p'_2)\gamma^\mu u_{r_1}(p_1))(\bar{u}_{r'_1}(p'_1)\gamma^\nu u_{r_2}(p_2)) .$$
$$(15.5.30)$$

The minus sign that appears here is due to the fact that, in the exchange term, one needs an odd number of anticommutations to bring the creation and annihilation operators into the same order as in the direct term. If, in (15.5.29) one inserts $i D_{F\mu\nu}(x_1 - x_2) = i \int d^4k \frac{-g_{\mu\nu} e^{-ik(x_1-x_2)}}{k^2 + i\epsilon}$, then, after carrying out the integrations, one obtains

$$\langle f | S^{(2)} | i \rangle = (2\pi)^4 \delta^{(4)}(p'_1 + p'_2 - p_1 - p_2)$$
$$\times \left(\frac{m^4}{V^4 E_{\mathbf{p}_1} E_{\mathbf{p}'_1} E_{\mathbf{p}_2} E_{\mathbf{p}'_2}} \right)^{\frac{1}{2}} (\mathcal{M}_a + \mathcal{M}_b) , \qquad (15.5.31)$$

where the matrix elements are given, for the graph 15.9a, by

$$\mathcal{M}_a = -e^2 \bar{u}(p'_1)\gamma^\mu u(p_1) i D_{F\mu\nu}(p_2 - p'_2)\bar{u}(p'_2)\gamma^\nu u(p_2) \qquad (15.5.32a)$$

and for the graph 15.9b, by

$$\mathcal{M}_b = e^2 \bar{u}(p'_2)\gamma^\mu u(p_1) i D_{F\mu\nu}(p_2 - p'_1)\bar{u}(p'_1)\gamma^\nu u(p_2) . \qquad (15.5.32b)$$

The δ function in (15.5.31) expresses the conservation of the total four-momentum of the two particles. Since, in the matrix element $\mathcal{M}_a$ for the direct scattering, the photon propagator, for example, has the argument $k \equiv p_2 - p'_2 = p'_1 - p_1$, the four-momentum of the particles is conserved at every vertex. Hereby, we fix the orientation of the photon momentum of the internal line to be from right to left. In principle, the orientation of the photon momentum is arbitrary since $D_{F\mu\nu}(k) = D_{F\mu\nu}(-k)$; however, one must select an orientation so that one can monitor the conservation of momentum at the vertices. The Feynman diagrams in momentum space are shown in Fig. 15.10.

The Feynman rules can now be extended as follows: To every internal photon line with momentum argument k, with end points at the vertices γ^μ and γ^ν, assign the propagator $i D_{F\mu\nu}(k) = i \frac{-g_{\mu\nu}}{k^2 + i\epsilon}$.

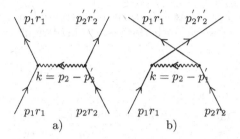

Fig. 15.10. Feynman diagrams in momentum space for electron–electron scattering a) direct scattering, b) exchange scattering

We now turn to the evaluation of the matrix element (15.5.32a), which is now lengthier than for Mott scattering. Instead of going through the details of the calculation, we refer the reader to the problems and the supplementary remarks at the end of this section and discuss the final result. The relation between the differential scattering cross-section and the matrix element $\mathcal{M}$ in the center-of-mass system for two fermions with mass m_1 and m_2 is given, according to Eq. (15.5.59), by

$$\left.\frac{d\sigma}{d\Omega}\right|_{\text{CM}} = \frac{1}{(4\pi)^2}\frac{m_1 m_2}{E_{\text{tot}}}|\mathcal{M}|^2, \tag{15.5.33}$$

where E_{tot} is the total energy. Inserting the results (15.5.37)–(15.5.43) into (15.5.33), one obtains for the scattering cross-section in the center-of-mass frame (Fig. 15.11) the Møller formula (1932)

$$\frac{d\sigma}{d\Omega} = \frac{\alpha^2(2E^2 - m^2)^2}{4E^2(E^2 - m^2)^2}$$
$$\times \left(\frac{4}{\sin^4\vartheta} - \frac{3}{\sin^2\vartheta} + \frac{(E^2 - m^2)^2}{(2E^2 - m^2)^2}\left(1 + \frac{4}{\sin^4\vartheta}\right)\right). \tag{15.5.34}$$

In the nonrelativistic limit, $E^2 \approx m^2$, $v^2 = (E^2 - m^2)/E^2$, this yields:

$$\left.\frac{d\sigma}{d\Omega}\right|_{\text{nr}} = \left(\frac{\alpha}{m}\right)^2\frac{1}{4v^4}\left(\frac{1}{\sin^4\frac{\vartheta}{2}} + \frac{1}{\cos^4\frac{\vartheta}{2}} - \frac{1}{\sin^2\frac{\vartheta}{2}\cos^2\frac{\vartheta}{2}}\right), \tag{15.5.35}$$

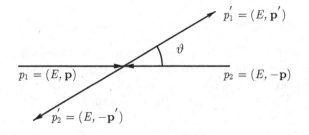

Fig. 15.11. Kinematics of the scattering of two identical particles in the center-of-mass system

a formula that was originally derived by Mott (1930). It is instructive to compare this result (15.5.35) with the classical Rutherford-scattering formula

$$\frac{d\sigma}{d\Omega} = \frac{\alpha^2 m^2}{16|\mathbf{p}|^4} \left\{ \frac{1}{\sin^4 \frac{\vartheta}{2}} + \frac{1}{\cos^4 \frac{\vartheta}{2}} \right\}. \tag{15.5.36}$$

This classical formula contains the familiar $\sin^{-4} \frac{\vartheta}{2}$ term, but also an additional $\cos^{-4} \frac{\vartheta}{2}$ term, since here we are considering the scattering of two (identical) electrons. If one observes the scattering at a particular angle ϑ, then the probability of observing the electron incident from the left is proportional to $\sin^{-4} \frac{\vartheta}{2}$. The probability that the electron incident from the right is scattered into this direction is, as can be seen from symmetry considerations, proportional to

$$\sin^{-4}\left(\frac{\pi - \vartheta}{2} \right) = \cos^{-4} \frac{\vartheta}{2}.$$

Classically, these probabilities simply add, which leads to (15.5.36). In the quantum-mechanical result (15.5.35), however, an additional term arises due to the interference between the two electrons. In quantum mechanics it is the two amplitudes, corresponding to the Feynman diagrams 15.9a and 15.9b, that are added. The scattering cross-section is then obtained as the absolute magnitude squared. The minus sign in the interference term results from the Fermi statistics; for bosons one obtains a plus sign.

In the extreme relativistic case, $\frac{E}{m} \to \infty$, from (15.5.34) we have

$$\begin{aligned}
\left. \frac{d\sigma}{d\Omega} \right|_{\text{er}} &= \frac{\alpha^2}{E^2} \left(\frac{4}{\sin^4 \vartheta} - \frac{2}{\sin^2 \vartheta} + \frac{1}{4} \right) \\
&= \frac{\alpha^2}{4E^2} \left(\frac{1}{\sin^4 \frac{\vartheta}{2}} + \frac{1}{\cos^4 \frac{\vartheta}{2}} + 1 \right) \\
&= \frac{\alpha^2}{4E^2} \frac{(3 + \cos^2 \vartheta)^2}{\sin^4 \vartheta}.
\end{aligned} \tag{15.5.37}$$

Supplement: Calculation of the differential scattering cross-section for electron–electron scattering. For the scattering cross-section, we need

$$|\mathcal{M}|^2 = |\mathcal{M}_a|^2 + |\mathcal{M}_b|^2 + 2\mathrm{Re}\mathcal{M}_a\mathcal{M}_b^* \tag{15.5.38}$$

from (15.5.32a,b) with $iD_{F\mu\nu}(k) = \frac{-ig_{\mu\nu}}{k^2 + i\varepsilon}$. We assume an unpolarized electron beam that scatters from likewise unpolarized electrons and, furthermore, that the polarization of the scattered particles is not registered; this implies the summation $\frac{1}{4}\sum_{r_1}\sum_{r_2}\sum_{r_1'}\sum_{r_2'} \equiv \frac{1}{4}\sum_{r_i, r_i'}$. For the first term in (15.5.38) we obtain

$$\overline{|\mathcal{M}_a|^2} = \frac{e^2}{4} \sum_{r_i,r_i'} \bar{u}_{r_1'}(p_1')\gamma^\mu u_{r_1}(p_1)\bar{u}_{r_2'}(p_2')\gamma_\mu u_{r_2}(p_2)$$

$$\times\ \bar{u}_{r_1}(p_1)\gamma^\nu u_{r_1'}(p_1')\bar{u}_{r_2}(p_2)\gamma_\nu u_{r_2'}(p_2')\frac{1}{[(p_1'-p_1)^2]^2}$$

$$= \frac{e^4}{4} \sum_{r_i,r_i'} \bar{u}_{r_1}(p_1)\gamma^\nu u_{r_1'}(p_1')\bar{u}_{r_1'}(p_1')\gamma^\mu u_{r_1}(p_1)$$

$$\times\ \bar{u}_{r_2}(p_2)\gamma_\nu u_{r_2'}(p_2')\bar{u}_{r_2'}(p_2')\gamma_\mu u_{r_2}(p_2)\frac{1}{[(p_1'-p_1)^2]^2} \tag{15.5.39}$$

$$= \frac{e^4}{4}\mathrm{Tr}\left(\gamma_\nu\frac{p_1'+m}{2m}\gamma_\mu\frac{p_1+m}{2m}\right)$$

$$\times\ \mathrm{Tr}\left(\gamma^\nu\frac{p_2'+m}{2m}\gamma^\mu\frac{p_2+m}{2m}\right)\frac{1}{[(p_1'-p_1)^2]^2} \quad.$$

The second term of (15.5.38) is obtained by exchanging the momenta p_1' and p_2' in $|\mathcal{M}_a|^2$:

$$\overline{|\mathcal{M}_b|^2} = \overline{|\mathcal{M}_a|^2}\,(p_1' \leftrightarrow p_2') \ , \tag{15.5.40}$$

and the third is

$$\overline{\mathrm{Re}(\mathcal{M}_a\mathcal{M}_b^*)} = \frac{e^2}{4}\sum_{r_i,r_i'}\frac{1}{(p_1'-p_2)^2(p_2'-p_1)^2}$$

$$\times\ \mathrm{Re}\left[\bar{u}_{r_1'}(p_1')\gamma_\mu u_{r_1}(p_1)\bar{u}_{r_2'}(p_2')\gamma^\mu u_{r_2}(p_2)\right.$$

$$\left.\times\ \bar{u}_{r_1}(p_1)\gamma^\nu u_{r_1'}(p_1')\bar{u}_{r_2}(p_2)\gamma_\nu u_{r_2'}(p_2')\right]$$

$$= -\frac{e^4}{4}\frac{1}{(p_1'-p_2)^2(p_2'-p_1)^2}$$

$$\times\ \mathrm{Tr}\left(\gamma_\nu\frac{p_1'+m}{2m}\gamma_\mu\frac{p_1+m}{2m}\gamma^\nu\frac{p_2'+m}{2m}\gamma^\mu\frac{p_2+m}{2m}\right) \ . \tag{15.5.41}$$

In the final expression here, it is possible to omit the Re since its argument is already real. There still remains the evaluation of the traces: For Eq. (15.5.39) one needs

$$\mathrm{Tr}\left(\gamma_\nu(p_1'+m)\gamma_\mu(p_1+m)\right) = 4(g_{\mu\nu}m^2 + p_{1\mu}p_{1\nu}' + p_{1\nu}p_{1\mu}' - g_{\mu\nu}p_1'\cdot p_1) \ . \tag{15.5.42}$$

In Eq. (15.5.41) we encounter

$$\gamma_\nu(p_1'+m)\gamma_\mu(p_1+m)\gamma^\nu = -2p_1\gamma_\mu p_1' + 4m(p_{1\mu}+p_{1\mu}') - 2m^2\gamma_\mu \tag{15.5.43}$$

and

$$\text{Tr}\,(\gamma_\nu(\not{p}_1' + m)\gamma_\mu(\not{p}_1 + m)\gamma^\nu(\not{p}_2' + m)\gamma^\mu(\not{p}_2 + m))$$
$$= \text{Tr}\,((-2\not{p}_1\gamma_\mu\not{p}_1' + 4m(p_1 + p_1')_\mu - 2m^2\gamma_\mu)(\not{p}_2' + m)\gamma^\mu(\not{p}_2 + m))$$
$$= \text{Tr}\Big(-2\not{p}_1(4p_1' \cdot p_2' - 2m\not{p}_1')(\not{p}_2 + m) + 4m(\not{p}_2' + m)(\not{p}_1 + \not{p}_1')(\not{p}_2 + m)$$
$$- 2m^2(-2\not{p}_2' + 4m)(\not{p}_2 + m)\Big)$$
$$= 16(-2p_1 \cdot p_2\,p_1' \cdot p_2' + m^2 p_1 \cdot p_1' + m^2(p_1 + p_1') \cdot (p_2 + p_2') + m^2 p_2 \cdot p_2' - 2m^4)\ .$$
$$(15.5.44)$$

The formulas (15.5.38)–(15.5.44) were used in going from (15.5.31) to (15.5.34).

*15.5.3.2 Scattering Cross-Section and S-Matrix Element

In many applications, it is important to have a general relation between the scattering cross-section and the relevant S-matrix element.

We consider the scattering of two particles with four-momenta $p_i = (E_i, \mathbf{p}_i)$, $i = 1, 2$, which react to yield a final state containing n particles with momenta $p_f' = (E_f', \mathbf{p}_f')$, $f = 1, \ldots, n$. For brevity, we suppress the polarization states. The S-matrix element for the transition from the initial state $|i\rangle$ into the final state $|f\rangle$ has the form

$$\langle f|\,S\,|i\rangle = \delta_{fi} + (2\pi)^4\delta^{(4)}(\sum p_f' - \sum p_i)$$
$$\times \prod_i \left(\frac{1}{2VE_i}\right)^{1/2} \prod_f \left(\frac{1}{2VE_f'}\right)^{1/2} \prod_{\substack{\text{external}\\\text{fermions}}} (2m)^{1/2}\mathcal{M}\ . \quad (15.5.45)$$

The final product $\prod_{\text{external fermions}}(2m)^{1/2}$ stems from the normalization factor in (15.3.12a,b) and contributes a factor $(2m)^{1/2}$ for every external fermion, whereby the masses can be different. The amplitude factor $\mathcal{M} = \sum_{n=1}^\infty \mathcal{M}^{(n)}$ is the sum over all orders of perturbation theory, where $\mathcal{M}^{(n)}$ stems from the term $S^{(n)}$. The four-dimensional δ function is obtained for an infinite time interval and an infinite normalization volume. As was the case for Mott scattering, it is convenient to consider a finite time interval T as well as a finite volume. One then has

$$(2\pi)^4\delta^{(4)}(\sum p_f' - \sum p_i)$$
$$\rightarrow \lim_{T\to\infty, V\to\infty} \int_{-T/2}^{T/2} dt \int_V d^3x\, e^{ix(\sum p_f' - \sum p_i)} \quad (15.5.46)$$

and

$$\left(\lim_{T \to \infty, V \to \infty} \int_{-T/2}^{T/2} dt \int_V d^3x \, e^{ix(\sum p'_f - \sum p_i)} \right)^2$$

$$= TV(2\pi)^4 \delta^{(4)} \left(\sum p'_f - \sum p_i \right).$$

(15.5.47)

This leads to a transition rate or, in other words, a transition probability per unit time, of

$$w_{fi} = \frac{|S_{fi}|^2}{T}$$

$$= V(2\pi)^4 \delta^{(4)} \left(\sum p'_f - \sum p_i \right) \left(\prod_i \frac{1}{2VE_i} \right) \left(\prod_f \frac{1}{2VE'_f} \right)$$

$$\times \prod_{\substack{\text{external} \\ \text{fermions}}} (2m) \, |\mathcal{M}|^2 \,.$$

(15.5.48)

w_{fi} is the transition rate into a particular final state f. The transition rate into a volume element in momentum space $\prod_l d^3p'_l$ is obtained by multiplying (15.5.48) by the number of states in this element

$$\frac{|S_{fi}|^2}{T} \prod_f V \frac{d^3p'_f}{(2\pi)^3} \,.$$

(15.5.49)

The scattering cross-section is the ratio of the transition rate to the incident flux. In differential form this implies

$$d\sigma = \frac{|S_{fi}|^2}{T} \frac{V}{v_{\text{rel}}} \prod_f V \frac{d^3p'_f}{(2\pi)^3}$$

$$= (2\pi)^4 \delta^{(4)} \left(\sum p'_f - \sum p_i \right) \frac{1}{4E_1 E_2 v_{\text{rel}}}$$

$$\times \prod_{\substack{\text{external} \\ \text{fermions}}} (2m) \prod_f \frac{d^3p'_f}{(2\pi)^3 E'_f} |\mathcal{M}|^2$$

(15.5.50)

$$\equiv \frac{1}{4E_1 E_2 v_{\text{rel}}} \prod_{\substack{\text{external} \\ \text{fermions}}} (2m) \, |\mathcal{M}|^2 d\Phi_n \,.$$

The normalization is chosen such that the volume V contains one particle and the incident flux equals v_{rel}/V, with the relative velocity v_{rel}. In the center-of-mass frame ($\mathbf{p}_2 = -\mathbf{p}_1$), the relative velocity of the two incident particles is

$$v_{\text{rel}} = \frac{|\mathbf{p}_1|}{E_1} + \frac{|\mathbf{p}_2|}{E_2} = |\mathbf{p}_1| \frac{E_1 + E_2}{E_1 E_2} \,.$$

(15.5.51)

In the laboratory frame, where particle 2 is assumed to be at rest, we have $\mathbf{p}_2 = 0$, and the relative velocity is

$$v_{\text{rel}} = \frac{|\mathbf{p}_1|}{E_1} \, . \tag{15.5.52}$$

In the calculation of the scattering cross-sections, one encounters phase-space factors of the outgoing particles

$$d\Phi_n \equiv (2\pi)^4 \delta^{(4)} \left(\sum p'_f - p_1 - p_2 \right) \prod_f \frac{d^3 p'_f}{(2\pi)^3 2E'_f} \, . \tag{15.5.53}$$

If one is interested in the cross-section of the transition into a certain region of phase space, one must integrate over the remaining variables. Since the total four-momentum is conserved, the momenta $\mathbf{p}'_1, \dots \mathbf{p}'_n$ cannot all be independent variables. We consider the important special case of two outgoing particles

$$d\Phi_2 = (2\pi)^4 \delta^{(4)}(p'_1 + p'_2 - p_1 - p_2) \frac{d^3 p'_1}{(2\pi)^3 2E'_1} \frac{d^3 p'_2}{(2\pi)^3 2E'_2} \, . \tag{15.5.54}$$

The integration over $\mathbf{p}'_2$ yields[8]:

$$d\Phi_2 = \frac{1}{(2\pi)^2} \delta(E'_1 + E'_2 - E_1 - E_2) \frac{d^3 p'_1}{4E'_1 E'_2}$$

$$= \frac{\delta(E'_1 + E'_2 - E_1 - E_2) p'^2_1 dp'_1 d\Omega'_1}{16\pi^2 E'_1 E'_2} \, , \tag{15.5.55}$$

where, in this equation, $E'_2 \equiv E_{\mathbf{p}'_2 = \mathbf{p}_1 + \mathbf{p}_2 - \mathbf{p}'_1}$. The further integration over $p_1 \equiv |\mathbf{p}'_1|$ yields[9]:

$$d\Phi_2 = \frac{|\mathbf{p}'_1|^2}{16\pi^2 E'_1 E'_2 \frac{\partial(E'_1 + E'_2)}{\partial |\mathbf{p}'_1|}} d\Omega'_1 \, . \tag{15.5.56}$$

In the center-of-mass system we have $\mathbf{p}'_2 = -\mathbf{p}'_1$. From the relation

$$E'^2_f = m'^2_f + \mathbf{p}'^2_f, \quad f = 1, 2 \tag{15.5.57}$$

it follows that

$$\frac{\partial(E'_1 + E'_2)}{\partial |\mathbf{p}'_1|} = |\mathbf{p}'_1| \left(\frac{1}{E'_1} + \frac{1}{E'_2} \right) = |\mathbf{p}'_1| \frac{E'_1 + E'_2}{E'_1 E'_2} \, . \tag{15.5.58}$$

---

[8] The phase space changes in size when going from (15.5.54) to (15.5.56), since finally we are calculating the cross-section per element of solid angle $d\Omega'_1$, independently of $|\mathbf{p}'_1|$ and $\mathbf{p}'_2$. The notation $d\Phi$ is retained throughout.

[9] $\delta(f(x)) = \sum_{x_0} \frac{1}{|f'(x_0)|} \delta(x - x_0)$, where the sum extends over all (simple) zeros of $f(x)$.

Inserting this into (15.5.56) and (15.5.58), and inserting (15.5.52) into (15.5.50), one obtains the differential scattering cross-section in the center-of-mass frame as

$$\frac{d\sigma}{d\Omega}\bigg|_{CM} = \frac{1}{4}\frac{1}{(4\pi)^2(E_1+E_2)^2}\frac{|\mathbf{p}_1'|}{|\mathbf{p}_1|}\prod_{\substack{\text{external}\\\text{fermions}}}(2m_{\text{Fermi}})|\mathcal{M}|^2. \qquad (15.5.59)$$

This is the relationship we are seeking between the differential scattering cross-section and the amplitude $\mathcal{M}$. The special case of electron–electron scattering was analyzed in Sect. 15.5.3.1.

15.5.3.3 Compton Scattering

Compton scattering refers to the scattering of a photon from a free electron. In practice, the electrons are frequently bound but the high energy of the photons often means that they can still be considered as effectively free[10]. In this scattering process

$$e^- + \gamma \longrightarrow e^- + \gamma$$

the initial state contains an electron and a photon, as does the final state. In second-order perturbation theory, Wick's theorem yields two contributions, each with one contraction of two Fermi operators ψ and $\bar\psi$. The two Feynman diagrams are shown in Fig. 15.12. From these we can directly deduce a further Feynman rule. To each internal fermion line there corresponds a propagator $iS_F(p) = \frac{i}{\not{p}-m+i\epsilon}$. The two diagrams (b) and (c) are topologically equivalent: it is sufficient to consider just one of them.

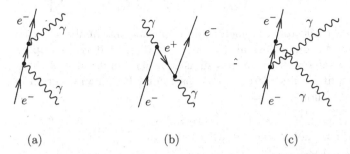

     (a)   (b)   (c)

Fig. 15.12. Compton scattering (a) A photon is absorbed and then emitted. (b) The photon first creates an e^+e^- pair. This diagram is topologically equivalent to (c), where first a photon is emitted, and only afterwards is the incident photon absorbed by the electron

---

[10] The great historical significance of the Compton effect for the evolution of quantum mechanics was described in QM I, Sect. 1.2.1.3.

Remark. In connection with the calculation of the photon propagator in the Coulomb gauge (Sect. 14.5), it was asserted that the photon propagator appears only in the combination $j_\mu D_F^{\mu\nu} j_\nu$. We now explain this in second-order perturbation theory, for which the contribution to the S matrix is

$$S^{(2)} = \frac{(-\mathrm{i})^2}{2!} \int \int d^4x\, d^4x'\, T\left(j^\mu(x)A_\mu(x)j^\nu(x')A_\nu(x')\right)$$

$$= \frac{(-\mathrm{i})^2}{2!} \int \int d^4x\, d^4x'\, T\left(j^\mu(x)j^\nu(x')\right) T\left(A_\mu(x)A_\nu(x')\right) ,$$

since the electron and photon operators commute with one another. The contraction of the two photon fields yields:

$$\int d^4x\, d^4x'\, T\left(j^\mu(x)j^\nu(x')\right) D_{F\mu\nu}(x-x') = \int d^4k\, T\left(j^\mu(k)j^\nu(k)\right) D_{F\mu\nu}(k) ,$$

and, on account of the continuity equation $j^\mu(k)k_\mu = 0$, the part of $D_{F\mu\nu}(k)$ which we call redundant in (E.26c) makes no contribution.

15.5.4 Feynman Rules of Quantum Electrodynamics

In our analysis of scattering processes in Sects. 15.5.2 and 15.5.3, we were able to apply Wick's theorem to derive the most important elements of the Feynman rules which associate analytical expressions to the Feynman diagrams. We summarize these rules in the list below and in Fig. 15.13.

For given initial and final states $|i\rangle$ and $|f\rangle$, the S-matrix element has the form

$$\langle f|\, S\, |i\rangle$$

$$= \delta_{fi} + \left[(2\pi)^4 \delta^{(4)}(P_f - P_i) \left(\prod_{\substack{\text{ext.} \\ \text{fermion}}} \sqrt{\frac{m}{VE}} \right) \left(\prod_{\substack{\text{ext.} \\ \text{photon}}} \sqrt{\frac{1}{2V|\mathbf{k}|}} \right) \right] \mathcal{M} ,$$

where P_i and P_f are the total momenta of the initial and final states. In order to determine $\mathcal{M}$, one draws all topologically distinct diagrams up to the desired order in the interaction and sums over the amplitudes of these diagrams. The amplitude associated with a particular Feynman diagram is itself determined as follows:

1. One assigns a factor of $-\mathrm{i}e\gamma^\mu$ to every vertex point.
2. For every internal photon line one writes a factor $\mathrm{i}D_{F\mu\nu}(k) = \mathrm{i}\frac{-g^{\mu\nu}}{k^2+\mathrm{i}\epsilon}$.
3. For every internal fermion line one writes $\mathrm{i}S_F(p) = \mathrm{i}\frac{1}{\not{p}-m+\mathrm{i}\epsilon}$.
4. To the external lines one assigns the following free spinors and polarization vectors:
   incoming electron: $u_r(p)$
   outgoing electron: $\bar{u}_r(p)$
   incoming positron: $\bar{w}_r(p)$
   outgoing positron: $w_r(p)$

Feynman rules of quantum electrodynamics in momentum space

External lines:

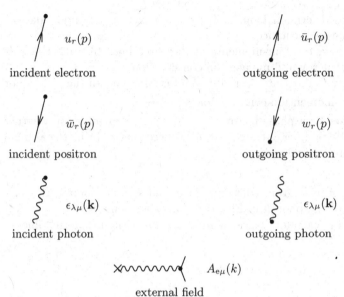

$u_r(p)$

incident electron

$\bar{u}_r(p)$

outgoing electron

$\bar{w}_r(p)$

incident positron

$w_r(p)$

outgoing positron

$\epsilon_{\lambda\mu}(\mathbf{k})$

incident photon

$\epsilon_{\lambda\mu}(\mathbf{k})$

outgoing photon

$A_{e\mu}(k)$

external field

Internal lines:

$iS_F(p) = i\frac{1}{\not{p}-m+i\epsilon}$

p

internal electron line

k

$iD_F^{\mu\nu}(k) = i\frac{-g^{\mu\nu}}{k^2+i\epsilon}$

μ ν

internal photon line

p'

k

$-ie\gamma^\mu$

p

vertex

Fig. 15.13. The Feynman rules of quantum electrodynamics. The end points where the external lines and propagators can be attached to a vertex are indicated by dots

incoming photon: $\epsilon_{\lambda\mu}(\mathbf{k})$

outgoing photon: $\epsilon_{\lambda\mu}(\mathbf{k})$

5. The spinor factors (γ matrices, S_F propagators, four-spinors) are ordered for each fermion line such that reading them from right to left amounts to following the arrows along the fermion lines.

6. For each closed fermion loop, multiply by a factor (-1) and take the trace over the spinor indices.

7. At every vertex, the four-momenta of the three lines that meet at this point satisfy energy and momentum conservation.

8. It is necessary to integrate over all free internal momenta (i.e., those not fixed by four-momentum conservation): $\int \frac{d^4q}{(2\pi)^4}$.

9. One multiplies by a phase factor $\delta_p = 1(\text{or} - 1)$, depending on whether an even or odd number of transpositions is necessary to bring the fermion operators into normal order.

An example of a diagram containing a closed fermion loop is the self-energy diagram of the photon in Fig. 15.21, another being the vacuum diagram of Fig. 15.14. Vacuum diagrams are diagrams without external lines.

Fig. 15.14. The vacuum diagram of lowest order

The minus sign for a closed fermion loop has the following origin: Starting from the part of the T-product which leads to the closed loop $T\left(\ldots \bar{\psi}(x_1)A(x_1)\psi(x_1)\ldots\bar{\psi}(x_2)A(x_2)\psi(x_2)\ldots\bar{\psi}(x_f)A(x_f)\psi(x_f)\ldots\right)$, one has to permute $\bar{\psi}(x_1)$ with an odd number of fermion fields to arrive at the arrangement $\ldots A(x_1)\psi(x_1)\bar{\psi}(x_2)A(x_2)\psi(x_2)\ldots\bar{\psi}(x_f)A(x_f)\psi(x_1)\ldots$ leading to the sequence of propagators $\underline{\psi(x_1)\bar{\psi}(x_2)}\ldots\underline{\psi(x_f)\bar{\psi}(x_1)}$ with a minus sign.

*15.6 Radiative Corrections

We will now describe a few other typical elements of Feynman diagrams, which lead in scattering processes to higher-order corrections in the charge. These corrections go by the general name of *radiative corrections*. If, in electron–electron scattering, for example, one includes higher order Feynman diagrams, one obtains correction terms in powers of the fine-structure constant α. Some of these diagrams are of a completely new form, whereas others can be shown to be modfications of the electron propagator, the photon propagator, or the electron–electron–photon vertex. In this section we will investigate the latter elements of Feynman diagrams. The intention is to give the reader a general, qualitative impression of the way in which higher corrections are calculated, and of regularization and renormalization. We will not attempt to present detailed quantitative calculations.[11]

15.6.1 The Self-Energy of the Electron

15.6.1.1 Self-Energy and the Dyson Equation

As observed earlier, an electron interacts with its own radiation field. It can emit and reabsorb photons. These photons, which are described by perturbation-theoretical contributions of higher order, modify the propagation properties of the electron. If, for example, a fermion line within some diagram is replaced by the diagram shown in Fig. 15.15b, this means that the propagator as a whole, (a)+(b), becomes

$$S_F(p) \to S_F'(p) = S_F(p) + S_F(p)\Sigma(p)S_F(p) \,. \tag{15.6.1}$$

(a)   (b)

Fig. 15.15. Replacement of a fermion propagator (a) by two propagators with inclusion of the self-energy (b)

The bubble consisting of a photon and a fermion line (Fig. 15.15b) is called the *self-energy* $\Sigma(p)$. The corresponding analytical expression is given in Eq. (15.6.4). Summing over processes of this type

$$\begin{aligned}
S_F'(p) &= S_F(p) + S_F(p)\Sigma(p)S_F(p) \\
&\quad + S_F(p)\Sigma(p)S_F(p)\Sigma(p)S_F(p) + \dots \\
&= S_F(p) + S_F(p)\Sigma(p)\Big(S_F(p) + S_F(p)\Sigma(p)S_F(p) + \dots\Big),
\end{aligned}$$
$$\tag{15.6.2a}$$

one obtains the *Dyson equation*

$$S_F'(p) = S_F(p) + S_F(p)\Sigma(p)S_F'(p) \,, \tag{15.6.2b}$$

---

[11] See, e.g., J. M. Jauch and F. Rohrlich, *The Theory of Photons and Electrons* or J. D. Bjorken and S. D. Drell, *Relativistic Quantum Mechanics*, McGraw-Hill, New York, 1964, p. 153

with the solution

$$S'_F(p) = \frac{1}{(S_F(p))^{-1} - \Sigma(p)} .$$
(15.6.3)

Hence, the self-energy $\Sigma(p)$ and its associated *self-energy diagram* give, among other things, a correction to the mass and a modification of the particle's energy, the latter being the source of the name "self-energy". The diagrammatic representation of (15.6.2a) and (15.6.2b) is given in Fig. 15.16. To distinguish it from the free (or bare) propagator $S_F(p)$, the propagator

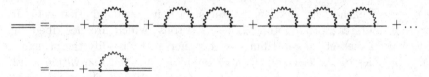

Fig. 15.16. Diagrammatic representation of the Dyson equation (15.6.2a,b). The propagator $S'_F(p)$ is represented by the double line

$S'_F(p)$ is called the interacting (dressed) propagator. It is represented diagrammatically by a double line.

A few self-energy diagrams of higher order have already been given in Fig. 15.2. In general, a part of a diagram is called a self-energy contribution when it is linked to the rest of the diagram only by two $S_F(p)$ propagators. A *proper* self-energy diagram (also one-particle irreducible) is one that cannot be separated into two parts by cutting a single $S_F(p)$ line. Otherwise, one has an *improper* self-energy diagram. The self-energy diagrams in Fig. 15.2 are all proper ones. The second summand in the first line of Fig. 15.16 contains a proper self-energy part; all the others are improper. The Dyson equation (15.6.2b) can be extended to arbitrarily high orders; then, $\Sigma(p)$ in (15.6.2b) and (15.6.3) consists of the sum of all proper self-energy diagrams.

The analytical expression corresponding to the lowest order self-energy diagram of Fig. 15.17, which is also contained as a part of Fig. 15.15, reads according to the Feynman rules

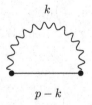

$p - k$

Fig. 15.17. Lowest (proper) self-energy diagram of the electron

$$-i\Sigma(p) = \frac{(-ie)^2}{(2\pi)^4} \int d^4k \, iD_{F\mu\nu}(k)\gamma^\mu iS_F(p-k)\gamma^\nu$$

$$= \frac{e^2}{(2\pi)^4} \int d^4k \frac{1}{k^2 + i\epsilon} \frac{2\not{p} - 2\not{k} - 4m}{(p-k)^2 - m^2 + i\epsilon}. \qquad (15.6.4)$$

This integral is ultraviolet divergent; it diverges logarithmically at the upper limit.

The bare (free) propagator $\frac{i}{\not{p}-m+i\epsilon}$ has a pole at the bare mass $\not{p} = m$, i.e., $\frac{i}{\not{p}-m+i\epsilon} = \frac{i(\not{p}+m)}{p^2-m^2+i\epsilon}$ has a pole at $p^2 = m^2$. Correspondingly, the interacting propagator that follows from (15.6.3)

$$iS_F'(p) = \frac{i}{\not{p} - m - \Sigma(p) + i\epsilon}. \qquad (15.6.5)$$

will possess a pole at a different, *physical* or *renormalized* mass

$$m_R = m + \delta m. \qquad (15.6.6)$$

The mass of the electron is modified by the emission and reabsorption of virtual photons (see, e.g., the diagrams in Fig. 15.2). We rewrite (15.6.5) with the help of (15.6.6),

$$iS_F'(p) = \frac{i}{\not{p} - m_R - \Sigma(p) + \delta m + i\epsilon}. \qquad (15.6.7)$$

15.6.1.2 The Physical and the Bare Mass, Mass Renormalization

It will prove convenient to redefine the fermion part of the Lagrangian density, and likewise that of the Hamiltonian,

$$\mathcal{L} \equiv \bar{\psi}(i\not{\partial} - m)\psi - e\bar{\psi}\not{A}\psi$$
$$= \bar{\psi}(i\not{\partial} - m_R)\psi - e\bar{\psi}\not{A}\psi + \delta m \, \bar{\psi}\psi, \qquad (15.6.8)$$

so that the free Lagrangian density contains the physical mass. This then takes account of the fact that the nonlinear interaction in $\mathcal{L}_1$ modifies the bare mass m, and that the resulting physical mass m_R, which is observed in experiment, differs from m according to (15.6.6). Individual particles that are widely separated from one another and noninteracting, as is the case for scattering processes before and after the scattering event, possess the physical mass m_R. According to Eq. (15.6.8), the Hamiltonian density contains, in addition to $e\bar{\psi}\not{A}\psi$, the further perturbation term $-\delta m \, \bar{\psi}\psi$. The δm has to be determined such that the combined effect of the two terms in the modified interaction part,

$$\mathcal{H}_I = e\bar{\psi}\not{A}\psi - \delta m \, \bar{\psi}\psi, \qquad (15.6.9)$$

produces no change in the physical electron mass. The perturbation term $-\delta m \, \bar{\psi}\psi$ is represented diagrammatically in Fig. 15.18. It has the form of a

$-\delta m$

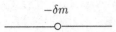

Fig. 15.18. Feynman diagram for the mass counter term $-\delta m\,\bar\psi\psi$

vertex with two lines. Subtraction of the term $-\delta m\,\bar\psi\psi$, has the result that the "bare" particles of the thus redefined Lagrangian density have the same mass, and, hence, the same energy spectrum, as the physical particles, namely m_R. Every self-energy term of the form shown in Fig. 15.19a is accompanied by a mass counter term (b), which cancels the k-independent contribution from (a). In higher orders of e, there are further proper self-energy diagrams to be considered, and δm contains higher-order corrections in e. For the redefined Lagrangian density (15.6.8) and the interaction Hamiltonian density (15.6.9) the propagator also has the form (15.6.7), where the self-energy $\Sigma(p)$

$$-i\Sigma(p) = -e^2 \int \frac{d^4k}{(2\pi)^4} \frac{-i}{k^2+i\epsilon}\gamma_\nu \frac{i}{\not{p}-\not{k}-m_R+i\epsilon}\gamma^\nu \qquad (15.6.10)$$

differs from (15.6.4) only in the appearance of the mass m_R. The mass shift δm is obtained from the condition that the sum of the third and fourth terms in the denominator of (15.6.7) produces no change in the (physical) mass, i.e., that $iS_F'(p)$ has a pole at $\not{p}=m_R$:

$$\Sigma(p)|_{\not{p}=m_R} = \delta m \quad . \qquad (15.6.11)$$

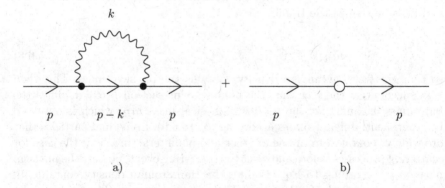

Fig. 15.19. The lowest self-energy contributions according to the Lagrangian density (15.6.8) or (15.6.9). (a) Self-energy as in Eq. (15.6.4) with $m \to m_R$; (b) mass counter term resulting from the mass correction in (15.6.9)

15.6.1.3 Regularization and Charge Renormalization

Since the integrand in (15.6.10) falls off only as k^{-3}, the integral is ultraviolet divergent. Thus, in order to determine the physical effects associated with $\Sigma(p)$, one needs to carry out a regularization which makes the integral finite. One possibility is to replace the photon propagator by

$$\frac{1}{k^2 + i\epsilon} \longrightarrow \frac{1}{k^2 - \lambda^2 + i\epsilon} - \frac{1}{k^2 - \Lambda^2 + i\epsilon} \,. \tag{15.6.12}$$

Here, Λ is a large cut-off wave vector: for $k \ll \Lambda$ the propagator is unchanged and for $k \gg \Lambda$ it falls off as k^{-4}, such that $\Sigma(p)$ becomes finite. In the limit $\Lambda \to \infty$ one has the original QED. In addition, λ is an artificial photon mass that is introduced so as to avoid infrared divergences, and which will eventually be set to zero. With the regularization (15.6.12), $\Sigma(p)$ becomes finite. It will be helpful to expand $\Sigma(p)$ in powers of $(\not{p} - m_R)$,

$$\Sigma(p) = A - (\not{p} - m_R)B + (\not{p} - m_R)^2 \Sigma_f(p) \,. \tag{15.6.13}$$

From (15.6.10) one sees that the p-independent coefficients A and B diverge logarithmically in Λ, whereas $\Sigma_f(p)$ is finite and independent of Λ. If one multiplies $\Sigma(p)$ from the left and right by spinors for the mass m_R, only the constant A remains. If one considers $\frac{\partial \Sigma(p)}{\partial p_\mu}$, and again multiplies from the left and right by the spinors, then only $-\gamma^\mu B$ remains. We will need this result later in connection with the Ward identity.

The result of the explicit calculation[11] is: According to Eq. (15.6.11), the mass shift δm is obtained as

$$\delta m = A = \frac{3 m_R \alpha}{2\pi} \log \frac{\Lambda}{m_R} \,, \tag{15.6.14}$$

which is logarithmically divergent. The coefficient B is

$$B = \frac{\alpha}{4\pi} \log \frac{\Lambda^2}{m_R^2} - \frac{\alpha}{2\pi} \log \frac{m_R^2}{\lambda^2}. \tag{15.6.15}$$

The explicit form of the finite function $\Sigma_f(p)$ will not be needed here. It follows that

$$\begin{aligned}
iS_F'(p) &= \frac{i}{(\not{p} - m_R)\left[1 + B - (\not{p} - m_R)\Sigma_f(p)\right]} \\
&= \frac{i}{(\not{p} - m_R)(1 + B)(1 - (\not{p} - m_R)\Sigma_f(p)) + \mathcal{O}(\alpha^2)} \\
&= \frac{iZ_2}{(\not{p} - m_R)(1 - (\not{p} - m_R)\Sigma_f(p)) + \mathcal{O}(\alpha^2)}
\end{aligned} \tag{15.6.16}$$

with

$$Z_2^{-1} \equiv 1 + B = 1 + \frac{\alpha}{4\pi}\left(\log\frac{\Lambda^2}{m_R^2} - 2\log\frac{m_R^2}{\lambda^2}\right) \ . \tag{15.6.17}$$

The quantity Z_2 is known as the wave function renormalization constant.

Now, a propagator connects two vertices, each of which contributes a factor e. Hence, the factor Z_2 can be split into two factors of $\sqrt{Z_2}$ and, taking into account the two fermions entering at each vertex, one can redefine the value of the charge

$$e_R' = Z_2 e \equiv (1 - B)e \ . \tag{15.6.18}$$

Here, e_R' is the preliminary *renormalized charge*. In the following, we will undertake two further renormalizations. The electron propagator that remains after the renormalization has the form

$$i\tilde{S}_F'(p) = Z_2^{-1}iS_F'(p)$$
$$= \frac{i}{(\not{p} - m_R)(1 - (\not{p} - m_R)\Sigma_f(p)) + \mathcal{O}(\alpha^2)} \tag{15.6.19}$$

and is finite.

15.6.1.4 Renormalization of External Electron Lines

The diagram 15.20a contains a self-energy insertion in an external electron line. This, together with the mass counter term of Fig. 15.20b, leads to the following modification of the spinor of the incident electron:

$$u_r(p) \to u_r(p) + \frac{i}{\not{p} - m_R + i\epsilon}\left(i(\not{p} - m_R)B - i(\not{p} - m_R)^2\Sigma_f(p)\right)u_r$$
$$\to \left(1 - \frac{i}{\not{p} - m_R + i\epsilon}(\not{p} - m_R)B\right)u_r(p) \ , \tag{15.6.20}$$

since the last term in the first line vanishes on account of $(\not{p} - m_R)u_r(p) = 0$. The expression in the second line is undetermined, as can be seen, either by allowing the two operators to cancel with one another, or by applying $(\not{p} - m_R)$ to $u_r(p)$. By switching the interaction on and off adiabatically,

$$\mathcal{H}_I = \zeta(t)e\,\bar{\psi}\gamma_\mu\psi A^\mu - \zeta(t)^2\delta m\,\bar{\psi}\psi, \tag{15.6.21}$$

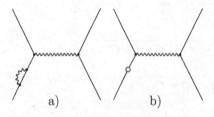

a)   b)

Fig. 15.20. (a) A diagram with inclusion of the self-energy in an external fermion line. (b) Mass counter term

where $\lim_{t \to \pm\infty} \zeta(t) = 0$, and $\zeta(0) = 1$, Eq. (15.6.20) is replaced by a well-defined mathematical expression, with the result

$$u_r(p) \overset{!}{\to} u_r(p)\sqrt{1-B} \,. \tag{15.6.22}$$

This means that the external lines, like the internal ones, also yield a factor $\sqrt{1-B}$ in the renormalization of the charge. Thus, Eq. (15.6.18) also holds for vertices with external lines:

$$e \to e_R' = (1 - B)e \,.$$

Apart from the factor $Z_2^{1/2}$, which goes into the charge renormalization, there are no radiative corrections in the external electron lines. The result (15.6.22) is to be expected intuitively for the following reasons: (i) Even an external electron must have been emitted somewhere and is thus an internal electron in some larger process. It thus yields a factor of $\sqrt{1-B}$ at every vertex. (ii) The transition from S_F' to $\tilde{S}_F'$ in Eq. (15.6.19) can be regarded as a replacement of the field ψ by a renormalized field $\psi_R = Z_2^{-1/2}\psi + \ldots$, or $Z_2^{1/2}\psi_R = \psi + \ldots$. From this, one also sees that Z_2 represents the probability of finding in a physical electron state one bare electron.

15.6.2 Self-Energy of the Photon, Vacuum Polarization

The lowest contribution to the photon self-energy is represented in Fig. 15.21. This diagram makes a contribution to the photon propagator. The photon creates a virtual electron–positron pair, which subsequently recombines to yield a photon once more. Since the virtual electron–positron pair has a fluctuating dipole moment that can be polarized by an electric field, one speaks in this context of *vacuum polarization*.

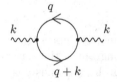

Fig. 15.21. Vacuum polarization: A photon decays into an electron–positron pair which recombines to a photon

According to the Feynman rules, the analytical expression equivalent to Fig. 15.21 is

$$\Pi_{\mu\nu}(k, m_R) = \int \frac{d^4q}{(2\pi)^4}(-1)$$

$$\times \operatorname{Tr}\left((-ie\gamma_\mu)\frac{i}{\not{q} + \not{k} - m_R + i\epsilon}(-ie\gamma_\nu)\frac{i}{\not{q} - m_R + i\epsilon}\right). \tag{15.6.23}$$

At first sight, this expression would appear to diverge quadratically at the upper limit. However, due to the gauge invariance, the ultraviolet contributions are in fact only logarithmically divergent.

Regularizing the expression by cutting off the integral at a wave vector Λ would violate gauge invariance. One thus regularizes (15.6.23) using the Pauli–Villars method[11], by replacing $\Pi_{\mu\nu}(k, m_R)$ by $\Pi^R_{\mu\nu}(k, m_R) \equiv \Pi_{\mu\nu}(k, m_R) - \sum_i C_i \Pi_{\mu\nu}(k, M_i)$, where the M_i are large additional fictitious fermion masses, and the coefficients satisfy $\sum_i C_i = 1$, $\sum_i C_i M_i^2 = m_R^2$. The final result only involves $\log \frac{M^2}{m_R^2} \equiv \sum_i C_i \log \frac{M_i^2}{m_R^2}$.

Finally, because of the vacuum polarization self-energy contributions, the photon propagator for small k takes the form

$$iD'_{\mu\nu}(k) = -\frac{ig_{\mu\nu}}{k^2 + i\epsilon} Z_3 \left(1 - \frac{\alpha}{\pi m_R^2} \left(\frac{1}{15} - \frac{1}{40} \left(\frac{k^2}{m_R^2} \right) \right) \right) , \qquad (15.6.24)$$

where

$$Z_3 \equiv 1 - C = 1 - \frac{\alpha}{3\pi} \log \frac{M^2}{m_R^2} \qquad (15.6.25)$$

is the photon field renormalization constant. This factor also leads to a renormalization of the charge

$$e_R''^2 \equiv Z_3 e'^2 \approx \left(1 - \frac{\alpha}{3\pi} \log \frac{M^2}{m_R^2} \right) e^2 . \qquad (15.6.26)$$

The photon propagator that remains after charge renormalization, for small k, has the form

$$i\tilde{D}'_{F\mu\nu}(k) = Z_3^{-1} iD'_{F\mu\nu}(k)$$
$$= \frac{-ig_{\mu\nu}}{k^2 + i\epsilon} \left(1 - \frac{\alpha k^2}{\pi m_R^2} \left(\frac{1}{15} - \frac{1}{40} \left(\frac{k^2}{m_R^2} \right) \right) \right) . \qquad (15.6.27)$$

15.6.3 Vertex Corrections

We now proceed to the discussion of vertex corrections. The divergences that occur here can again be removed by renormalization. A diagram of the type shown in Fig. 15.22a contains two fermion and one photon line; it thus has the same structure as the vertex $\bar{\psi}\gamma^\mu A_\mu \psi$ in Fig. 15.22b. For diagrams of this kind, one thus speaks of vertex corrections. The diagram 15.22a represents the lowest (lowest power in e) vertex correction. This diagram also yields the leading contribution to the anomalous magnetic moment of the electron. The amplitude for the diagram, without the external lines, is given by

$$\Lambda_\mu(p', p) = (-ie)^2 \int \frac{d^4k}{(2\pi)^4} \frac{-i}{k^2 + i\epsilon}$$
$$\times \gamma_\nu \frac{i}{p\!\!\!/' - k\!\!\!/ - m_R + i\epsilon} \gamma_\mu \frac{i}{p\!\!\!/ - k\!\!\!/ - m_R + i\epsilon} \gamma^\nu . \qquad (15.6.28)$$

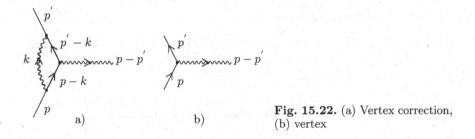

a)   b)

Fig. 15.22. (a) Vertex correction, (b) vertex

$\Lambda_\mu(p',p)$ is logarithmically divergent and is regularized in the following by replacing the photon propagator as specified in Eq. (15.6.12). One can split $\Lambda_\mu(p',p)$ into a component that diverges in the limit $\Lambda \to \infty$ and a component that remains finite. We first consider Λ_μ multiplied from the left and right by two spinors corresponding to the mass m_R, to yield $\bar{u}_{r'}(P)\Lambda_\mu(P,P)u_r(P)$. We denote the momentum of spinors such as these, which correspond to real particles, by P. Due to Lorentz invariance, this expression can only be proportional to γ_μ and to P_μ. With the help of the Gordon identity (10.1.5), one can replace a P^μ dependence by γ^μ, so that one has

$$\bar{u}_{r'}(P)\Lambda_\mu(P,P)u_r(P) = L\bar{u}_{r'}(P)\gamma_\mu u_r(P) \qquad (15.6.29)$$

with a constant L that remains to be determined. For general four-vectors p, p', we separate $\Lambda_\mu(p',p)$ in the following way:

$$\Lambda_\mu(p',p) = L\gamma_\mu + \Lambda_\mu^f(p',p) \quad . \qquad (15.6.30)$$

Whereas L diverges in the limit $\Lambda \to \infty$, the term $\Lambda_\mu^f(p',p)$ remains finite. In order to see this, we expand the fermion part of (15.6.28) in terms of the deviation of the momentum vectors p and p' from the momentum P of free physical particles used in (15.6.29):

$$\left(\frac{1}{\not{P} - \not{k} - m_R + i\epsilon} - \frac{1}{\not{P} - \not{k} - m_R + i\epsilon}(\not{p}' - \not{P})\frac{1}{\not{P} - \not{k} - m_R + i\epsilon} + \cdots \right)$$

$$\times \gamma_\mu \left(\frac{1}{\not{P} - \not{k} - m_R + i\epsilon} \right.$$

$$\left. - \frac{1}{\not{P} - \not{k} - m_R + i\epsilon}(\not{p} - \not{P})\frac{1}{\not{P} - \not{k} - m_R + i\epsilon} + \cdots \right) .$$

$$(15.6.31)$$

The divergence in (15.6.28) stems from the leading term (the product of the first terms in the brackets in (15.6.31)); this yields $L\gamma^\mu$, whilst the remaining terms are finite.

The first term in (15.6.30), together with γ_μ, leads to the replacement $\gamma_\mu \to \gamma_\mu(1 + L)$ and produces a further renormalization of the charge

$$e_R = (1 + L)e''_R \equiv Z_1^{-1} e''_R . \tag{15.6.32}$$

We need not pursue the calculation of L any further since, as will be generally shown, it is related to the constant B introduced in (15.6.13) and (15.6.15), and in the charge renormalization cancels with it.

15.6.4 The Ward Identity and Charge Renormalization

Taken together, the various renormalization factors for the charge yield

$$e \to e_R = \sqrt{1 - C}\,(1 - B)(1 + L)e . \tag{15.6.33}$$

The factor $\sqrt{1 - C}$ comes from the vacuum polarization (Fig. 15.21), the factor $1 - B$ from the wave function renormalization of the electron (Fig. 15.15), and the factor $1 + L$ from the vertex renormalization. However, in quantum electrodynamics, it turns out that the coefficients B and L are equal. To demonstrate this, we write the self-energy of the electron (15.6.10) in the form

$$\Sigma(p) = ie^2 \int \frac{d^4 k}{(2\pi)^4} D_F(k) \gamma_\nu S_F(p - k) \gamma^\nu , \tag{15.6.34}$$

and the vertex function (15.6.28) as

$$\Lambda_\mu(p', p) = e^2 \int \frac{d^4 k}{(2\pi)^4} D_F(k) \gamma_\nu S_F(p' - k) \gamma_\mu S_F(p - k) \gamma_\nu . \tag{15.6.35}$$

We now make use of the relation

$$\frac{\partial S_F(p)}{\partial p^\mu} = -S_F(p) \gamma_\mu S_F(p), \tag{15.6.36}$$

which is obtained by differentiating

$$S_F(p) S_F^{-1}(p) = 1 \tag{15.6.37}$$

with respect to p^μ, i.e.,

$$\frac{\partial S_F(p)}{\partial p^\mu} S_F^{-1}(p) + S_F(p) \frac{\partial}{\partial p^\mu}(\not{p} - m_R) = 0 , \tag{15.6.38}$$

and then multiplying by $S_F(p)$ from the right. Equation (15.6.36) states that the insertion of a vertex γ_μ in an internal electron line, without energy transfer, is equivalent to the differentiation of the electron propagator with respect to p^μ (Fig. 15.23). With the help of this identity, we can write the vertex function (15.6.35) in the limit of equal momenta as

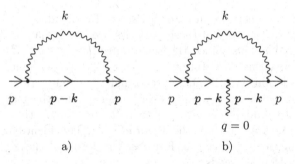

Fig. 15.23. Diagrammatic representation of the Ward identity. (a) Self-energy diagram. (b) The differentiation is equivalent to inserting into the fermion line a vertex for a photon with zero momentum

$$
\begin{aligned}
L\gamma_\mu &= \lim_{p' \to p} \Lambda_\mu(p',p)\bigg|_{\not{p}'=m_R,\ \not{p}=m_R} \\
&= -ie^2 \int \frac{d^4k}{(2\pi)^4} D_F(k)\gamma_\nu \frac{\partial S_F(p-k)}{\partial(p-k)^\mu}\gamma^\nu \\
&= -ie^2 \int \frac{d^4k}{(2\pi)^4} D_F(k)\gamma_\nu \frac{\partial S_F(p-k)}{\partial p^\mu}\gamma^\nu \ .
\end{aligned}
\tag{15.6.39}
$$

On the other hand, from the definition of B in (15.6.13), one obtains

$$
\begin{aligned}
\bar{u}_{r'}(p)B\gamma_\mu u_r(p) &= \bar{u}_{r'}(p)\left(\frac{-\partial \Sigma(p)}{\partial p^\mu}\right)u_r(p) \\
&= \bar{u}_{r'}(p)\left(e^2\int \frac{d^4k}{(2\pi)^4} D_F(k)\gamma_\nu \frac{\partial S_F(p-k)}{\partial p^\mu}\gamma^\nu\right)u_r(p) \\
&= \bar{u}_{r'}(p)L\gamma_\mu u_r(p) \ ,
\end{aligned}
\tag{15.6.40}
$$

from which it follows that

$$
B = L \ . \tag{15.6.41}
$$

This relation implies

$$
(1-B)(1+L) = 1 + \mathcal{O}(\alpha^2) \ , \tag{15.6.42}
$$

so that the charge renormalization simplifies to

$$
e \to e_R = \sqrt{1-C}\,e \equiv Z_3^{1/2}e \ . \tag{15.6.43}
$$

The renormalized charge e_R is equal to the experimentally measured charge $e_R^2 \equiv \frac{4\pi}{137}$. The bare charge e^2 is, according to (15.6.26), larger than e_R^2.

The factors arising from the renormalization of the vertex and of the wave function of the fermion cancel one another. It follows from this result that the charge renormalization is independent of the type of fermion considered.

In particular, it is the same for electrons and muons. Hence, for the identical bare charges, the renormalized charges of these particles are also equal, as is the case for electrons and muons. Since the renormalization factors (Z factors) depend on the mass, this last statement would not hold without such cancellation. The prediction that the charge renormalization only arises from the photon field renormalization is valid at every order of perturbation theory. The relation (15.6.36) and its generalization to higher orders, together with its implication (15.6.41), are known as the *Ward identity*. This identity is a general consequence of gauge invariance. Expressed in terms of the Z factors, the Ward identity (15.6.41) reads

$$Z_1 = Z_2 \ .$$

Remarks:

(i) We add a remark here concerning the form of the radiative corrections for the electron–electron scattering that was treated to leading order in Sects. 15.5.3.1 and 15.5.3.2. We will confine ourselves to the direct scattering. The leading diagram is shown in Fig. 15.24a. Taking this as the starting point, one obtains diagrams that contain self-energy insertions in internal (b) and external (d) lines, and vertex corrections (c). These are taken into account by charge renormalization and by the replacements $D \to \tilde{D}'$ and $\gamma_\mu \to (\gamma_\mu + \Lambda_\mu^f)$, as was briefly sketched above. The diagram

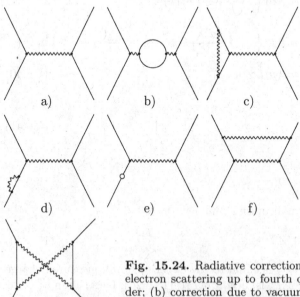

Fig. 15.24. Radiative corrections to the direct electron–electron scattering up to fourth order in e. (a) Second order; (b) correction due to vacuum polarization; (c) vertex correction; (d) self-energy insertion in an external line; (f) and (g) two further diagrams

(e) stems from the mass counter term $-\delta m \bar{\psi}\psi$. In addition to these diagrams there are two further diagrams (f,g) that make finite contributions in second order.

(ii) Quantum electrodynamics in four space–time dimensions is renormalizable since, at every order of perturbation theory, all divergences can be removed by means of a finite number of reparameterizations (renormalization constants δm, Z_1, Z_2, and Z_3).

15.6.5 Anomalous Magnetic Moment of the Electron

An interesting consequence of the radiative corrections is their effect on the magnetic moment of the electron. This we elucidate by considering the scattering in an external electromagnetic potential A_e^μ. In the interaction (15.6.9) the field operator A^μ is thus replaced by $A^\mu + A_e^\mu$. The process to first order is shown in Fig. 15.25a. The corresponding analytical expression is

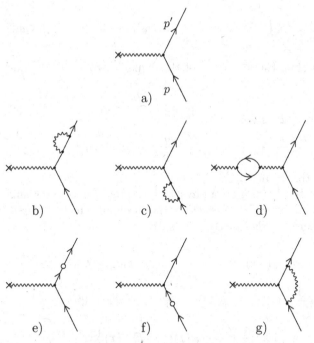

Fig. 15.25. Radiative corrrections of second order for the QED vertex with two fermions and an external potential A_e^μ

$$- ie_R \bar{u}_{r'}(p') \not{A}_e(p' - p) u_r(p)$$

$$= -\frac{ie_R}{2m_R} \bar{u}_{r'}(p') \left[(p' + p)^\mu + i\sigma^{\mu\nu}(p' - p)_\nu \right] u_r(p) A_{e\mu}(p' - p) ,$$

$$\text{(15.6.44)}$$

where we have used the Gordon identity (10.1.5). In anticipation of the charge renormalization (see below), we have already inserted the renormalized charge here. The second term in the square brackets is the transition amplitude for the scattering of a spin-$\frac{1}{2}$ particle with the magnetic moment $\frac{e_R}{2m_R} = -\frac{\hat{e}_0}{2m_R}$, where $\hat{e}_0$ is the elementary charge[12]; i.e., the gyromagnetic ratio is $g = 2$. The processes of higher order are shown in Fig. 15.25b–g. The self-energy insertions (b,c) and the contribution C from vacuum polarization (d) and L coming from the vertex correction (g) lead to charge renormalization, i.e., in (15.6.44) one has, instead of e, the physical charge e_R. Furthermore, the diagrams (d) and (g) yield finite corrections. For the spin-dependent scattering, only the vertex correction $\Lambda_\mu^f(p', p)$ is important. For (15.6.30)[11] the calculation yields

$$\Lambda_\mu^f(p', p) = \gamma_\mu \frac{\alpha}{3\pi} \frac{q^2}{m_R^2} \left(\log \frac{m_R}{\lambda} - \frac{3}{8} \right) + \frac{\alpha}{8\pi m_R} [\not{q} , \gamma_\mu] \qquad \text{(15.6.45)}$$

with $q = p' - p$. By adding the last term of this equation to (15.6.44), one obtains

$$-ie_R \bar{u}_{r'}(p')(\gamma_\mu + \frac{i\alpha}{2\pi} \frac{\sigma_{\mu\nu} q^\nu}{2m_R}) u_r(p) A_e^\mu(q) \qquad \text{(15.6.46)}$$

$$= -ie_R \bar{u}_{r'}(p') \left[\frac{(p + p')_\mu}{2m_R} + (1 + \frac{\alpha}{2\pi}) \frac{i\sigma_{\mu\nu} q^\nu}{2m_R} \right] u_r(p) A_e^\mu(q) .$$

In coordinate space, the term $i\sigma_{\mu\nu} q^\nu A_e^\mu$ has the form $-\sigma_{\mu\nu} \partial^\nu A_e^\mu(x) = -\frac{1}{2}\sigma_{\mu\nu} F^{\mu\nu}$. In order to be able to give a physical interpretation of the result (15.6.46), we consider an effective interaction Hamiltonian which, in first-order perturbation theory, yields exactly (15.6.46):

$$\mathcal{H}^{\text{eff}} \equiv e_R \int d^3x \left\{ \bar{\psi}(x)\gamma_\mu\psi(x)A_e^\mu(x) + \frac{\alpha}{2\pi} \frac{1}{2m_R} \bar{\psi}(x)\sigma_{\mu\nu}\psi(x)\partial^\nu A_e^\mu(x) \right\}$$

$$= e_R \int d^3x \left\{ \frac{i}{2m_R} \left(\bar{\psi}(x) \left(\partial_\mu\psi(x) \right) - \left(\partial_\mu\bar{\psi}(x) \right) \psi(x) \right) A_e^\mu(x) \right.$$

$$\left. + \left(1 + \frac{\alpha}{2\pi} \right) \frac{1}{4m_R} \bar{\psi}(x)\sigma_{\mu\nu}\psi F_e^{\mu\nu}(x) \right\} .$$

$$\text{(15.6.47)}$$

Here, we have again used the Gordon identity. The first term after the second equals sign represents a convective current. The second term, in the case of

---

a constant magnetic field, can be interpreted as a magnetic dipole energy. Since this can, through the substitution $F^{12} = B^3$, $F^{23} = B^1$, $F^{31} = B^2$, $\sigma_{12} = \Sigma_3$, etc., be brought into the form

$$-\mathbf{B} \left(\frac{e_R}{2m_R} \left(1 + \frac{\alpha}{2\pi} \right) 2 \int d^3x \bar{\psi}(x) \frac{\boldsymbol{\Sigma}}{2} \psi(x) \right) \equiv -\mathbf{B}\boldsymbol{\mu} . \tag{15.6.48}$$

For slow electrons, the upper components of the spinors are significantly larger than the lower ones. In this nonrelativistic limit, the magnetic moment of a single electron is, according to (15.6.48), effectively given by

$$\frac{e_R}{2m_R} \left(1 + \frac{\alpha}{2\pi} \right) 2 \frac{\boldsymbol{\sigma}}{2} , \tag{15.6.49}$$

where $\boldsymbol{\sigma}$ are the 2×2 Pauli matrices. The contribution to (15.6.49) proportional to the fine-structure constant is referred to as the anomalous magnetic moment of the electron. It should be stressed, however, that (15.6.47) does not represent a fundamental interaction: It merely serves to describe the second-order radiative correction within first-order perturbation theory. From (15.6.49) one obtains the modification of the g factor

$$\frac{g - 2}{2} = \frac{\alpha}{2\pi} = 0.00116 .$$

When one includes corrections of order α^2 and α^3, which arise from higher-order diagrams, one finds the value

$$\frac{g - 2}{2} = 0.0011596524(\pm 4),$$

which is in impressive agreement with the experimental value of

$$0.00115965241(\pm 20) .$$

The increase in the magnetic moment of the electron can be understood qualitatively as follows. The electron is continually emitting and reabsorbing photons and is thus surrounded by a cloud of photons. Thus, a certain amount of the electron's energy, and therefore mass, resides with these photons. Hence, the charge-to-mass ratio of the electron is effectively increased and this reveals itself in a measurement of the magnetic moment. In the diagram 15.25g, the electron emits a photon before interacting with the external magnetic field. The correction is proportional to the emission probability and, thus, to the fine-structure constant α.

Problems

15.1 Confirm the expression (15.2.17) for the propagator $\phi(x_1)\phi(x_2)$, instead of starting from (15.2.16), by evaluating (15.2.15) directly.

15.2 The interaction of the complex Klein–Gordon field with the radiation field reads, according to Eq. (F.7), to first order in $A_\mu(x)$

$$\mathcal{H}_I(x) = j^\mu(x)A_\mu(x) \,,$$

where $j^\mu = -ie : \frac{\partial\phi^\dagger}{\partial x_\mu}\phi - \frac{\partial\phi}{\partial x_\mu}\phi^\dagger :$ is the current density.

Calculate the differential scattering cross-section for the scattering from a nucleus with charge Z. Establish the result

$$\frac{d\sigma}{d\Omega} = \frac{(\alpha Z)^2}{4E^2 v^4 \sin^4\frac{\vartheta}{2}} \,.$$

15.3 Show that for fermions

$$\langle e^-, \mathbf{p}| \, j^\mu(x) \, |e^-, \mathbf{p}\rangle = \frac{p^\mu}{V E_\mathbf{p}} \,,$$

where $j^\mu(x)$ is the current-density operator, $|e^-, \mathbf{p}\rangle = b^\dagger_{\mathbf{p},r} |0\rangle$, and $E_\mathbf{p} = \sqrt{\mathbf{p}^2 + m^2}$.

15.4 Verify Eq. (15.5.39).

15.5 Verify Eqs. (15.5.42) and (15.5.43).

15.6 a) With the help of the Feynman rules, give the analytical expression for the transition amplitude corresponding to the Feynman diagrams of Compton scattering, Fig. 15.12a,b.

b) Derive these expressions by making use of Wick's theorem.

Bibliography for Part III

A.I. Achieser and W.B. Berestezki, *Quantum Electrodynamics*, Consultants Bureau Inc., New York, 1957

I. Aitchison and A. Hey, *Gauge Theories in Particle Physics*, Adam Hilger, Bristol, 1982

J.D. Bjorken and S.D. Drell, *Relativistic Quantum Mechanics*, McGraw-Hill, New York, 1964

J.D. Bjorken and S.D. Drell, *Relativistic Quantum Fields*, McGraw-Hill, New York, 1965

N.N. Bogoliubov and D.V. Shirkov, *Quantum Fields*, The Benjamin/Cummings Publishing Company, Inc., London, 1983; and *Introduction to the Theory of Quantized Fields*, 3rd edition, John Wiley & Sons, New York, 1980

S.J. Chang, *Introduction to Quantum Field Theory*, Lecture Notes in Physics Vol. 29, World Scientific, Singapore, 1990

K. Huang, *Quarks, Leptons and Gauge Fields*, World Scientific, Singapore, 1982

C. Itzykson and J.-B. Zuber, *Quantum Field Theory*, McGraw Hill, New York, 1980

J.M. Jauch and F. Rohrlich, *The Theory of Photons and Electrons*, 2nd ed., Springer, New York, 1976

G. Källen, *Elementary Particle Physics*, Addison Wesley, Reading, 1964

G. Kane, *Modern Elementary Particle Physics*, Addison Wesley, Redwood City, 1987

F. Mandl and G. Shaw, *Quantum Field Theory*, John Wiley & Sons, Chichester, 1984

O. Nachtmann, *Elementary Particle Physics*, Springer, Heidelberg, 1990

Yu.V. Novozhilov, *Introduction to Elementary Particle Theory*, Pergamon Press, Oxford, 1975

D.H. Perkins, *Introduction to High Energy Physics*, Addison Wesley, Reading, 1987

S.S. Schweber, *An Introduction to Relativistic Quantum Field Theory*, Harper & Row, New York, 1961.

J.C. Taylor, *Gauge Theories of Weak Interactions*, Cambridge Univ. Press, Cambridge, 1976

S. Weinberg, *The Quantum Theory of Fields*, Cambridge University Press, Cambridge, 1995

Appendix

A Alternative Derivation of the Dirac Equation

Here, we shall give an alternative derivation of the Dirac equation. In so doing, we will also deduce the Pauli equation as well as a decomposition of the Dirac equation that is related to the Weyl equations for massless spin-$\frac{1}{2}$ particles.

Our starting point is the nonrelativistic kinetic energy

$$H = \frac{\mathbf{p}^2}{2m} \rightarrow \frac{1}{2m}\left(\frac{\hbar}{i}\nabla\right)^2 . \tag{A.1}$$

Provided there is no external magnetic field, instead of this Hamiltonian, one can use the completely equivalent form

$$H = \frac{1}{2m}(\boldsymbol{\sigma}\cdot\mathbf{p})(\boldsymbol{\sigma}\cdot\mathbf{p}) \tag{A.2}$$

as can be established from the identity

$$(\boldsymbol{\sigma}\cdot\mathbf{a})(\boldsymbol{\sigma}\cdot\mathbf{b}) = \mathbf{a}\cdot\mathbf{b} + i\boldsymbol{\sigma}\cdot(\mathbf{a}\times\mathbf{b}) .$$

If one starts from (A.1) when introducing the coupling to the magnetic field, it is, in addition, necessary to add the coupling of the electron spin to the magnetic field "by hand". Alternatively, one can start with (A.2) and write the Hamiltonian with magnetic field as

$$\begin{aligned}
H &= \frac{1}{2m}\boldsymbol{\sigma}\cdot\left(\mathbf{p} - \frac{e}{c}\mathbf{A}\right)\boldsymbol{\sigma}\cdot\left(\mathbf{p} - \frac{e}{c}\mathbf{A}\right) \\
&= \frac{1}{2m}\left(\mathbf{p} - \frac{e}{c}\mathbf{A}\right)^2 + \frac{i}{2m}\boldsymbol{\sigma}\cdot\left[\left(\mathbf{p} - \frac{e}{c}\mathbf{A}\right)\times\left(\mathbf{p} - \frac{e}{c}\mathbf{A}\right)\right] \\
&= \frac{1}{2m}\left(\mathbf{p} - \frac{e}{c}\mathbf{A}\right)^2 - \frac{e\hbar}{2mc}\boldsymbol{\sigma}\cdot\mathbf{B} .
\end{aligned} \tag{A.3}$$

Here, we have made use of the rearrangements that lead from (5.3.29) to (5.3.29′). In this way, one obtains the Pauli equation with the correct Landé factor $g = 2$.

We now wish to establish the relativistic generalization of this equation. To this end, we start with the relativistic energy–momentum relation

$$\frac{E^2}{c^2} - \mathbf{p}^2 = (mc)^2 ,\tag{A.4}$$

which we rewrite as

$$\left(\frac{E}{c} - \boldsymbol{\sigma} \cdot \mathbf{p}\right)\left(\frac{E}{c} + \boldsymbol{\sigma} \cdot \mathbf{p}\right) = (mc)^2 .\tag{A.5}$$

According to the correspondence principle ($E \rightarrow i\hbar\frac{\partial}{\partial t}$, $\mathbf{p} \rightarrow -i\hbar\boldsymbol{\nabla}$), the quantum-mechanical relation is

$$\left(i\hbar\frac{\partial}{\partial t\,c} + \boldsymbol{\sigma}i\hbar\boldsymbol{\nabla}\right)\left(i\hbar\frac{\partial}{\partial t\,c} - \boldsymbol{\sigma}i\hbar\boldsymbol{\nabla}\right)\phi = (mc)^2\phi ,\tag{A.6}$$

where ϕ is a two-component wave function (spinor). This equation was originally put forward by van der Waerden. In order to obtain a differential equation of first order in time, we introduce two *two-component spinors*

$$\phi^{(L)} = -\phi \quad \text{and} \quad \phi^{(R)} = -\frac{1}{mc}\left(i\hbar\frac{\partial}{\partial x_0} - i\hbar\boldsymbol{\sigma}\cdot\boldsymbol{\nabla}\right)\phi^{(L)} .$$

The last equation, defining $\phi^{(R)}$, together with the remaining differential equation from (A.6), yields:

$$\begin{aligned}\left(i\hbar\frac{\partial}{\partial x_0} - i\hbar\boldsymbol{\sigma}\cdot\boldsymbol{\nabla}\right)\phi^{(L)} &= -mc\phi^{(R)}\\[6pt]\left(i\hbar\frac{\partial}{\partial x_0} + i\hbar\boldsymbol{\sigma}\cdot\boldsymbol{\nabla}\right)\phi^{(R)} &= -mc\phi^{(L)} .\end{aligned}\tag{A.7}$$

The notation $\phi^{(L)}$ and $\phi^{(R)}$ refers to the fact that, in the limit $m \rightarrow 0$, these functions represent left- and right-handed polarized states (i.e., spins antiparallel and parallel to the momentum). In order to make the connection to the Dirac equation, we write $\boldsymbol{\sigma}\boldsymbol{\nabla} \equiv \sigma^i\partial_i$ and form the difference and the sum of the two equations (A.7)

$$\begin{aligned}i\hbar\frac{\partial}{\partial x_0}&\left(\phi^{(R)} - \phi^{(L)}\right) + i\hbar\sigma^i\partial_i\left(\phi^{(R)} + \phi^{(L)}\right)\\&- mc\left(\phi^{(R)} - \phi^{(L)}\right) = 0\\[6pt]-i\hbar\frac{\partial}{\partial x_0}&\left(\phi^{(R)} + \phi^{(L)}\right) - i\hbar\sigma^i\partial_i\left(\phi^{(R)} - \phi^{(L)}\right)\\&- mc\left(\phi^{(R)} + \phi^{(L)}\right) = 0 .\end{aligned}\tag{A.8}$$

Combining the two-component spinors into the bispinor

$$\psi = \begin{pmatrix}\phi^{(R)} - \phi^{(L)}\\\phi^{(R)} + \phi^{(L)}\end{pmatrix}\tag{A.9a}$$

yields

$$\left(i\hbar\gamma^0 \frac{\partial}{\partial x_0} + i\hbar\gamma^i \partial_i - mc \right) \psi = 0 , \tag{A.9b}$$

with

$$\gamma^0 = \begin{pmatrix} \mathbb{1} & 0 \\ 0 & -\mathbb{1} \end{pmatrix} \quad , \quad \gamma^i = \begin{pmatrix} 0 & \sigma^i \\ -\sigma^i & 0 \end{pmatrix} . \tag{A.9c}$$

We thus obtain the standard representation of the Dirac equation.

B Dirac Matrices

B.1 Standard Representation

$$\gamma^0 = \begin{pmatrix} \mathbb{1} & 0 \\ 0 & -\mathbb{1} \end{pmatrix} , \qquad \gamma^i = \begin{pmatrix} 0 & \sigma^i \\ -\sigma^i & 0 \end{pmatrix} , \qquad \gamma^5 = \begin{pmatrix} 0 & \mathbb{1} \\ \mathbb{1} & 0 \end{pmatrix}$$

$$\beta = \begin{pmatrix} \mathbb{1} & 0 \\ 0 & -\mathbb{1} \end{pmatrix} , \qquad \alpha^i = \begin{pmatrix} 0 & \sigma^i \\ \sigma^i & 0 \end{pmatrix}$$

Chirality operator : γ^5

$$(\gamma^5)^2 = \mathbb{1}$$

$$\{\gamma^5, \gamma^\mu\} = 0$$

$$\not{a}\not{b} = a \cdot b - ia^\mu b^\nu \sigma_{\mu\nu} , \quad \not{a} \equiv \gamma_\mu a^\mu$$

$$\sigma_{\mu\nu} = \frac{i}{2} [\gamma_\mu, \gamma_\nu]$$

$$\sigma_{\mu\nu} = -\sigma_{\nu\mu}$$

$$\gamma^\mu \gamma_\mu = 4 , \quad \gamma^\mu \gamma^\nu \gamma_\mu = -2\gamma^\nu$$

$$\gamma^\mu \gamma^\nu \gamma^\rho \gamma_\mu = 4g^{\nu\rho} , \quad \gamma^\mu \gamma^\nu \gamma^\rho \gamma^\sigma \gamma_\mu = -2\gamma^\sigma \gamma^\rho \gamma^\nu$$

B.2 Chiral Representation

$$\gamma^0 = \beta = \begin{pmatrix} 0 & -\mathbb{1} \\ -\mathbb{1} & 0 \end{pmatrix} , \quad \alpha = \begin{pmatrix} \sigma & 0 \\ 0 & -\sigma \end{pmatrix} , \quad \gamma = \begin{pmatrix} 0 & \sigma \\ -\sigma & 0 \end{pmatrix} ,$$

$$\sigma_{0i} = \frac{i}{2} [\gamma_0, \gamma_i] = -i\alpha_i = \frac{1}{i} \begin{pmatrix} \sigma^i & 0 \\ 0 & -\sigma^i \end{pmatrix}$$

$$\sigma_{ij} = \frac{i}{2} [\gamma_i, \gamma_j] = -\frac{i}{2} [\alpha_i, \alpha_j] = \epsilon^{ijk} \begin{pmatrix} \sigma^k & 0 \\ 0 & \sigma^k \end{pmatrix}$$

B.3 Majorana Representations

$$\gamma^0 = \begin{pmatrix} 0 & \sigma^2 \\ \sigma^2 & 0 \end{pmatrix}, \; \gamma^1 = i \begin{pmatrix} \sigma^3 & 0 \\ 0 & \sigma^3 \end{pmatrix}, \; \gamma^2 = \begin{pmatrix} 0 & -\sigma^2 \\ \sigma^2 & 0 \end{pmatrix}, \; \gamma^3 = -i \begin{pmatrix} \sigma^1 & 0 \\ 0 & \sigma^1 \end{pmatrix}$$

or

$$\gamma_0 = \begin{pmatrix} 0 & \sigma^2 \\ \sigma^2 & 0 \end{pmatrix}, \; \gamma_1 = i \begin{pmatrix} 0 & \sigma^1 \\ \sigma^1 & 0 \end{pmatrix}, \; \gamma_2 = i \begin{pmatrix} 1 & 0 \\ 0 & -1 \end{pmatrix}, \; \gamma_3 = i \begin{pmatrix} 0 & \sigma^3 \\ \sigma^3 & 0 \end{pmatrix}$$

C Projection Operators for the Spin

C.1 Definition

We define here the spin projection operator and summarize its properties. Since this projection operator contains the Dirac matrix γ^5 (Eq. 6.2.48), we give the following useful representation of γ^5

$$\gamma^5 = i\gamma^0\gamma^1\gamma^2\gamma^3 = -\frac{i}{4!}\epsilon^{\mu\nu\rho\sigma}\gamma_\mu\gamma_\nu\gamma_\rho\gamma_\sigma = -\frac{i}{4!}\epsilon_{\mu\nu\rho\sigma}\gamma^\mu\gamma^\nu\gamma^\rho\gamma^\sigma \quad . \tag{C.1}$$

Here, $\epsilon^{\mu\nu\rho\sigma}$ is the totally antisymmetric tensor of fourth rank:

$$\epsilon^{\mu\nu\rho\sigma} = \begin{cases} 1 & \text{for even permutations of 0123} \\ -1 & \text{for odd permutations of 0123} \\ 0 & \text{otherwise}. \end{cases} \tag{C.2}$$

The *spin projection operator* is defined by

$$P(n) = \frac{1}{2}(1 + \gamma_5 \slashed{n}) \quad . \tag{C.3}$$

Here, $\slashed{n} = \gamma^\mu n_\mu$, and n_μ is a space-like unit vector satisfying $n^2 = n^\mu n_\mu = -1$ and $n_\mu k^\mu = 0$. In the rest frame, these two vectors are denoted by $\check{n}^\mu$ and $\check{k}^\mu$ and have the form $\check{n} = (0, \check{n})$ and $\check{k} = (m, \mathbf{0})$.

C.2 Rest Frame

For the *special case* where $n \equiv n_{(3)} \equiv (0, 0, 0, 1)$ is a unit vector in the positive z direction, one obtains

$$P(n_{(3)}) = \frac{1}{2}(1 + \gamma_5\gamma_3) = \frac{1}{2}\begin{pmatrix} 1 + \sigma^3 & 0 \\ 0 & 1 - \sigma^3 \end{pmatrix}, \tag{C.4}$$

since $\gamma_5(-\gamma^3) = -\begin{pmatrix} 0 & 1 \\ 1 & 0 \end{pmatrix}\begin{pmatrix} 0 & \sigma^3 \\ -\sigma^3 & 0 \end{pmatrix} = \begin{pmatrix} \sigma^3 & 0 \\ 0 & -\sigma^3 \end{pmatrix}$. The effect of the projection operator $P(n_{(3)})$ on the spinors of particles at rest (Eq. (6.3.3) or (6.3.11a,b) for $\mathbf{k} = 0$) is thus given by

$$P(n_{(3)}) \begin{cases} u_1(m, \mathbf{0}) \\ u_2(m, \mathbf{0}) \end{cases} = \begin{cases} u_1(m, \mathbf{0}) \\ 0 \end{cases}$$

$$P(n_{(3)}) \begin{cases} v_1(m, \mathbf{0}) \\ v_2(m, \mathbf{0}) \end{cases} = \begin{cases} 0 \\ v_2(m, \mathbf{0}) \, . \end{cases} \tag{C.5}$$

Equation (C.5) implies that, in the rest frame, $P(n)$ projects onto eigenstates of $\frac{1}{2}\mathbf{\Sigma} \cdot \mathbf{n}$, with the eigenvalue $+\frac{1}{2}$ for positive energy states and the eigenvalue $-\frac{1}{2}$ for negative energy states.

In Problem 6.15 the following properties of $P(n)$ and of the projection operators $\Lambda_\pm(k)$ acting on spinors of positive and negative energy have already been demonstrated:

$$[\Lambda_\pm(k), P(n)] = 0$$

$$\Lambda_+(k)P(n) + \Lambda_-(k)P(n) + \Lambda_+(k)P(-n) + \Lambda_-(k)P(-n) = \mathbb{1} \tag{C.6}$$

$$\mathrm{Tr}\Lambda_\pm(k)P(\pm n) = 1 \, .$$

C.3 General Significance of the Projection Operator $P(n)$

We will now investigate the effect of $P(n)$ for a general space-like unit vector n, which thus obeys $n^2 = -1$ and $n \cdot k = 0$. Useful quantities for this purpose are the vector

$$W_\mu = -\frac{1}{2}\gamma_5\gamma_\mu \slashed{k} \tag{C.7a}$$

and the scalar product

$$W \cdot n = -\frac{1}{4}\epsilon_{\mu\nu\rho\sigma}n^\mu k^\nu \sigma^{\rho\sigma} \, , \tag{C.7b}$$

which can also be written as

$$W \cdot n = -\frac{1}{2}\gamma_5\slashed{n}\slashed{k} \, . \tag{C.7c}$$

The equivalence of these two expressions can be seen most easily by transforming into a frame of reference in which k is purely time-like ($k = (k^0, 0, 0, 0)$) and hence n, on account of $n \cdot k = 0$, purely space-like ($n = (0, n^1, 0, 0)$). In this rest frame, the right-hand side of (C.7b) becomes

$$-\frac{1}{4}\epsilon_{10\rho\sigma}n^1 k^0 \sigma^{\rho\sigma} = -\frac{1}{4}\epsilon_{10\rho\sigma}n^1 k^0 i\gamma^\rho\gamma^\sigma$$

$$= -\frac{1}{4}(\epsilon_{1023}n^1 k^0 i\gamma^2\gamma^3 + \epsilon_{1032}n^1 k^0 i\gamma^3\gamma^2)$$

$$= -\frac{i}{2}n^1 k^0 \gamma^2\gamma^3$$

and for the right-hand side of (C.7c) we have

$$-\frac{1}{2}\gamma_5 \not{n} \not{k} = -\frac{i}{2}\gamma^0\gamma^1\gamma^2\gamma^3(-n^1\gamma^1)k_0\gamma^0 = -\frac{i}{2}n^1 k^0\gamma^2\gamma^3 \,,$$

thus demonstrating that they are identical.
In the *rest frame* the vector (C.7a) has the spatial components

$$\mathbf{W} = -\frac{1}{2}\gamma_5\boldsymbol{\gamma}\gamma^0 k^0 = +\frac{1}{2}\gamma_5\gamma^0\boldsymbol{\gamma}m = \frac{m}{2}\boldsymbol{\Sigma} \,, \tag{C.8}$$

where we have put $k^0 = m$. Assuming that n is directed along the z axis, i.e.,
$n = n_{(3)} \equiv (0,0,0,1)$, it follows from (C.8) that

$$W \cdot n = \frac{m}{2}\Sigma^3 \,. \tag{C.9}$$

The plane waves in the rest frame are eigenvectors of $-\frac{W \cdot n_{(3)}}{m} = \frac{1}{2}\Sigma^3$:

$$\frac{1}{2}\Sigma^3 u_1(m, \mathbf{k}=0) = \quad \frac{1}{2}u_1(m, \mathbf{k}=0)$$

$$\frac{1}{2}\Sigma^3 u_2(m, \mathbf{k}=0) = \quad -\frac{1}{2}u_2(m, \mathbf{k}=0)$$

$$\frac{1}{2}\Sigma^3 v_1(m, \mathbf{k}=0) = \quad \frac{1}{2}v_1(m, \mathbf{k}=0) \tag{C.10}$$

$$\frac{1}{2}\Sigma^3 v_2(m, \mathbf{k}=0) = -\frac{1}{2}v_2(m, \mathbf{k}=0) \,.$$

After carrying out a Lorentz transformation from $(m, \mathbf{k}=0)$ to $(k^0, \mathbf{k})$, we
have

$$-\frac{W \cdot n}{m} = \frac{1}{2m}\gamma_5 \not{n}\not{k} \,,$$

where n is the transform of $n_{(3)}$. The equations (C.10) then transform into
eigenvalue equations for $u_r(k)$ and $v_r(k)$

$$-\frac{W \cdot n}{m}u_r(k) = \frac{1}{2m}\gamma_5\not{n}\not{k}u_r(k) = \frac{1}{2}\gamma_5\not{n}u_r(k)$$

$$= \pm\frac{1}{2}u_r(k) \quad \text{for} \quad r = \begin{cases} 1 \\ 2 \end{cases}$$

$$-\frac{W \cdot n}{m}v_r(k) = \frac{1}{2m}\gamma_5\not{n}\not{k}v_r(k) = -\frac{1}{2}\gamma_5\not{n}v_r(k) \tag{C.11}$$

$$= \pm\frac{1}{2}v_r(k) \quad \text{for} \quad r = \begin{cases} 1 \\ 2 \end{cases} \,,$$

where, after the first and the second equals signs, we have made use of (C.7c)
and of $\not{k}u_r(k) = mu_r(k)$ and $\not{k}v_r(k) = -mv_r(k)$, respectively. Finally, after
the third equals sign we obtain the right-hand side of (C.10). The action of

$\gamma_5 \not{n}$ on the $u_r(k)$ and $v_r(k)$ is apparent from (C.11) and it is likewise evident that

$$P(n) = \frac{1}{2}(\mathbb{1} + \gamma_5 \not{n}) \tag{C.12a}$$

is a projection operator onto $u_1(k)$ and $v_2(k)$ and that

$$P(-n) = \frac{1}{2}(\mathbb{1} - \gamma_5 \not{n}) \tag{C.12b}$$

is a projection operator onto $u_2(k)$ and $v_1(k)$.

Let n be an arbitrary space-like vector, with $n \cdot k = 0$ and let $\check{n}$ be the corresponding vector in the rest frame. Then, $P(n)$ projects onto spinors $u(k, n)$ that are polarized along $+\check{n}$ in the rest frame, and onto those $v(k, n)$ that are polarized along $-\check{n}$ in the rest frame. We have the eigenvalue equations

$$\begin{aligned} \boldsymbol{\Sigma} \cdot \check{n}\, u(\check{k}, \check{n}) &= u(\check{k}, \check{n}) \\ \boldsymbol{\Sigma} \cdot \check{n}\, v(\check{k}, \check{n}) &= -v(\check{k}, \check{n}) \ . \end{aligned} \tag{C.13}$$

The vectors k and n are related to their counterparts $\check{k}$ and $\check{n}$ in the rest frame by a Lorentz transformation Λ: $k^\mu = \Lambda^\mu_{\ \nu} \check{k}^\nu$ with $\check{k}^\nu = (m, 0, 0, 0)$ and $n^\mu = \Lambda^\mu_{\ \nu} \check{n}^\nu$ with $\check{n}^\nu = (0, \mathbf{n})$. The inverse relation reads $\check{n}^\nu = \Lambda^\nu_{\ \mu} n^\mu$. As is usual in the present context, we have used the notation $u(k, n)$ and $v(k, n)$ for the spinors. These are related to the $u_r(k)$ and $v_r(k)$ used previously by

$$\begin{aligned} u_1(k) &= u(k, n), u_2(k) = u(k, -n) \\ v_1(k) &= v(k, -n), v_2(k) = v(k, n) \ , \end{aligned} \tag{C.14}$$

where $n = \Lambda n_{(3)}$ with $n_{(3)} = (0, 0, 0, 1)$.

We now consider a unit vector n_k, whose spatial part is parallel to $\mathbf{k}$:

$$n_k = \left(\frac{|\mathbf{k}|}{m}, \frac{k^0}{m} \frac{\mathbf{k}}{|\mathbf{k}|} \right) \ . \tag{C.15}$$

Trivially, this satisfies

$$n_k^2 = \frac{\mathbf{k}^2}{m^2} - \frac{k_0^2}{m^2} = -1 \quad \text{and} \quad n_k \cdot k = \frac{|\mathbf{k}| k_0}{m} - \frac{k_0}{m} \frac{\mathbf{k}^2}{|\mathbf{k}|} = 0 \ .$$

We now show that the combined effect of the projection operator $P(n_k)$ and the projection operators $\Lambda_\pm(k)$ on spinors with positive and negative energy can be represented by

$$P(n_k)\Lambda_\pm(k) = \left(\mathbb{1} \pm \frac{\boldsymbol{\Sigma} \cdot \mathbf{k}}{|\mathbf{k}|} \right) \Lambda_\pm(k) \ . \tag{C.16}$$

To prove this relation, one starts from the definitions

$$P(n_k)\Lambda_\pm(k) = \frac{1}{2}(\mathbb{1} + \gamma_5 \not{n}_k)\frac{\pm \not{k} + m}{2m}$$

and rearranges as follows:

$$\gamma_5 \not{n}_k \frac{\pm \not{k} + m}{2m} = \gamma_5 \not{n}_k \frac{\pm \not{k} + m}{2m}\frac{\pm \not{k} + m}{2m} = \left(\frac{1}{2}\gamma_5 \not{n}_k \pm \gamma_5 \not{n}_k \frac{\not{k}}{2m}\right)\frac{\pm \not{k} + m}{2m} \ .$$

This yields:

$$\frac{1}{2}\gamma_5 \not{n}_k \frac{\pm \not{k} + m}{2m} = \pm \gamma_5 \not{n}_k \frac{\not{k}}{2m}\frac{\pm \not{k} + m}{2m}$$

which gives us the intermediate result

$$P(n_k)\Lambda_\pm(k) = \frac{1}{2}(\mathbb{1} \pm \gamma_5 \not{n}_k \frac{\not{k}}{m})\frac{\pm \not{k} + m}{2m} \ . \tag{C.17}$$

We then proceed by writing

$$\gamma_5 \not{n}_k \not{k} = \gamma_5(\underbrace{n_k \cdot k}_{=0} - \mathrm{i}n_k^\mu \sigma_{\mu\nu}k^\nu)$$

$$= -\mathrm{i}\gamma_5(n_k^0 \sigma_{0j}k^j + n_k^j \sigma_{j0}k^0) = \mathrm{i}\gamma_5 \sigma_{0j}(\frac{|\mathbf{k}|}{m}k^j - \frac{k^0}{m}\frac{k^j}{|\mathbf{k}|}k^0)$$

$$= \mathrm{i}\gamma_5 \frac{m}{|\mathbf{k}|}\sigma_{0j}k^j = \gamma_5 \frac{m}{|\mathbf{k}|}\gamma^0 \gamma^j k^j \ .$$

Here, we have used $n_k \cdot k = 0$ and also the fact that the purely spatial components make no contribution, due to the antisymmetry of σ^{ij}. By considering, as an example, the $j = 3$ component of $\gamma_5\gamma^0\gamma^j$:

$$\gamma_5\gamma^0\gamma^3 = -\mathrm{i}\gamma^1\gamma^2(\gamma^3)^2 = \mathrm{i}\gamma^1\gamma^2 = \sigma^{12} = \Sigma^3 \ ,$$

the assertion (C.16) is confirmed. Equation (C.16) reveals the following property of the projection operator $P(n_k)$: The operator $P(n_k)$ projects states with positive energy onto states with positive helicity, and states with negative energy onto states with negative helicity. Analogously, we have

$$P(-n_k)\Lambda_\pm(k) = \frac{1}{2}(\mathbb{1} \mp \boldsymbol{\Sigma} \cdot \frac{\mathbf{k}}{|\mathbf{k}|})\Lambda_\pm \ ;$$

thus $P(-n_k)$ projects spinors with positive energy onto spinors with negative helicity and spinors with negative energy onto spinors with positive helicity.

D The Path-Integral Representation of Quantum Mechanics

We start from the Schrödinger equation

$$i\hbar \frac{\partial}{\partial t} |\psi, t\rangle = H |\psi, t\rangle \qquad (D.1)$$

with the Hamiltonian

$$H = \frac{1}{2m} p^2 + V , \qquad (D.2)$$

and let the eigenstates of H be $|n\rangle$. Assuming that $\lim_{x \to \pm\infty} V(x) = \infty$, we know that the eigenvalues of H are discrete. In the coordinate representation, the eigenstates of H are the wave functions $\psi_n(x) = \langle x|n\rangle$, where $|x\rangle$ is the position eigenstate with position x. The discussion that follows will be based on the Schrödinger representation. If, at time 0 the particle is in the position state $|y\rangle$, then at time t its state is $e^{-iHt/\hbar} |y\rangle$. The probability amplitude that at time t the particle is located at x is given by

$$G(y, 0|x, t) = \langle x| e^{-itH/\hbar} |y\rangle . \qquad (D.3)$$

We call $G(y, 0|x, t)$ the Green's function. It satisfies the initial condition $G(y, 0|x, 0) = \delta(y - x)$. Inserting the closure relation $\mathbb{1} = \sum_n |n\rangle \langle n|$ in (D.3),

$$G(y, 0|x, t) = \sum_{n,m} \langle x|n\rangle \langle n| e^{-itH/\hbar} |m\rangle \langle m|y\rangle ,$$

we obtain the coordinate representation of the Green's function

$$G(y, 0|x, t) = \sum_n e^{-itE_n/\hbar} \psi_n(x) \psi_n^*(y). \qquad (D.4)$$

By dividing the time interval $[0, t]$ into N parts (Fig. D.1), whereby increasing N yields ever smaller time differences $\Delta t = \frac{t}{N}$, we may express the Green's function as follows:

$$\begin{aligned} G(y, 0|x, t) &= \langle x| e^{-iH\Delta t/\hbar} \dots e^{-iH\Delta t/\hbar} |y\rangle \\ &= \int dz_1 \dots \int dz_{N-1} \langle z_N| e^{-iH\Delta t/\hbar} |z_{N-1}\rangle \dots \langle z_1| e^{-iH\Delta t/\hbar} |z_0\rangle , \end{aligned} \qquad (D.5)$$

Fig. D.1. Discretization of the time interval $[0, t]$, with the z_i of the identity operators introduced in (D.5) ($z_0 = y$, $z_N = x$)

where we have introduced the identity operators $\mathbb{1} = \int dz_i \, |z_i\rangle \, \langle z_i|$. We then have

$$e^{-iH\Delta t/\hbar} = e^{-i\frac{\Delta t}{\hbar}\frac{V(x)}{2}} e^{-i\frac{\Delta t p^2}{2\hbar m}} e^{-\frac{i\Delta t}{\hbar}\frac{V(x)}{2}} + \mathcal{O}((\Delta t))^2. \tag{D.6}$$

We now determine the necessary matrix elements

$$
\begin{aligned}
\langle \xi | e^{-i\frac{p^2}{2m}\frac{\Delta t}{\hbar}} | \xi' \rangle &= \int \frac{dk}{2\pi} \, \langle \xi | k \rangle \, e^{-i\frac{(k\hbar)^2}{2m}\frac{\Delta t}{\hbar}} \, \langle k | \xi' \rangle \\
&= \int \frac{dk}{2\pi} \, e^{ik(\xi-\xi')} e^{-i\frac{(k\hbar)^2}{2m}\frac{\Delta t}{\hbar}} \\
&= \int \frac{dk}{2\pi} \, e^{-\frac{i\Delta t\hbar}{2m}\left(k - \frac{\xi-\xi'}{2\Delta t\hbar/2m}\right)^2 + i(\xi-\xi')^2 \frac{m}{2\Delta t\hbar}} \\
&= \left(\frac{-im}{2\pi\hbar\Delta t}\right)^{1/2} e^{\frac{im}{2\hbar\Delta t}(\xi-\xi')^2}.
\end{aligned}
\tag{D.7}
$$

In the first step the completeness relation for the momentum eigenfunctions was inserted twice. From (D.6) and (D.7) it follows that

$$
\begin{aligned}
&\langle \xi | e^{-iH\Delta t/\hbar} | \xi' \rangle \\
&= \exp\left[-i\frac{\Delta t}{\hbar}\left(\frac{V(\xi)+V(\xi')}{2} - \frac{m(\xi-\xi')^2}{2(\Delta t)^2}\right)\right] \left(\frac{-im}{2\pi\hbar\Delta t}\right)^{\frac{1}{2}}
\end{aligned}
\tag{D.8}
$$

and finally, for the Green's function,

$$
\begin{aligned}
G(y,0|x,t) = \int dz_1 \ldots \int dz_{N-1} \\
\times \exp\Bigg[\frac{i\Delta t}{\hbar}\sum_{n=1}^{N}\left\{\frac{m(z_n-z_{n-1})^2}{2(\Delta t)^2} - \frac{V(z_n)+V(z_{n-1})}{2}\right\} \\
+ \frac{N}{2}\log\frac{-im}{2\pi\hbar\Delta t}\Bigg].
\end{aligned}
$$

In the limit $N \to \infty$ this yields the *Feynman path-integral representation*[1]

$$G(y,0|x,t) = \int \mathcal{D}[z] \exp\frac{i}{\hbar}\int_0^t dt' \left\{\frac{m\dot{z}(t')^2}{2} - V(z(t'))\right\}, \tag{D.9}$$

where

$$\mathcal{D}[z] = \lim_{N\to\infty}\left(\frac{-im}{2\pi\hbar\Delta t}\right)^{\frac{N}{2}}\prod_{n=1}^{N-1}dz_n \tag{D.10}$$

[1] R. P. Feynman and A. R. Hibbs, *Quantum Mechanics and Path Integrals*, McGraw-Hill, New York, 1965; G. Parisi, *Statistical Field Theory*, Addison-Wesley, 1988, p.234

and

$$z_n = z\left(\frac{nt}{N}\right).$$

The probability amplitude for the transition from y to x after a time t is given as the sum of the amplitudes of all possible trajectories from y to x, each being given the weight $\exp\frac{i}{\hbar}\int_0^t dt'\, L(\dot{z}, z)$ and where

$$L(\dot{z}, z) = \frac{m\dot{z}(t)^2}{2} - V(z(t))$$

is the classical Lagrangian. The phase of the probability amplitude is just the classical action. In the limit $\hbar \to 0$, the main contribution to the functional integral comes from the neighborhood of the trajectory whose phase is stationary. This is just the classical trajectory.

E Covariant Quantization of the Electromagnetic Field, the Gupta–Bleuler Method

E.1 Quantization and the Feynman Propagator

In the main text, we treated the radiation field in the Coulomb gauge. This has the advantage that only the two transverse photons occur. To determine the propagator, however, one has to combine the photon contributions with the Coulomb interaction in order to obtain the final covariant expression. In this appendix, we describe an alternative and explicitly covariant quantization of the radiation field by means of the Gupta–Bleuler method[2]. In the covariant theory, one begins with

$$\mathcal{L}_L = -\frac{1}{2}(\partial_\nu A_\mu)(\partial^\nu A^\mu) - j_\mu A^\mu \ . \tag{E.1}$$

The components of the momentum conjugate to A^μ are

$$\Pi_L^\mu = \frac{\partial \mathcal{L}_L}{\partial \dot{A}^\mu} = -\dot{A}_\mu \ . \tag{E.2}$$

From the Lagrangian density (E.1), we obtain the field equations

$$\Box A^\mu(x) = j^\mu(x) \ . \tag{E.3}$$

---

[2] Detailed presentations of the Gupta–Bleuler method can be found in S.N. Gupta, *Quantum Electrodynamics* Gordon and Breach, New York, 1977; C. Itzykson and J.-B. Zuber, *Quantum Field Theory* McGraw Hill, New York, 1980; F. Mandl and G. Shaw, *Quantum Field Theory*, J. Wiley, Chichester 1984; J.M. Jauch and F. Rohrlich, *The Theory of Photons and Electrons*, 2nd ed., Springer, New York, 1976, Sect.6.3.

These are only equivalent to the Maxwell equations when the four-potential $A^\mu(x)$ satisfies the gauge condition

$$\partial_\mu A^\mu(x) = 0 . \tag{E.4}$$

The most general solution of the free field equations $(j^\mu = 0)$ is given by the linear superposition[3]

$$
\begin{aligned}
A^\mu(x) &= A^{\mu\,+}(x) + A^{\mu\,-}(x) \\
&= \sum_{\mathbf{k},r} \left(\frac{1}{2V|\mathbf{k}|}\right)^{1/2} \left(\epsilon_r^\mu(\mathbf{k})a_r(\mathbf{k})e^{-ikx} + \epsilon_r^\mu(\mathbf{k})a_r^\dagger(\mathbf{k})e^{ikx}\right) .
\end{aligned}
\tag{E.5}
$$

The four polarization vectors obey the orthogonality and completeness relations

$$\epsilon_r(\mathbf{k})\epsilon_s(\mathbf{k}) \equiv \epsilon_{r\mu}(\mathbf{k})\epsilon_s^\mu(\mathbf{k}) = -\zeta_r\delta_{rs}, \quad r,s = 0,1,2,3 \tag{E.6a}$$

$$\sum_r \zeta_r \epsilon_r^\mu(\mathbf{k})\epsilon_r^\nu(\mathbf{k}) = -g^{\mu\nu} , \tag{E.6b}$$

where

$$\zeta_0 = -1 \quad , \quad \zeta_1 = \zeta_2 = \zeta_3 = 1 . \tag{E.6c}$$

On occasion it is useful to employ the special polarization vectors

$$\epsilon_0^\mu(\mathbf{k}) = n^\mu \equiv (1,0,0,0) \tag{E.7a}$$

$$\epsilon_r^\mu(\mathbf{k}) = (0, \boldsymbol{\epsilon}_{\mathbf{k},r}) \qquad r = 1,2,3 , \tag{E.7b}$$

where $\boldsymbol{\epsilon}_{\mathbf{k},1}$ and $\boldsymbol{\epsilon}_{\mathbf{k},2}$ are unit vectors that are orthogonal, both to one another and to $\mathbf{k}$, and

$$\boldsymbol{\epsilon}_{\mathbf{k},3} = \mathbf{k}/|\mathbf{k}| . \tag{E.7c}$$

This implies

$$\mathbf{n} \cdot \boldsymbol{\epsilon}_{\mathbf{k},r} = 0 , \qquad\qquad\qquad r = 1,2 \tag{E.7d}$$

$$\boldsymbol{\epsilon}_{\mathbf{k},r}\boldsymbol{\epsilon}_{\mathbf{k},s} = \delta_{rs} , \qquad\qquad r,s = 1,2,3 . \tag{E.7e}$$

The longitudinal vector can also be expressed in the form

$$\epsilon_3^\mu(\mathbf{k}) = \frac{k^\mu - (kn)n^\mu}{\left((kn)^2 - k^2\right)^{1/2}} . \tag{E.7f}$$

---

[3] To distinguish them from the polarization vectors $\boldsymbol{\epsilon}_{\mathbf{k},\lambda}$ and creation and annihilation operators $a_{\mathbf{k}\lambda}^\dagger$ and $a_{\mathbf{k}\lambda}$ $(\lambda = 1,2)$ of the main text, where the Coulomb gauge was employed, we denote the polarization vectors in the covariant representation by $\epsilon_r^\mu(\mathbf{k})$, and the corresponding creation and annihilation operators by $a_r^\dagger(\mathbf{k})$ and $a_r(\mathbf{k})$.

The four vectors describe

$\epsilon_1^{\mu}, \epsilon_2^{\mu}$ transverse polarization

ϵ_3^{μ} longitudinal polarization

ϵ_0^{μ} scalar or time-like polarization .

The covariant simultaneous canonical commutation relations for the radiation field read:

$$[A^{\mu}(\mathbf{x},t), A^{\nu}(\mathbf{x}',t)] = 0, \quad \left[\dot{A}^{\mu}(\mathbf{x},t), \dot{A}^{\nu}(\mathbf{x}',t)\right] = 0$$
$$\left[A^{\mu}(\mathbf{x},t), \dot{A}^{\nu}(\mathbf{x}',t)\right] = -ig^{\mu\nu}\delta(\mathbf{x}-\mathbf{x}') . \tag{E.8}$$

The commutation relations are the same as those for the massless Klein–Gordon field, but with the additional factor $-g^{\mu\nu}$. The zero component has the opposite sign to the spatial components.

One can thus obtain the propagators directly from those for the Klein–Gordon equation:

$$[A^{\mu}(\mathbf{x}), A^{\nu}(\mathbf{x}')] = iD^{\mu\nu}(x-x') \tag{E.9a}$$

$$D^{\mu\nu}(x-x') = ig^{\mu\nu}\int \frac{d^4k}{(2\pi)^3}\delta(k^2)\epsilon(k_0)e^{-ikx} \tag{E.9b}$$

$$\langle 0| T\left(A^{\mu}(x)A^{\nu}(x')\right)|0\rangle = iD_F^{\mu\nu}(x-x') \tag{E.10a}$$

$$D_F^{\mu\nu}(x-x') = -g^{\mu\nu}\int \frac{d^4k\, e^{-ikx}}{k^2+i\epsilon} . \tag{E.10b}$$

By inverting (E.5) and using (E.8), one obtains the commutation relations for the creation and annihilation operators

$$[a_r(\mathbf{k}), a_s^{\dagger}(\mathbf{k}')] = \zeta_r\delta_{rs}\delta_{\mathbf{kk}'} , \;\; \zeta_0 = -1 , \;\; \zeta_1 = \zeta_2 = \zeta_3 = 1 ,$$
$$[a_r(\mathbf{k}), a_s(\mathbf{k}')] = [a_r^{\dagger}(\mathbf{k}), a_s^{\dagger}(\mathbf{k}')] = 0 . \tag{E.11}$$

E.2 The Physical Significance of Longitudinal and Scalar Photons

For the components $1, 2, 3$ (i.e., the two transverse, and the longitudinal, photons) one has, according to Eq. (E.11), the usual commutation relations, whereas for the scalar photon ($r = 0$) the roles of the creation and annihilation operators seem to be reversed. The *vacuum state* $|0\rangle$ is defined by

$$a_r(\mathbf{k})|0\rangle = 0 \quad \text{for all } \mathbf{k} \text{ and } r = 0, 1, 2, 3 , \tag{E.12}$$

i.e.,

$$A^{\mu\,+}(x)|0\rangle = 0$$

for all x. One-photon states have the form

$$|\mathbf{q}s\rangle = a_s^\dagger(\mathbf{q})\,|0\rangle \;. \tag{E.13}$$

The Hamiltonian is obtained from (E.1) as

$$H = \int d^3x \;:\; \left(\Pi_L^\mu(x)\dot{A}_\mu(x) - \mathcal{L}(x) \right) \;:\; . \tag{E.14}$$

Inserting (E.2) and the expansion (E.5) into this Hamiltonian yields

$$H = \sum_{r,\mathbf{k}} |\mathbf{k}|\,\zeta_r\,a_r^\dagger(\mathbf{k})a_r(\mathbf{k}) \;. \tag{E.15}$$

One may be concerned that the energy might not be positive definite, because of $\zeta_0 = -1$. However, because of the commutation relation (E.11) the energy is indeed positive definite

$$
\begin{aligned}
H\,|\mathbf{q},s\rangle &= \sum_{r,\mathbf{k}} |\mathbf{k}|\,\zeta_r\,a_r^\dagger(\mathbf{k})a_r(\mathbf{k})a_s^\dagger(\mathbf{q})\,|0\rangle \\
&= |\mathbf{q}|\,a_s^\dagger(\mathbf{q})\,|0\rangle \quad, \qquad s = 0,1,2,3 \quad.
\end{aligned}
\tag{E.16}
$$

Correspondingly, one defines the occupation-number operator

$$\hat{n}_{r\mathbf{k}} = \zeta_r a_r^\dagger(\mathbf{k})a_r(\mathbf{k}) \;. \tag{E.17}$$

For the norm of the states, one finds

$$\langle \mathbf{q}s|\mathbf{q}s\rangle = \langle 0|\,a_s(\mathbf{q})a_s^\dagger(\mathbf{q})\,|0\rangle = \zeta_s\,\langle 0|0\rangle = \zeta_s \;. \tag{E.18}$$

In the Gupta–Bleuler theory, the norm of a state with a scalar photon is negative. More generally, every state with an odd number of scalar photons has a negative norm. However, the Lorentz condition ensures that, essentially, the scalar photons are eliminated from all physical effects. In combination with the longitudinal photons, they merely lead to the Coulomb interaction between charged particles.

For the theory to be really equivalent to the Maxwell equations, we still need to satisfy the Lorentz condition (E.4). In the quantized theory, however, it is not possible to impose the Lorentz condition as an operator identity. If one were to attempt this, Eq. (E.9a) would imply that

$$[\partial_\mu A^\mu(x), A^\nu(x')] = \mathrm{i}\partial_\mu D^{\mu\nu}(x - x') \tag{E.19}$$

must vanish. However, from (E.10b) we know that this is not the case. Gupta and Bleuler replaced the Lorentz condition by a condition[4] on the states

$$\partial_\mu A^{\mu+}(x) \, |\Psi\rangle = 0 \,. \tag{E.20a}$$

This also gives

$$\langle\Psi| \, \partial_\mu A^{\mu-}(x) = 0 \tag{E.20b}$$

and thus

$$\langle\Psi| \, \partial_\mu A^\mu(x) \, |\Psi\rangle = 0 \,. \tag{E.21}$$

It is thereby guaranteed that the Maxwell equations are always satisfied in the classical limit.

The subsidiary condition (E.20a) affects only the longitudinal- and scalar-photon states since the polarization vectors of the transverse photons are orthogonal to k. From (E.20a), (E.5), and (E.6), it follows for all $\mathbf{k}$ that

$$(a_3(\mathbf{k}) - a_0(\mathbf{k})) \, |\Psi\rangle = 0 \,. \tag{E.22}$$

Equation (E.22) amounts to a restriction on the allowed combinations of excitations of scalar and longitudinal photons. If $|\Psi\rangle$ satisfies the condition (E.22), the expectation value of the term with the corresponding wave vector in the Hamiltonian is

$$\begin{aligned}
\langle\Psi| \, a_3^\dagger(\mathbf{k}) a_3(\mathbf{k}) &- a_0^\dagger(\mathbf{k}) a_0(\mathbf{k}) \, |\Psi\rangle \\
&= \langle\Psi| \, a_3^\dagger(\mathbf{k}) a_3(\mathbf{k}) - a_0^\dagger(\mathbf{k}) a_0(\mathbf{k}) - a_0^\dagger(\mathbf{k})(a_3(\mathbf{k}) - a_0(\mathbf{k})) \, |\Psi\rangle \\
&= \langle\Psi| \, (a_3^\dagger(\mathbf{k}) - a_0^\dagger(\mathbf{k})) a_3(\mathbf{k}) \, |\Psi\rangle = 0 \,.
\end{aligned} \tag{E.23}$$

Thus, with (E.15), we have

$$\langle\Psi| \, H \, |\Psi\rangle = \langle\Psi| \sum_{\mathbf{k}} \sum_{r=1,2} |\mathbf{k}| \, a_r^\dagger(\mathbf{k}) a_r(\mathbf{k}) \, |\Psi\rangle \,, \tag{E.24}$$

---

[4] As already stated prior to Eq.(E.19), the Lorentz condition cannot be imposed as an operator condition, and cannot even be imposed as a condition on the states in the form

$$\partial_\mu A^\mu(x) \, |\Psi\rangle = 0 \,. \tag{E.20c}$$

For the vacuum state, Eq. (E.20c) would yield

$$\partial_\mu A^\mu(x) \, |\Psi_0\rangle = \partial_\mu A^{\mu-}(x) \, |\Psi_0\rangle = 0 \,.$$

Multiplication of the middle expression by $A^+(y)$ yields $A_\mu^+(y) \partial^\nu A_\nu^-(x) |\Psi_0\rangle = \frac{\partial}{\partial x_\nu}(A_\mu^+(y) A_\nu^-(x)) |\Psi_0\rangle = \frac{\partial}{\partial x_\nu}([A_\mu^+(y), A_\nu^-(x)] + A_\nu^-(x) A_\mu^+(y)) |\Psi_0\rangle = \frac{\partial}{\partial x_\nu} i g_{\mu\nu} D^+(y-x) |\Psi_0\rangle \neq 0$, which constitutes a contradiction. Thus the Lorentz condition can only be imposed in the weaker form (E.20a).

so that only the two transverse photons contribute to the expectation value of the Hamiltonian. From the structure of the remaining observables $\mathbf{P}, \mathbf{J}$, etc., one sees that this is also the case for the expectation values of these observables. Thus for free fields, in observable quantities only transverse photons occur, as is the case for the Coulomb gauge. The excitation of scalar and longitudinal photons obeying the subsidiary condition (E.20a) leads, in the absence of charges, to no observable consequences. One can show that the excitation of such photons leads merely to a transformation to another gauge that also satisfies the Lorentz condition. It is thus simplest to take as the vacuum state the state containing no photons.

When charges are present, the longitudinal and scalar photons provide the Coulomb interaction between the charges and thus appear as virtual particles in intermediate states. However, the initial and final states still contain only transverse photons.

E.3 The Feynman Photon Propagator

We now turn to a more detailed analysis of the photon propagator. For this we utilize the equation

$$g^{\mu\nu} = -\sum_r \zeta_r \epsilon_r^\mu(\mathbf{k}) \epsilon_r^\nu(\mathbf{k}) \tag{E.6b}$$

and insert the specific choice (E.7a-c) for the polarization vector tetrad into the Fourier transform of (E.10b):

$$D_F^{\mu\nu}(k) = \frac{1}{k^2 + i\epsilon} \left\{ \sum_{r=1,2} \epsilon_r^\mu(\mathbf{k}) \epsilon_r^\nu(\mathbf{k}) \right. \\ \left. + \frac{\left(k^\mu - (k \cdot n)n^\mu\right)\left(k^\nu - (k \cdot n)n^\nu\right)}{(kn)^2 - k^2} - n^\mu n^\nu \right\} . \tag{E.25}$$

The first term on the right-hand side represents the exchange of transverse photons

$$D_{F,\text{trans}}^{\mu\nu}(k) = \frac{1}{k^2 + i\epsilon} \sum_{r=1,2} \epsilon_r^\mu(\mathbf{k}) \epsilon_r^\nu(\mathbf{k}) . \tag{E.26a}$$

We divide the remainder of the expression, i.e., the second and third terms, into two parts:

$$D_{F,\text{Coul}}^{\mu\nu}(k) = \frac{1}{k^2 + i\epsilon} \left\{ \frac{(kn)^2 n^\mu n^\nu}{(kn)^2 - k^2} - n^\mu n^\nu \right\} \\ = \frac{k^2}{k^2 + i\epsilon} \frac{n^\mu n^\nu}{(kn)^2 - k^2} = \frac{n^\mu n^\nu}{(kn)^2 - k^2} \\ = \frac{n^\mu n^\nu}{\mathbf{k}^2} \tag{E.26b}$$

and

$$D^{\mu\nu}_{F,\text{red}}(k) = \frac{1}{k^2 + i\epsilon} \left[\frac{k^\mu k^\nu - (kn)(k^\mu n^\nu + n^\mu k^\nu)}{(kn)^2 - k^2} \right] . \tag{E.26c}$$

In coordinate space, $D^{\mu\nu}_{F,\text{Coul}}$ reads:

$$
\begin{aligned}
D^{\mu\nu}_{F,\text{Coul}}(x) &= n^\mu n^\nu \int \frac{d^3k\,dk^0}{(2\pi)^4} e^{-ikx} \frac{1}{|\mathbf{k}|^2} \\
&= g^{\mu 0} g^{\nu 0} \int \frac{d^3k\, e^{i\mathbf{k}x}}{|\mathbf{k}|^2} \int dk^0 e^{ik^0 x^0} \\
&= g^{\mu 0} g^{\nu 0} \frac{1}{4\pi|\mathbf{x}|} \delta(x^0) .
\end{aligned}
\tag{E.26b'}
$$

This part of the propagator represents the instantaneous Coulomb inter-action. The longitudinal and scalar photons thus yield the instantaneous Coulomb interaction between charged particles. In the Coulomb gauge only transverse photons occurred. The scalar potential was not a dynamical degree of freedom and was determined through Eq. (14.2.2) by the charge density of the particles (the charge density of the Dirac field). In the covariant quantiza-tion, the longitudinal and scalar (time-like) components were also quantized. The Coulomb interaction now no longer occurs explicitly in the theory, but is contained as the exchange of scalar and longitudinal photons in the prop-agator of the theory (in going from (E.25) to (E.26b) it is not only the third term of (E.25) that contributes, but also a part of the second term). The re-maining term $D^{\mu\nu}_{F,\text{red}}$ makes no physical contribution and is thus redundant, as can be seen from the structure of perturbation theory (see the Remark in Sect. 15.5.3.3),

$$
\begin{aligned}
&\int d^4x \int d^4x'\, j_{1\mu}(x) D^{\mu\nu}_F(x - x') j_{2\nu}(x') \\
&= \int d^4k\, j_{1\mu}(k) D^{\mu\nu}_F(k) j_{2\nu}(k) .
\end{aligned}
\tag{E.27}
$$

Since the current density is conserved,

$$\partial_\mu j^\mu = 0 \quad \text{and hence} \quad j_\mu k^\mu = 0 , \tag{E.28}$$

the term $D^{\mu\nu}_{F,\text{red}}$, comprising terms proportional to k^μ or k^ν, makes no con-tribution.

E.4 Conserved Quantities

From the free Lagrangian density corrresponding to (E.1),

$$\mathcal{L}_L = -\frac{1}{2} (\partial^\nu A_\mu)(\partial_\nu A^\mu) = -\frac{1}{2} A_{\mu,\nu} A^{\mu,\nu} , \tag{E.29}$$

according to (12.4.1), one obtains for the energy–momentum tensor

$$T^{\mu\nu} = -A_{\sigma}^{,\mu} A^{\sigma,\nu} - g^{\mu\nu}\mathcal{L}_L \ , \tag{E.30a}$$

and hence the energy and momentum densities

$$T^{00} = -\frac{1}{2}(\dot{A}^{\nu}\dot{A}_{\nu} + \partial_k A^{\nu}\partial_k A^{\nu}) \tag{E.30b}$$

$$T^{0k} = -\dot{A}_{\nu}\partial^k A^{\nu} \ . \tag{E.30c}$$

Furthermore, from (12.4.21), one obtains the angular-momentum tensor

$$M^{\mu\nu\sigma} = -A^{\nu,\mu}A^{\sigma} + A^{\sigma,\mu}A^{\nu} + x^{\nu}T^{\mu\sigma} - x^{\sigma}T^{\mu\nu} \tag{E.31a}$$

having the spin contribution

$$S^{\mu\nu\sigma} = -A^{\sigma}A^{\nu,\mu} + A^{\nu}A^{\sigma,\mu} \tag{E.31b}$$

from which one finally establishes the spin three-vector

$$\mathbf{S} = \mathbf{A}(x) \times \dot{\mathbf{A}}(x) \ . \tag{E.31c}$$

The vector product of the polarization vectors of the transverse photons $\epsilon_1(\mathbf{k}) \times \epsilon_2(\mathbf{k})$ equals $\mathbf{k}/|\mathbf{k}|$, and hence the value of the spin, is 1 with only two possible orientations, parallel or antiparallel to the wave vector. In this context it is instructive to make the transition from the two creation and annihilation operators $a_1^{\dagger}(\mathbf{k})$ and $a_2^{\dagger}(\mathbf{k})$ (or $a_1(\mathbf{k})$ and $a_2(\mathbf{k})$) to the creation and annihilation operators for helicity eigenstates.

F Coupling of Charged Scalar Mesons to the Electromagnetic Field

The Lagrangian density for the complex Klein–Gordon field is, according to (13.2.1),

$$\mathcal{L}_{\mathrm{KG}} = \left(\partial_{\mu}\phi^{\dagger}\right)\left(\partial^{\mu}\phi\right) - m^2\phi^{\dagger}\phi \ . \tag{F.1a}$$

In order to obtain the coupling to the radiation field, one has to make the replacement $\partial^{\mu} \rightarrow \partial^{\mu} + ieA^{\mu}$. The resulting covariant Lagrangian density, including the Lagrangian density of the electromagnetic field

$$\mathcal{L}_{\mathrm{rad}} = -\frac{1}{2}\left(\partial^{\nu}A_{\mu}\right)\left(\partial_{\nu}A^{\mu}\right) \tag{F.1b}$$

reads:

$$\mathcal{L} = -\frac{1}{2}\left(\partial^{\nu}A_{\mu}\right)\left(\partial_{\nu}A^{\mu}\right) - \left(\frac{\partial\phi^{\dagger}}{\partial x^{\mu}} - ieA_{\mu}\phi^{\dagger}\right)\left(\frac{\partial\phi}{\partial x_{\mu}} + ieA^{\mu}\phi\right) - m^2\phi^{\dagger}\phi \ . \tag{F.2}$$

The equations of motion for the vector potential are obtained from

$$-\frac{\partial}{\partial x^\nu}\frac{\partial \mathcal{L}}{\partial A^\mu_{,\nu}} = \Box A_\mu = -\frac{\partial \mathcal{L}}{\partial A^\mu} \; . \tag{F.3}$$

By differentiating with respect to $\phi^\dagger$, one obtains the Klein–Gordon equation in the presence of an electromagnetic field. Defining the electromagnetic current density

$$j_\mu = -\frac{\partial \mathcal{L}}{\partial A^\mu} \; , \tag{F.4}$$

one obtains

$$j_\mu = -\mathrm{i}e\left(\left(\frac{\partial \phi^\dagger}{\partial x^\mu} - \mathrm{i}eA_\mu\phi^\dagger\right)\phi - \phi^\dagger\left(\frac{\partial \phi}{\partial x^\mu} + \mathrm{i}eA_\mu\phi\right)\right) \; , \tag{F.5}$$

which, by virtue of the equations of motion, is conserved.

The Lagrangian density (F.6) can be separated into the Lagrangian density of the free Klein–Gordon field $\mathcal{L}_{\mathrm{KG}}$, that of the free radiation field $\mathcal{L}_{\mathrm{rad}}$, and an interaction Lagrangian density $\mathcal{L}_1$,

$$\mathcal{L} = \left(\partial_\mu\phi^\dagger\right)\left(\partial^\mu\phi\right) - m^2\phi^\dagger\phi - \frac{1}{2}\left(\partial^\nu A_\mu\right)\left(\partial_\nu A^\mu\right) + \mathcal{L}_1 \; , \tag{F.6}$$

where

$$\mathcal{L}_1 = \mathrm{i}e\left(\frac{\partial \phi^\dagger}{\partial x^\mu}\phi - \phi^\dagger\frac{\partial \phi}{\partial x^\mu}\right)A^\mu + e^2 A_\mu A^\mu \phi^\dagger\phi \; . \tag{F.7}$$

The occurrence of the term $e^2 A_\mu A^\mu \phi^\dagger\phi$ is characteristic for the Klein–Gordon field and corresponds, in the nonrelativistic limit, to the $\mathbf{A}^2$ term in the Schrödinger equation. From (F.7) one obtains for the interaction Hamiltonian density which enters the S matrix (15.3.4)[5] for charged particles

$$\mathcal{H}_I(x) = -\mathrm{i}e\left(\phi^\dagger(x)\frac{\partial \phi}{\partial x^\mu} - \frac{\partial \phi^\dagger}{\partial x^\mu}\phi(x)\right)A^\mu(x) - e^2\phi^\dagger(x)\phi(x)A^\mu(x)A_\mu(x) \; . \tag{F.8}$$

---

[5] P.T. Matthews, Phys. Rev. **76**, 684L (1949); **76**, 1489 (1949); S.S. Schweber, *An Introduction to Relativistic Quantum Field Theory*, Harper & Row, New York, 1961, p.482; C. Itzykson and J.-B. Zuber, *Quantum Field Theory*, McGraw Hill, New York, 1980, p.285

Index